Graduate Texts in Mathematics 244

Graduate Texts in Mathematics

(*continued after index*)

J.A. Bondy   U.S.R. Murty

Graph Theory

Springer

J.A. Bondy
Université de Lyon
Université Lyon 1
France
and
Université de Paris
Université Paris 6
France

U.S.R. Murty
University of Waterloo
Canada

Graduate Texts in Mathematics series ISSN: 0072-5285
ISBN: 978-1-846996-690-0   e-ISBN: 978-1-84628-970-5
DOI: 10.1007/978-1-84628-970-5

Library of Congress Control Number: 2007940370

Mathematics Subject Classification (2000): 05C; 68R10

Dedication

To the memory of our dear friends and mentors

CLAUDE BERGE   PAUL ERDŐS   BILL TUTTE

Dedication

To the memory of our dear friends and mentors

Preface

For more than one hundred years, the development of graph theory was inspired and guided mainly by the Four-Colour Conjecture. The resolution of the conjecture by K. Appel and W. Haken in 1976, the year in which our first book *Graph Theory with Applications* appeared, marked a turning point in its history. Since then, the subject has experienced explosive growth, due in large measure to its role as an essential structure underpinning modern applied mathematics. Computer science and combinatorial optimization, in particular, draw upon and contribute to the development of the theory of graphs. Moreover, in a world where communication is of prime importance, the versatility of graphs makes them indispensable tools in the design and analysis of communication networks.

Building on the foundations laid by Claude Berge, Paul Erdős, Bill Tutte, and others, a new generation of graph-theorists has enriched and transformed the subject by developing powerful new techniques, many borrowed from other areas of mathematics. These have led, in particular, to the resolution of several longstanding conjectures, including Berge's Strong Perfect Graph Conjecture and Kneser's Conjecture, both on colourings, and Gallai's Conjecture on cycle coverings.

One of the dramatic developments over the past thirty years has been the creation of the theory of graph minors by G. N. Robertson and P. D. Seymour. In a long series of deep papers, they have revolutionized graph theory by introducing an original and incisive way of viewing graphical structure. Developed to attack a celebrated conjecture of K. Wagner, their theory gives increased prominence to embeddings of graphs in surfaces. It has led also to polynomial-time algorithms for solving a variety of hitherto intractable problems, such as that of finding a collection of pairwise-disjoint paths between prescribed pairs of vertices.

A technique which has met with spectacular success is the probabilistic method. Introduced in the 1940s by Erdős, in association with fellow Hungarians A. Rényi and P. Turán, this powerful yet versatile tool is being employed with ever-increasing frequency and sophistication to establish the existence or nonexistence of graphs, and other combinatorial structures, with specified properties.

As remarked above, the growth of graph theory has been due in large measure to its essential role in the applied sciences. In particular, the quest for efficient algorithms has fuelled much research into the structure of graphs. The importance of spanning trees of various special types, such as breadth-first and depth-first trees, has become evident, and tree decompositions of graphs are a central ingredient in the theory of graph minors. Algorithmic graph theory borrows tools from a number of disciplines, including geometry and probability theory. The discovery by S. Cook in the early 1970s of the existence of the extensive class of seemingly intractable $\mathcal{NP}$-complete problems has led to the search for efficient approximation algorithms, the goal being to obtain a good approximation to the true value. Here again, probabilistic methods prove to be indispensable.

The links between graph theory and other branches of mathematics are becoming increasingly strong, an indication of the growing maturity of the subject. We have already noted certain connections with topology, geometry, and probability. Algebraic, analytic, and number-theoretic tools are also being employed to considerable effect. Conversely, graph-theoretical methods are being applied more and more in other areas of mathematics. A notable example is Szemerédi's regularity lemma. Developed to solve a conjecture of Erdős and Turán, it has become an essential tool in additive number theory, as well as in extremal conbinatorics. An extensive account of this interplay can be found in the two-volume *Handbook of Combinatorics*.

It should be evident from the above remarks that graph theory is a flourishing discipline. It contains a body of beautiful and powerful theorems of wide applicability. The remarkable growth of the subject is reflected in the wealth of books and monographs now available. In addition to the *Handbook of Combinatorics*, much of which is devoted to graph theory, and the three-volume treatise on combinatorial optimization by Schrijver (2003), destined to become a classic, one can find monographs on colouring by Jensen and Toft (1995), on flows by Zhang (1997), on matching by Lovász and Plummer (1986), on extremal graph theory by Bollobás (1978), on random graphs by Bollobás (2001) and Janson et al. (2000), on probabilistic methods by Alon and Spencer (2000) and Molloy and Reed (1998), on topological graph theory by Mohar and Thomassen (2001), on algebraic graph theory by Biggs (1993), and on digraphs by Bang-Jensen and Gutin (2001), as well as a good choice of textbooks. Another sign is the significant number of new journals dedicated to graph theory.

The present project began with the intention of simply making minor revisions to our earlier book. However, we soon came to the realization that the changing face of the subject called for a total reorganization and enhancement of its contents. As with *Graph Theory with Applications*, our primary aim here is to present a coherent introduction to the subject, suitable as a textbook for advanced undergraduate and beginning graduate students in mathematics and computer science. For pedagogical reasons, we have concentrated on topics which can be covered satisfactorily in a course. The most conspicuous omission is the theory of graph minors, which we only touch upon, it being too complex to be accorded an adequate treatment. We have maintained as far as possible the terminology and notation of our earlier book, which are now generally accepted.

Particular care has been taken to provide a systematic treatment of the theory of graphs without sacrificing its intuitive and aesthetic appeal. Commonly used proof techniques are described and illustrated. Many of these are to be found in insets, whereas others, such as search trees, network flows, the regularity lemma and the local lemma, are the topics of entire sections or chapters. The exercises, of varying levels of difficulty, have been designed so as to help the reader master these techniques and to reinforce his or her grasp of the material. Those exercises which are needed for an understanding of the text are indicated by a star. The more challenging exercises are separated from the easier ones by a dividing line.

A second objective of the book is to serve as an introduction to research in graph theory. To this end, sections on more advanced topics are included, and a number of interesting and challenging open problems are highlighted and discussed in some detail. These and many more are listed in an appendix.

Despite this more advanced material, the book has been organized in such a way that an introductory course on graph theory may be based on the first few sections of selected chapters. Like number theory, graph theory is conceptually simple, yet gives rise to challenging unsolved problems. Like geometry, it is visually pleasing. These two aspects, along with its diverse applications, make graph theory an ideal subject for inclusion in mathematical curricula.

We have sought to convey the aesthetic appeal of graph theory by illustrating the text with many interesting graphs — a full list can be found in the index. The cover design, taken from Chapter 10, depicts simultaneous embeddings on the projective plane of K_6 and its dual, the Petersen graph.

A Web page for the book is available at

http://blogs.springer.com/bondyandmurty

The reader will find there hints to selected exercises, background to open problems, other supplementary material, and an inevitable list of errata. For instructors wishing to use the book as the basis for a course, suggestions are provided as to an appropriate selection of topics, depending on the intended audience.

We are indebted to many friends and colleagues for their interest in and help with this project. Tommy Jensen deserves a special word of thanks. He read through the entire manuscript, provided numerous unfailingly pertinent comments, simplified and clarified several proofs, corrected many technical errors and linguistic infelicities, and made valuable suggestions. Others who went through and commented on parts of the book include Noga Alon, Roland Assous, Xavier Buchwalder, Genghua Fan, Frédéric Havet, Bill Jackson, Stephen Locke, Zsolt Tuza, and two anonymous readers. We were most fortunate to benefit in this way from their excellent knowledge and taste.

Colleagues who offered advice or supplied exercises, problems, and other helpful material include Michael Albertson, Marcelo de Carvalho, Joseph Cheriyan, Roger Entringer, Herbert Fleischner, Richard Gibbs, Luis Goddyn, Alexander Kelmans,

Henry Kierstead, László Lovász, Cláudio Lucchesi, George Purdy, Dieter Rautenbach, Bruce Reed, Bruce Richmond, Neil Robertson, Alexander Schrijver, Paul Seymour, Miklós Simonovits, Balázs Szegedy, Robin Thomas, Stéphan Thomassé, Carsten Thomassen, and Jacques Verstraëte. We thank them all warmly for their various contributions. We are grateful also to Martin Crossley for allowing us to use (in Figure 10.24) drawings of the Möbius band and the torus taken from his book Crossley (2005).

Facilities and support were kindly provided by Maurice Pouzet at Université Lyon 1 and Jean Fonlupt at Université Paris 6. The glossary was prepared using software designed by Nicola Talbot of the University of East Anglia. Her promptly-offered advice is much appreciated. Finally, we benefitted from a fruitful relationship with Karen Borthwick at Springer, and from the technical help provided by her colleagues Brian Bishop and Frank Ganz.

We are dedicating this book to the memory of our friends Claude Berge, Paul Erdős, and Bill Tutte. It owes its existence to their achievements, their guiding hands, and their personal kindness.

J. A. Bondy and U. S. R. Murty

September 2007

Preface to the second printing

We have taken the opportunity of this second printing to correct a number of errors. Most of these have already been indicated on the blog of the book. We have also made minor modifications to a few exercises, reclassified some easy exercises as hard or vice-versa, or shifted them to new sections or even new chapters. Eight exercises in Section 11.2 have been moved to their proper place in Section 11.1, for example. In addition, several new exercises have been included. *Consequently, many exercise numbers have changed from the first printing.*

More importantly, we have moved some material (the Gallai-Milgram Theorem and Berge's Path Partition Conjecture) from Section 19.2 to Sections 12.1 and 14.1, respectively, and slightly remodelled Sections 14.5 and 14.6.

J. A. Bondy and U. S. R. Murty

June 2008

Contents

1

Graphs

Contents

1.1 Graphs and Their Representation

DEFINITIONS AND EXAMPLES

Many real-world situations can conveniently be described by means of a diagram consisting of a set of points together with lines joining certain pairs of these points.

For example, the points could represent people, with lines joining pairs of friends; or the points might be communication centres, with lines representing communication links. Notice that in such diagrams one is mainly interested in whether two given points are joined by a line; the manner in which they are joined is immaterial. A mathematical abstraction of situations of this type gives rise to the concept of a graph.

A *graph* G is an ordered pair $(V(G), E(G))$ consisting of a set $V(G)$ of *vertices* and a set $E(G)$, disjoint from $V(G)$, of *edges*, together with an *incidence function* ψ_G that associates with each edge of G an unordered pair of (not necessarily distinct) vertices of G. If e is an edge and u and v are vertices such that $\psi_G(e) = \{u, v\}$, then e is said to *join* u and v, and the vertices u and v are called the *ends* of e. We denote the numbers of vertices and edges in G by $v(G)$ and $e(G)$; these two basic parameters are called the *order* and *size* of G, respectively.

Two examples of graphs should serve to clarify the definition. For notational simplicity, we write uv for the unordered pair $\{u, v\}$.

Example 1.

$$G = (V(G), E(G))$$

where

$$V(G) = \{u, v, w, x, y\}$$
$$E(G) = \{a, b, c, d, e, f, g, h\}$$

and ψ_G is defined by

$$\psi_G(a) = uv \quad \psi_G(b) = uu \quad \psi_G(c) = vw \quad \psi_G(d) = wx$$
$$\psi_G(e) = vx \quad \psi_G(f) = wx \quad \psi_G(g) = ux \quad \psi_G(h) = xy$$

Example 2.

$$H = (V(H), E(H))$$

where

$$V(H) = \{v_0, v_1, v_2, v_3, v_4, v_5\}$$
$$E(H) = \{e_1, e_2, e_3, e_4, e_5, e_6, e_7, e_8, e_9, e_{10}\}$$

and ψ_H is defined by

$$\psi_H(e_1) = v_1v_2 \quad \psi_H(e_2) = v_2v_3 \quad \psi_H(e_3) = v_3v_4 \quad \psi_H(e_4) = v_4v_5 \quad \psi_H(e_5) = v_5v_1$$
$$\psi_H(e_6) = v_0v_1 \quad \psi_H(e_7) = v_0v_2 \quad \psi_H(e_8) = v_0v_3 \quad \psi_H(e_9) = v_0v_4 \quad \psi_H(e_{10}) = v_0v_5$$

DRAWINGS OF GRAPHS

Graphs are so named because they can be represented graphically, and it is this graphical representation which helps us understand many of their properties. Each vertex is indicated by a point, and each edge by a line joining the points representing its ends. Diagrams of G and H are shown in Figure 1.1. (For clarity, vertices are represented by small circles.)

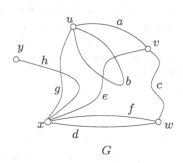
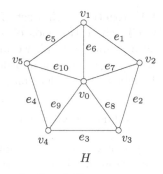

Fig. 1.1. Diagrams of the graphs G and H

There is no single correct way to draw a graph; the relative positions of points representing vertices and the shapes of lines representing edges usually have no significance. In Figure 1.1, the edges of G are depicted by curves, and those of H by straight-line segments. A diagram of a graph merely depicts the incidence relation holding between its vertices and edges. However, we often draw a diagram of a graph and refer to it as the graph itself; in the same spirit, we call its points 'vertices' and its lines 'edges'.

Most of the definitions and concepts in graph theory are suggested by this graphical representation. The ends of an edge are said to be *incident* with the edge, and *vice versa*. Two vertices which are incident with a common edge are *adjacent*, as are two edges which are incident with a common vertex, and two distinct adjacent vertices are *neighbours*. The set of neighbours of a vertex v in a graph G is denoted by $N_G(v)$.

An edge with identical ends is called a *loop*, and an edge with distinct ends a *link*. Two or more links with the same pair of ends are said to be *parallel edges*. In the graph G of Figure 1.1, the edge b is a loop, and all other edges are links; the edges d and f are parallel edges.

Throughout the book, the letter G denotes a graph. Moreover, when there is no scope for ambiguity, we omit the letter G from graph-theoretic symbols and write, for example, V and E instead of $V(G)$ and $E(G)$. In such instances, we denote the numbers of vertices and edges of G by n and m, respectively.

A graph is *finite* if both its vertex set and edge set are finite. In this book, we mainly study finite graphs, and the term 'graph' always means 'finite graph'. The graph with no vertices (and hence no edges) is the *null graph*. Any graph with just one vertex is referred to as *trivial*. All other graphs are *nontrivial*. We admit the null graph solely for mathematical convenience. Thus, unless otherwise specified, all graphs under discussion should be taken to be nonnull.

A graph is *simple* if it has no loops or parallel edges. The graph H in Example 2 is simple, whereas the graph G in Example 1 is not. Much of graph theory is concerned with the study of simple graphs.

A set V, together with a set E of two-element subsets of V, defines a simple graph (V, E), where the ends of an edge uv are precisely the vertices u and v. Indeed, in any simple graph we may dispense with the incidence function ψ by renaming each edge as the unordered pair of its ends. In a diagram of such a graph, the labels of the edges may then be omitted.

SPECIAL FAMILIES OF GRAPHS

Certain types of graphs play prominent roles in graph theory. A *complete graph* is a simple graph in which any two vertices are adjacent, an *empty graph* one in which no two vertices are adjacent (that is, one whose edge set is empty). A graph is *bipartite* if its vertex set can be partitioned into two subsets X and Y so that every edge has one end in X and one end in Y; such a partition (X, Y) is called a *bipartition* of the graph, and X and Y its *parts*. We denote a bipartite graph G with bipartition (X, Y) by $G[X, Y]$. If $G[X, Y]$ is simple and every vertex in X is joined to every vertex in Y, then G is called a *complete bipartite graph*. A *star* is a complete bipartite graph $G[X, Y]$ with $|X| = 1$ or $|Y| = 1$. Figure 1.2 shows diagrams of a complete graph, a complete bipartite graph, and a star.

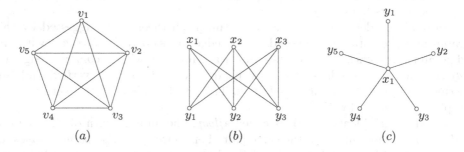

Fig. 1.2. (a) A complete graph, (b) a complete bipartite graph, and (c) a star

A *path* is a simple graph whose vertices can be arranged in a linear sequence in such a way that two vertices are adjacent if they are consecutive in the sequence, and are nonadjacent otherwise. Likewise, a *cycle* on three or more vertices is a simple graph whose vertices can be arranged in a cyclic sequence in such a way that two vertices are adjacent if they are consecutive in the sequence, and are nonadjacent otherwise; a cycle on one vertex consists of a single vertex with a loop, and a cycle on two vertices consists of two vertices joined by a pair of parallel edges. The *length* of a path or a cycle is the number of its edges. A path or cycle of length k is called a *k-path* or *k-cycle*, respectively; the path or cycle is *odd* or *even* according to the parity of k. A 3-cycle is often called a *triangle*, a 4-cycle a *quadrilateral*, a 5-cycle a *pentagon*, a 6-cycle a *hexagon*, and so on. Figure 1.3 depicts a 3-path and a 5-cycle.

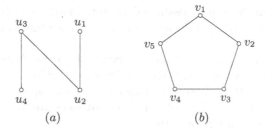

(a)   (b)

Fig. 1.3. (a) A path of length three, and (b) a cycle of length five

A graph is *connected* if, for every partition of its vertex set into two nonempty sets X and Y, there is an edge with one end in X and one end in Y; otherwise the graph is *disconnected*. In other words, a graph is disconnected if its vertex set can be partitioned into two nonempty subsets X and Y so that no edge has one end in X and one end in Y. (It is instructive to compare this definition with that of a bipartite graph.) Examples of connected and disconnected graphs are displayed in Figure 1.4.

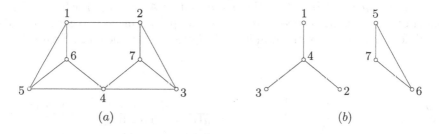

(a)   (b)

Fig. 1.4. (a) A connected graph, and (b) a disconnected graph

As observed earlier, examples of graphs abound in the real world. Graphs also arise naturally in the study of other mathematical structures such as polyhedra, lattices, and groups. These graphs are generally defined by means of an adjacency rule, prescribing which unordered pairs of vertices are edges and which are not. A number of such examples are given in the exercises at the end of this section and in Section 1.3.

For the sake of clarity, we observe certain conventions in representing graphs by diagrams: we do not allow an edge to intersect itself, nor let an edge pass through a vertex that is not an end of the edge; clearly, this is always possible. However, two edges may intersect at a point that does not correspond to a vertex, as in the drawings of the first two graphs in Figure 1.2. A graph which can be drawn in the plane in such a way that edges meet only at points corresponding to their common ends is called a *planar graph*, and such a drawing is called a *planar embedding* of the graph. For instance, the graphs G and H of Examples 1 and 2 are both planar, even though there are crossing edges in the particular drawing of G shown in Figure 1.1. The first two graphs in Figure 1.2, on the other hand, are not planar, as proved later.

Although not all graphs are planar, every graph can be drawn on some surface so that its edges intersect only at their ends. Such a drawing is called an *embedding* of the graph on the surface. Figure 1.21 provides an example of an embedding of a graph on the torus. Embeddings of graphs on surfaces are discussed in Chapter 3 and, more thoroughly, in Chapter 10.

INCIDENCE AND ADJACENCY MATRICES

Although drawings are a convenient means of specifying graphs, they are clearly not suitable for storing graphs in computers, or for applying mathematical methods to study their properties. For these purposes, we consider two matrices associated with a graph, its incidence matrix and its adjacency matrix.

Let G be a graph, with vertex set V and edge set E. The *incidence matrix* of G is the $n \times m$ matrix $\mathbf{M}_G := (m_{ve})$, where m_{ve} is the number of times $(0, 1, \text{or } 2)$ that vertex v and edge e are incident. Clearly, the incidence matrix is just another way of specifying the graph.

The *adjacency matrix* of G is the $n \times n$ matrix $\mathbf{A}_G := (a_{uv})$, where a_{uv} is the number of edges joining vertices u and v, each loop counting as two edges. Incidence and adjacency matrices of the graph G of Figure 1.1 are shown in Figure 1.5.

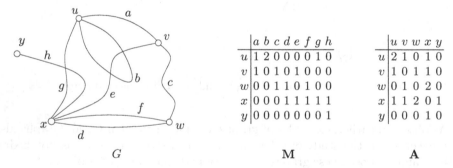

	a b c d e f g h
u	1 2 0 0 0 0 1 0
v	1 0 1 0 1 0 0 0
w	0 0 1 1 0 1 0 0
x	0 0 0 1 1 1 1 1
y	0 0 0 0 0 0 0 1

	u v w x y
u	2 1 0 1 0
v	1 0 1 1 0
w	0 1 0 2 0
x	1 1 2 0 1
y	0 0 0 1 0

G **M**   **A**

Fig. 1.5. Incidence and adjacency matrices of a graph

Because most graphs have many more edges than vertices, the adjacency matrix of a graph is generally much smaller than its incidence matrix and thus requires less storage space. When dealing with simple graphs, an even more compact representation is possible. For each vertex v, the neighbours of v are listed in some order. A list $(N(v) : v \in V)$ of these lists is called an *adjacency list* of the graph. Simple graphs are usually stored in computers as adjacency lists.

When G is a bipartite graph, as there are no edges joining pairs of vertices belonging to the same part of its bipartition, a matrix of smaller size than the adjacency matrix may be used to record the numbers of edges joining pairs of vertices. Suppose that $G[X, Y]$ is a bipartite graph, where $X := \{x_1, x_2, \ldots, x_r\}$ and $Y := \{y_1, y_2, \ldots, y_s\}$. We define the *bipartite adjacency matrix* of G to be the $r \times s$ matrix $\mathbf{B}_G = (b_{ij})$, where b_{ij} is the number of edges joining x_i and y_j.

VERTEX DEGREES

The *degree* of a vertex v in a graph G, denoted by $d_G(v)$, is the number of edges of G incident with v, each loop counting as two edges. In particular, if G is a simple graph, $d_G(v)$ is the number of neighbours of v in G. A vertex of degree zero is called an *isolated vertex*. We denote by $\delta(G)$ and $\Delta(G)$ the minimum and maximum degrees of the vertices of G, and by $d(G)$ their *average degree*, $\frac{1}{n} \sum_{v \in V} d(v)$. The following theorem establishes a fundamental identity relating the degrees of the vertices of a graph and the number of its edges.

Theorem 1.1 *For any graph G,*

$$\sum_{v \in V} d(v) = 2m \tag{1.1}$$

Proof Consider the incidence matrix $\mathbf{M}$ of G. The sum of the entries in the row corresponding to vertex v is precisely $d(v)$. Therefore $\sum_{v \in V} d(v)$ is just the sum of all the entries in $\mathbf{M}$. But this sum is also $2m$, because each of the m column sums of $\mathbf{M}$ is 2, each edge having two ends. $\square$

Corollary 1.2 *In any graph, the number of vertices of odd degree is even.*

Proof Consider equation (1.1) modulo 2. We have

$$d(v) \equiv \begin{cases} 1 \pmod 2 & \text{if } d(v) \text{ is odd,} \\ 0 \pmod 2 & \text{if } d(v) \text{ is even.} \end{cases}$$

Thus, modulo 2, the left-hand side is congruent to the number of vertices of odd degree, and the right-hand side is zero. The number of vertices of odd degree is therefore congruent to zero modulo 2. $\square$

A graph G is *k-regular* if $d(v) = k$ for all $v \in V$; a *regular graph* is one that is k-regular for some k. For instance, the complete graph on n vertices is $(n-1)$-regular, and the complete bipartite graph with k vertices in each part is k-regular. For $k = 0, 1$ and 2, k-regular graphs have very simple structures and are easily characterized (Exercise 1.1.5). By contrast, 3-regular graphs can be remarkably complex. These graphs, also referred to as *cubic* graphs, play a prominent role in graph theory. We present a number of interesting examples of such graphs in the next section.

PROOF TECHNIQUE: COUNTING IN TWO WAYS

In proving Theorem 1.1, we used a common proof technique in combinatorics, known as *counting in two ways*. It consists of considering a suitable matrix and computing the sum of its entries in two different ways: firstly as the sum of its row sums, and secondly as the sum of its column sums. Equating these two quantities results in an identity. In the case of Theorem 1.1, the matrix we considered was the incidence matrix of G. In order to prove the identity of Exercise 1.1.9a, the appropriate matrix to consider is the bipartite adjacency matrix of the bipartite graph $G[X, Y]$. In both these cases, the choice of the appropriate matrix is fairly obvious. However, in some cases, making the right choice requires ingenuity.

Note that an upper bound on the sum of the column sums of a matrix is clearly also an upper bound on the sum of its row sums (and *vice versa*). The method of counting in two ways may therefore be adapted to establish inequalities. The proof of the following proposition illustrates this idea.

Proposition 1.3 *Let $G[X, Y]$ be a bipartite graph without isolated vertices such that $d(x) \geq d(y)$ for all $xy \in E$, where $x \in X$ and $y \in Y$. Then $|X| \leq |Y|$, with equality if and only if $d(x) = d(y)$ for all $xy \in E$.*

Proof The first assertion follows if we can find a matrix with $|X|$ rows and $|Y|$ columns in which each row sum is one and each column sum is at most one. Such a matrix can be obtained from the bipartite adjacency matrix $\mathbf{B}$ of $G[X, Y]$ by dividing the row corresponding to vertex x by $d(x)$, for each $x \in X$. (This is possible since $d(x) \neq 0$.) Because the sum of the entries of $\mathbf{B}$ in the row corresponding to x is $d(x)$, all row sums of the resulting matrix $\widetilde{\mathbf{B}}$ are equal to one. On the other hand, the sum of the entries in the column of $\widetilde{\mathbf{B}}$ corresponding to vertex y is $\sum 1/d(x)$, the sum being taken over all edges xy incident to y, and this sum is at most one because $1/d(x) \leq 1/d(y)$ for each edge xy, by hypothesis, and because there are $d(y)$ edges incident to y.

The above argument may be expressed more concisely as follows.

$$|X| = \sum_{\substack{x \in X \\ xy \in E}} \sum_{y \in Y} \frac{1}{d(x)} = \sum_{\substack{x \in X \\ y \in Y}} \sum_{xy \in E} \frac{1}{d(x)} \leq \sum_{\substack{x \in X \\ y \in Y}} \sum_{xy \in E} \frac{1}{d(y)} = \sum_{y \in Y} \sum_{\substack{x \in X \\ xy \in E}} \frac{1}{d(y)} = |Y|$$

Furthermore, if $|X| = |Y|$, the middle inequality must be an equality, implying that $d(x) = d(y)$ for all $xy \in E$. $\qquad \square$

An application of this proof technique to a problem about geometric configurations is described in Exercise 1.3.15.

Exercises

1.1.1 Let G be a simple graph. Show that $m \leq \binom{n}{2}$, and determine when equality holds.

1.1.2 Let $G[X, Y]$ be a simple bipartite graph, where $|X| = r$ and $|Y| = s$.

a) Show that $m \leq rs$.
b) Deduce that $m \leq n^2/4$.
c) Describe the simple bipartite graphs G for which equality holds in (b).

⋆**1.1.3** Show that:

a) every path is bipartite,
b) a cycle is bipartite if and only if its length is even.

1.1.4 Show that, for any graph G, $\delta(G) \leq d(G) \leq \Delta(G)$.

1.1.5 For $k = 0, 1, 2$, characterize the k-regular graphs.

1.1.6

a) Show that, in any group of two or more people, there are always two who have exactly the same number of friends within the group.
b) Describe a group of five people, any two of whom have exactly one friend in common. Can you find a group of four people with this same property?

1.1.7 n-CUBE
The n-cube Q_n ($n \geq 1$) is the graph whose vertex set is the set of all n-tuples of 0s and 1s, where two n-tuples are adjacent if they differ in precisely one coordinate.

a) Draw Q_1, Q_2, Q_3, and Q_4.
b) Determine $v(Q_n)$ and $e(Q_n)$.
c) Show that Q_n is bipartite for all $n \geq 1$.

1.1.8 The *boolean lattice* BL_n ($n \geq 1$) is the graph whose vertex set is the set of all subsets of $\{1, 2, \ldots, n\}$, where two subsets X and Y are adjacent if their symmetric difference has precisely one element.

a) Draw BL_1, BL_2, BL_3, and BL_4.
b) Determine $v(BL_n)$ and $e(BL_n)$.
c) Show that BL_n is bipartite for all $n \geq 1$.

⋆**1.1.9** Let $G[X, Y]$ be a bipartite graph.

a) Show that $\sum_{v \in X} d(v) = \sum_{v \in Y} d(v)$.
b) Deduce that if G is k-regular, with $k \geq 1$, then $|X| = |Y|$.

★1.1.10 k-PARTITE GRAPH
A k-*partite graph* is one whose vertex set can be partitioned into k subsets, or *parts*, in such a way that no edge has both ends in the same part. (Equivalently, one may think of the vertices as being colourable by k colours so that no edge joins two vertices of the same colour.) Let G be a simple k-partite graph with parts of sizes $a_1, a_2, \ldots, a_k$. Show that $m \leq \frac{1}{2} \sum_{i=1}^{k} a_i(n - a_i)$.

★1.1.11 TURÁN GRAPH
A k-partite graph is *complete* if any two vertices in different parts are adjacent. A simple complete k-partite graph on n vertices whose parts are of equal or almost equal sizes (that is, $\lfloor n/k \rfloor$ or $\lceil n/k \rceil$) is called a *Turán graph* and denoted $T_{k,n}$.

a) Show that $T_{k,n}$ has more edges than any other simple complete k-partite graph on n vertices.
b) Determine $e(T_{k,n})$.

1.1.12

a) Show that if G is simple and $m > \binom{n-1}{2}$, then G is connected.
b) For $n > 1$, find a disconnected simple graph G with $m = \binom{n-1}{2}$.

1.1.13

a) Show that if G is simple and $\delta > \frac{1}{2}(n - 2)$, then G is connected.
b) For n even, find a disconnected $\frac{1}{2}(n - 2)$-regular simple graph.

1.1.14 For a simple graph G, show that the diagonal entries of both $\mathbf{A}^2$ and $\mathbf{MM}^t$ (where $\mathbf{M}^t$ denotes the transpose of $\mathbf{M}$) are the degrees of the vertices of G.

1.1.15 Show that the rank over $GF(2)$ of the incidence matrix of a graph G is at most $n - 1$, with equality if and only if G is connected.

1.1.16 DEGREE SEQUENCE
If G has vertices $v_1, v_2, \ldots, v_n$, the sequence $(d(v_1), d(v_2), \ldots, d(v_n))$ is called a *degree sequence* of G. Let $\mathbf{d} := (d_1, d_2, \ldots, d_n)$ be a nonincreasing sequence of nonnegative integers, that is, $d_1 \geq d_2 \geq \cdots \geq d_n \geq 0$. Show that:

a) there is a graph with degree sequence $\mathbf{d}$ if and only if $\sum_{i=1}^{n} d_i$ is even,
b) there is a loopless graph with degree sequence $\mathbf{d}$ if and only if $\sum_{i=1}^{n} d_i$ is even and $d_1 \leq \sum_{i=2}^{n} d_i$.

1.1.17 COMPLEMENT OF A GRAPH
Let G be a simple graph. The *complement* $\overline{G}$ of G is the simple graph whose vertex set is V and whose edges are the pairs of nonadjacent vertices of G.

a) Express the degree sequence of $\overline{G}$ in terms of the degree sequence of G.
b) Show that if G is disconnected, then $\overline{G}$ is connected. Is the converse true?

————————⁌⁌————————

1.1.18 GRAPHIC SEQUENCE
A sequence $\mathbf{d} = (d_1, d_2, \ldots, d_n)$ is *graphic* if there is a simple graph with degree sequence $\mathbf{d}$. Show that:

a) the sequences $(7, 6, 5, 4, 3, 3, 2)$ and $(6, 6, 5, 4, 3, 3, 1)$ are not graphic,
b) if $\mathbf{d} = (d_1, d_2, \ldots, d_n)$ is graphic and $d_1 \geq d_2 \geq \cdots \geq d_n$, then $\sum_{i=1}^{n} d_i$ is even and

$$\sum_{i=1}^{k} d_i \leq k(k-1) + \sum_{i=k+1}^{n} \min\{k, d_i\}, \quad 1 \leq k \leq n$$

(Erdős and Gallai (1960) showed that these necessary conditions for a sequence to be graphic are also sufficient.)

1.1.19 Let $\mathbf{d} = (d_1, d_2, \ldots, d_n)$ be a nonincreasing sequence of nonnegative integers. Set $\mathbf{d}' := (d_2 - 1, d_3 - 1, \ldots, d_{d_1+1} - 1, d_{d_1+2}, \ldots, d_n)$.

a) Show that $\mathbf{d}$ is graphic if and only if $\mathbf{d}'$ is graphic.
b) Using (a), describe an algorithm which accepts as input a nonincreasing sequence $\mathbf{d}$ of nonnegative integers, and returns either a simple graph with degree sequence $\mathbf{d}$, if such a graph exists, or else a proof that $\mathbf{d}$ is not graphic.
(V. HAVEL AND S.L. HAKIMI)

1.1.20 Let S be a set of n points in the plane, the distance between any two of which is at least one. Show that there are at most $3n$ pairs of points of S at distance exactly one.

1.1.21 EIGENVALUES OF A GRAPH
Recall that the eigenvalues of a square matrix $\mathbf{A}$ are the roots of its characteristic polynomial $\det(\mathbf{A} - x\mathbf{I})$. An *eigenvalue* of a graph is an eigenvalue of its adjacency matrix. Likewise, the *characteristic polynomial* of a graph is the characteristic polynomial of its adjacency matrix. Show that:

a) every eigenvalue of a graph is real,
b) every rational eigenvalue of a graph is integral.

1.1.22

a) Let G be a k-regular graph. Show that:
   i) $\mathbf{M}\mathbf{M}^t = \mathbf{A} + k\mathbf{I}$, where $\mathbf{I}$ is the $n \times n$ identity matrix,
   ii) k is an eigenvalue of G, with corresponding eigenvector $\mathbf{1}$, the n-vector in which each entry is 1.
b) Let G be a complete graph of order n. Denote by $\mathbf{J}$ the $n \times n$ matrix all of whose entries are 1. Show that:
   i) $\mathbf{A} = \mathbf{J} - \mathbf{I}$,
   ii) the eigenvalues of $\mathbf{J}$ are 0 (with multiplicity $n - 1$) and n.
c) Derive from (b) the eigenvalues of a complete graph and their multiplicities, and determine the corresponding eigenspaces.

1.1.23 Let G be a simple graph.

a) Show that $\overline{G}$ has adjacency matrix $\mathbf{J} - \mathbf{I} - \mathbf{A}$.
b) Suppose now that G is k-regular.
    i) Deduce from Exercise 1.1.22 that $n - k - 1$ is an eigenvalue of $\overline{G}$, with corresponding eigenvector $\mathbf{1}$.
    ii) Show that if λ is an eigenvalue of G different from k, then $-1 - \lambda$ is an eigenvalue of $\overline{G}$, with the same multiplicity. (Recall that eigenvectors corresponding to distinct eigenvalues of a real symmetric matrix are orthogonal.)

1.1.24 Show that:

a) no eigenvalue of a graph G has absolute value greater than Δ,
b) if G is a connected graph and Δ is an eigenvalue of G, then G is regular,
c) if G is a connected graph and $-\Delta$ is an eigenvalue of G, then G is both regular and bipartite.

1.1.25 STRONGLY REGULAR GRAPH
A simple graph G which is neither empty nor complete is said to be *strongly regular* with parameters (v, k, λ, μ) if:

▷ $v(G) = v$,
▷ G is k-regular,
▷   any two adjacent vertices of G have λ common neighbours,
▷   any two nonadjacent vertices of G have μ common neighbours.

Let G be a strongly regular graph with parameters (v, k, λ, μ). Show that:

a) $\overline{G}$ is strongly regular,
b) $k(k - \lambda - 1) = (v - k - 1)\mu$,
c) $\mathbf{A}^2 = k\,\mathbf{I} + \lambda\,\mathbf{A} + \mu\,(\mathbf{J} - \mathbf{I} - \mathbf{A})$.

1.2 Isomorphisms and Automorphisms

ISOMORPHISMS

Two graphs G and H are *identical*, written $G = H$, if $V(G) = V(H)$, $E(G) = E(H)$, and $\psi_G = \psi_H$. If two graphs are identical, they can clearly be represented by identical diagrams. However, it is also possible for graphs that are not identical to have essentially the same diagram. For example, the graphs G and H in Figure 1.6 can be represented by diagrams which look exactly the same, as the second drawing of H shows; the sole difference lies in the labels of their vertices and edges. Although the graphs G and H are not identical, they do have identical structures, and are said to be isomorphic.

In general, two graphs G and H are *isomorphic*, written $G \cong H$, if there are bijections $\theta : V(G) \to V(H)$ and $\phi : E(G) \to E(H)$ such that $\psi_G(e) = uv$ if and only if $\psi_H(\phi(e)) = \theta(u)\theta(v)$; such a pair of mappings is called an *isomorphism* between G and H.

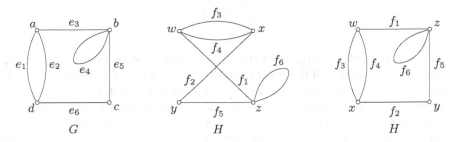

Fig. 1.6. Isomorphic graphs

In order to show that two graphs are isomorphic, one must indicate an isomorphism between them. The pair of mappings (θ, ϕ) defined by

$$\theta := \begin{pmatrix} a & b & c & d \\ w & z & y & x \end{pmatrix} \qquad \phi := \begin{pmatrix} e_1 & e_2 & e_3 & e_4 & e_5 & e_6 \\ f_3 & f_4 & f_1 & f_6 & f_5 & f_2 \end{pmatrix}$$

is an isomorphism between the graphs G and H in Figure 1.6.

In the case of simple graphs, the definition of isomorphism can be stated more concisely, because if (θ, ϕ) is an isomorphism between simple graphs G and H, the mapping ϕ is completely determined by θ; indeed, $\phi(e) = \theta(u)\theta(v)$ for any edge $e = uv$ of G. Thus one may define an isomorphism between two simple graphs G and H as a bijection $\theta : V(G) \to V(H)$ which preserves adjacency (that is, the vertices u and v are adjacent in G if and only if their images $\theta(u)$ and $\theta(v)$ are adjacent in H).

Consider, for example, the graphs G and H in Figure 1.7.

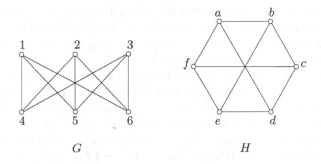

Fig. 1.7. Isomorphic simple graphs

The mapping

$$\theta := \begin{pmatrix} 1 & 2 & 3 & 4 & 5 & 6 \\ b & d & f & c & e & a \end{pmatrix}$$

is an isomorphism between G and H, as is

$$\theta' := \begin{pmatrix} 1\ 2\ 3\ 4\ 5\ 6 \\ a\ c\ e\ d\ f\ b \end{pmatrix}$$

Isomorphic graphs clearly have the same numbers of vertices and edges. On the other hand, equality of these parameters does not guarantee isomorphism. For instance, the two graphs shown in Figure 1.8 both have eight vertices and twelve edges, but they are not isomorphic. To see this, observe that the graph G has four mutually nonadjacent vertices, v_1, v_3, v_6, and v_8. If there were an isomorphism θ between G and H, the vertices $\theta(v_1)$, $\theta(v_3)$, $\theta(v_6)$, and $\theta(v_8)$ of H would likewise be mutually nonadjacent. But it can readily be checked that no four vertices of H are mutually nonadjacent. We deduce that G and H are not isomorphic.

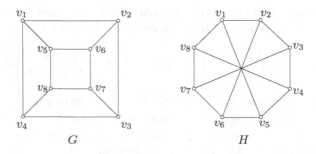

Fig. 1.8. Nonisomorphic graphs

It is clear from the foregoing discussion that if two graphs are isomorphic, then they are either identical or differ merely in the names of their vertices and edges, and thus have the same structure. Because it is primarily in structural properties that we are interested, we often omit labels when drawing graphs; formally, we may define an *unlabelled graph* as a representative of an equivalence class of isomorphic graphs. We assign labels to vertices and edges in a graph mainly for the purpose of referring to them (in proofs, for instance).

Up to isomorphism, there is just one complete graph on n vertices, denoted K_n. Similarly, given two positive integers m and n, there is a unique complete bipartite graph with parts of sizes m and n (again, up to isomorphism), denoted $K_{m,n}$. In this notation, the graphs in Figure 1.2 are K_5, $K_{3,3}$, and $K_{1,5}$, respectively. Likewise, for any positive integer n, there is a unique path on n vertices and a unique cycle on n vertices. These graphs are denoted P_n and C_n, respectively. The graphs depicted in Figure 1.3 are P_4 and C_5.

TESTING FOR ISOMORPHISM

Given two graphs on n vertices, it is certainly possible in principle to determine whether they are isomorphic. For instance, if G and H are simple, one could just consider each of the $n!$ bijections between $V(G)$ and $V(H)$ in turn, and check whether it is an isomorphism between the two graphs. If the graphs happen to be isomorphic, an isomorphism might (with luck) be found quickly. On the other hand, if they are not isomorphic, one would need to check all $n!$ bijections to discover this fact. Unfortunately, even for moderately small values of n (such as $n = 100$), the number $n!$ is unmanageably large (indeed, larger than the number of particles in the universe!), so this 'brute force' approach is not feasible. Of course, if the graphs are not regular, the number of bijections to be checked will be smaller, as an isomorphism must map each vertex to a vertex of the same degree (Exercise 1.2.1a). Nonetheless, except in particular cases, this restriction does not serve to reduce their number sufficiently. Indeed, no efficient generally applicable procedure for testing isomorphism is known. However, by employing powerful group-theoretic methods, Luks (1982) devised an efficient isomorphism-testing algorithm for cubic graphs and, more generally, for graphs of bounded maximum degree.

There is another important matter related to algorithmic questions such as graph isomorphism. Suppose that two simple graphs G and H are isomorphic. It might not be easy to find an isomorphism between them, but once such an isomorphism θ has been found, it is a simple matter to verify that θ is indeed an isomorphism: one need merely check that, for each of the $\binom{n}{2}$ pairs uv of vertices of G, $uv \in E(G)$ if and only if $\theta(u)\theta(v) \in E(H)$. On the other hand, if G and H happen not to be isomorphic, how can one verify this fact, short of checking all possible bijections between $V(G)$ and $V(H)$? In certain cases, one might be able to show that G and H are not isomorphic by isolating some structural property of G that is not shared by H, as we did for the graphs G and H of Figure 1.8. However, in general, verifying that two nonisomorphic graphs are indeed not isomorphic seems to be just as hard as determining in the first place whether they are isomorphic or not.

AUTOMORPHISMS

An *automorphism* of a graph is an isomorphism of the graph to itself. In the case of a simple graph, an automorphism is just a permutation α of its vertex set which preserves adjacency: if uv is an edge then so is $\alpha(u)\alpha(v)$.

The automorphisms of a graph reflect its symmetries. For example, if u and v are two vertices of a simple graph, and if there is an automorphism α which maps u to v, then u and v are alike in the graph, and are referred to as *similar* vertices. Graphs in which all vertices are similar, such as the complete graph K_n, the complete bipartite graph $K_{n,n}$ and the n-cube Q_n, are called *vertex-transitive*. Graphs in which no two vertices are similar are called *asymmetric*; these are the graphs which have only the identity permutation as automorphism (see Exercise 1.2.14).

Particular drawings of a graph may often be used to display its symmetries. As an example, consider the three drawings shown in Figure 1.9 of the *Petersen graph*, a graph which turns out to have many special properties. (We leave it as an exercise (1.2.5) that they are indeed drawings of one and the same graph.) The first drawing shows that the five vertices of the outer pentagon are similar (under rotational symmetry), as are the five vertices of the inner pentagon. The third drawing exhibits six similar vertices (under reflective or rotational symmetry), namely the vertices of the outer hexagon. Combining these two observations, we conclude that all ten vertices of the Petersen graph are similar, and thus that the graph is vertex-transitive.

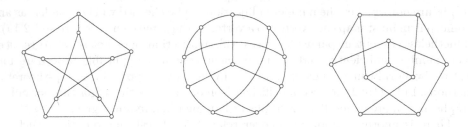

Fig. 1.9. Three drawings of the Petersen graph

We denote the set of all automorphisms of a graph G by $\mathrm{Aut}(G)$, and their number by $\mathrm{aut}(G)$. It can be verified that $\mathrm{Aut}(G)$ is a group under the operation of composition (Exercise 1.2.9). This group is called the *automorphism group* of G. The automorphism group of K_n is the symmetric group S_n, consisting of all permutations of its vertex set. In general, for any simple graph G on n vertices, $\mathrm{Aut}(G)$ is a subgroup of S_n. For instance, the automorphism group of C_n is D_n, the dihedral group on n elements (Exercise 1.2.10).

LABELLED GRAPHS

As we have seen, the edge set E of a simple graph $G = (V, E)$ is usually considered to be a subset of $\binom{V}{2}$, the set of all 2-subsets of V; edge labels may then be omitted in drawings of such graphs. A simple graph whose vertices are labelled, but whose edges are not, is referred to as a *labelled simple graph*. If $|V| = n$, there are $2^{\binom{n}{2}}$ distinct subsets of $\binom{V}{2}$, so $2^{\binom{n}{2}}$ labelled simple graphs with vertex set V. We denote by $\mathcal{G}_n$ the set of labelled simple graphs with vertex set $V := \{v_1, v_2, \ldots, v_n\}$. The set $\mathcal{G}_3$ is shown in Figure 1.10.

A *priori*, there are $n!$ ways of assigning the labels $v_1, v_2, \ldots, v_n$ to the vertices of an unlabelled simple graph on n vertices. But two of these will yield the same labelled graph if there is an automorphism of the graph mapping one labelling to the other. For example, all six labellings of K_3 result in the same element of $\mathcal{G}_3$, whereas the six labellings of P_3 yield three distinct labelled graphs, as shown in Figure 1.10. The number of distinct labellings of a given unlabelled simple graph G on n vertices is, in fact, $n!/\mathrm{aut}(G)$ (Exercise 1.2.15). Consequently,

$$\sum_G \frac{n!}{\mathrm{aut}(G)} = 2^{\binom{n}{2}}$$

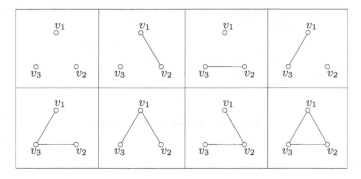

Fig. 1.10. The eight labelled graphs on three vertices

where the sum is over all unlabelled simple graphs on n vertices. In particular, the number of unlabelled simple graphs on n vertices is at least

$$\left\lceil \frac{2^{\binom{n}{2}}}{n!} \right\rceil \tag{1.2}$$

For small values of n, this bound is not particularly good. For example, there are four unlabelled simple graphs on three vertices, but the bound (1.2) is just two. Likewise, the number of unlabelled simple graphs on four vertices is eleven (Exercise 1.2.6), whereas the bound given by (1.2) is three. Nonetheless, when n is large, this bound turns out to be a good approximation to the actual number of unlabelled simple graphs on n vertices because the vast majority of graphs are asymmetric (see Exercise 1.2.15d).

Exercises

1.2.1

a) Show that any isomorphism between two graphs maps each vertex to a vertex of the same degree.
b) Deduce that isomorphic graphs necessarily have the same (nonincreasing) degree sequence.

1.2.2 Show that the graphs in Figure 1.11 are not isomorphic (even though they have the same degree sequence).

1.2.3 Let G be a connected graph. Show that every graph which is isomorphic to G is connected.

1.2.4 Determine:

a) the number of isomorphisms between the graphs G and H of Figure 1.7,

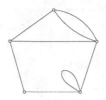 

Fig. 1.11. Nonisomorphic graphs

b) the number of automorphisms of each of these graphs.

$\star$**1.2.5** Show that the three graphs in Figure 1.9 are isomorphic.

1.2.6 Draw:

a) all the labelled simple graphs on four vertices,
b) all the unlabelled simple graphs on four vertices,
c) all the unlabelled simple cubic graphs on eight or fewer vertices.

1.2.7 Show that the n-cube Q_n and the boolean lattice BL_n (defined in Exercises 1.1.7 and 1.1.8) are isomorphic.

1.2.8 Show that two simple graphs G and H are isomorphic if and only if there exists a permutation matrix $\mathbf{P}$ such that $\mathbf{A}_H = \mathbf{P}\mathbf{A}_G\mathbf{P}^t$.

1.2.9 Show that $\mathrm{Aut}(G)$ is a group under the operation of composition.

1.2.10

a) Show that, for $n \geq 2$, $\mathrm{Aut}(P_n) \cong S_2$ and $\mathrm{Aut}(C_n) = D_n$, the dihedral group on n elements (where $\cong$ denotes isomorphism of groups; see, for example, Herstein (1996)).
b) Determine the automorphism group of the complete bipartite graph $K_{m,n}$.

1.2.11 Show that, for any simple graph G, $\mathrm{Aut}(G) = \mathrm{Aut}(\overline{G})$.

1.2.12 Consider the subgroup Γ of S_3 with elements $(1)(2)(3)$, (123), and (132).

a) Show that there is no simple graph whose automorphism group is Γ.
b) Find a simple graph whose automorphism group is isomorphic to Γ.
   (Frucht (1938) showed that every abstract group is isomorphic to the automorphism group of some simple graph.)

1.2.13 ORBITS OF A GRAPH

a) Show that similarity is an equivalence relation on the vertex set of a graph.
b) The equivalence classes with respect to similarity are called the *orbits* of the graph. Determine the orbits of the graphs in Figure 1.12.

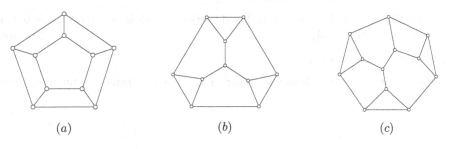

(a) (b) (c)

Fig. 1.12. Determine the orbits of these graphs (Exercise 1.2.13)

1.2.14

a) Show that there is no asymmetric simple graph on five or fewer vertices.
b) For each $n \geq 6$, find an asymmetric simple graph on n vertices.

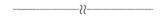

1.2.15 Let G and H be isomorphic members of $\mathcal{G}_n$, let θ be an isomorphism between G and H, and let α be an automorphism of G.

a) Show that $\theta\alpha$ is an isomorphism between G and H.
b) Deduce that the set of all isomorphisms between G and H is the coset $\theta\text{Aut}(G)$ of $\text{Aut}(G)$.
c) Deduce that the number of labelled graphs isomorphic to G is equal to $n!/\text{aut}(G)$.
d) Erdős and Rényi (1963) have shown that almost all simple graphs are asymmetric (that is, the proportion of simple graphs on n vertices that are asymmetric tends to one as n tends to infinity). Using this fact, deduce from (c) that the number of unlabelled graphs on n vertices is asymptotically equal to $2^{\binom{n}{2}}/n!$ (G. PÓLYA)

1.2.16 SELF-COMPLEMENTARY GRAPH
A simple graph is *self-complementary* if it is isomorphic to its complement. Show that:

a) each of the graphs P_4 and C_5 (shown in Figure 1.3) is self-complementary,
b) every self-complementary graph is connected,
c) if G is self-complementary, then $n \equiv 0, 1 \pmod 4$,
d) every self-complementary graph on $4k + 1$ vertices has a vertex of degree $2k$.

1.2.17 EDGE-TRANSITIVE GRAPH
A simple graph is *edge-transitive* if, for any two edges uv and xy, there is an automorphism α such that $\alpha(u)\alpha(v) = xy$.

a) Show that the Petersen graph is edge-transitive.
b) Find a graph which is vertex-transitive but not edge-transitive.

c) Show that any graph without isolated vertices which is edge-transitive but not vertex-transitive is bipartite. (E. DAUBER)

1.2.18 THE FOLKMAN GRAPH

a) Show that the graph shown in Figure 1.13a is edge-transitive but not vertex-transitive.

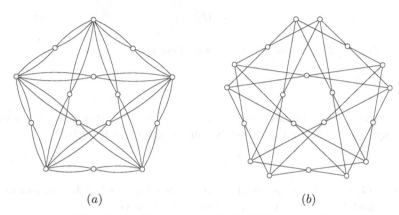

(a)   (b)

Fig. 1.13. Construction of the Folkman graph

b) The *Folkman graph*, depicted in Figure 1.13b, is the 4-regular graph obtained from the graph of Figure 1.13a by replacing each vertex v of degree eight by two vertices of degree four, both of which have the same four neighbours as v. Show that the Folkman graph is edge-transitive but not vertex-transitive. (J. FOLKMAN)

1.2.19 GENERALIZED PETERSEN GRAPH
Let k and n be positive integers, with $n > 2k$. The *generalized Petersen graph* $P_{k,n}$ is the simple graph with vertices $x_1, x_2, \ldots, x_n, y_1, y_2, \ldots, y_n$, and edges $x_i x_{i+1}, y_i y_{i+k}, x_i y_i$, $1 \le i \le n$, indices being taken modulo n. (Note that $P_{2,5}$ is the Petersen graph.)

a) Draw the graphs $P_{2,7}$ and $P_{3,8}$.
b) Which of these two graphs are vertex-transitive, and which are edge-transitive?

1.2.20 Show that if G is simple and the eigenvalues of $\mathbf{A}$ are distinct, then every automorphism of G is of order one or two. (A. MOWSHOWITZ)

1.3 Graphs Arising from Other Structures

As remarked earlier, interesting graphs can often be constructed from geometric and algebraic objects. Such constructions are often quite straightforward, but in some instances they rely on experience and insight.

POLYHEDRAL GRAPHS

A *polyhedral graph* is the 1-skeleton of a polyhedron, that is, the graph whose vertices and edges are just the vertices and edges of the polyhedron, with the same incidence relation. In particular, the five platonic solids (the tetrahedron, the cube, the octahedron, the dodecahedron, and the icosahedron) give rise to the five *platonic graphs* shown in Figure 1.14. For classical polyhedra such as these, we give the graph the same name as the polyhedron from which it is derived.

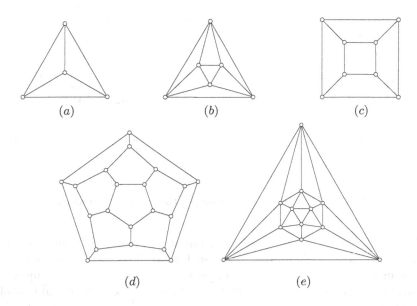

Fig. 1.14. The five platonic graphs: (a) the tetrahedron, (b) the octahedron, (c) the cube, (d) the dodecahedron, (e) the icosahedron

SET SYSTEMS AND HYPERGRAPHS

A *set system* is an ordered pair $(V, \mathcal{F})$, where V is a set of elements and $\mathcal{F}$ is a family of subsets of V. Note that when $\mathcal{F}$ consists of pairs of elements of V, the set system $(V, \mathcal{F})$ is a loopless graph. Thus set systems can be thought of as generalizations of graphs, and are usually referred to as *hypergraphs*, particularly when one seeks to extend properties of graphs to set systems (see Berge (1973)). The elements of V are then called the *vertices* of the hypergraph, and the elements of $\mathcal{F}$ its *edges* or *hyperedges*. A hypergraph is *k-uniform* if each edge is a k-set (a set of k elements). As we show below, set systems give rise to graphs in two principal ways: incidence graphs and intersection graphs.

Many interesting examples of hypergraphs are provided by geometric configurations. A *geometric configuration* $(P, \mathcal{L})$ consists of a finite set P of elements called *points*, and a finite family $\mathcal{L}$ of subsets of P called *lines*, with the property that at most one line contains any given pair of points. Two classical examples of geometric configurations are the *Fano plane* and the *Desargues configuration*. These two configurations are shown in Figure 1.15. In both cases, each line consists of three points. These configurations thus give rise to 3-uniform hypergraphs; the *Fano hypergraph* has seven vertices and seven edges, the *Desargues hypergraph* ten vertices and ten edges.

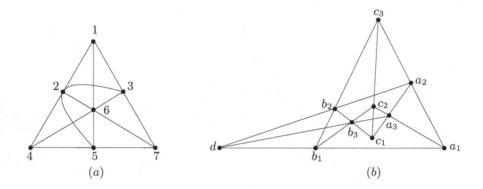

Fig. 1.15. (a) The Fano plane, and (b) the Desargues configuration

The Fano plane is the simplest of an important family of geometric configurations, the *projective planes* (see Exercise 1.3.13). The Desargues configuration arises from a well-known theorem in projective geometry. Other examples of interesting geometric configurations are described in Coxeter (1950) and Godsil and Royle (2001).

INCIDENCE GRAPHS

A natural graph associated with a set system $H = (V, \mathcal{F})$ is the bipartite graph $G[V, \mathcal{F}]$, where $v \in V$ and $F \in \mathcal{F}$ are adjacent if $v \in F$. This bipartite graph G is called the *incidence graph* of the set system H, and the bipartite adjacency matrix of G the *incidence matrix* of H; these are simply alternative ways of representing a set system. Incidence graphs of geometric configurations often give rise to interesting bipartite graphs; in this context, the incidence graph is sometimes called the *Levi graph* of the configuration. The incidence graph of the Fano plane is shown in Figure 1.16. This graph is known as the *Heawood graph*.

INTERSECTION GRAPHS

With each set system $(V, \mathcal{F})$ one may associate its *intersection graph*. This is the graph whose vertex set is $\mathcal{F}$, two sets in $\mathcal{F}$ being adjacent if their intersection is nonempty. For instance, when V is the vertex set of a simple graph G and $\mathcal{F} := E$,

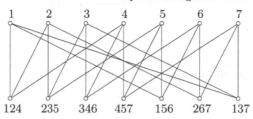

Fig. 1.16. The incidence graph of the Fano plane: the Heawood graph

the edge set of G, the intersection graph of $(V, \mathcal{F})$ has as vertices the edges of G, two edges being adjacent if they have an end in common. For historical reasons, this graph is known as the *line graph* of G and denoted $L(G)$. Figure 1.17 depicts a graph and its line graph.

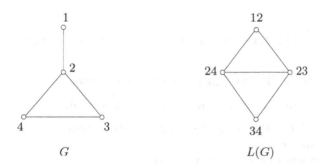

Fig. 1.17. A graph and its line graph

It can be shown that the intersection graph of the Desargues configuration is isomorphic to the line graph of K_5, which in turn is isomorphic to the complement of the Petersen graph (Exercise 1.3.2). As for the Fano plane, its intersection graph is isomorphic to K_7, because any two of its seven lines have a point in common.

The definition of the line graph $L(G)$ may be extended to all loopless graphs G as being the graph with vertex set E in which two vertices are joined by just as many edges as their number of common ends in G.

When $V = \mathbb{R}$ and $\mathcal{F}$ is a set of closed intervals of $\mathbb{R}$, the intersection graph of $(V, \mathcal{F})$ is called an *interval graph*. Examples of practical situations which give rise to interval graphs can be found in the book by Berge (1973). Berge even wrote a detective story whose resolution relies on the theory of interval graphs; see Berge (1995).

It should be evident from the above examples that graphs are implicit in a wide variety of structures. Many such graphs are not only interesting in their own right but also serve to provide insight into the structures from which they arise.

Exercises

1.3.1

a) Show that the graph in Figure 1.18 is isomorphic to the Heawood graph (Figure 1.16).

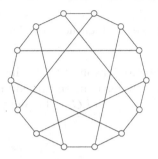

Fig. 1.18. Another drawing of the Heawood graph

b) Deduce that the Heawood graph is vertex-transitive.

1.3.2 Show that the following three graphs are isomorphic:

▷ the intersection graph of the Desargues configuration,
▷ the line graph of K_5,
▷ the complement of the Petersen graph.

1.3.3 Show that the line graph of $K_{3,3}$ is self-complementary.

1.3.4 Show that neither of the graphs displayed in Figure 1.19 is a line graph.

1.3.5 Let $H := (V, \mathcal{F})$ be a hypergraph. The number of edges incident with a vertex v of H is its *degree*, denoted $d(v)$. A *degree sequence* of H is a vector $\mathbf{d} := (d(v) : v \in V)$. Let $\mathbf{M}$ be the incidence matrix of H and $\mathbf{d}$ the corresponding degree sequence of H. Show that the sum of the columns of $\mathbf{M}$ is equal to $\mathbf{d}$.

1.3.6 Let $H := (V, \mathcal{F})$ be a hypergraph. For $v \in V$, let $\mathcal{F}_v$ denote the set of edges of H incident to v. The *dual* of H is the hypergraph H^* whose vertex set is $\mathcal{F}$ and whose edges are the sets $\mathcal{F}_v$, $v \in V$.

Fig. 1.19. Two graphs that are not line graphs

a) How are the incidence graphs of H and H^* related?
b) Show that the dual of H^* is isomorphic to H.
c) A hypergraph is *self-dual* if it is isomorphic to its dual. Show that the Fano and Desargues hypergraphs are self-dual.

1.3.7 HELLY PROPERTY

A family of sets has the *Helly Property* if the members of each pairwise intersecting subfamily have an element in common.

a) Show that the family of closed intervals on the real line has the Helly Property.
(E. HELLY)

b) Deduce that the graph in Figure 1.20 is not an interval graph.

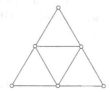

Fig. 1.20. A graph that is not an interval graph

1.3.8 KNESER GRAPH

Let m and n be positive integers, where $n > 2m$. The *Kneser graph* $KG_{m,n}$ is the graph whose vertices are the m-subsets of an n-set S, two such subsets being adjacent if and only if their intersection is empty. Show that:

a) $KG_{1,n} \cong K_n$, $n \geq 3$,
b) $KG_{2,n}$ is isomorphic to the complement of $L(K_n)$, $n \geq 5$.

1.3.9 Let G be a simple graph with incidence matrix $\mathbf{M}$.

a) Show that the adjacency matrix of its line graph $L(G)$ is $\mathbf{M}^t\mathbf{M} - 2\mathbf{I}$, where $\mathbf{I}$ is the $m \times m$ identity matrix.
b) Using the fact that $\mathbf{M}^t\mathbf{M}$ is positive-semidefinite, deduce that:
 i) each eigenvalue of $L(G)$ is at least -2,
 ii) if the rank of $\mathbf{M}$ is less than m, then -2 is an eigenvalue of $L(G)$.

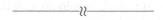

1.3.10

a) Consider the following two matrices $\mathbf{B}$ and $\mathbf{C}$, where x is an indeterminate, $\mathbf{M}$ is an arbitrary $n \times m$ matrix, and $\mathbf{I}$ is an identity matrix of the appropriate dimension.

$$\mathbf{B} := \begin{bmatrix} \mathbf{I} & \mathbf{M} \\ \mathbf{M}^t & x\mathbf{I} \end{bmatrix} \qquad \mathbf{C} := \begin{bmatrix} x\mathbf{I} & -\mathbf{M} \\ \mathbf{0} & \mathbf{I} \end{bmatrix}$$

By equating the determinants of **BC** and **CB**, derive the identity

$$\det(x\mathbf{I} - \mathbf{M}^t\mathbf{M}) = x^{m-n}\det(x\mathbf{I} - \mathbf{M}\mathbf{M}^t)$$

b) Let G be a simple k-regular graph with $k \geq 2$. By appealing to Exercise 1.3.9 and using the above identity, establish the following relationship between the characteristic polynomials of $L(G)$ and G.

$$\det(\mathbf{A}_{L(G)} - x\mathbf{I}) = (-1)^{m-n}(x+2)^{m-n}\det(\mathbf{A}_G - (x+2-k)\mathbf{I})$$

c) Deduce that:
  i) to each eigenvalue $\lambda \neq -k$ of G, there corresponds an eigenvalue $\lambda + k - 2$ of $L(G)$, with the same multiplicity,
  ii) -2 is an eigenvalue of $L(G)$ with multiplicity $m - n + r$, where r is the multiplicity of the eigenvalue $-k$ of G. (If $-k$ is not an eigenvalue of G then $r = 0$.)
  (H. SACHS)

1.3.11

a) Using Exercises 1.1.22 and 1.3.10, show that the eigenvalues of $L(K_5)$ are

$$(6, 1, 1, 1, 1, -2, -2, -2, -2, -2)$$

b) Applying Exercise 1.1.23, deduce that the Petersen graph has eigenvalues

$$(3, 1, 1, 1, 1, 1, -2, -2, -2, -2)$$

1.3.12 SPERNER'S LEMMA

Let T be a triangle in the plane. A subdivision of T into triangles is *simplicial* if any two of the triangles which intersect have either a vertex or an edge in common. Consider an arbitrary simplicial subdivision of T into triangles. Assign the colours red, blue, and green to the vertices of these triangles in such a way that each colour is missing from one side of T but appears on the other two sides. (Thus, in particular, the vertices of T are assigned the colours red, blue, and green in some order.)

a) Show that the number of triangles in the subdivision whose vertices receive all three colours is odd.   (E. SPERNER)

b) Deduce that there is always at least one such triangle.

(Sperner's Lemma, generalized to n-dimensional simplices, is the key ingredient in a proof of Brouwer's Fixed Point Theorem: *every continuous mapping of a closed n-disc to itself has a fixed point*; see Bondy and Murty (1976).)

1.3.13 FINITE PROJECTIVE PLANE

A *finite projective plane* is a geometric configuration $(P, \mathcal{L})$ in which:

  i) any two points lie on exactly one line,
  ii) any two lines meet in exactly one point,

iii) there are four points no three of which lie on a line.

(Condition (iii) serves only to exclude two trivial configurations — the *pencil*, in which all points are collinear, and the *near-pencil*, in which all but one of the points are collinear.)

a) Let $(P, \mathcal{L})$ be a finite projective plane. Show that there is an integer $n \geq 2$ such that $|P| = |\mathcal{L}| = n^2 + n + 1$, each point lies on $n + 1$ lines, and each line contains $n + 1$ points (the instance $n = 2$ being the Fano plane). This integer n is called the *order* of the projective plane.
b) How many vertices has the incidence graph of a finite projective plane of order n, and what are their degrees?

1.3.14 Consider the nonzero vectors in $\mathbb{F}^3$, where $\mathbb{F} = GF(q)$ and q is a prime power. Define two of these vectors to be *equivalent* if one is a multiple of the other. One can form a finite projective plane $(P, \mathcal{L})$ of order q by taking as points and lines the $(q^3 - 1)/(q - 1) = q^2 + q + 1$ equivalence classes defined by this equivalence relation and defining a point (a, b, c) and line (x, y, z) to be incident if $ax + by + cz = 0$ (in $GF(q)$). This plane is denoted $PG_{2,q}$.

a) Show that $PG_{2,2}$ is isomorphic to the Fano plane.
b) Construct $PG_{2,3}$.

1.3.15 THE DE BRUIJN–ERDŐS THEOREM

a) Let $G[X, Y]$ be a bipartite graph, each vertex of which is joined to at least one, but not all, vertices in the other part. Suppose that $d(x) \geq d(y)$ for all $xy \notin E$. Show that $|Y| \geq |X|$, with equality if and only if $d(x) = d(y)$ for all $xy \notin E$ with $x \in X$ and $y \in Y$.
b) Deduce the following theorem.
Let $(P, \mathcal{L})$ be a geometric configuration in which any two points lie on exactly one line and not all points lie on a single line. Then $|\mathcal{L}| \geq |P|$. Furthermore, if $|\mathcal{L}| = |P|$, then $(P, \mathcal{L})$ is either a finite projective plane or a near-pencil.
(N.G. DE BRUIJN AND P. ERDŐS)

1.3.16 Show that:

a) the line graphs $L(K_n)$, $n \geq 4$, and $L(K_{n,n})$, $n \geq 2$, are strongly regular,
b) the *Shrikhande graph*, displayed in Figure 1.21 (where vertices with the same label are to be identified), is strongly regular, with the same parameters as those of $L(K_{4,4})$, but is not isomorphic to $L(K_{4,4})$.

1.3.17

a) Show that:
   i) $\mathrm{Aut}(L(K_n)) \not\cong \mathrm{Aut}(K_n)$ for $n = 2$ and $n = 4$,
   ii) $\mathrm{Aut}(L(K_n)) \cong \mathrm{Aut}(K_n)$ for $n = 3$ and $n \geq 5$.
b) Appealing to Exercises 1.2.11 and 1.3.2, deduce that the automorphism group of the Petersen graph is isomorphic to the symmetric group S_5.

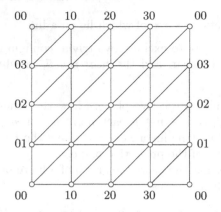

Fig. 1.21. An embedding of the Shrikhande graph on the torus

1.3.18 CAYLEY GRAPH

Let Γ be a group, and let S be a set of elements of Γ not including the identity element. Suppose, furthermore, that the inverse of every element of S also belongs to S. The *Cayley graph* of Γ with respect to S is the graph $\mathrm{CG}(\Gamma, S)$ with vertex set Γ in which two vertices x and y are adjacent if and only if $xy^{-1} \in S$. (Note that, because S is closed under taking inverses, if $xy^{-1} \in S$, then $yx^{-1} \in S$.)

a) Show that the n-cube is a Cayley graph.
b) Let G be a Cayley graph $\mathrm{CG}(\Gamma, S)$ and let x be an element of Γ.
   i) Show that the mapping α_x defined by the rule that $\alpha_x(y) := yx$ is an automorphism of G.
   ii) Deduce that every Cayley graph is vertex-transitive.
c) By considering the Petersen graph, show that not every vertex-transitive graph is a Cayley graph.

1.3.19 CIRCULANT

A *circulant* is a Cayley graph $\mathrm{CG}(\mathbb{Z}_n, S)$, where $\mathbb{Z}_n$ is the additive group of integers modulo n. Let p be a prime, and let i and j be two nonzero elements of $\mathbb{Z}_p$.

a) Show that $\mathrm{CG}(\mathbb{Z}_p, \{i, -i\}) \cong \mathrm{CG}(\mathbb{Z}_p, \{j, -j\})$.
b) Determine when $\mathrm{CG}(\mathbb{Z}_p, \{1, -1, i, -i\}) \cong \mathrm{CG}(\mathbb{Z}_p, \{1, -1, j, -j\})$.

1.3.20 PALEY GRAPH

Let q be a prime power, $q \equiv 1 \pmod{4}$. The *Paley graph* PG_q is the graph whose vertex set is the set of elements of the field $GF(q)$, two vertices being adjacent if their difference is a nonzero square in $GF(q)$.

a) Draw PG_5, PG_9, and PG_{13}.
b) Show that these three graphs are self-complementary.
c) Let a be a nonsquare in $GF(q)$. By considering the mapping $\theta : GF(q) \to GF(q)$ defined by $\theta(x) := ax$, show that PG_q is self-complementary for all q.

1.4 Constructing Graphs from Other Graphs

We have already seen a couple of ways in which we may associate with each graph another graph: the complement (in the case of simple graphs) and the line graph. If we start with two graphs G and H rather than just one, a new graph may be defined in several ways. For notational simplicity, we assume that G and H are simple, so that each edge is an unordered pair of vertices; the concepts described here can be extended without difficulty to the general context.

UNION AND INTERSECTION

Two graphs are *disjoint* if they have no vertex in common, and *edge-disjoint* if they have no edge in common. The most basic ways of combining graphs are by union and intersection. The *union* of simple graphs G and H is the graph $G \cup H$ with vertex set $V(G) \cup V(H)$ and edge set $E(G) \cup E(H)$. If G and H are disjoint, we refer to their union as a *disjoint union*, and generally denote it by $G + H$. These operations are associative and commutative, and may be extended to an arbitrary number of graphs. It can be seen that a graph is disconnected if and only if it is a disjoint union of two (nonnull) graphs. More generally, every graph G may be expressed uniquely (up to order) as a disjoint union of connected graphs (Exercise 1.4.1). These graphs are called the *connected components*, or simply the *components*, of G. The number of components of G is denoted $c(G)$. (The null graph has the anomalous property of being the only graph without components.)

The *intersection* $G \cap H$ of G and H is defined analogously. (Note that when G and H are disjoint, their intersection is the null graph.) Figure 1.22 illustrates these concepts. The graph $G \cup H$ shown in Figure 1.22 has just one component, whereas the graph $G \cap H$ has two components.

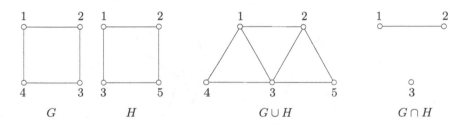

Fig. 1.22. The union and intersection of two graphs

CARTESIAN PRODUCT

There are also several ways of forming from two graphs a new graph whose vertex set is the cartesian product of their vertex sets. These constructions are consequently referred to as 'products'. We now describe one of them.

The *cartesian product* of simple graphs G and H is the graph $G \square H$ whose vertex set is $V(G) \times V(H)$ and whose edge set is the set of all pairs $(u_1, v_1)(u_2, v_2)$ such that either $u_1 u_2 \in E(G)$ and $v_1 = v_2$, or $v_1 v_2 \in E(H)$ and $u_1 = u_2$. Thus, for each edge $u_1 u_2$ of G and each edge $v_1 v_2$ of H, there are four edges in $G \square H$, namely $(u_1, v_1)(u_2, v_1)$, $(u_1, v_2)(u_2, v_2)$, $(u_1, v_1)(u_1, v_2)$, and $(u_2, v_1)(u_2, v_2)$ (see Figure 1.23a); the notation used for the cartesian product reflects this fact. More generally, the cartesian product $P_m \square P_n$ of two paths is the $(m \times n)$-*grid*. An example is shown in Figure 1.23b.

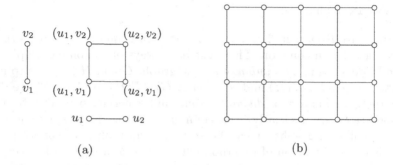

(a)   (b)

Fig. 1.23. (a) The cartesian product $K_2 \square K_2$, and (b) the (5×4)-grid

For $n \geq 3$, the cartesian product $C_n \square K_2$ is a polyhedral graph, the *n-prism*; the 3-prism, 4-prism, and 5-prism are commonly called the *triangular prism*, the *cube*, and the *pentagonal prism* (see Figure 1.24). The cartesian product is arguably the most basic of graph products. There exist a number of others, each arising naturally in various contexts. We encounter several of these in later chapters.

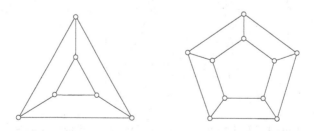

Fig. 1.24. The triangular and pentagonal prisms

Exercises

1.4.1 Show that every graph may be expressed uniquely (up to order) as a disjoint union of connected graphs.

1.4.2 Show that the rank over $GF(2)$ of the incidence matrix of a graph G is $n - c$.

1.4.3 Show that the cartesian product is both associative and commutative.

1.4.4 Find an embedding of the cartesian product $C_m \square C_n$ on the torus.

1.4.5

a) Show that the cartesian product of two vertex-transitive graphs is vertex-transitive.
b) Give an example to show that the cartesian product of two edge-transitive graphs need not be edge-transitive.

1.4.6

a) Let G be a self-complementary graph and let P be a path of length three disjoint from G. Form a new graph H from $G \cup P$ by joining the first and third vertices of P to each vertex of G. Show that H is self-complementary.
b) Deduce (by appealing to Exercise 1.2.16) that there exists a self-complementary graph on n vertices if and only if $n \equiv 0, 1 \pmod 4$.

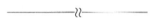

1.5 Directed Graphs

Although many problems lend themselves to graph-theoretic formulation, the concept of a graph is sometimes not quite adequate. When dealing with problems of traffic flow, for example, it is necessary to know which roads in the network are one-way, and in which direction traffic is permitted. Clearly, a graph of the network is not of much use in such a situation. What we need is a graph in which each link has an assigned orientation, namely a directed graph.

Formally, a *directed graph* D is an ordered pair $(V(D), A(D))$ consisting of a set $V := V(D)$ of *vertices* and a set $A := A(D)$, disjoint from $V(D)$, of *arcs*, together with an *incidence function* ψ_D that associates with each arc of D an ordered pair of (not necessarily distinct) vertices of D. If a is an arc and $\psi_D(a) = (u, v)$, then a is said to *join* u to v; we also say that u *dominates* v. The vertex u is the *tail* of a, and the vertex v its *head*; they are the two *ends* of a. Occasionally, the orientation of an arc is irrelevant to the discussion. In such instances, we refer to the arc as an *edge* of the directed graph. The number of arcs in D is denoted by $a(D)$. The vertices which dominate a vertex v are its *in-neighbours*, those which are dominated by the vertex its *outneighbours*. These sets are denoted by $N_D^-(v)$ and $N_D^+(v)$, respectively.

For convenience, we abbreviate the term 'directed graph' to *digraph*. A *strict* digraph is one with no loops or parallel arcs (arcs with the same head and the same tail).

With any digraph D, we can associate a graph G on the same vertex set simply by replacing each arc by an edge with the same ends. This graph is the *underlying graph* of D, denoted $G(D)$. Conversely, any graph G can be regarded as a digraph, by replacing each of its edges by two oppositely oriented arcs with the same ends; this digraph is the *associated digraph* of G, denoted $D(G)$. One may also obtain a digraph from a graph G by replacing each edge by just one of the two possible arcs with the same ends. Such a digraph is called an *orientation* of G. We occasionally use the symbol $\overrightarrow{G}$ to specify an orientation of G (even though a graph generally has many orientations). An orientation of a simple graph is referred to as an *oriented graph*. One particularly interesting instance is an orientation of a complete graph. Such an oriented graph is called a *tournament*, because it can be viewed as representing the results of a round-robin tournament, one in which each team plays every other team (and there are no ties).

Digraphs, like graphs, have a simple pictorial representation. A digraph is represented by a diagram of its underlying graph together with arrows on its edges, each arrow pointing towards the head of the corresponding arc. The four unlabelled tournaments on four vertices are shown in Figure 1.25 (see Exercise 1.5.3a).

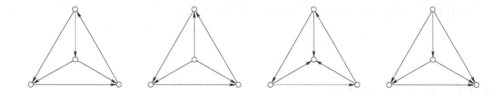

Fig. 1.25. The four unlabelled tournaments on four vertices

Every concept that is valid for graphs automatically applies to digraphs too. For example, the *degree* of a vertex v in a digraph D is simply the degree of v in $G(D)$, the underlying graph of D.[1] Likewise, a digraph is said to be *connected* if its underlying graph is connected.[2] But there are concepts in which orientations play an essential role. For instance, the *indegree* $d_D^-(v)$ of a vertex v in D is the number of arcs with head v, and the *outdegree* $d_D^+(v)$ of v is the number of arcs with tail v. The minimum indegree and outdegree of D are denoted by $\delta^-(D)$ and $\delta^+(D)$, respectively; likewise, the maximum indegree and outdegree of D are

---

[1] *In such cases, we employ the same notation as for graphs (with G replaced by D). Thus the degree of v in D is denoted by $d_D(v)$. These instances of identical notation are recorded only once in the glossaries, namely for graphs.*

[2] *The index includes only those definitions for digraphs which differ substantively from their analogues for graphs. Thus the term 'connected digraph' does not appear there, only 'connected graph'.*

denoted $\Delta^-(D)$ and $\Delta^+(D)$, respectively. A digraph is k-*diregular* if each indegree and each outdegree is equal to k. A vertex of indegree zero is called a *source*, one of outdegree zero a *sink*. A *directed path* or *directed cycle* is an orientation of a path or cycle in which each vertex dominates its successor in the sequence. There is also a notion of connectedness in digraphs which takes directions into account, as we shall see in Chapter 2.

Two special digraphs are shown in Figure 1.26. The first of these is a 2-diregular digraph, the second a 3-diregular digraph (see Bondy (1978)); we adopt here the convention of representing two oppositely oriented arcs by an edge. These digraphs can both be constructed from the Fano plane (Exercise 1.5.8). They also possess other unusual properties, to be described in Chapter 2.

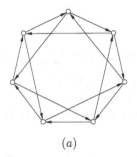 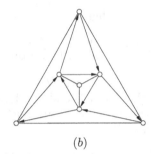

(a)   (b)

Fig. 1.26. (a) the Koh–Tindell digraph, and (b) a directed analogue of the Petersen graph

Further examples of interesting digraphs can be derived from other mathematical structures, such as groups. For example, there is a natural directed analogue of a Cayley graph. If Γ is a group, and S a subset of Γ not including the identity element, the *Cayley digraph* of Γ with respect to S is the digraph, denoted $\mathrm{CD}(\Gamma, S)$, whose vertex set is Γ and in which vertex x dominates vertex y if and only if $xy^{-1} \in S$. A *directed circulant* is a Cayley digraph $\mathrm{CD}(\mathbb{Z}_n, S)$, where $\mathbb{Z}_n$ is the group of integers modulo n. The *Koh–Tindell digraph* of Figure 1.26a is a directed circulant based on $\mathbb{Z}_7$.

With each digraph D, one may associate another digraph, $\overleftarrow{D}$, obtained by reversing each arc of D. The digraph $\overleftarrow{D}$ is called the *converse* of D. Because the converse of the converse is just the original digraph, the converse of a digraph can be thought of as its 'directional dual'. This point of view gives rise to a simple yet useful principle.

PRINCIPLE OF DIRECTIONAL DUALITY
Any statement about a digraph has an accompanying 'dual' statement, obtained by applying the statement to the converse of the digraph and reinterpreting it in terms of the original digraph.

For instance, the sum of the indegrees of the vertices of a digraph is equal to the total number of arcs (Exercise 1.5.2). Applying the Principle of Directional Duality, we immediately deduce that the sum of the outdegrees is also equal to the number of arcs.

Apart from the practical aspect mentioned earlier, assigning suitable orientations to the edges of a graph is a convenient way of exploring properties of the graph, as we shall see in Chapter 6.

Exercises

1.5.1 How many orientations are there of a labelled graph G?

$\star$**1.5.2** Let D be a digraph.

  a) Show that $\sum_{v \in V} d^-(v) = m$.
  b) Using the Principle of Directional Duality, deduce that $\sum_{v \in V} d^+(v) = m$.

1.5.3 Two digraphs D and D' are *isomorphic*, written $D \cong D'$, if there are bijections $\theta : V(D) \to V(D')$ and $\phi : A(D) \to A(D')$ such that $\psi_D(a) = (u, v)$ if and only if $\psi_{D'}(\phi(a)) = (\theta(u), \theta(v))$. Such a pair of mappings is called an *isomorphism* between D and D'.

  a) Show that the four tournaments in Figure 1.25 are pairwise nonisomorphic, and that these are the only ones on four vertices, up to isomorphism.
  b) How many tournaments are there on five vertices, up to isomorphism?

1.5.4

  a) Define the notions of vertex-transitivity and arc-transitivity for digraphs.
  b) Show that:
     i) every vertex-transitive digraph is diregular,
     ii) the Koh–Tindell digraph (Figure 1.26a) is vertex-transitive but not arc-transitive.

1.5.5 A digraph is *self-converse* if it is isomorphic to its converse. Show that both digraphs in Figure 1.26 are self-converse.

1.5.6 INCIDENCE MATRIX OF A DIGRAPH
Let D be a digraph with vertex set V and arc set A. The *incidence matrix* of D (with respect to given orderings of its vertices and arcs) is the $n \times m$ matrix $\mathbf{M}_D := (m_{va})$, where

$$
m_{va} = \begin{cases} 1 & \text{if arc } a \text{ is a link and vertex } v \text{ is the tail of } a \\ -1 & \text{if arc } a \text{ is a link and vertex } v \text{ is the head of } a \\ 0 & \text{otherwise} \end{cases}
$$

Let $\mathbf{M}$ be the incidence matrix of a connected digraph D. Show that the rank of $\mathbf{M}$ is $n - 1$.

⋆**1.5.7** TOTALLY UNIMODULAR MATRIX
A matrix is *totally unimodular* if each of its square submatrices has determinant equal to 0, +1, or −1. Let $\mathbf{M}$ be the incidence matrix of a digraph.

a) Show that $\mathbf{M}$ is totally unimodular.   (H. POINCARÉ)
b) Deduce that the matrix equation $\mathbf{Mx} = \mathbf{b}$ has a solution in integers provided that it is consistent and the vector $\mathbf{b}$ is integral.

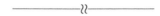

1.5.8 Describe how the two digraphs in Figure 1.26 can be constructed from the Fano plane.

1.5.9 PALEY TOURNAMENT
Let q be a prime power, $q \equiv 3 \pmod 4$. The *Paley tournament* PT_q is the tournament whose vertex set is the set of elements of the field $GF(q)$, vertex i dominating vertex j if and only if $j - i$ is a nonzero square in $GF(q)$.

a) Draw PT_3, PT_7, and PT_{11}.
b) Show that these three digraphs are self-converse.

1.5.10 STOCKMEYER TOURNAMENT
For a nonzero integer k, let pow (k) denote the greatest integer p such that 2^p divides k, and set odd $(k) := k/2^p$. (For example, pow $(12) = 2$ and odd $(12) = 3$, whereas pow $(-1) = 0$ and odd $(-1) = -1$.) The *Stockmeyer tournament* ST_n, where $n \geq 1$, is the tournament whose vertex set is $\{1, 2, 3, \ldots, 2^n\}$ in which vertex i dominates vertex j if odd $(j - i) \equiv 1 \pmod 4$.

a) Draw ST_2 and ST_3.
b) Show that ST_n is both self-converse and asymmetric (that is, has no nontrivial automorphisms).   (P.K. STOCKMEYER)

1.5.11 ARC-TRANSITIVE GRAPH
An undirected graph G is *arc-transitive* if its associated digraph $D(G)$ is arc-transitive. (Equivalently, G is arc-transitive if, given any two ordered pairs (x, y) and (u, v) of adjacent vertices, there exists an automorphism of G which maps (x, y) to (u, v).)

a) Show that any graph which is arc-transitive is both vertex-transitive and edge-transitive.
b) Let G be a k-regular graph which is both vertex-transitive and edge-transitive, but not arc-transitive. Show that k is even. (An example of such a graph with $k = 4$ may be found in Godsil and Royle (2001).)

1.5.12 ADJACENCY MATRIX OF A DIGRAPH
The *adjacency matrix* of a digraph D is the $n \times n$ matrix $\mathbf{A}_D = (a_{uv})$, where a_{uv} is the number of arcs in D with tail u and head v. Let $\mathbf{A}$ be the adjacency matrix of a tournament on n vertices. Set $\mathbf{B} := \mathbf{A} - \mathbf{A}^T$. Show that rank $\mathbf{B} = n - 1$ if n is odd and rank $\mathbf{B} = n$ if n is even.

1.6 Infinite Graphs

As already mentioned, the graphs studied in this book are assumed to be finite. There is, however, an extensive theory of graphs defined on infinite sets of vertices and/or edges. Such graphs are known as *infinite graphs*. An infinite graph is *countable* if both its vertex and edge sets are countable. Figure 1.27 depicts three well-known countable graphs, the *square lattice*, the *triangular lattice*, and the *hexagonal lattice*.

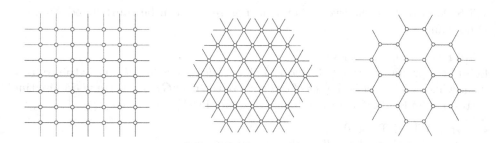

Fig. 1.27. The square, triangular and hexagonal lattices

Most notions that are valid for finite graphs are either directly applicable to infinite graphs or else require some simple modification. Whereas the definition of the degree of a vertex is essentially the same as for finite graphs (with 'number' replaced by 'cardinality'), there are two types of infinite path, one having an initial but no terminal vertex (called a *one-way infinite path*), and one having neither initial nor terminal vertices (called a *two-way infinite path*); the square lattice is the cartesian product of two two-way infinite paths. However, certain concepts for finite graphs have no natural 'infinite' analogue, the cycle for instance (although, in some circumstances, a two-way infinite path may be regarded as an infinite cycle).

While the focus of this book is on finite graphs, we include occasional remarks and exercises on infinite graphs, mainly to illustrate the differences between finite and infinite graphs. Readers interested in pursuing the topic are referred to the survey article by Thomassen (1983a) or the book by Diestel (2005), which includes a chapter on infinite graphs.

Exercises

1.6.1 Locally Finite Graph
An infinite graph is *locally finite* if every vertex is of finite degree. Give an example of a locally finite graph in which no two vertices have the same degree.

1.6.2 For each positive integer d, describe a simple infinite planar graph with minimum degree d. (We shall see, in Chapter 10, that every simple finite planar graph has a vertex of degree at most five.)

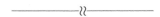

1.6.3 Give an example of a self-complementary infinite graph.

1.6.4 Unit Distance Graph
The *unit distance graph* on a subset V of $\mathbb{R}^2$ is the graph with vertex set V in which two vertices (x_1, y_1) and (x_2, y_2) are adjacent if their euclidean distance is equal to 1, that is, if $(x_1 - x_2)^2 + (y_1 - y_2)^2 = 1$. When $V = \mathbb{Q}^2$, this graph is called the *rational* unit distance graph, and when $V = \mathbb{R}^2$, the *real* unit distance graph. (Note that these are both infinite graphs.)

a) Let V be a finite subset of the vertex set of the infinite 2-dimensional integer lattice (see Figure 1.27), and let d be an odd positive integer. Denote by G the graph with vertex set V in which two vertices (x_1, y_1) and (x_2, y_2) are adjacent if their euclidean distance is equal to d. Show that G is bipartite.
b) Deduce that the rational unit distance graph is bipartite.
c) Show, on the other hand, that the real unit distance graph is not bipartite.

1.7 Related Reading

History of Graph Theory

An attractive account of the history of graph theory up to 1936, complete with annotated extracts from pivotal papers, can be found in Biggs et al. (1986). The first book on graph theory was published by König (1936). It led to the development of a strong school of graph theorists in Hungary which included P. Erdős and T. Gallai. Also in the thirties, H. Whitney published a series of influential articles (see Whitney (1992)).

As with every branch of mathematics, graph theory is best learnt by doing. The book *Combinatorial Problems and Exercises* by Lovász (1993) is highly recommended as a source of stimulating problems and proof techniques. A general guide to solving problems in mathematics is the very readable classic *How to Solve It* by Pólya (2004). The delightful *Proofs from the Book* by Aigner and Ziegler (2004) is a compilation of beautiful proofs in mathematics, many of which treat combinatorial questions.

2

Subgraphs

Contents

2.1 Subgraphs and Supergraphs

EDGE AND VERTEX DELETION

Given a graph G, there are two natural ways of deriving smaller graphs from G. If e is an edge of G, we may obtain a graph on $m - 1$ edges by deleting e from G but leaving the vertices and the remaining edges intact. The resulting graph is denoted by $G \setminus e$. Similarly, if v is a vertex of G, we may obtain a graph on $n - 1$ vertices by deleting from G the vertex v together with all the edges incident with v. The resulting graph is denoted by $G - v$. These operations of *edge deletion* and *vertex deletion* are illustrated in Figure 2.1.

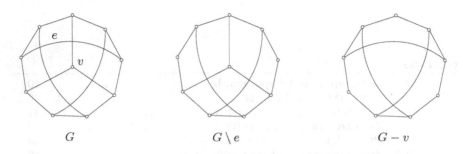

$$G \qquad\qquad G \setminus e \qquad\qquad G - v$$

Fig. 2.1. Edge-deleted and vertex-deleted subgraphs of the Petersen graph

The graphs $G \setminus e$ and $G - v$ defined above are examples of subgraphs of G. We call $G \setminus e$ an *edge-deleted subgraph*, and $G - v$ a *vertex-deleted subgraph*. More generally, a graph F is called a *subgraph* of a graph G if $V(F) \subseteq V(G)$, $E(F) \subseteq E(G)$, and ψ_F is the restriction of ψ_G to $E(F)$. We then say that G *contains* F or that F is *contained in* G, and write $G \supseteq F$ or $F \subseteq G$, respectively. Any subgraph F of G can be obtained by repeated applications of the basic operations of edge and vertex deletion; for instance, by first deleting the edges of G not in F and then deleting the vertices of G not in F. Note that the null graph is a subgraph of every graph.

We remark in passing that in the special case where G is vertex-transitive, all vertex-deleted subgraphs of G are isomorphic. In this case, the notation $G - v$ is used to denote any vertex-deleted subgraph. Likewise, we write $G \setminus e$ to denote any edge-deleted subgraph of an edge-transitive graph G.

A *copy* of a graph F in a graph G is a subgraph of G which is isomorphic to F. Such a subgraph is also referred to as an F-*subgraph* of G; for instance, a K_3-subgraph is a triangle in the graph. An *embedding* of a graph F in a graph G is an isomorphism between F and a subgraph of G. For each copy of F in G, there are $\mathrm{aut}(F)$ embeddings of F in G.

A *supergraph* of a graph G is a graph H which contains G as a subgraph, that is, $H \supseteq G$. Note that any graph is both a subgraph and a supergraph of itself.

All other subgraphs F and supergraphs H are referred to as *proper*; we then write $F \subset G$ or $H \supset G$, respectively.

The above definitions apply also to digraphs, with the obvious modifications.

In many applications of graph theory, one is interested in determining if a given graph has a subgraph or supergraph with prescribed properties. The theorem below provides a sufficient condition for a graph to contain a cycle. In later chapters, we study conditions under which a graph contains a long path or cycle, or a complete subgraph of given order. Although supergraphs with prescribed properties are encountered less often, they do arise naturally in the context of certain applications. One such is discussed in Chapter 16 (see also Exercises 2.2.17 and 2.2.26).

Theorem 2.1 *Let G be a graph in which all vertices have degree at least two. Then G contains a cycle.*

Proof If G has a loop, it contains a cycle of length one, and if G has parallel edges, it contains a cycle of length two. So we may assume that G is simple.

Let $P := v_0 v_1 \ldots v_{k-1} v_k$ be a longest path in G. Because the degree of v_k is at least two, it has a neighbour v different from v_{k-1}. If v is not on P, the path $v_0 v_1 \ldots v_{k-1} v_k v$ contradicts the choice of P as a longest path. Therefore, $v = v_i$, for some i, $0 \leq i \leq k - 2$, and $v_i v_{i+1} \ldots v_k v_i$ is a cycle in G. $\qquad\square$

MAXIMALITY AND MINIMALITY

The proof of Theorem 2.1 proceeded by first selecting a longest path in the graph, and then finding a cycle based on this path. Of course, from a purely mathematical point of view, this is a perfectly sound approach. The graph, being finite, must certainly have a longest path. However, if we wished to actually find a cycle in our graph by tracing through the steps of the proof, we would first have to find such a path, and this turns out to be a very hard task in general (in a sense to be made precise in Chapter 8). Fortunately, the very same proof remains valid if 'longest path' is replaced by 'maximal path', a maximal path being one that cannot be extended to a longer path from either end. Moreover, a maximal path is easily found: one simply starts at any vertex and grows a path until it can no longer be extended either way. For reasons such as this, the concepts of maximality and minimality (of subgraphs) turn out to be rather important.

Let $\mathcal{F}$ be a family of subgraphs of a graph G. A member F of $\mathcal{F}$ is *maximal* in $\mathcal{F}$ if no member of $\mathcal{F}$ properly contains F; likewise, F is *minimal* in $\mathcal{F}$ if no member of $\mathcal{F}$ is properly contained in F. When $\mathcal{F}$ consists of the set of all paths of G, we simply refer to a maximal member of $\mathcal{F}$ as a *maximal path* of G. We use similar terminology for describing maximal and minimal members of other special families of subgraphs. For instance, when $\mathcal{F}$ is the set of all connected subgraphs of G, the maximal members of $\mathcal{F}$ are simply its components (Exercise 2.1.1). Similarly, because an odd cycle is not bipartite, but each of its proper subgraphs is bipartite (Exercise 1.1.3), the odd cycles of a graph are its minimal nonbipartite subgraphs (see Figure 2.2b). Indeed, as we shall see, the odd cycles are the only minimal nonbipartite subgraphs.

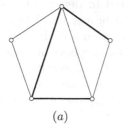
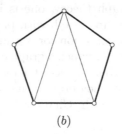
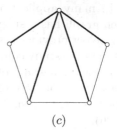

(a)   (b)   (c)

Fig. 2.2. (a) A maximal path, (b) a minimal nonbipartite subgraph, and (c) a maximal bipartite subgraph

The notions of maximality and minimality should not be confused with those of maximum and minimum cardinality. Every cycle in a graph is a maximal cycle, because no cycle is contained in another; by the same token, every cycle is a minimal cycle. On the other hand, by a maximum cycle of a graph we mean one of maximum length, that is, a longest cycle, and by a minimum cycle we mean one of minimum length. In a graph G which has at least one cycle, the length of a longest cycle is called its *circumference* and the length of a shortest cycle its *girth*.

ACYCLIC GRAPHS AND DIGRAPHS

A graph is *acyclic* if does not contain a cycle. It follows from Theorem 2.1 that an acyclic graph must have a vertex of degree less than two. In fact, every nontrivial acyclic graph has at least two vertices of degree less than two (Exercise 2.1.2).

Analogously, a digraph is *acyclic* if it has no directed cycle. One particularly interesting class of acyclic digraphs are those associated with partially ordered sets. A *partially ordered set*, or for short *poset*, is an ordered pair $P = (X, \prec)$, where X is a set and $\prec$ is a *partial order* on X, that is, an irreflexive, antisymmetric, and transitive binary relation. Two elements u and v of X are *comparable* if either $u \prec v$ or $v \prec u$, and *incomparable* otherwise. A set of pairwise comparable elements in P is a *chain*, a set of pairwise incomparable elements an *antichain*.

One can form a digraph $D := D(P)$ from a poset $P = (X, \prec)$ by taking X as the set of vertices, (u, v) being an arc of D if and only if $u \prec v$. This digraph is acyclic and transitive, where *transitive* here means that (u, w) is an arc whenever both (u, v) and (v, w) are arcs. (It should be emphasized that, despite its name, this notion of transitivity in digraphs has no connection whatsoever with the group-theoretic notions of vertex-transitivity and edge-transitivity defined earlier.) Conversely, to every strict acyclic transitive digraph D there corresponds a poset P on the vertex set of D. An acyclic tournament is frequently referred to as a *transitive tournament.* It can be seen that chains in P correspond to transitive subtournaments of D.

PROOF TECHNIQUE: THE PIGEONHOLE PRINCIPLE

If $n + 1$ letters are distributed among n pigeonholes, at least two of them will end up in the same pigeonhole. This is known as the *Pigeonhole Principle*, and is a special case of a simple statement concerning multisets (sets with repetitions allowed) of real numbers.
Let $S = (a_1, a_2, \ldots, a_n)$ be a multiset of real numbers and let a denote their average. Clearly, the minimum of the a_i is no larger than a, and the maximum of the a_i is at least as large as a. Thus, if all the elements of S are integers, we may assert that there is an element that is no larger than $\lfloor a \rfloor$, and also one that is at least as large as $\lceil a \rceil$. The Pigeonhole Principle merely amounts to saying that if n integers sum to $n + 1$ or more, one of them is at least $\lceil (n + 1)/n \rceil = 2$.
Exercise 1.1.6a is a simple example of a statement that can be proved by applying this principle. As a second application, we establish a sufficient condition for the existence of a quadrilateral in a graph, due to Reiman (1958).
Theorem 2.2 *Any simple graph G with $\sum_{v \in V} \binom{d(v)}{2} > \binom{n}{2}$ contains a quadrilateral.*

Proof Denote by p_2 the number of paths of length two in G, and by $p_2(v)$ the number of such paths whose central vertex is v. Clearly, $p_2(v) = \binom{d(v)}{2}$. As each path of length two has a unique central vertex, $p_2 = \sum_{v \in V} p_2(v) = \sum_{v \in V} \binom{d(v)}{2}$. On the other hand, each such path also has a unique pair of ends. Therefore the set of all paths of length two can be partitioned into $\binom{n}{2}$ subsets according to their ends. The hypothesis $\sum_{v \in V} \binom{d(v)}{2} > \binom{n}{2}$ now implies, by virtue of the Pigeonhole Principle, that one of these subsets contains two or more paths; that is, there exist two paths of length two with the same pair of ends. The union of these paths is a quadrilateral. □

Exercises

⋆**2.1.1** Show that the maximal connected subgraphs of a graph are its components.

⋆**2.1.2**

a) Show that every nontrivial acyclic graph has at least two vertices of degree less than two.
b) Deduce that every nontrivial connected acyclic graph has at least two vertices of degree one. When does equality hold?

2.1.3

a) Show that if $m \geq n$, then G contains a cycle.
b) For each positive integer n, find an acyclic graph with n vertices and $n - 1$ edges.

2.1.4

a) Show that every simple graph G contains a path of length δ.
b) For each $k \geq 0$, find a simple graph G with $\delta = k$ which contains no path of length greater than k.

2.1.5

a) Show that every simple graph G with $\delta \geq 2$ contains a cycle of length at least $\delta + 1$.
b) For each $k \geq 2$, find a simple graph G with $\delta = k$ which contains no cycle of length greater than $k + 1$.

2.1.6 Show that every simple graph has a vertex x and a family of $\lfloor \frac{1}{2}d(x) \rfloor$ cycles any two of which meet only in the vertex x.

2.1.7

a) Show that the Petersen graph has girth five and circumference nine.
b) How many cycles are there of length k in this graph, for $5 \leq k \leq 9$?

2.1.8

a) Show that a k-regular graph of girth four has at least $2k$ vertices.
b) For $k \geq 2$, determine all k-regular graphs of girth four on exactly $2k$ vertices.

2.1.9

a) Show that a k-regular graph of girth five has at least $k^2 + 1$ vertices.
b) Determine all k-regular graphs of girth five on exactly $k^2 + 1$ vertices, $k = 2, 3$.

2.1.10 Show that the incidence graph of a finite projective plane has girth six.

$\star$**2.1.11** A *topological sort* of a digraph D is an linear ordering of its vertices such that, for every arc a of D, the tail of a precedes its head in the ordering.

a) Show that every acyclic digraph has at least one source and at least one sink.
b) Deduce that a digraph admits a topological sort if and only if it is acyclic.

2.1.12 Show that every strict acyclic digraph contains an arc whose reversal results in an acyclic digraph.

2.1.13 Let D be a strict digraph. Setting $k := \max \{\delta^-, \delta^+\}$, show that:

a) D contains a directed path of length at least k,
b) if $k > 0$, then D contains a directed cycle of length at least $k + 1$.

2.1.14

a) Let G be a graph all of whose vertex-deleted subgraphs are isomorphic. Show that G is vertex-transitive.

b) Let G be a graph all of whose edge-deleted subgraphs are isomorphic. Is G necessarily edge-transitive?

2.1.15 Using Theorem 2.2 and the Cauchy–Schwarz Inequality[1], show that a simple graph G contains a quadrilateral if $m > \frac{1}{4}n(\sqrt{4n-3}+1)$.   (I. Reiman)

2.1.16 Triangle-Free Graph

A *triangle-free* graph is one which contains no triangles. Let G be a simple triangle-free graph.

a) Show that $d(x) + d(y) \leq n$ for all $xy \in E$.

b) Deduce that $\sum_{v \in V} d(v)^2 \leq mn$.

c) Applying the Cauchy–Schwarz Inequality[1], deduce that $m \leq n^2/4$.

(W. Mantel)

d) For each positive integer n, find a simple triangle-free graph G with $m = \lfloor n^2/4 \rfloor$.

2.1.17

a) Let G be a triangle-free graph with $\delta > 2n/5$. Show that G is bipartite.

b) For $n \equiv 0 \pmod 5$, find a nonbipartite triangle-free graph with $\delta = 2n/5$.

(B. Andrásfai, P. Erdős, and V.T. Sós)

2.1.18 Let G be a simple graph with $v(G) = kp$ and $\delta(G) \geq kq$. Show that G has a subgraph F with $v(F) = p$ and $\delta(F) \geq q$.   (C.St.J.A. Nash-Williams)

2.1.19 Show that the Kneser graph $KG_{m,n}$ has no odd cycle of length less than $n/(n-2m)$.

$\star$**2.1.20** Let K_n be a complete graph whose edges are coloured red or blue. Call a subgraph of this graph *monochromatic* if all of its edges have the same colour, and *bichromatic* if edges of both colours are present.

a) Let v be a vertex of K_n. Show that the number of bichromatic 2-paths in K_n whose central vertex is v is at most $(n-1)^2/4$. When does equality hold?

b) Deduce that the total number of bichromatic 2-paths in K_n is at most $n(n-1)^2/4$.

c) Observing that each bichromatic triangle contains exactly two bichromatic 2-paths, deduce that the number of monochromatic triangles in K_n is at least $n(n-1)(n-5)/24$. When does equality hold?   (A.W. Goodman)

d) How many monochromatic triangles must there be, at least, when $n = 5$ and when $n = 6$?

---

[1] $\sum_{i=1}^{n} a_i^2 \sum_{i=1}^{n} b_i^2 \geq \left(\sum_{i=1}^{n} a_i b_i\right)^2$ for real numbers $a_i, b_i, 1 \leq i \leq n$.

2.1.21 Let T be a tournament on n vertices, and let v be a vertex of T.

a) Show that the number of directed 2-paths in T whose central vertex is v is at most $(n-1)^2/4$. When does equality hold?
b) Deduce that the total number of directed 2-paths in T is at most $n(n-1)^2/4$.
c) Observing that each transitive triangle contains exactly one directed 2-path and that each directed triangle contains exactly three directed 2-paths, deduce that the number of directed triangles in T is at most $\frac{1}{4}\binom{n+1}{3}$. When does equality hold?

$\star$**2.1.22** Let $P = (X, \prec)$ be a poset. Show that the maximum number of elements in a chain of P is equal to the minimum number of antichains into which X can be partitioned.

(L. MIRSKY)

2.1.23 GEOMETRIC GRAPH
A *geometric graph* is a graph embedded in the plane in such a way that each edge is a line segment. Let G be a geometric graph in which any two edges intersect (possibly at an end).

a) Show that G has at most n edges.
b) For each $n \geq 3$, find an example of such a graph G with n edges.

(H. HOPF AND E. PANNWITZ)

2.2 Spanning and Induced Subgraphs

SPANNING SUBGRAPHS

A *spanning subgraph* of a graph G is a subgraph obtained by edge deletions only, in other words, a subgraph whose vertex set is the entire vertex set of G. If S is the set of deleted edges, this subgraph of G is denoted $G \setminus S$. Observe that every simple graph is a spanning subgraph of a complete graph.

Spanning supergraphs are defined analogously. The inverse operation to edge deletion is *edge addition*. Adding a set S of edges to a graph G yields a *spanning supergraph* of G, denoted $G + S$. By starting with a disjoint union of two graphs G and H and adding edges joining every vertex of G to every vertex of H, one obtains the *join* of G and H, denoted $G \vee H$. The join $C_n \vee K_1$ of a cycle C_n and a single vertex is referred to as a *wheel* with n *spokes* and denoted W_n. (The graph H of Figure 1.1 is the wheel W_5.) One may also add a set X of vertices to a graph, resulting in a supergraph of G denoted $G + X$.

Certain types of spanning subgraph occur frequently in applications of graph theory and, for historical reasons, have acquired special names. For example, spanning paths and cycles are called *Hamilton paths* and *Hamilton cycles*, respectively, and spanning k-regular subgraphs are referred to as *k-factors*. Rédei's Theorem (Theorem 2.3, see inset) tells us that every tournament has a directed Hamilton path. Not every tournament (on three or more vertices) has a directed Hamilton cycle, however. Indeed, the transitive tournament has no directed cycles at all.

PROOF TECHNIQUE: INDUCTION

One of the most widely used proof techniques in mathematics is the *Principle of Mathematical Induction*. Suppose that, for each nonnegative integer i, we have a mathematical statement S_i. One may prove that all assertions in the sequence $(S_0, S_1, \ldots)$ are true by:

▷   directly verifying S_0 (the *basis* of the induction),
▷   for each integer $n \geq 1$, deducing that S_n is true (the *inductive step*) from the assumption that S_{n-1} is true (the *inductive hypothesis*).

The justification for this technique is provided by the principle that each nonempty subset of $\mathbb{N}$ has a minimal element: if not all S_i were true, the set $\{i \in \mathbb{N} : S_i \text{ is false}\}$ would be a nonempty subset of $\mathbb{N}$, and would therefore have a minimal element n. Thus S_{n-1} would be true and S_n false.

We shall come across many examples of inductive proofs throughout the book. Here, as a simple illustration of the technique, we prove a basic theorem on tournaments due to Rédei (1934).

Theorem 2.3   RÉDEI'S THEOREM
Every tournament has a directed Hamilton path.

Proof   Clearly, the trivial tournament (on one vertex) has a directed Hamilton path. Assume that, for some integer $n \geq 2$, every tournament on $n - 1$ vertices has a directed Hamilton path. Let T be a tournament on n vertices and let $v \in V(T)$. The digraph $T' := T - v$ is a tournament on $n-1$ vertices. By the inductive hypothesis, T' has a directed Hamilton path $P' := (v_1, v_2, \ldots, v_{n-1})$. If (v, v_1) is an arc of T, the path $(v, v_1, v_2, \ldots, v_{n-1})$ is a directed Hamilton path of T. Similarly, if (v_{n-1}, v) is an arc of T, the path $(v_1, v_2, \ldots, v_{n-1}, v)$ is a directed Hamilton path of T. Because T is a tournament, v is adjacent to each vertex of P', so we may assume that both (v_1, v) and (v, v_{n-1}) are arcs of T. It follows that there exists an integer i, $1 \leq i < n - 1$, such that both (v_i, v) and (v, v_{i+1}) are arcs of T. But now $P := (v_1, \ldots, v_i, v, v_{i+1}, \ldots, v_{n-1})$ is a directed Hamilton path of T.   □

Inductive proofs may be presented in a variety of ways. The above proof, for example, may be recast as a 'longest path' proof. We take P to be a longest directed path in the tournament T. Assuming that P is not a directed Hamilton path, we then obtain a contradiction by showing that T has a directed path longer than P (Exercise 2.2.4).

Graph-theoretical statements generally assert that all graphs belonging to some well-defined class possess a certain property. Any 'proof' that fails to cover all cases is false. This is a common mistake in attempts to prove statements of this sort by induction. Another common error is neglecting to verify the basis of the induction. For an example of how not to use induction, see Exercise 2.2.21.

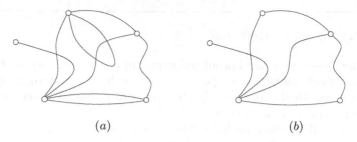

Fig. 2.3. (a) A graph and (b) its underlying simple graph

Nonetheless, Camion (1959) proved that every tournament in which any vertex can be reached from any other vertex by means of a directed path does indeed have a directed Hamilton cycle (Exercise 3.4.12a).

By deleting from a graph G all loops and, for every pair of adjacent vertices, all but one link joining them, we obtain a simple spanning subgraph called the *underlying simple graph* of G. Up to isomorphism, each graph has a unique underlying simple graph. Figure 2.3 shows a graph and its underlying simple graph.

Given spanning subgraphs $F_1 = (V, E_1)$ and $F_2 = (V, E_2)$ of a graph $G = (V, E)$, we may form the spanning subgraph of G whose edge set is the symmetric difference $E_1 \triangle E_2$ of E_1 and E_2. This graph is called the *symmetric difference* of F_1 and F_2, and denoted $F_1 \triangle F_2$. Figure 2.4 shows the symmetric difference of two spanning subgraphs of a graph on five vertices.

INDUCED SUBGRAPHS

A subgraph obtained by vertex deletions only is called an *induced subgraph*. If X is the set of vertices deleted, the resulting subgraph is denoted by $G - X$. Frequently, it is the set $Y := V \setminus X$ of vertices which remain that is the focus of interest. In such cases, the subgraph is denoted by $G[Y]$ and referred to as the subgraph of G *induced by* Y. Thus $G[Y]$ is the subgraph of G whose vertex set is Y and whose edge set consists of all edges of G which have both ends in Y.

The following theorem, due to Erdős (1964/1965), tells us that every graph has a induced subgraph whose minimum degree is relatively large.

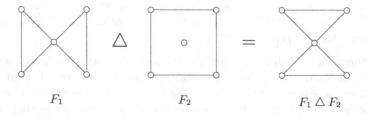

$$F_1 \qquad\qquad F_2 \qquad\qquad F_1 \triangle F_2$$

Fig. 2.4. The symmetric difference of two graphs

PROOF TECHNIQUE: CONTRADICTION

A common approach to proving graph-theoretical statements is to proceed by assuming that the stated assertion is false and analyse the consequences of that assumption so as to arrive at a contradiction. As a simple illustration of this method, we present an interesting and very useful result due to Erdős (1965).

Theorem 2.4 *Every loopless graph G contains a spanning bipartite subgraph F such that $d_F(v) \geq \frac{1}{2}d_G(v)$ for all $v \in V$.*

Proof Let G be a loopless graph. Certainly, G has spanning bipartite subgraphs, one such being the empty spanning subgraph. Let $F := F[X, Y]$ be a spanning bipartite subgraph of G with the greatest possible number of edges. We claim that F satisfies the required property. Suppose not. Then there is some vertex v for which

$$d_F(v) < \frac{1}{2}d_G(v) \tag{2.1}$$

Without loss of generality, we may suppose that $v \in X$. Consider the spanning bipartite subgraph F' whose edge set consists of all edges of G with one end in $X \setminus \{v\}$ and the other in $Y \cup \{v\}$. The edge set of F' is the same as that of F except for the edges of G incident to v; those which were in F are not in F', and those which were not in F are in F'. We thus have:

$$e(F') = e(F) - d_F(v) + (d_G(v) - d_F(v)) = e(F) + (d_G(v) - 2d_F(v)) > e(F)$$

the inequality following from (2.1). But this contradicts the choice of F. It follows that F does indeed have the required property.   □

The method of contradiction is merely a convenient way of presenting the idea underlying the above proof. Implicit in the proof is an algorithm which finds, in any graph, a spanning bipartite subgraph with the stated property: one starts with any spanning bipartite subgraph and simply moves vertices between parts so as to achieve the desired objective (see also Exercises 2.2.2 and 2.2.20).

Theorem 2.5 *Every graph with average degree at least $2k$, where k is a positive integer, has an induced subgraph with minimum degree at least $k + 1$.*

Proof Let G be a graph with average degree $d(G) \geq 2k$, and let F be an induced subgraph of G with the largest possible average degree and, subject to this, the smallest number of vertices. We show that $\delta(F) \geq k + 1$. This is clearly true if $v(F) = 1$, since then $\delta(F) = d(F) \geq d(G)$, by the choice of F. We may therefore assume that $v(F) > 1$.

Suppose, by way of contradiction, that $d_F(v) \leq k$ for some vertex v of F. Consider the vertex-deleted subgraph $F' := F - v$. Note that F' is also an induced subgraph of G. Moreover

$$d(F') = \frac{2e(F')}{v(F')} \geq \frac{2(e(F) - k)}{v(F) - 1} \geq \frac{2e(F) - d(G)}{v(F) - 1} \geq \frac{2e(F) - d(F)}{v(F) - 1} = d(F)$$

Because $v(F') < v(F)$, this contradicts the choice of F. Therefore $\delta(F) \geq k + 1$. $\square$

The bound on the minimum degree given in Theorem 2.5 is sharp (Exercise 3.1.6).

Subgraphs may also be induced by sets of edges. If S is a set of edges, the *edge-induced subgraph* $G[S]$ is the subgraph of G whose edge set is S and whose vertex set consists of all ends of edges of S. Any edge-induced subgraph $G[S]$ can be obtained by first deleting the edges in $E \setminus S$ and then deleting all resulting isolated vertices; indeed, an edge-induced subgraph is simply a subgraph without isolated vertices.

WEIGHTED GRAPHS AND SUBGRAPHS

When graphs are used to model practical problems, one often needs to take into account additional factors, such as costs associated with edges. In a communications network, for example, relevant factors might be the cost of transmitting data along a link, or of constructing a new link between communication centres. Such situations are modelled by weighted graphs.

With each edge e of G, let there be associated a real number $w(e)$, called its *weight*. Then G, together with these weights on its edges, is called a *weighted graph*, and denoted (G, w). One can regard a weighting $w : E \to \mathbb{R}$ as a vector whose coordinates are indexed by the edge set E of G; the set of all such vectors is denoted by $\mathbb{R}^E$ or, when the weights are rational numbers, by $\mathbb{Q}^E$.

If F is a subgraph of a weighted graph, the *weight* $w(F)$ of F is the sum of the weights on its edges, $\sum_{e \in E(F)} w(e)$. Many optimization problems amount to finding, in a weighted graph, a subgraph of a certain type with minimum or maximum weight. Perhaps the best known problem of this kind is the following one.

A travelling salesman wishes to visit a number of towns and then return to his starting point. Given the travelling times between towns, how should he plan his itinerary so that he visits each town exactly once and minimizes his total travelling time? This is known as the *Travelling Salesman Problem*. In graph-theoretic terms, it can be phrased as follows.

Problem 2.6   THE TRAVELLING SALESMAN PROBLEM (TSP)
   GIVEN: *a weighted complete graph* (G, w),
   FIND: *a minimum-weight Hamilton cycle of* G.

Note that it suffices to consider the TSP for complete graphs because nonadjacent vertices can be joined by edges whose weights are prohibitively high. We discuss this problem, and others of a similar flavour, in Chapters 6 and 8, as well as in later chapters.

Exercises

2.2.1 Let G be a graph on n vertices and m edges and c components.

a) How many spanning subgraphs has G?
b) How many edges need to be added to G to obtain a connected spanning supergraph?

$\star$**2.2.2**

a) Deduce from Theorem 2.4 that every loopless graph G contains a spanning bipartite subgraph F with $e(F) \geq \frac{1}{2}e(G)$.
b) Describe an algorithm for finding such a subgraph by first arranging the vertices in a linear order and then assigning them, one by one, to either X or Y, using a simple rule.

2.2.3 Determine the number of 1-factors in each of the following graphs: (a) the Petersen graph, (b) the pentagonal prism, (c) K_{2n}, (d) $K_{n,n}$.

2.2.4 Give a proof of Theorem 2.3 by means of a longest path argument.

(D. KÖNIG AND P. VERESS)

2.2.5

a) Show that every Hamilton cycle of the k-prism uses either exactly two consecutive edges linking the two k-cycles or else all of them.
b) How many Hamilton cycles are there in the pentagonal prism?

2.2.6 Show that there is a Hamilton path between two vertices in the Petersen graph if and only if these vertices are nonadjacent.

2.2.7
Which grids have Hamilton paths, and which have Hamilton cycles?

2.2.8 Give an example to show that the following simple procedure, known as the *Greedy Heuristic*, is not guaranteed to solve the Travelling Salesman Problem.

▷ Select an arbitrary vertex v.
▷ Starting with the trivial path v, grow a Hamilton path one edge at a time, choosing at each iteration an edge of minimum weight between the terminal vertex of the current path and a vertex not on this path.
▷ Form a Hamilton cycle by adding the edge joining the two ends of the Hamilton path.

2.2.9 Let G be a graph on n vertices and m edges.

a) How many induced subgraphs has G?
b) How many edge-induced subgraphs has G?

2.2.10 Show that every shortest cycle in a simple graph is an induced subgraph.

⋆**2.2.11** Show that if G is simple and connected, but not complete, then G contains an induced path of length two.

⋆**2.2.12** Let P and Q be distinct paths in a graph G with the same initial and terminal vertices. Show that $P \cup Q$ contains a cycle by considering the subgraph $G[E(P) \triangle E(Q)]$ and appealing to Theorem 2.1.

2.2.13

a) Show that any two longest paths in a connected graph have a vertex in common.
b) Deduce that if P is a longest path in a connected graph G, then no path in $G - V(P)$ is as long as P.

2.2.14 Give a constructive proof of Theorem 2.5.

2.2.15

a) Show that an induced subgraph of a line graph is itself a line graph.
b) Deduce that no line graph can contain either of the graphs in Figure 1.19 as an induced subgraph.
c) Show that these two graphs are minimal with respect to the above property. Can you find other such graphs? (There are nine in all.)

2.2.16

a) Show that an induced subgraph of an interval graph is itself an interval graph.
b) Deduce that no interval graph can contain the graph in Figure 1.20 as an induced subgraph.
c) Show that this graph is minimal with respect to the above property.

2.2.17 Let G be a bipartite graph of maximum degree k.

a) Show that there is a k-regular bipartite graph H which contains G as an induced subgraph.
b) Show, moreover, that if G is simple, then there exists such a graph H which is simple.

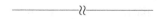

2.2.18

a) Show that if $m \geq n + 4$, then G contains two edge-disjoint cycles. (L. Pósa)
b) For each integer $n \geq 6$, find a graph with n vertices and $n + 3$ edges which does not contain two edge-disjoint cycles.

2.2.19 Chord of a Cycle
A *chord* of a cycle C in a graph G is an edge in $E(G) \setminus E(C)$ both of whose ends lie on C. Let G be a simple graph with $m \geq 2n - 3$, where $n \geq 4$. Show that G contains a cycle with at least one chord. (L. Pósa)

2.2.20 Let G be a simple connected graph.

a) Show that there is an ordering $v_1, v_2, \ldots, v_n$ of V such that at least $\frac{1}{2}(n-1)$ vertices v_j are adjacent to an odd number of vertices v_i with $i < j$.
b) By starting with such an ordering and adopting the approach outlined in Exercise 2.2.2b, deduce that G has a bipartite subgraph with at least $\frac{1}{2}m + \frac{1}{4}(n-1)$ edges.   (C. EDWARDS; P. ERDŐS)

2.2.21 Read the 'Theorem' and 'Proof' given below, and then answer the questions which follow.

'Theorem'. Let G be a simple graph with $\delta \geq n/2$, where $n \geq 3$. Then G has a Hamilton cycle.

'Proof'. By induction on n. The 'Theorem' is true for $n = 3$, because $G = K_3$ in this case. Suppose that it holds for $n = k$, where $k \geq 3$. Let G' be a simple graph on k vertices in which $\delta \geq k/2$, and let C' be a Hamilton cycle of G'. Form a graph G on $k+1$ vertices in which $\delta \geq (k+1)/2$ by adding a new vertex v and joining v to at least $(k+1)/2$ vertices of G'. Note that v must be adjacent to two consecutive vertices, u and w, of C'. Replacing the edge uw of C' by the path uvw, we obtain a Hamilton cycle C of G. Thus the 'Theorem' is true for $n = k+1$. By the Principle of Mathematical Induction, it is true for all $n \geq 3$.   □

a) Is the 'Proof' correct?
b) If you claim that the 'Proof' is incorrect, give reasons to support your claim.
c) Can you find any graphs for which the 'Theorem' fails? Does the existence or nonexistence of such graphs have any relationship to the correctness or incorrectness of the 'Proof'?   (D.R. WOODALL)

2.2.22

a) Let D be an oriented graph with minimum outdegree k, where $k \geq 1$.
   i) Show that D has a vertex x whose indegree and outdegree are both at least k.
   ii) Let D' be the digraph obtained from D by deleting $N^-(x) \cup \{x\}$ and adding an arc (u, v) from each vertex u of the set $N^{--}(x)$ of in-neighbours of $N^-(x)$ to each vertex v of $N^+(x)$, if there was no such arc in D. Show that D' is a strict digraph with minimum outdegree k.
b) Deduce, by induction on n, that every strict digraph D with minimum outdegree k, where $k \geq 1$, contains a directed cycle of length at most $2n/k$.
                                     (V. CHVÁTAL AND E. SZEMERÉDI)

2.2.23 The *complement* $\overline{D}$ of a strict digraph D is its complement in $D(K_n)$. Let $D = (V, A)$ be a strict digraph and let P be a directed Hamilton path of D. Form a bipartite graph $B[\mathcal{F}, S_n]$, where $\mathcal{F}$ is the family of spanning subgraphs of D each component of which is a directed path and S_n is the set of permutations of V, a subgraph $F \in \mathcal{F}$ being adjacent in B to a permutation $\sigma \in S_n$ if and only if $\sigma(F) \subseteq \sigma(D) \cap P$.

a) Which vertices $F \in \mathcal{F}$ are of odd degree in B?

b) Describe a bijection between the vertices $\sigma \in S_n$ of odd degree in B and the directed Hamilton paths of $\overline{D}$.

c) Deduce that $h(D) \equiv h(\overline{D}) \pmod 2$, where $h(D)$ denotes the number of directed Hamilton paths in D.

2.2.24 Let D be a tournament, and let (x, y) be an arc of D. Set $D^- := D \setminus (x, y)$ and $D^+ := D + (y, x)$.

a) Describe a bijection between the directed Hamilton paths of D^- and those of $\overline{D^+}$.

b) Deduce from Exercise 2.2.23 that $h(D^-) \equiv h(D^+) \pmod 2$.

c) Consider the tournament D' obtained from D on reversing the arc (x, y). Show that $h(D') = h(D^+) - h(D) + h(D^-)$.

d) Deduce that $h(D') \equiv h(D) \pmod 2$.

e) Conclude that every tournament has an odd number of directed Hamilton paths. (L. Rédei)

2.2.25

a) Let S be a set of n points in the plane, the distance between any two of which is at most one. Show that there are at most n pairs of points of S at distance exactly one. (P. Erdős)

b) For each $n \geq 3$, describe such a set S for which the number of pairs of points at distance exactly one is n.

2.2.26 Let G be a simple graph on n vertices and m edges, with minimum degree δ and maximum degree Δ.

a) Show that there is a simple Δ-regular graph H which contains G as an induced subgraph.

b) Let H be such a graph, with $v(H) = n + r$. Show that:
   i) $r \geq \Delta - \delta$,
   ii) $r\Delta \equiv n\Delta \pmod 2$,
   iii) $r\Delta \geq n\Delta - 2m \geq r\Delta - r(r - 1)$.

(Erdős and Kelly (1967) showed that if r is the smallest positive integer which satisfies the above three conditions, then there does indeed exist a simple Δ-regular graph H on $n + r$ vertices which contains G as an induced subgraph.)

2.2.27 Let G be a simple graph on n vertices, where $n \geq 4$, and let k be an integer, $2 \leq k \leq n - 2$. Suppose that all induced subgraphs of G on k vertices have the same number of edges. Show that G is either empty or complete.

2.3 Modifying Graphs

We have already discussed some simple ways of modifying a graph, namely deleting or adding vertices or edges. Here, we describe several other local operations on graphs. Although they do not give rise to subgraphs or supergraphs, it is natural and convenient to introduce them here.

VERTEX IDENTIFICATION AND EDGE CONTRACTION

To *identify* nonadjacent vertices x and y of a graph G is to replace these vertices by a single vertex incident to all the edges which were incident in G to either x or y. We denote the resulting graph by $G \,/\, \{x, y\}$ (see Figure 2.5a). To *contract* an edge e of a graph G is to delete the edge and then (if the edge is a link) identify its ends. The resulting graph is denoted by $G \,/\, e$ (see Figure 2.5b).

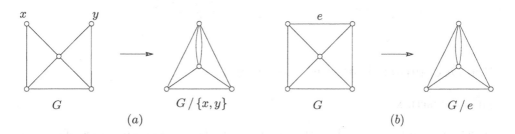

Fig. 2.5. (a) Identifying two vertices, and (b) contracting an edge

VERTEX SPLITTING AND EDGE SUBDIVISION

The inverse operation to edge contraction is vertex splitting. To *split* a vertex v is to replace v by two adjacent vertices, v' and v'', and to replace each edge incident to v by an edge incident to either v' or v'' (but not both, unless it is a loop at v), the other end of the edge remaining unchanged (see Figure 2.6a). Note that a vertex of positive degree can be split in several ways, so the resulting graph is not unique in general.

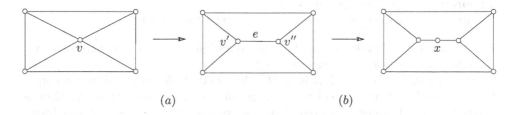

Fig. 2.6. (a) Splitting a vertex, and (b) subdividing an edge

A special case of vertex splitting occurs when exactly one link, or exactly one end of a loop, is assigned to either v' or v''. The resulting graph can then be viewed as having been obtained by subdividing an edge of the original graph, where to *subdivide* an edge e is to delete e, add a new vertex x, and join x to the ends of e (when e is a link, this amounts to replacing e by a path of length two, as in Figure 2.6b).

Exercises

2.3.1

a) Show that $c(G \,/\, e) = c(G)$ for any edge e of a graph G.
b) Let G be an acyclic graph, and let $e \in E$.
   i) Show that $G \,/\, e$ is acyclic.
   ii) Deduce that $m = n - c$.

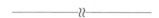

2.4 Decompositions and Coverings

DECOMPOSITIONS

A *decomposition* of a graph G is a family $\mathcal{F}$ of edge-disjoint subgraphs of G such that

$$\cup_{F \in \mathcal{F}} E(F) = E(G) \tag{2.2}$$

If the family $\mathcal{F}$ consists entirely of paths or entirely of cycles, we call $\mathcal{F}$ a *path decomposition* or *cycle decomposition* of G.

Every loopless graph has a trivial path decomposition, into paths of length one. On the other hand, not every graph has a cycle decomposition. Observe that if a graph has a cycle decomposition $\mathcal{C}$, the degree of each vertex is twice the number of cycles of $\mathcal{C}$ to which it belongs, so is even. A graph in which each vertex has even degree is called an *even graph*. Thus, a graph which admits a cycle decomposition is necessarily even. Conversely, as was shown by Veblen (1912/13), every even graph admits a cycle decomposition.

Theorem 2.7   VEBLEN'S THEOREM
A graph admits a cycle decomposition if and only if it is even.

Proof   We have already shown that the condition of evenness is necessary. We establish the converse by induction on $e(G)$.

Suppose that G is even. If G is empty, then $E(G)$ is decomposed by the empty family of cycles. If not, consider the subgraph F of G induced by its vertices of positive degree. Because G is even, F also is even, so every vertex of F has degree two or more. By Theorem 2.1, F contains a cycle C. The subgraph $G' := G \setminus E(C)$ is even, and has fewer edges than G. By induction, G' has a cycle decomposition $\mathcal{C}'$. Therefore G has the cycle decomposition $\mathcal{C} := \mathcal{C}' \cup \{C\}$. □

There is a corresponding version of Veblen's Theorem for digraphs (see Exercise 2.4.2).

PROOF TECHNIQUE: LINEAR INDEPENDENCE

Algebraic techniques can occasionally be used to solve problems where combinatorial methods fail. Arguments involving the ranks of appropriately chosen matrices are particularly effective. Here, we illustrate this technique by giving a simple proof, due to Tverberg (1982), of a theorem of Graham and Pollak (1971) on decompositions of complete graphs into complete bipartite graphs. There are many ways in which a complete graph can be decomposed into complete bipartite graphs. For example, K_4 may be decomposed into six copies of K_2, into three copies of $K_{1,2}$, into the stars $K_{1,1}$, $K_{1,2}$, and $K_{1,3}$, or into $K_{2,2}$ and two copies of K_2. What Graham and Pollak showed is that, no matter how K_n is decomposed into complete bipartite graphs, there must be at least $n - 1$ of them in the decomposition. Observe that this bound can always be achieved, for instance by decomposing K_n into the stars $K_{1,k}$, $1 \le k \le n - 1$.

Theorem 2.8 *Let* $\mathcal{F} := \{F_1, F_2, \ldots, F_k\}$ *be a decomposition of* K_n *into complete bipartite graphs. Then* $k \ge n - 1$.

Proof Let $V := V(K_n)$ and let F_i have bipartition (X_i, Y_i), $1 \le i \le k$. Consider the following system of $k + 1$ homogeneous linear equations in the variables x_v, $v \in V$:

$$\sum_{v \in V} x_v = 0, \qquad \sum_{v \in X_i} x_v = 0, \ \ 1 \le i \le k$$

Suppose that $k < n - 1$. Then this system, consisting of fewer than n equations in n variables, has a solution $x_v = c_v$, $v \in V$, with $c_v \ne 0$ for at least one $v \in V$. Thus

$$\sum_{v \in V} c_v = 0 \text{ and } \sum_{v \in X_i} c_v = 0, \ \ 1 \le i \le k$$

Because $\mathcal{F}$ is a decomposition of K_n,

$$\sum_{vw \in E} c_v c_w = \sum_{i=1}^{k} \left(\sum_{v \in X_i} c_v \right) \left(\sum_{w \in Y_i} c_w \right)$$

Therefore

$$0 = \left(\sum_{v \in V} c_v \right)^2 = \sum_{v \in V} c_v^2 + 2 \sum_{i=1}^{k} \left(\sum_{v \in X_i} c_v \right) \left(\sum_{w \in Y_i} c_w \right) = \sum_{v \in V} c_v^2 > 0$$

a contradiction. We conclude that $k \ge n - 1$. $\square$

Further proofs based on linear independence arguments are outlined in Exercises 2.4.10 and 14.2.15.

COVERINGS

We now define the related concept of a covering. A *covering* or *cover* of a graph G is a family $\mathcal{F}$ of subgraphs of G, not necessarily edge-disjoint, satisfying (2.2). A covering is *uniform* if it covers each edge of G the same number of times; when this number is k, the covering is called a *k-cover*. A 1-cover is thus simply a decomposition. A 2-cover is usually called a *double cover*. If the family $\mathcal{F}$ consists entirely of paths or entirely of cycles, the covering is referred to as a *path covering* or *cycle covering*. Every graph which admits a cycle covering also admits a uniform cycle covering (Exercise 3.5.7).

The notions of decomposition and covering crop up frequently in the study of graphs. In Section 3.5, we discuss a famous unsolved problem concerning cycle coverings, the Cycle Double Cover Conjecture. The concept of covering is also useful in the study of another celebrated unsolved problem, the Reconstruction Conjecture (see Section 2.7, in particular Exercise 2.7.11).

Exercises

2.4.1 Let e be an edge of an even graph G. Show that G/e is even.

⋆**2.4.2** EVEN DIRECTED GRAPH
A digraph D is *even* if $d^-(v) = d^+(v)$ for each vertex $v \in V$. Prove the following directed version of Veblen's Theorem (2.7): *A directed graph admits a decomposition into directed cycles if and only if it is even.*

2.4.3 Find a decomposition of K_{13} into three copies of the circulant $\mathrm{CG}(\mathbb{Z}_{13}, \{1, -1, 5, -5\})$.

2.4.4 Let $\mathcal{C}$ be a cycle decomposition of a connected even graph G, not a cycle. Show that $G[E(G) \setminus E(C)]$ is connected for some cycle C in $\mathcal{C}$.

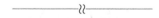

2.4.5

a) Show that K_n can be decomposed into copies of K_p only if $n - 1$ is divisible by $p - 1$ and $n(n - 1)$ is divisible by $p(p - 1)$. For which integers n do these two conditions hold when p is a prime?
b) For k a prime power, describe a decomposition of K_{k^2+k+1} into copies of K_{k+1}, based on a finite projective plane of order k.

2.4.6 Let n be a positive integer.

a) Describe a decomposition of K_{2n+1} into Hamilton cycles.
b) Deduce that K_{2n} admits a decomposition into Hamilton paths.

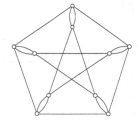

Fig. 2.7. The Petersen graph with a doubled 1-factor

⋆**2.4.7** Consider the graph obtained from the Petersen graph by replacing each of the five edges in a 1-factor by two parallel edges, as shown in Figure 2.7. Show that every cycle decomposition of this 4-regular graph includes a 2-cycle.

2.4.8   Let G be a connected graph with an even number of edges.

a) Show that G can be oriented so that the outdegree of each vertex is even.
b) Deduce that G admits a decomposition into paths of length two.

2.4.9 Show that every loopless digraph admits a decomposition into two acyclic digraphs.

2.4.10 Give an alternative proof of the de Bruijn–Erdős Theorem (see Exercise 1.3.15b) by proceeding as follows. Let $\mathbf{M}$ be the incidence matrix of a geometric configuration $(P, \mathcal{L})$ which has at least two lines and in which any two points lie on exactly one line.

a) Show that the columns of $\mathbf{M}$ span $\mathbb{R}^n$, where $n := |P|$.
b) Deduce that $\mathbf{M}$ has rank n.
c) Conclude that $|\mathcal{L}| \geq |P|$.

2.5 Edge Cuts and Bonds

Edge Cuts

Let X and Y be sets of vertices (not necessarily disjoint) of a graph $G = (V, E)$. We denote by $E[X, Y]$ the set of edges of G with one end in X and the other end in Y, and by $e(X, Y)$ their number. If $Y = X$, we simply write $E(X)$ and $e(X)$ for $E[X, X]$ and $e(X, X)$, respectively. When $Y = V \setminus X$, the set $E[X, Y]$ is called the *edge cut* of G associated with X, or the *coboundary* of X, and is denoted by $\partial(X)$; note that $\partial(X) = \partial(Y)$ in this case, and that $\partial(V) = \emptyset$. In this notation, a graph $G = (V, E)$ is bipartite if $\partial(X) = E$ for some subset X of V, and is connected if $\partial(X) \neq \emptyset$ for every nonempty proper subset X of V. The edge cuts of a graph are illustrated in Figure 2.8.

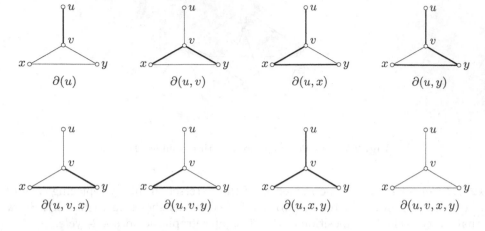

Fig. 2.8. The edge cuts of a graph

An edge cut $\partial(v)$ associated with a single vertex v is a *trivial* edge cut; this is simply the set of all links incident with v. If there are no loops incident with v, it follows that $|\partial(v)| = d(v)$. Accordingly, in the case of loopless graphs, we refer to $|\partial(X)|$ as the *degree* of X and denote it by $d(X)$.

The following theorem is a natural generalization of Theorem 1.1, the latter theorem being simply the case where $X = V$. Its proof is based on the technique of counting in two ways, and is left as an exercise (2.5.1a).

Theorem 2.9 *For any graph G and any subset X of V,*

$$|\partial(X)| = \sum_{v \in X} d(v) - 2e(X) \qquad \square$$

Veblen's Theorem (2.7) characterizes even graphs in terms of cycles. Even graphs may also be characterized in terms of edge cuts, as follows.

Theorem 2.10 *A graph G is even if and only if $|\partial(X)|$ is even for every subset X of V.*

Proof Suppose that $|\partial(X)|$ is even for every subset X of V. Then, in particular, $|\partial(v)|$ is even for every vertex v. But, as noted above, $\partial(v)$ is just the set of all links incident with v. Because loops contribute two to the degree, it follows that all degrees are even. Conversely, if G is even, then Theorem 2.9 implies that all edge cuts are of even cardinality. $\qquad \square$

The operation of symmetric difference of spanning subgraphs was introduced in Section 2.1. The following propositions show how edge cuts behave with respect to symmetric difference.

Proposition 2.11 *Let G be a graph, and let X and Y be subsets of V. Then*

$$\partial(X) \vartriangle \partial(Y) = \partial(X \vartriangle Y)$$

Proof Consider the Venn diagram, shown in Figure 2.9, of the partition of V

$$(X \cap Y, \quad X \setminus Y, \quad Y \setminus X, \quad \overline{X} \cap \overline{Y})$$

determined by the partitions $(X, \overline{X})$ and $(Y, \overline{Y})$, where $\overline{X} := V \setminus X$ and $\overline{Y} := V \setminus Y$. The edges of $\partial(X)$, $\partial(Y)$, and $\partial(X \bigtriangleup Y)$ between these four subsets of V are indicated schematically in Figure 2.10. It can be seen that $\partial(X) \bigtriangleup \partial(Y) = \partial(X \bigtriangleup Y)$. $\qquad\square$

	Y	$\overline{Y}$
X	$X \cap Y$	$X \setminus Y$
$\overline{X}$	$Y \setminus X$	$\overline{X} \cap \overline{Y}$

Fig. 2.9. Partition of V determined by the partitions $(X, \overline{X})$ and $(Y, \overline{Y})$

Corollary 2.12 *The symmetric difference of two edge cuts is an edge cut.* $\qquad\square$

We leave the proof of the second proposition to the reader (Exercise 2.5.1b).

Proposition 2.13 *Let F_1 and F_2 be spanning subgraphs of a graph G, and let X be a subset of V. Then*

$$\partial_{F_1 \bigtriangleup F_2}(X) = \partial_{F_1}(X) \bigtriangleup \partial_{F_2}(X) \qquad\qquad\square$$

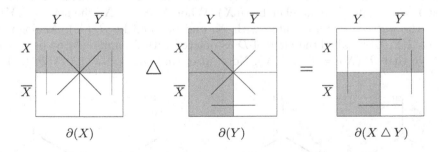

Fig. 2.10. The symmetric difference of two cuts

BONDS

A *bond* of a graph is a minimal nonempty edge cut, that is, a nonempty edge cut none of whose nonempty proper subsets is an edge cut. The bonds of the graph whose edge cuts are depicted in Figure 2.8 are shown in Figure 2.11.

The following two theorems illuminate the relationship between edge cuts and bonds. The first can be deduced from Proposition 2.11 (Exercise 2.5.1c). The second provides a convenient way to check when an edge cut is in fact a bond.

Theorem 2.14 *A set of edges of a graph is an edge cut if and only if it is a disjoint union of bonds.*   □

Theorem 2.15 *In a connected graph G, a nonempty edge cut $\partial(X)$ is a bond if and only if both $G[X]$ and $G[V \setminus X]$ are connected.*

Proof Suppose, first, that $\partial(X)$ is a bond, and let Y be a nonempty proper subset of X. Because G is connected, both $\partial(Y)$ and $\partial(X \setminus Y)$ are nonempty. It follows that $E[Y, X \setminus Y]$ is nonempty, for otherwise $\partial(Y)$ would be a nonempty proper subset of $\partial(X)$, contradicting the supposition that $\partial(X)$ is a bond. We conclude that $G[X]$ is connected. Likewise, $G[V \setminus X]$ is connected.

Conversely, suppose that $\partial(X)$ is not a bond. Then there is a nonempty proper subset Y of V such that $X \cap Y \neq \emptyset$ and $\partial(Y) \subset \partial(X)$. But this implies (see Figure 2.10) that $E[X \cap Y, X \setminus Y] = E[Y \setminus X, \overline{X} \cap \overline{Y}] = \emptyset$. Thus $G[X]$ is not connected if $X \setminus Y \neq \emptyset$. On the other hand, if $X \setminus Y = \emptyset$, then $\emptyset \subset Y \setminus X \subset V \setminus X$, and $G[V \setminus X]$ is not connected.   □

CUTS IN DIRECTED GRAPHS

If X and Y are sets of vertices (not necessarily disjoint) of a digraph $D = (V, A)$, we denote the set of arcs of D whose tails lie in X and whose heads lie in Y by $A(X, Y)$, and their number by $a(X, Y)$. This set of arcs is denoted by $A(X)$ when $Y = X$, and their number by $a(X)$. When $Y = V \setminus X$, the set $A(X, Y)$ is called the *outcut* of D associated with X, and denoted by $\partial^+(X)$. Analogously, the set $A(Y, X)$ is called the *incut* of D associated with X, and denoted by $\partial^-(X)$. Observe that $\partial^+(X) = \partial^-(V \setminus X)$. Note, also, that $\partial(X) = \partial^+(X) \cup \partial^-(X)$. In

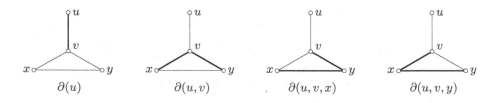

Fig. 2.11. The bonds of a graph

the case of loopless digraphs, we refer to $|\partial^+(X)|$ and $|\partial^-(X)|$ as the *outdegree* and *indegree* of X, and denote these quantities by $d^+(X)$ and $d^-(X)$, respectively.

A digraph D is called *strongly connected* or *strong* if $\partial^+(X) \neq \emptyset$ for every nonempty proper subset X of V (and thus $\partial^-(X) \neq \emptyset$ for every nonempty proper subset X of V, too).

Exercises

⋆**2.5.1**

a) Prove Theorem 2.9.
b) Prove Proposition 2.13.
c) Deduce Theorem 2.14 from Proposition 2.11.

⋆**2.5.2** Let D be a digraph, and let X be a subset of V.

a) Show that $|\partial^+(X)| = \sum_{v \in X} d^+(v) - a(X)$.
b) Suppose that D is even. Using the Principle of Directional Duality, deduce that $|\partial^+(X)| = |\partial^-(X)|$.
c) Deduce from (b) that every connected even digraph is strongly connected.

2.5.3 Let G be a graph, and let X and Y be subsets of V. Show that $\partial(X \cup Y) \bigtriangleup \partial(X \cap Y) = \partial(X \bigtriangleup Y)$.

⋆**2.5.4** Let G be a loopless graph, and let X and Y be subsets of V.

a) Show that:

$$d(X) + d(Y) = d(X \cup Y) + d(X \cap Y) + 2e(X \setminus Y, Y \setminus X)$$

b) Deduce the following *submodular inequality* for degrees of sets of vertices.

$$d(X) + d(Y) \geq d(X \cup Y) + d(X \cap Y)$$

c) State and prove a directed analogue of this submodular inequality.

⋆**2.5.5** An *odd graph* is one in which each vertex is of odd degree. Show that a graph G is odd if and only if $|\partial(X)| \equiv |X| \pmod 2$ for every subset X of V.

⋆**2.5.6** Show that each arc of a strong digraph is contained in a directed cycle.

2.5.7 DIRECTED BOND
A *directed bond* of a digraph is a bond $\partial(X)$ such that $\partial^-(X) = \emptyset$ (in other words, $\partial(X)$ is the outcut $\partial^+(X)$).

a) Show that an arc of a digraph is contained either in a directed cycle, or in a directed bond, but not both.   (G.J. MINTY)
b) Deduce that:
    i) a digraph is acyclic if and only if every bond is a directed bond,

ii) a digraph is strong if and only if no bond is a directed bond.

⋆**2.5.8** FEEDBACK ARC SET

A *feedback arc set* of a digraph D is a set S of arcs such that $D \setminus S$ is acyclic. Let S be a minimal feedback arc set of a digraph D. Show that there is a linear ordering of the vertices of D such that the arcs of S are precisely those arcs whose heads precede their tails in the ordering.

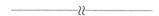

2.5.9 Let (D, w) be a weighted oriented graph. For $v \in V$, set $w^+(v) := \sum \{w(a) : a \in \partial^+(v)\}$. Suppose that $w^+(v) \geq 1$ for all $v \in V \setminus \{y\}$, where $y \in V$. Show that D contains a directed path of weight at least one, by proceeding as follows.

a) Consider an arc $(x, y) \in \partial^-(y)$ of maximum weight. Contract this arc to a vertex y', delete all arcs with tail y', and replace each pair $\{a, a'\}$ of multiple arcs (with head y') by a single arc of weight $w(a) + w(a')$, all other arcs keeping their original weights. Denote the resulting weighted digraph by (D', w'). Show that if D' contains a directed path of weight at least one, then so does D.

b) Deduce, by induction on V, that D contains a directed path of weight at least one.   (B. BOLLOBÁS AND A.D. SCOTT)

2.6 Even Subgraphs

By an *even subgraph* of a graph G we understand a *spanning* even subgraph of G, or frequently just the edge set of such a subgraph. Observe that the first two subgraphs in Figure 2.4 are both even, as is their symmetric difference. Indeed, it is an easy consequence of Proposition 2.13 that the symmetric difference of even subgraphs is always even.

Corollary 2.16 *The symmetric difference of two even subgraphs is an even subgraph.*

Proof Let F_1 and F_2 be even subgraphs of a graph G, and let X be a subset of V. By Proposition 2.13,

$$\partial_{F_1 \triangle F_2}(X) = \partial_{F_1}(X) \triangle \partial_{F_2}(X)$$

By Theorem 2.10, $\partial_{F_1}(X)$ and $\partial_{F_2}(X)$ are both of even cardinality, so their symmetric difference is too. Appealing again to Theorem 2.10, we deduce that $F_1 \triangle F_2$ is even.   □

As we show in Chapters 4 and 21, the even subgraphs of a graph play an important structural role. When discussing even subgraphs (and only in this context), by a *cycle* we mean the edge set of a cycle. By the same token, we use the term *disjoint cycles* to mean edge-disjoint cycles. With this convention, the cycles of a graph are its minimal nonempty even subgraphs, and Theorem 2.7 may be restated as follows.

Theorem 2.17 *A set of edges of a graph is an even subgraph if and only if it is a disjoint union of cycles.*   □

THE CYCLE AND BOND SPACES

Even subgraphs and edge cuts are related in the following manner.

Proposition 2.18 *In any graph, every even subgraph meets every edge cut in an even number of edges.*

Proof We first show that every cycle meets every edge cut in an even number of edges. Let C be a cycle and $\partial(X)$ an edge cut. Each vertex of C is either in X or in $V \setminus X$. As C is traversed, the number of times it crosses from X to $V \setminus X$ must be the same as the number of times it crosses from $V \setminus X$ to X. Thus $|E(C) \cap \partial(X)|$ is even.

By Theorem 2.17, every even subgraph is a disjoint union of cycles. It follows that every even subgraph meets every edge cut in an even number of edges.   □

We denote the set of all subsets of the edge set E of a graph G by $\mathcal{E}(G)$. This set forms a vector space of dimension m over $GF(2)$ under the operation of symmetric difference. We call $\mathcal{E}(G)$ the *edge space* of G. With each subset X of E, we may associate its *incidence vector* $\mathbf{f}_X$, where $f_X(e) = 1$ if $e \in X$ and $f_X(e) = 0$ if $e \notin X$. The function which maps X to $\mathbf{f}_X$ for all $X \subseteq E$ is an isomorphism from $\mathcal{E}$ to $(GF(2))^E$ (Exercise 2.6.2).

By Corollary 2.16, the set of all even subgraphs of a graph G forms a subspace $\mathcal{C}(G)$ of the edge space of G. We call this subspace the *cycle space* of G, because it is generated by the cycles of G. Likewise, by Corollary 2.12, the set of all edge cuts of G forms a subspace $\mathcal{B}(\mathcal{G})$ of $\mathcal{E}(G)$, called the *bond space* (Exercise 2.6.4a,b). Proposition 2.18 implies that these two subspaces are orthogonal. They are, in fact, orthogonal complements (Exercise 2.6.4c).

In Chapter 20, we extend the above concepts to arbitrary fields, in particular to the field of real numbers.

Exercises

2.6.1 Show that:

a) a graph G is even if and only if E is an even subgraph of G,
b) a graph G is bipartite if and only if E is an edge cut of G.

$\star$**2.6.2** Show that the edge space $\mathcal{E}(G)$ is a vector space over $GF(2)$ with respect to the operation of symmetric difference, and that it is isomorphic to $(GF(2))^E$.

2.6.3

a) Draw all the elements of the cycle and bond spaces of the wheel W_4.

b) How many elements are there in each of these two vector spaces?

★**2.6.4** Show that:

a) the cycles of a graph generate its cycle space,
b) the bonds of a graph generate its bond space,
c) the bond space of a graph G is the row space of its incidence matrix **M** over $GF(2)$, and the cycle space of G is its orthogonal complement.

2.6.5 How many elements are there in the cycle and bond spaces of a graph G?

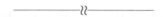

2.6.6 Show that every graph G has an edge cut $[X, Y]$ such that $G[X]$ and $G[Y]$ are even.

2.7 Graph Reconstruction

Two graphs G and H on the same vertex set V are called *hypomorphic* if, for all $v \in V$, their vertex-deleted subgraphs $G - v$ and $H - v$ are isomorphic. Does this imply that G and H are themselves isomorphic? Not necessarily: the graphs $2K_1$ and K_2, though not isomorphic, are clearly hypomorphic. However, these two graphs are the only known nonisomorphic pair of hypomorphic simple graphs, and it was conjectured in 1941 by Kelly (1942) (see also Ulam (1960)) that there are no other such pairs. This conjecture was reformulated by Harary (1964) in the more intuitive language of reconstruction. A *reconstruction* of a graph G is any

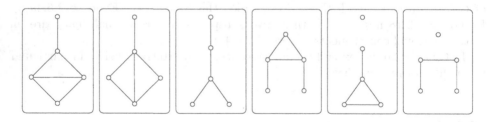

Fig. 2.12. The deck of a graph on six vertices

graph that is hypomorphic to G. We say that a graph G is *reconstructible* if every reconstruction of G is isomorphic to G, in other words, if G can be 'reconstructed' up to isomorphism from its vertex-deleted subgraphs. Informally, one may think of the (unlabelled) vertex-deleted subgraphs as being presented on cards, one per card. The problem of reconstructing a graph is then that of determining the graph from its *deck* of cards. The reader is invited to reconstruct the graph whose deck of six cards is shown in Figure 2.12.

THE RECONSTRUCTION CONJECTURE

Conjecture 2.19 *Every simple graph on at least three vertices is reconstructible.*

The Reconstruction Conjecture has been verified by computer for all graphs on up to ten vertices by McKay (1977). In discussing it, we implicitly assume that our graphs have at least three vertices.

One approach to the Reconstruction Conjecture is to show that it holds for various classes of graphs. A class of graphs is *reconstructible* if every member of the class is reconstructible. For instance, regular graphs are easily shown to be reconstructible (Exercise 2.7.5). One can also prove that disconnected graphs are reconstructible (Exercise 2.7.11). Another approach is to prove that specific parameters are reconstructible. We call a graphical parameter *reconstructible* if the parameter takes the same value on all reconstructions of G. A fundamental result of this type was obtained by Kelly (1957). For graphs F and G, we adopt the notation of Lauri and Scapellato (2003) and use $\binom{G}{F}$ to denote the number of copies of F in G. For instance, if $F = K_2$, then $\binom{G}{F} = e(G)$, if $F = G$, then $\binom{G}{F} = 1$, and if $v(F) > v(G)$, then $\binom{G}{F} = 0$.

Lemma 2.20   KELLY'S LEMMA
For any two graphs F and G such that $v(F) < v(G)$, the parameter $\binom{G}{F}$ is reconstructible.

Proof   Each copy of F in G occurs in exactly $v(G) - v(F)$ of the vertex-deleted subgraphs $G - v$ (namely, whenever the vertex v is not present in the copy). Therefore

$$\binom{G}{F} = \frac{1}{v(G) - v(F)} \sum_{v \in V} \binom{G - v}{F}$$

Since the right-hand side of this identity is reconstructible, so too is the left-hand side. □

Corollary 2.21 *For any two graphs F and G such that $v(F) < v(G)$, the number of subgraphs of G that are isomorphic to F and include a given vertex v is reconstructible.*

Proof   This number is $\binom{G}{F} - \binom{G-v}{F}$, which is reconstructible by Kelly's Lemma. □

Corollary 2.22 *The size and the degree sequence are reconstructible parameters.*

Proof Take $F = K_2$ in Kelly's Lemma and Corollary 2.21, respectively.   □

An edge analogue of the Reconstruction Conjecture was proposed by Harary (1964). A graph is *edge-reconstructible* if it can be reconstructed up to isomorphism from its edge-deleted subgraphs.

THE EDGE RECONSTRUCTION CONJECTURE

Conjecture 2.23 *Every simple graph on at least four edges is edge-reconstructible.*

Note that the bound on the number of edges is needed on account of certain small counterexamples (see Exercise 2.7.2). The notions of *edge reconstructibility* of classes of graphs and of graph parameters are defined in an analogous manner to those of reconstructibility, and there is an edge version of Kelly's Lemma, whose proof we leave as an exercise (Exercise 2.7.13a).

Lemma 2.24 KELLY'S LEMMA: EDGE VERSION
For any two graphs F and G such that $e(F) < e(G)$, the parameter $\binom{G}{F}$ is edge reconstructible.   □

Because edge-deleted subgraphs are much closer to the original graph than are vertex-deleted subgraphs, it is intuitively clear (but not totally straightforward to prove) that the Edge Reconstruction Conjecture is no harder than the Reconstruction Conjecture (Exercise 2.7.14). Indeed, a number of approaches have been developed which are effective for edge reconstruction, but not for vertex reconstruction. We describe below one of these approaches, Möbius Inversion.

PROOF TECHNIQUE: MÖBIUS INVERSION

We discussed earlier the proof technique of counting in two ways. Here, we present a more subtle counting technique, that of *Möbius Inversion*. This is a generalization of the *Inclusion-Exclusion Formula*, a formula which expresses the cardinality of the union of a family of sets $\{A_i : i \in T\}$ in terms of the cardinalities of intersections of these sets:

$$| \cup_{i \in T} A_i | = \sum_{\emptyset \subset X \subseteq T} (-1)^{|X|-1} |\cap_{i \in X} A_i| \qquad (2.3)$$

the case of two sets being the formula $|A \cup B| = |A| + |B| - |A \cap B|$.

MÖBIUS INVERSION (CONTINUED)

Theorem 2.25   THE MÖBIUS INVERSION FORMULA
Let $f : 2^T \to \mathbb{R}$ be a real-valued function defined on the subsets of a finite set T. Define the function $g : 2^T \to \mathbb{R}$ by

$$g(S) := \sum_{S \subseteq X \subseteq T} f(X) \tag{2.4}$$

Then, for all $S \subseteq T$,

$$f(S) = \sum_{S \subseteq X \subseteq T} (-1)^{|X|-|S|} g(X) \tag{2.5}$$

Remark. Observe that (2.4) is a linear transformation of the vector space of real-valued functions defined on 2^T. The Möbius Inversion Formula (2.5) simply specifies the inverse of this transformation.

Proof   By the Binomial Theorem,

$$\sum_{S \subseteq X \subseteq Y} (-1)^{|X|-|S|} = \sum_{|S| \leq |X| \leq |Y|} \binom{|Y|-|S|}{|X|-|S|} (-1)^{|X|-|S|} = (1-1)^{|Y|-|S|}$$

which is equal to 0 if $S \subset Y$, and to 1 if $S = Y$. Therefore,

$$
\begin{aligned}
f(S) &= \sum_{S \subseteq Y \subseteq T} f(Y) \sum_{S \subseteq X \subseteq Y} (-1)^{|X|-|S|} \\
&= \sum_{S \subseteq X \subseteq T} (-1)^{|X|-|S|} \sum_{X \subseteq Y \subseteq T} f(Y) = \sum_{S \subseteq X \subseteq T} (-1)^{|X|-|S|} g(X) \qquad \square
\end{aligned}
$$

We now show how the Möbius Inversion Formula can be applied to the problem of edge reconstruction. This highly effective approach was introduced by Lovász (1972c) and refined successively by Müller (1977) and Nash-Williams (1978).

The idea is to count the mappings between two simple graphs G and H on the same vertex set V according to the intersection of the image of G with H. Each such mapping is determined by a permutation σ of V, which one extends to $G = (V, E)$ by setting $\sigma(G) := (V, \sigma(E))$, where $\sigma(E) := \{\sigma(u)\sigma(v) : uv \in E\}$. For each spanning subgraph F of G, we consider the permutations of G which map the edges of F onto edges of H and the remaining edges of G onto edges of $\overline{H}$. We denote their number by $|G \to H|_F$, that is:

$$|G \to H|_F := |\{\sigma \in S_n : \sigma(G) \cap H = \sigma(F)\}|$$

In particular, if $F = G$, then $|G \to H|_F$ is simply the number of embeddings of G in H, which we denote for brevity by $|G \to H|$, and if F is empty, $|G \to H|_F$ is the number of embeddings of G in the complement of H; that is, $|G \to \overline{H}|$. These concepts are illustrated in Figure 2.13 for all spanning subgraphs F of G when $G = K_1 + K_{1,2}$ and $H = 2K_2$. Observe that, for any subgraph F of G,

$$\sum_{F \subseteq X \subseteq G} |G \to H|_X = |F \to H| \qquad (2.6)$$

and that

$$|F \to H| = \text{aut}(F) \binom{H}{F} \qquad (2.7)$$

where $\text{aut}(F)$ denotes the number of automorphisms of F, because the subgraph F of G can be mapped onto each copy of F in H in $\text{aut}(F)$ distinct ways.

Lemma 2.26   Nash-Williams' Lemma
Let G be a graph, F a spanning subgraph of G, and H an edge reconstruction of G that is not isomorphic to G. Then

$$|G \to G|_F - |G \to H|_F = (-1)^{e(G)-e(F)}\text{aut}(G) \qquad (2.8)$$

Proof   By (2.6) and (2.7),

$$\sum_{F \subseteq X \subseteq G} |G \to H|_X = \text{aut}(F) \binom{H}{F}$$

We invert this identity by applying the Möbius Inversion Formula (identifying each spanning subgraph of G with its edge set), to obtain:

$$|G \to H|_F = \sum_{F \subseteq X \subseteq G} (-1)^{e(X)-e(F)}\text{aut}(X) \binom{H}{X}$$

Therefore,

$$|G \to G|_F - |G \to H|_F = \sum_{F \subseteq X \subseteq G} (-1)^{e(X)-e(F)}\text{aut}(X) \left(\binom{G}{X} - \binom{H}{X} \right)$$

Because H is an edge reconstruction of G, we have $\binom{G}{X} = \binom{H}{X}$ for every proper spanning subgraph X of G, by the edge version of Kelly's Lemma (2.24). Finally, $\binom{G}{G} = 1$, whereas $\binom{H}{G} = 0$ since $e(H) = e(G)$ and $H \not\cong G$. $\square$

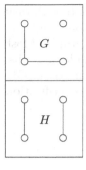

	F	○ ○ ○ ○	○─○ ○ ○	○ ○ ○──○	○─○ ○──○
H	$\|G \to G\|_F$	10	6	6	2
	$\|G \to H\|_F$	8	8	8	0

Fig. 2.13. Counting mappings

MÖBIUS INVERSION (CONTINUED)

Theorem 2.27 *A graph G is edge reconstructible if there exists a spanning subgraph F of G such that either of the following two conditions holds.*

(i) $|G \to H|_F$ takes the same value for all edge reconstructions H of G,
(ii) $|F \to G| < 2^{e(G)-e(F)-1}\mathrm{aut}(G)$.

Proof Let H be an edge reconstruction of G. If condition (i) holds, the left-hand side of (2.8) is zero whereas the right-hand side is nonzero. The inequality of condition (ii) is equivalent, by (2.6), to the inequality

$$\sum_{F \subseteq X \subseteq G} |G \to G|_X < 2^{e(G)-e(F)-1}\mathrm{aut}(G)$$

But this implies that $|G \to G|_X < \mathrm{aut}(G)$ for some spanning subgraph X of G such that $e(G) - e(X)$ is even, and identity (2.8) is again violated (with $F := X$). Thus, in both cases, Nash-Williams' Lemma implies that H is isomorphic to G. □

Choosing F as the empty graph in Theorem 2.27 yields two sufficient conditions for the edge reconstructibility of a graph in terms of its edge density, due to Lovász (1972) and Müller (1977), respectively (Exercise 2.7.8).

Corollary 2.28 *A graph G is edge reconstructible if either $m > \frac{1}{2}\binom{n}{2}$ or $2^{m-1} > n!$* □

Two other applications of the Möbius Inversion Formula to graph theory are given in Exercises 2.7.17 and 14.7.12. For further examples, see Whitney (1932b). Theorem 2.25 was extended by Rota (1964) to the more general context of partially ordered sets.

It is natural to formulate corresponding conjectures for digraphs (see Harary (1964)). Tools such as Kelly's Lemma apply to digraphs as well, and one might be led to believe that the story is much the same here as for undirected graphs. Most surprisingly, this is not so. Several infinite families of nonreconstructible digraphs, and even nonreconstructible tournaments, were constructed by Stockmeyer (1981) (see Exercise 2.7.18). One such pair is shown in Figure 2.14. We leave its verification to the reader (Exercise 2.7.9).

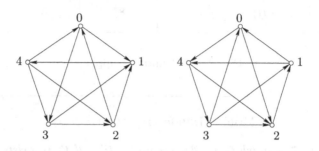

Fig. 2.14. A pair of nonreconstructible tournaments

We remark that there also exist infinite families of nonreconstructible hypergraphs (see Exercise 2.7.10 and Kocay (1987)) and nonreconstructible infinite graphs (see Exercise 4.2.10). Further information on graph reconstruction can be found in the survey articles by Babai (1995), Bondy (1991), and Ellingham (1988), and in the book by Lauri and Scapellato (2003).

Exercises

2.7.1 Find two nonisomorphic graphs on six vertices whose decks both include the first five cards displayed in Figure 2.12.   (P.K. STOCKMEYER)

2.7.2 Find a pair of simple graphs on two edges, and also a pair of simple graphs on three edges, which are not edge reconstructible.

2.7.3 Two dissimilar vertices u and v of a graph G are called *pseudosimilar* if the vertex-deleted subgraphs $G - u$ and $G - v$ are isomorphic.

a) Find a pair of pseudosimilar vertices in the graph of Figure 2.15.
b) Construct a connected acyclic graph with a pair of pseudosimilar vertices.
                    (F. HARARY AND E.M. PALMER)

2.7.4 A class $\mathcal{G}$ of graphs is *recognizable* if, for each graph $G \in \mathcal{G}$, every reconstruction of G also belongs to $\mathcal{G}$. The class $\mathcal{G}$ is *weakly reconstructible* if, for each graph $G \in \mathcal{G}$, every reconstruction of G that belongs to $\mathcal{G}$ is isomorphic to G. Show that a class of graphs is reconstructible if and only if it is both recognizable and weakly reconstructible.

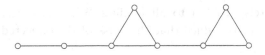

Fig. 2.15. A graph containing a pair of pseudosimilar vertices (Exercise 2.7.3)

2.7.5

a) Show that regular graphs are both recognizable and weakly reconstructible.
b) Deduce that this class of graphs is reconstructible.

2.7.6

a) Let G be a connected graph on at least two vertices, and let P be a maximal path in G, starting at x and ending at y. Show that $G - x$ and $G - y$ are connected.
b) Deduce that a graph on at least three vertices is connected if and only if at least two vertex-deleted subgraphs are connected.
c) Conclude that the class of disconnected graphs is recognizable.

2.7.7 Verify identity (2.6) for the graphs G and H of Figure 2.13, and for all spanning subgraphs F of G.

$\star$**2.7.8** Deduce Corollary 2.28 from Theorem 2.27.

2.7.9 Show that the two tournaments displayed in Figure 2.14 form a pair of nonreconstructible tournaments.   (P.K. STOCKMEYER)

2.7.10 Consider the hypergraphs G and H with vertex set $V := \{1, 2, 3, 4, 5\}$ and respective edge sets

$$\mathcal{F}(G) := \{123, 125, 135, 234, 345\} \quad \text{and} \quad \mathcal{F}(H) := \{123, 135, 145, 234, 235\}$$

Show that (G, H) is a nonreconstructible pair.

———————⁂———————

2.7.11 Let G be a graph, and let $\mathcal{F} := (F_1, F_2, \ldots, F_k)$ be a sequence of graphs (not necessarily distinct). A *covering* of G by $\mathcal{F}$ is a sequence $(G_1, G_2, \ldots, G_k)$ of subgraphs of G such that $G_i \cong F_i$, $1 \leq i \leq k$, and $\cup_{i=1}^{k} G_i = G$. We denote the number of coverings of G by $\mathcal{F}$ by $c(\mathcal{F}, G)$. For example, if $\mathcal{F} := (K_2, K_{1,2})$, the coverings of G by $\mathcal{F}$ for each graph G such that $c(\mathcal{F}, G) > 0$ are as indicated in Figure 2.16 (where the edge of K_2 is shown as a dotted line).

a) Show that, for any graph G and any sequence $\mathcal{F} := (F_1, F_2, \ldots, F_k)$ of graphs such that $v(F_i) < v(G)$, $1 \leq i \leq k$, the parameter

$$\sum_{X} c(\mathcal{F}, X) \binom{G}{X}$$

is reconstructible, where the sum extends over all unlabelled graphs X such that $v(X) = v(G)$.   (W.L. KOCAY)

b) Applying Exercise 2.7.11a to all families $\mathcal{F} := (F_1, F_2, \ldots, F_k)$ such that $\sum_{i=1}^{k} v(F_i) = v(G)$, deduce that the class of disconnected graphs is weakly reconstructible.

c) Applying Exercise 2.7.6c, conclude that this class is reconstructible.

(P.J.KELLY)

2.7.12 Let G and H be two graphs on the same vertex set V, where $|V| \geq 4$. Suppose that $G - \{x, y\} \cong H - \{x, y\}$ for all $x, y \in V$. Show that $G \cong H$.

⋆**2.7.13**

a) Prove the edge version of Kelly's Lemma (Lemma 2.24).

b) Using the edge version of Kelly's Lemma, show that the number of isolated vertices is edge reconstructible.

c) Deduce that the Edge Reconstruction Conjecture is valid for all graphs provided that it is valid for all graphs without isolated vertices.

2.7.14

a) By applying Exercise 2.7.11a, show that the (vertex) deck of any graph without isolated vertices is edge reconstructible.

b) Deduce from Exercise 2.7.13c that the Edge Reconstruction Conjecture is true if the Reconstruction Conjecture is true.   (D.L. GREENWELL)

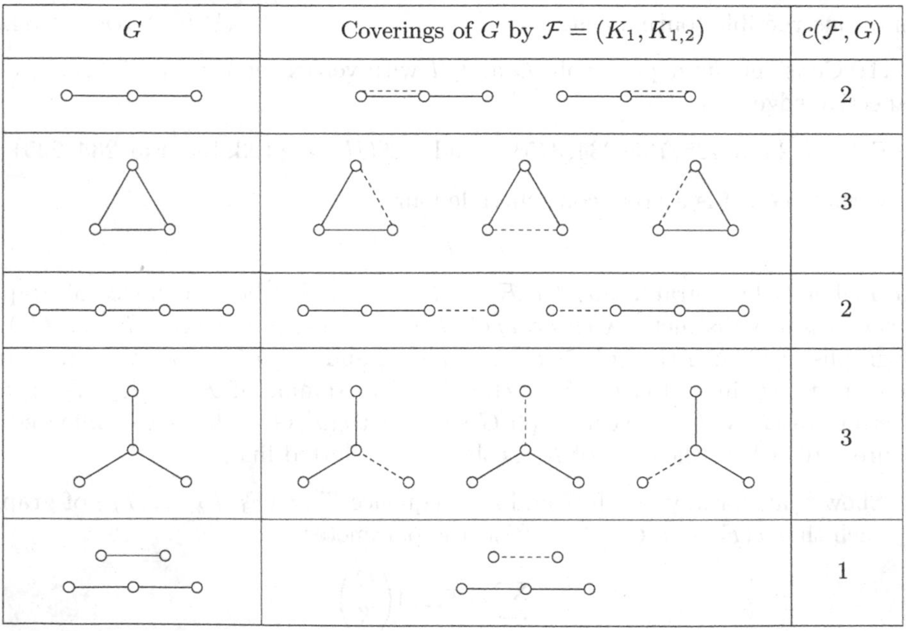

G	Coverings of G by $\mathcal{F} = (K_1, K_{1,2})$	$c(\mathcal{F}, G)$
		2
		3
		2
		3
		1

Fig. 2.16. Covering a graph by a sequence of graphs (Exercise 2.7.11)

2.7.15 Let $\{A_i : i \in T\}$ be a family of sets. For $S \subseteq T$, define $f(S) := |(\cap_{i \in S} A_i) \setminus (\cup_{i \in T \setminus S} A_i)|$ and $g(S) := |\cap_{i \in S} A_i|$, where, by convention, $\cap_{i \in \emptyset} A_i = \cup_{i \in T} A_i$.

a) Show that $g(S) = \sum_{S \subseteq X \subseteq T} f(X)$.

b) Deduce from the Möbius Inversion Formula (2.5) that

$$\sum_{\emptyset \subseteq X \subseteq T} (-1)^{|X|} |\cap_{i \in X} A_i| = 0.$$

c) Show that this identity is equivalent to the Inclusion–Exclusion Formula (2.3).

2.7.16 Use the Binomial Theorem to establish the Inclusion-Exclusion Formula (2.3) directly, without appealing to Möbius Inversion.

2.7.17 Consider the lower-triangular matrix $\mathbf{A}_n$ whose rows and columns are indexed by the isomorphism types of the graphs on n vertices, listed in increasing order of size, and whose (X, Y) entry is $\binom{X}{Y}$.

a) Compute $\mathbf{A}_3$ and $\mathbf{A}_4$.

b) For $k \in \mathbb{Z}$, show that the (X, Y) entry of $(\mathbf{A}_n)^k$ is $k^{e(X)-e(Y)} \binom{X}{Y}$.

(V.B. Mnukhin)

2.7.18 Consider the Stockmeyer tournament ST_n, defined in Exercise 1.5.10.

a) Show that each vertex-deleted subgraph of ST_n is self-converse.

b) Denote by $odd(ST_n)$ and $even(ST_n)$ the subtournaments of ST_n induced by its odd and even vertices, respectively. For $n \geq 1$, show that $odd(ST_n) \cong ST_{n-1} \cong even(ST_n)$.

c) Deduce, by induction on n, that $ST_n - k \cong ST_n - (2^n - k + 1)$ for all $k \in V(ST_n)$.

(W. Kocay)

d) Consider the following two tournaments obtained from ST_n by adding a new vertex 0. In one of these tournaments, 0 dominates the odd vertices and is dominated by the even vertices; in the other, 0 dominates the even vertices and is dominated by the odd vertices. Show that these two tournaments on $2^n + 1$ vertices form a pair of nonreconstructible digraphs.

(P.K. Stockmeyer)

2.7.19 To *switch* a vertex of a simple graph is to exchange its sets of neighbours and non-neighbours. The graph so obtained is called a *switching* of the graph. The collection of switchings of a graph G is called the (switching) *deck* of G. A graph is *switching-reconstructible* if every graph with the same deck as G is isomorphic to G.

a) Find four pairs of graphs on four vertices which are not switching-reconstruct-ible.

b) Let G be a graph with n odd. Consider the collection $\mathcal{G}$ consisting of the n^2 graphs in the decks of the graphs which comprise the deck of G.

    i) Show that G is the only graph which occurs an odd number of times in $\mathcal{G}$.

    ii) Deduce that G is switching-reconstructible.

c) Let G be a graph with $n \equiv 2 \pmod{4}$. Show that G is switching-reconstructible.

(R.P. Stanley; N. Alon)

2.8 Related Reading

PATH AND CYCLE DECOMPOSITIONS

Veblen's Theorem (2.7) tells us that every even graph can be decomposed into cycles, but it says nothing about the number of cycles in the decomposition. One may ask how many or how few cycles there can be in a cycle decomposition of a given even graph. These questions are not too hard to answer in special cases, such as when the graph is complete (see Exercises 2.4.5 and 2.4.6a). Some forty years ago, G. Hajós conjectured that *every simple even graph on n vertices admits a decomposition into at most $(n-1)/2$ cycles* (see Lovász (1968b)). Surprisingly little progress has been made on this simply stated problem. An analogous conjecture on path decompositions was proposed by T. Gallai at about the same time (see Lovász (1968b)), namely that *every simple connected graph on n vertices admits a decomposition into at most $(n+1)/2$ paths*. This bound is sharp if all the degrees are odd, because in any path decomposition each vertex must be an end of at least one path. Lovász (1968b) established the truth of Gallai's conjecture in this case (see also Donald (1980)).

LEGITIMATE DECKS

In the Reconstruction Conjecture (2.19), the deck of vertex-deleted subgraphs of a graph is supplied, the goal being to determine the graph. A natural problem, arguably even more fundamental, is to characterize such decks. A family $\mathcal{G} := \{G_1, G_2, \ldots, G_n\}$ of n graphs, each of order $n-1$, is called a *legitimate deck* if there is at least one graph G with vertex set $\{v_1, v_2, \ldots, v_n\}$ such that $G_i \cong G - v_i$, $1 \le i \le n$. The *Legitimate Deck Problem* asks for a characterization of legitimate decks. This problem was raised by Harary (1964). It was shown by Harary et al. (1982) and Mansfield (1982) that the problem of recognizing whether a deck is legitimate is as hard (in a sense to be discussed in Chapter 8) as that of deciding whether two graphs are isomorphic.

The various counting arguments deployed to attack the Reconstruction Conjecture provide natural necessary conditions for legitimacy. For instance, the proof of Kelly's Lemma (2.20) tells us that if $\mathcal{G}$ is the deck of a graph G, then $\binom{G}{F} = \sum_{i=1}^{n} \binom{G_i}{F}/(n - v(F))$ for every graph F on fewer than n vertices. Because the left-hand side is an integer, $\sum_{i=1}^{n} \binom{G_i}{F}$ must be a multiple of $n - v(F)$. It is not hard to come up with an illegitimate deck which passes this test. Indeed, next to nothing is known on the Legitimate Deck Problem. A more general problem would be to characterize, for a fixed integer k, the vectors $(\binom{G}{F} : v(F) = k)$, where G ranges over all graphs on n vertices. Although trivial for $k = 2$, the problem is unsolved already for $k = 3$ and appears to be very hard. Even determining the minimum number of triangles in a graph on n vertices with a specified number of edges is a major challenge (see Razborov (2006), where a complex asymptotic formula, derived by highly nontrivial methods, is given).

ULTRAHOMOGENEOUS GRAPHS

A simple graph is said to be k-*ultrahomogeneous* if any isomorphism between two of its isomorphic induced subgraphs on k or fewer vertices can be extended to an automorphism of the entire graph. It follows directly from the definition that every graph is 0-ultrahomogeneous, that 1-ultrahomogeneous graphs are the same as vertex-transitive graphs, and that complements of k-ultrahomogeneous graphs are k-ultrahomogeneous.

Cameron (1980) showed that any graph which is 5-ultrahomogeneous is k-ultrahomogeneous for all k. Thus it is of interest to classify the k-ultrahomogeneous graphs for $1 \leq k \leq 5$. The 5-ultrahomogeneous graphs were completely described by Gardiner (1976). They are the self-complementary graphs C_5 and $L(K_{3,3})$, and the Turán graphs $T_{k,rk}$, for all $k \geq 1$ and $r \geq 1$, as well as their complements. These graphs all have rather simple structures. There is, however, a remarkable 4-ultrahomogeneous graph. It arises from a very special geometric configuration, discovered by Schläfli (1858), consisting of twenty-seven lines on a cubic surface, and is known as the *Schläfli graph*. Here is a description due to Chudnovsky and Seymour (2005).

The vertex set of the graph is $\mathbb{Z}_3^3$, two distinct vertices (a,b,c) and (a',b',c') being joined by an edge if $a' = a$ and either $b' = b$ or $c' = c$, or if $a' = a+1$ and $b' \neq c$. This construction results in a 16-regular graph on twenty-seven vertices. The subgraph induced by the sixteen neighbours of a vertex of the Schläfli graph is isomorphic to the complement of the *Clebsch graph*, shown in Figure 12.12. In turn, the subgraph induced by the neighbour set of a vertex of the complement of the Clebsch graph is isomorphic to the complement of the Petersen graph. Thus, one may conclude that the Clebsch graph is 3-ultrahomogeneous and that the Petersen graph is 2-ultrahomogeneous. By employing the classification theorem for finite simple groups, Buczak (1980) showed that the the the Schläfli graph and its complement are the only two graphs which are 4-ultrahomogeneous without being 5-ultrahomogeneous.

The notion of ultrahomogeneity may be extended to infinite graphs. The countable random graph G described in Exercise 13.2.18 has the property that if F and F' are isomorphic induced subgraphs of G, then any isomorphism between F and F' can be extended to an automorphism of G. Further information about ultrahomogeneous graphs may be found in Cameron (1983) and Devillers (2002).

3
Connected Graphs

Contents

3.1 Walks and Connection

WALKS

In Section 1.1, the notion of connectedness was defined in terms of edge cuts. Here, we give an alternative definition based on the notion of a walk in a graph.

A *walk* in a graph G is a sequence $W := v_0 e_1 v_1 \ldots v_{\ell-1} e_\ell v_\ell$, whose terms are alternately vertices and edges of G (not necessarily distinct), such that v_{i-1} and v_i are the ends of e_i, $1 \leq i \leq \ell$. (We regard loops as giving rise to distinct walks with the same sequence, because they may be traversed in either sense. Thus if e is a loop incident with a vertex v, we count the walk vev not just once, but twice.) If $v_0 = x$ and $v_\ell = y$, we say that W *connects* x to y and refer to W as an *xy-walk*. The vertices x and y are called the *ends* of the walk, x being its *initial vertex* and y its *terminal vertex*; the vertices $v_1, \ldots, v_{\ell-1}$ are its *internal vertices*. The integer

ℓ (the number of edge terms) is the *length* of W. An *x-walk* is a walk with initial vertex x. If u and v are two vertices of a walk W, where u precedes v on W, the subsequence of W starting with u and ending with v is denoted by uWv and called the *segment* of W from u to v. The notation uWv is also used simply to signify a uv-walk W.

In a simple graph, a walk $v_0 e_1 v_1 \ldots v_{\ell-1} e_\ell v_\ell$ is determined, and is commonly specified, by the sequence $v_0 v_1 \ldots v_\ell$ of its vertices. Indeed, even if a graph is not simple, we frequently refer to a sequence of vertices in which consecutive terms are adjacent vertices as a 'walk'. In such cases, it should be understood that the discussion is valid for any walk with that vertex sequence. This convention is especially useful in discussing paths, which may be viewed as walks whose vertices (and edges) are distinct.

A walk in a graph is *closed* if its initial and terminal vertices are identical, and is a *trail* if all its edge terms are distinct. A closed trail of positive length whose initial and internal vertices are distinct is simply the sequence of vertices and edges of a cycle. Reciprocally, with any cycle one may associate a closed trail whose terms are just the vertices and edges of the cycle. Even though this correspondence is not one-to-one (the trail may start and end at any vertex of the cycle, and traverse it in either sense), we often specify a cycle by describing an associated closed trail and refer to that trail as the cycle itself.

CONNECTION

Connectedness of pairs of vertices in a graph G is an equivalence relation on V. Clearly, each vertex x is connected to itself by the trivial walk $W := x$; also, if x is connected to y by a walk W, then y is connected to x by the walk $\overleftarrow{W}$ obtained on reversing the sequence W; finally, for any three vertices, x, y, and z of G, if xWy and $yW'z$ are walks, the sequence $xWyW'z$, obtained by concatenating W and W' at y, is a walk; thus, if x is connected to y and y is connected to z, then x is connected to z. The equivalence classes determined by this relation of connectedness are simply the vertex sets of the components of G (Exercise 3.1.3).

If there is an xy-walk in a graph G, then there is also an xy-path (Exercise 3.1.1). The length of a shortest such path is called the *distance* between x and y and denoted $d_G(x,y)$. If there is no path connecting x and y (that is, if x and y lie in distinct components of G), we set $d_G(x,y) := \infty$.

We may extend the notion of an xy-path to paths connecting subsets X and Y of V. An (X,Y)-*path* is a path which starts at a vertex of X, ends at a vertex of Y, and whose internal vertices belong to neither X nor Y; if F_1 and F_2 are subgraphs of a graph G, we write (F_1, F_2)-*path* instead of $(V(F_1), V(F_2))$-path. A useful property of connected graphs is that any two nonempty sets of vertices (or subgraphs) are connected by such a path (Exercise 3.1.4).

PROOF TECHNIQUE: EIGENVALUES

We saw in Chapter 2 how certain problems can be solved by making use of arguments involving linear independence. Another powerful linear algebraic tool involves the computation of eigenvalues of appropriate matrices. Although this technique is suitable only for certain rather special problems, it is remarkably effective when applicable. Here is an illustration.

A *friendship graph* is a simple graph in which any two vertices have exactly one common neighbour. By using a clever mixture of graph-theoretical and eigenvalue arguments, Erdős et al. (1966) proved that all friendship graphs have a very simple structure.

Theorem 3.1   THE FRIENDSHIP THEOREM
Let G be a simple graph in which any two vertices (people) have exactly one common neighbour (friend). Then G has a vertex of degree $n-1$ (a politician, everyone's friend).

Proof   Suppose the theorem false, and let G be a friendship graph with $\Delta < n - 1$. Let us show first of all that G is regular. Consider two nonadjacent vertices x and y. For each neighbour v of x, denote by $f(v)$ the unique common neighbour of v and y. Since x and $f(v)$ have a unique common neighbour, namely v, the mapping $f : N(x) \to N(y)$ is one-to-one. Thus $d(x) = |N(x)| \leq |N(y)| = d(y)$. Likewise, $d(y) \leq d(x)$, and we conclude that $d(x) = d(y)$. Thus any two nonadjacent vertices of G have the same degree; equivalently, any two adjacent vertices of $\overline{G}$ have the same degree.

In order to prove that G is regular, it therefore suffices to show that $\overline{G}$ is connected. But $\overline{G}$ has no singleton component, because $\delta(\overline{G}) = n - 1 - \Delta(G) > 0$, and cannot have two components of order two or more, because G would then contain a 4-cycle, thus two vertices with two common neighbours. Therefore G is k-regular for some positive integer k. Moreover, by counting the number of 2-paths in G in two ways, we have $n\binom{k}{2} = \binom{n}{2}$; that is, $n = k^2 - k + 1$.

Let $\mathbf{A}$ be the adjacency matrix of G. Then (Exercise 3.1.2) $\mathbf{A}^2 = \mathbf{J} + (k-1)\mathbf{I}$, where $\mathbf{J}$ is the $n \times n$ matrix all of whose entries are 1, and $\mathbf{I}$ is the $n \times n$ identity matrix. Because the eigenvalues of $\mathbf{J}$ are 0, with multiplicity $n - 1$, and n, with multiplicity 1, the eigenvalues of $\mathbf{A}^2$ are $k - 1$, with multiplicity $n - 1$, and $n + k - 1 = k^2$, with multiplicity 1. The graph G therefore has eigenvalues $\pm\sqrt{k - 1}$, with total multiplicity $n - 1$, and k, with multiplicity 1 (see Exercise 1.1.22a).

Because G is simple, the sum of its eigenvalues, the trace of $\mathbf{A}$, is zero. Thus $t\sqrt{k - 1} = k$ for some integer t. But this implies that $k = 2$ and $n = 3$, contradicting the assumption that $\Delta < n - 1$. $\square$

Further applications of eigenvalues are outlined in Exercises 3.1.11 and 3.1.12.

The above notions apply equally to digraphs. If $W := v_0 a_1 v_1 \ldots v_{\ell-1} a_\ell v_\ell$ is a walk in a digraph, an arc a_i of W is a *forward arc* if v_{i-1} is the tail of a_i and v_i is its head, and a *reverse arc* if v_i is the tail of a_i and v_{i-1} its head. The sets of forward and reverse arcs of W are denoted by W^+ and W^-, respectively. Walks in which all arcs are forward arcs, called directed walks, are discussed in Section 3.4.

Connectedness plays an essential role in applications of graph theory. For example, the graph representing a communications network needs to be connected for communication to be possible between all vertices. Connectedness also plays a basic role in theoretical considerations. For instance, in developing an algorithm to determine whether a given graph is planar, we may restrict our attention to connected graphs, because a graph is planar if and only if each of its components is planar.

Exercises

$\star$**3.1.1** If there is an xy-walk in a graph G, show that there is also an xy-path in G.

3.1.2 Let G be a graph with vertex set V and adjacency matrix $\mathbf{A} = (a_{uv})$. Show that the number of uv-walks of length k in G is the (u, v) entry of $\mathbf{A}^k$.

$\star$**3.1.3** Show that the equivalence classes determined by the relation of connectedness between vertices are precisely the vertex sets of the components of the graph.

$\star$**3.1.4** Show that a graph G is connected if and only if there is an (X, Y)-path in G for any two nonempty subsets X and Y of V.

3.1.5 Show that, in any graph G, the distance function satisfies the *triangle inequality*: for any three vertices x, y, and z, $d(x, y) + d(y, z) \geq d(x, z)$.

3.1.6 POWER OF A GRAPH
The *kth power* of a simple graph $G = (V, E)$ is the graph G^k whose vertex set is V, two distinct vertices being adjacent in G^k if and only if their distance in G is at most k. The graph G^2 is referred to as the *square* of G, the graph G^3 as the *cube* of G. Consider P_n^k, the kth power of a path on n vertices, where $n > k^2 + k$. Show that:

a) $d(P_n^k) > 2k - 1$,
b) $\delta(F) \leq k$ for every induced subgraph F of P_n^k.

3.1.7 DIAMETER
The *diameter* of a graph G is the greatest distance between two vertices of G.

a) Let G be a simple graph of diameter greater than three. Show that $\overline{G}$ has diameter less than three.
b) Deduce that every self-complementary graph has diameter at most three.

c) For $k = 0, 1, 2, 3$, give an example of a self-complementary graph of diameter k, if there is one.

3.1.8 Show that if G is a simple graph of diameter two with $\Delta = n - 2$, then $m \geq 2n - 4$.

3.1.9 Show that the incidence graph of a finite projective plane has diameter three.

3.1.10 If the girth of a graph is at least $2k$, show that its diameter is at least k.

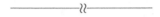

3.1.11

a) Let G_1 and G_2 be edge-disjoint copies of the Petersen graph on the same vertex set. Show that 2 is an eigenvalue of $G_1 \cup G_2$ by proceeding as follows.
   i) Observe that $\mathbf{1}$ is an eigenvector of both G_1 and G_2 corresponding to the eigenvalue 3.
   ii) Let S_1 and S_2 denote the eigenspaces of G_1 and G_2, respectively, corresponding to the eigenvalue 1. (Since 1 is an eigenvalue of the Petersen graph with multiplicity five, S_1 and S_2 are 5-dimensional subspaces of $\mathbb{R}^{10}$.) Using the fact that $\mathbf{1}$ is orthogonal to both S_1 and S_2, show that the dimension of $S_1 \cap S_2$ is at least one.
   iii) Noting that $\mathbf{A}_{G_1 \cup G_2} = \mathbf{A}_{G_1} + \mathbf{A}_{G_2}$, show that any nonzero vector in $S_1 \cap S_2$ is an eigenvector of $G_1 \cup G_2$ corresponding to the eigenvalue 2.
b) Appealing now to Exercises 1.3.2 and 1.3.11, conclude that K_{10} cannot be decomposed into three copies of the Petersen graph.   (A.J. SCHWENK)

3.1.12 MOORE GRAPH
A *Moore graph of diameter d* is a regular graph of diameter d and girth $2d + 1$. Consider a k-regular Moore graph G of diameter two.

a) Show that $n = k^2 + 1$.
b) Let $\mathbf{A}$ be the adjacency matrix of G and tr$(\mathbf{A})$ its trace.
   i) Show that tr$(\mathbf{A}) = 0$.
   ii) Evaluate the matrix $\mathbf{A}^2 + \mathbf{A}$, determine its eigenvalues and their multiplicities, and deduce the possible eigenvalues of $\mathbf{A}$ (but not their multiplicities).
   iii) Expressing tr$(\mathbf{A})$ in terms of the eigenvalues of $\mathbf{A}$ and their multiplicities, and noting that these multiplicities are necessarily integers, conclude that such a graph G can exist only if $k = 2, 3, 7$, or 57.
                      (A.J. HOFFMAN AND R.R. SINGLETON)
c) Find such a graph G for $k = 2$ and $k = 3$.

(A 7-regular example, the *Hoffman–Singleton graph*, discovered by Hoffman and Singleton (1960), is depicted in Figure 3.1; vertex i of P_j is joined to vertex $i + jk \pmod 5$ of Q_k. A 57-regular example would have 3250 vertices. No such graph is known.)

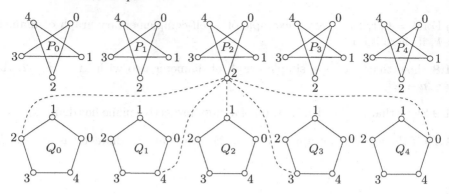

Fig. 3.1. The Hoffman–Singleton graph

3.1.13 CAGE

A k-regular graph of girth g with the least possible number of vertices is called a (k, g)-*cage*. A $(3, g)$-cage is often simply referred to as a g-*cage*. Let $f(k, g)$ denote the number of vertices in a (k, g)-cage. Observe that $f(2, g) = g$.

a) For $k \geq 3$, show that:
   i) $f(k, 2r) \geq (2(k-1)^r - 2)/(k-2)$,
   ii) $f(k, 2r+1) \geq (k(k-1)^r - 2)/(k-2)$.
b) Determine all g-cages, $g = 3, 4, 5, 6$.
c) Show that the incidence graph of a projective plane of order $k - 1$ is a $(k, 6)$-cage.

(Singleton (1966) showed, conversely, that any $(k, 6)$-cage of order $2(k^2 - k + 1)$ is necessarily the incidence graph of a projective plane of order $k - 1$.)

3.1.14 THE TUTTE–COXETER GRAPH

A highly symmetric cubic graph, known as *the Tutte–Coxeter graph*, is shown in Figure 3.2. Show that:

a) the Tutte–Coxeter graph is isomorphic to the bipartite graph $G[X, Y]$ derived from K_6 in the following manner. The vertices of X are the fifteen edges of K_6 and the vertices of Y are the fifteen 1-factors of K_6, an element e of X being adjacent to an element F of Y whenever e is an edge of the 1-factor F.

(H.S.M. COXETER)

b) the Tutte–Coxeter graph is an 8-cage.
   (Tutte (1947b) showed that this graph is, in fact, the unique 8-cage.)

3.1.15 t-ARC-TRANSITIVE GRAPH

A walk $(v_0, v_1, \ldots, v_t)$ in a graph such that $v_{i-1} \neq v_{i+1}$, for $1 \leq i \leq t - 1$, is called a t-*arc*. A simple connected graph G is t-*arc-transitive* if, given any two t-arcs $(v_0, v_1, \ldots, v_t)$ and $(w_0, w_1, \ldots, w_t)$, there is an automorphism of G which maps v_i to w_i, for $0 \leq i \leq t$. (Thus a 1-arc-transitive graph is the same as an arc-transitive graph, defined in Exercise 1.5.11.) Show that:

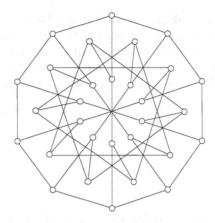

Fig. 3.2. The Tutte–Coxeter graph: the 8-cage

a) $K_{3,3}$ is 2-arc-transitive,
b) the Petersen graph is 3-arc-transitive,
c) the Heawood graph is 4-arc-transitive,
d) the Tutte–Coxeter graph is 5-arc-transitive.

(Tutte (1947b) showed that there are no t-arc-transitive cubic graphs when $t > 5$.)

3.2 Cut Edges

For any edge e of a graph G, it is easy to see that either $c(G \setminus e) = c(G)$ or $c(G \setminus e) = c(G) + 1$ (Exercise 3.2.1). If $c(G \setminus e) = c(G) + 1$, the edge e is called a *cut edge* of G. Thus a cut edge of a connected graph is one whose deletion results in a disconnected graph. More generally, the cut edges of a graph correspond to its bonds of size one (Exercise 3.2.2).

The graph in Figure 3.3 has three cut edges.

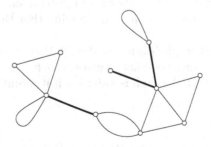

Fig. 3.3. The cut edges of a graph

If e is a cut edge of a graph G, its ends x and y belong to different components of $G \setminus e$, and so are not connected by a path in $G \setminus e$; equivalently, e lies in no cycle of G. Conversely, if $e = xy$ is not a cut edge of G, the vertices x and y belong to the same component of $G \setminus e$, so there is an xy-path P in $G \setminus e$, and $P + e$ is a cycle in G through e. Hence we have the following characterization of cut edges.

Proposition 3.2 *An edge e of a graph G is a cut edge if and only if e belongs to no cycle of G.* □

Exercises

$\star$**3.2.1** Show that if $e \in E$, then either $c(G \setminus e) = c(G)$ or $c(G \setminus e) = c(G) + 1$.

$\star$**3.2.2** Show that an edge e is a cut edge of a graph G if and only if $\{e\}$ is a bond of G.

3.2.3 Let G be a connected even graph. Show that:

a) G has no cut edge,
b) for any vertex $v \in V$, $c(G - v) \leq \frac{1}{2}d(v)$.

3.2.4 Let G be a k-regular bipartite graph with $k \geq 2$. Show that G has no cut edge.

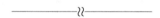

3.3 Euler Tours

A trail that traverses every edge of a graph is called an *Euler trail*, because Euler (1736) was the first to investigate the existence of such trails. In the earliest known paper on graph theory, he showed that it was impossible to cross each of the seven bridges of Königsberg once and only once during a walk through the town. A plan of Königsberg and the river Pregel is shown in Figure 3.4a. As can be seen, proving that such a walk is impossible amounts to showing that the graph in Figure 3.4b has no Euler trail.

A *tour* of a connected graph G is a closed walk that traverses each edge of G at least once, and an *Euler tour* one that traverses each edge exactly once (in other words, a closed Euler trail). A graph is *eulerian* if it admits an Euler tour.

FLEURY'S ALGORITHM

Let G be an eulerian graph, and let W be an Euler tour of G with initial and terminal vertex u. Each time a vertex v occurs as an internal vertex of W, two edges incident with v are accounted for. Since an Euler tour traverses each edge

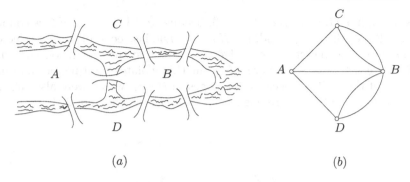

(a) (b)

Fig. 3.4. The bridges of Königsberg and their graph

exactly once, $d(v)$ is even for all $v \neq u$. Similarly, $d(u)$ is even, because W both starts and ends at u. Thus an eulerian graph is necessarily even.

The above necessary condition for the existence of an Euler tour in a connected graph also turns out to be sufficient. Moreover, there is a simple algorithm, due to Fleury (1883), which finds an Euler tour in an arbitrary connected even graph G (see also Lucas (1894)). Fleury's Algorithm constructs such a tour of G by tracing out a trail subject to the condition that, at any stage, a cut edge of the untraced subgraph F is taken only if there is no alternative.

Algorithm 3.3 FLEURY'S ALGORITHM

INPUT: a connected even graph G and a specified vertex u of G
OUTPUT: an Euler tour W of G starting (and ending) at u

1: *set $W := u$, $x := u$, $F := G$*
2: ***while** $\partial_F(x) \neq \emptyset$ **do***
3:     *choose an edge $e := xy \in \partial_F(x)$, where e is not a cut edge of F unless there is no alternative*
4:     *replace uWx by $uWxey$, x by y, and F by $F \setminus e$*
5: ***end while***
6: *return W*

Theorem 3.4 *If G is a connected even graph, the walk W returned by Fleury's Algorithm is an Euler tour of G.*

Proof The sequence W is initially a trail, and remains one throughout the procedure, because Fleury's Algorithm always selects an edge of F (that is, an as yet unchosen edge) which is incident to the terminal vertex x of W. Moreover, the algorithm terminates when $\partial_F(x) = \emptyset$, that is, when all the edges incident to the terminal vertex x of W have already been selected. Because G is even, we deduce that $x = u$; in other words, the trail W returned by the algorithm is a closed trail of G.

Suppose that W is not an Euler tour of G. Denote by X the set of vertices of positive degree in F when the algorithm terminates. Then $X \neq \emptyset$, and $F[X]$ is an even subgraph of G. Likewise $V \setminus X \neq \emptyset$, because $u \in V \setminus X$. Since G is connected, $\partial_G(X) \neq \emptyset$. On the other hand, $\partial_F(X) = \emptyset$. The last edge of $\partial_G(X)$ selected for inclusion in W was therefore a cut edge $e = xy$ of F at the time it was chosen, with $x \in X$ and $y \in V \setminus X$ (see Figure 3.5). But this violates the rule for choosing the next edge of the trail W, because the edges in $\partial_F(x)$, which were also candidates for selection at the time, were not cut edges of F, by Theorem 2.10. $\square$

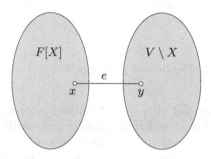

Fig. 3.5. Choosing a cut edge in Fleury's Algorithm

The validity of Fleury's Algorithm provides the following characterization of eulerian graphs.

Theorem 3.5 *A connected graph is eulerian if and only if it is even.* $\square$

Let now x and y be two distinct vertices of a graph G. Suppose that we wish to find an Euler xy-trail of G, if one exists. We may do so by adding a new edge e joining x and y. The graph G has an Euler trail connecting x and y if and only if $G+e$ has an Euler tour (Exercise 3.3.4). Thus Fleury's Algorithm may be adapted easily to find an Euler xy-trail in G, if one exists.

We remark that Fleury's Algorithm is an efficient algorithm, in a sense to be made precise in Chapter 8. When an edge is considered for inclusion in the current trail W, it must be examined to determine whether or not it is a cut edge of the remaining subgraph F. If it is not, it is appended to W right away. On the other hand, if it is found to be a cut edge of F, it remains a cut edge of F until it is eventually selected for inclusion in W; therefore, each edge needs to be examined only once. In Chapter 7, we present an efficient algorithm for determining whether or not an edge is a cut edge of a graph.

A comprehensive treatment of eulerian graphs and related topics can be found in Fleischner (1990, 1991).

Exercises

3.3.1 Which of the pictures in Figure 3.6 can be drawn without lifting one's pen from the paper and without tracing a line more than once?

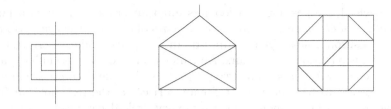

Fig. 3.6. Tracing pictures

3.3.2 If possible, give an example of an eulerian graph G with n even and m odd. Otherwise, explain why there is no such graph.

3.3.3 Give an alternative proof of Theorem 3.5 by appealing to Exercise 2.4.4.

$\star$**3.3.4** Let G be a graph with two distinct specified vertices x and y, and let $G + e$ be the graph obtained from G by the addition of a new edge e joining x and y.

a) Show that G has an Euler trail connecting x and y if and only if $G + e$ has an Euler tour.
b) Deduce that G has an Euler trail connecting x and y if and only if $d(x)$ and $d(y)$ are odd and $d(v)$ is even for all $v \in V \setminus \{x, y\}$.

3.3.5 Let G be a connected graph, and let X be the set of vertices of G of odd degree. Suppose that $|X| = 2k$, where $k \geq 1$.

a) Show that there are k edge-disjoint trails $Q_1, Q_2, \ldots, Q_k$ in G such that $E(G) = E(Q_1) \cup E(Q_2) \cup \ldots \cup E(Q_k)$.
b) Deduce that G contains k edge-disjoint paths connecting the vertices of X in pairs.

3.3.6 Let $W := v_0 e_1 v_1 e_2 v_2 \ldots e_m v_m$ be an Euler tour of a graph G, where $v_m = v_0$. Suppose that $v_i = v_0$, where $0 < i < m$. Show that $v_0 W v_i e_m v_{m-1} \overleftarrow{W} v_i$ is also an Euler tour of G.

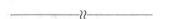

3.3.7 Let G be a nontrivial eulerian graph, and let $v \in V$. Show that each v-trail in G can be extended to an Euler tour of G if and only if $G - v$ is acyclic.

(O. ORE)

3.3.8 DOMINATING SUBGRAPH
A subgraph F of a graph G is *dominating* if every edge of G has at least one end in F. Let G be a graph with at least three edges. Show that $L(G)$ is hamiltonian if and only if G has a dominating eulerian subgraph.

(F. HARARY AND C.ST.J.A. NASH-WILLIAMS)

3.3.9 A cycle decomposition of a loopless eulerian graph G induces a family of pairs of edges of G, namely the consecutive pairs of edges in the cycles comprising the decomposition. Each edge thus appears in two pairs, and each trivial edge cut $\partial(v)$, $v \in V$, is partitioned into pairs. An Euler tour of G likewise induces a family of pairs of edges with these same two properties. A cycle decomposition and Euler tour are said to be *compatible* if, for all vertices v, the resulting partitions of $\partial(v)$ have no pairs in common. Show that every cycle decomposition of a loopless eulerian graph of minimum degree at least four is compatible with some Euler tour.   (A. KOTZIG)
(G. Sabidussi has conjectured that, conversely, every Euler tour of a loopless eulerian graph of minimum degree at least four is compatible with some cycle decomposition; see Appendix A.)

3.4 Connection in Digraphs

As we saw earlier, in Section 3.1, the property of connection in graphs may be expressed not only in terms of edge cuts but also in terms of walks. By the same token, the property of strong connection, defined in terms of outcuts in Section 2.5, may be expressed alternatively in terms of directed walks. This is an immediate consequence of Theorem 3.6 below.

A *directed walk* in a digraph D is an alternating sequence of vertices and arcs

$$W := (v_0, a_1, v_1, \ldots, v_{\ell-1}, a_\ell, v_\ell)$$

such that v_{i-1} and v_i are the tail and head of a_i, respectively, $1 \le i \le \ell$.[1] If x and y are the initial and terminal vertices of W, we refer to W as a *directed (x,y)-walk*. Directed trails, tours, paths, and cycles in digraphs are defined analogously. As for undirected graphs, the *(u,v)-segment* of a directed walk W, where u and v are two vertices of W, u preceding v, is the subsequence of W starting with u and ending with v, and is denoted uWv (the same notation as for undirected graphs).

We say that a vertex y is *reachable* from a vertex x if there is a directed (x,y)-path. The property of reachability can be expressed in terms of outcuts, as follows.

Theorem 3.6 *Let x and y be two vertices of a digraph D. Then y is reachable from x in D if and only if $\partial^+(X) \ne \emptyset$ for every subset X of V which contains x but not y.*

Proof Suppose, first, that y is reachable from x by a directed path P. Consider any subset X of V which contains x but not y. Let u be the last vertex of P which belongs to X and let v be its successor on P. Then $(u, v) \in \partial^+(X)$, so $\partial^+(X) \ne \emptyset$.

Conversely, suppose that y is not reachable from x, and let X be the set of vertices which are reachable from x. Then $x \in X$ and $y \notin X$. Furthermore, because no vertex of $V \setminus X$ is reachable from x, the outcut $\partial^+(X)$ is empty. $\square$

---
[1] Thus a walk in a graph corresponds to a directed walk in its associated digraph. This is consistent with our convention regarding the traversal of loops in walks.

In a digraph D, two vertices x and y are *strongly connected* if there is a directed (x, y)-walk and also a directed (y, x)-walk (that is, if each of x and y is reachable from the other). Just as connection is an equivalence relation on the vertex set of a graph, strong connection is an equivalence relation on the vertex set of a digraph (Exercise 3.4.1). The subdigraphs of D induced by the equivalence classes with respect to this relation are called the *strong components* of D. The strong components of the digraph shown in Figure 3.7a are indicated in Figure 3.7b. We leave it to the reader to verify that a digraph is strong if and only if it has exactly one strong component (Exercise 3.4.2).

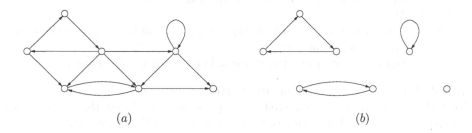

(a)   (b)

Fig. 3.7. (a) A digraph and (b) its strong components

A *directed Euler trail* is a directed trail which traverses each arc of the digraph exactly once, and a *directed Euler tour* is a directed tour with this same property. A digraph is *eulerian* if it admits a directed Euler tour. There is a directed version of Theorem 3.5, whose proof we leave as an exercise (3.4.8).

Theorem 3.7 *A connected digraph is eulerian if and only if it is even.*   □

Exercises

⋆**3.4.1** Show that strong connection is an equivalence relation on the vertex set of a digraph.

⋆**3.4.2** Show that a digraph is strong if and only if it has exactly one strong component.

⋆**3.4.3** Let C be a strong component of a digraph D, and let P be a directed path in D connecting two vertices of C. Show that P is contained in C.

3.4.4 Let D be a digraph with adjacency matrix $\mathbf{A} = (a_{uv})$. Show that the number of directed (u, v)-walks of length k in D is the (u, v) entry of $\mathbf{A}^k$.

3.4.5 Show that every tournament is either strong or can be transformed into a strong tournament by the reorientation of just one arc.

⋆**3.4.6** Condensation of a Digraph

a) Show that all the arcs linking two strong components of a digraph have their tails in one strong component (and their heads in the other).
b) The *condensation* $C(D)$ of a digraph D is the digraph whose vertices correspond to the strong components of D, two vertices of $C(D)$ being linked by an arc if and only if there is an arc in D linking the corresponding strong components, and with the same orientation. Draw the condensations of:
    i) the digraph of Figure 3.7a,
    ii) the four tournaments of Figure 1.25.
c) Show that the condensation of any digraph is acyclic.
d) Deduce that:
    i) every digraph has a *minimal* strong component, namely one that dominates no other strong component,
    ii) the condensation of any tournament is a transitive tournament.

3.4.7 A digraph is *unilateral* if any two vertices x and y are connected either by a directed (x, y)-path or by a directed (y, x)-path, or both. Show that a digraph is unilateral if and only if its condensation has a directed Hamilton path.

⋆**3.4.8** Prove Theorem 3.7.

3.4.9 De Bruijn–Good Digraph
The *de Bruijn–Good digraph* BG_n has as vertex set the set of all binary sequences of length n, vertex $a_1a_2 \ldots a_n$ being joined to vertex $b_1b_2 \ldots b_n$ if and only if $a_{i+1} = b_i$ for $1 \leq i \leq n - 1$. Show that BG_n is an eulerian digraph of order 2^n and directed diameter n.

3.4.10 De Bruijn–Good Sequence
A circular sequence $s_1s_2 \ldots s_{2^n}$ of zeros and ones is called a *de Bruijn–Good sequence* of order n if the 2^n subsequences $s_is_{i+1} \ldots s_{i+n-1}$, $1 \leq i \leq 2^n$ (where subscripts are taken modulo 2^n) are distinct, and so constitute all possible binary sequences of length n. For example, the sequence 00011101 is a de Bruijn–Good sequence of order three. Show how to derive such a sequence of any order n by considering a directed Euler tour in the de Bruijn–Good digraph BG_{n-1}.   (N.G. De Bruijn; I.J. Good)
(An application of de Bruijn–Good sequences can be found in Chapter 10 of Bondy and Murty (1976).)

⋆**3.4.11**

a) Show that a digraph which has a closed directed walk of odd length contains a directed odd cycle.
b) Deduce that a strong digraph which contains an odd cycle contains a directed odd cycle.

⋆**3.4.12** Show that:

a) every nontrivial strong tournament has a directed Hamilton cycle,
   (P. CAMION)
b) each vertex of a nontrivial strong tournament D is contained in a directed cycle of every length l, $3 \le l \le n$,
   (J.W. MOON)
c) each arc of an even tournament D is contained in a directed cycle of every length l, $3 \le l \le n$.
   (B. ALSPACH)

3.4.13 BALANCED DIGRAPH
A digraph D is *balanced* if $|d^+(v) - d^-(v)| \le 1$, for all $v \in V$. Show that every graph has a balanced orientation.

3.5 Cycle Double Covers

In this section, we discuss a beautiful conjecture concerning cycle coverings of graphs. In order for a graph to admit a cycle covering, each of its edges must certainly lie in some cycle. On the other hand, once this requirement is fulfilled, the set of all cycles of the graph clearly constitutes a covering. Thus, by Proposition 3.2, a graph admits a cycle covering if and only if it has no cut edge. We are interested here in cycle coverings which cover no edge too many times.

Recall that a *decomposition* is a covering in which each edge is covered exactly once. According to Veblen's Theorem (2.7), the only graphs which admit such cycle coverings are the even graphs. Thus, if a graph has vertices of odd degree, some edges will necessarily be covered more than once in a cycle covering. One is led to ask whether every graph without cut edges admits a cycle covering in which no edge is covered more than twice.

All the known evidence suggests that this is indeed so. For example, each of the platonic graphs (shown in Figure 1.14) has such a cycle covering consisting of its *facial cycles*, those which bound its regions, or *faces*, as in Figure 3.8. More generally, the same is true of all polyhedral graphs, and indeed of all planar graphs without cut edges, as we show in Chapter 10.

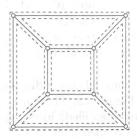

Fig. 3.8. A double covering of the cube by its facial cycles

In the example of Figure 3.8, observe that any five of the six facial cycles already constitute a cycle covering. Indeed, the covering shown, consisting of all six facial cycles each edge exactly twice. Such a covering is called a *cycle double cover* of the graph. It turns out that cycle coverings and cycle double covers are closely related.

Proposition 3.8 *If a graph has a cycle covering in which each edge is covered at most twice, then it has a cycle double cover.*

Proof Let $\mathcal{C}$ be a cycle covering of a graph G in which each edge is covered at most twice. The symmetric difference $\triangle\{E(C)|C \in \mathcal{C}\}$ of the edge sets of the cycles in $\mathcal{C}$ is then the set of edges of G which are covered just once by $\mathcal{C}$. Moreover, by Corollary 2.16, this set of edges is an even subgraph C' of G. By Veblen's Theorem (2.7), C' has a cycle decomposition $\mathcal{C}'$. It is now easily checked that $\mathcal{C} \cup \mathcal{C}'$ is a cycle double cover of G. □

Motivated by quite different considerations, Szekeres (1973) and Seymour (1979b) each put forward the conjecture that every graph without cut edges admits a cycle double cover.

THE CYCLE DOUBLE COVER CONJECTURE

Conjecture 3.9 *Every graph without cut edges has a cycle double cover.*

A graph has a cycle double cover if and only if each of its components has one. Thus, in order to prove the Cycle Double Cover Conjecture, it is enough to prove it for nontrivial connected graphs. Indeed, one may restrict one's attention even further, to *nonseparable graphs*. Roughly speaking, these are the connected graphs which cannot be obtained by piecing together two smaller connected graphs at a single vertex. (Nonseparable graphs are defined and discussed in Chapter 5.) In the case of planar graphs, the boundaries of the faces in any planar embedding are then cycles, as we show in Chapter 10, and these facial cycles constitute a cycle double cover of the graph. This suggests one natural approach to the Cycle Double Cover Conjecture: find a suitable embedding of the graph on some surface, an embedding in which each face is bounded by a cycle; the facial cycles then form a cycle double cover.

Consider, for example, the toroidal embeddings of the complete graph K_7 and the Petersen graph shown in Figure 3.9. The torus is represented here by a rectangle whose opposite sides are identified; identifying one pair of sides yields a cylinder, and identifying the two open ends of the cylinder results in a torus. In the embedding of K_7, there are fourteen faces, each bounded by a triangle; these triangles form a cycle double cover of K_7. In the embedding of the Petersen graph, there are five faces; three are bounded by cycles of length five (faces A, B, C), one by a cycle of length six (face D), and one by a cycle of length nine (face E). These five cycles constitute a cycle double cover of the Petersen graph.

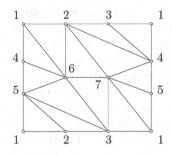
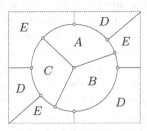

Fig. 3.9. Toroidal embeddings of (a) the complete graph K_7, and (b) the Petersen graph

The above approach to the Cycle Double Cover Conjecture, via surface embeddings, is supported by the following conjecture, which asserts that every loopless nonseparable graph can indeed be embedded in some surface in an appropriate fashion.

THE CIRCULAR EMBEDDING CONJECTURE

Conjecture 3.10 *Every loopless nonseparable graph can be embedded in some surface in such a way that each face in the embedding is bounded by a cycle.*

The origins of Conjecture 3.10 are uncertain. It was mentioned by W.T. Tutte (unpublished) in the mid-1960s, but was apparently already known at the time to several other graph-theorists, according to Robertson (2007). We discuss surface embeddings of graphs in greater detail in Chapter 10, and describe there a stronger conjecture on embeddings of graphs.

Apart from its intrinsic beauty, due to the simplicity of its statement and the fact that it applies to essentially all graphs, the Cycle Double Cover Conjecture is of interest because it is closely related to a number of other basic problems in graph theory, including the Circular Embedding Conjecture. We encounter several more in future chapters.

DOUBLE COVERS BY EVEN SUBGRAPHS

There is another attractive formulation of the Cycle Double Cover Conjecture, in terms of even subgraphs; here, by an even subgraph we mean the edge set of such a subgraph.

If a graph has a cycle covering, then it has a covering by even subgraphs because cycles are even subgraphs. Conversely, by virtue of Theorem 2.17, any covering by even subgraphs can be converted into a cycle covering by simply decomposing each even subgraph into cycles. It follows that a graph has a cycle double cover if and only if it has a double cover by even subgraphs. Coverings by even subgraphs therefore provide an alternative approach to the Cycle Double Cover Conjecture. If every graph without cut edges had a covering by at most two even subgraphs, such a covering would yield a cycle covering in which each edge was covered at most twice, thereby establishing the Cycle Double Cover Conjecture by virtue of Proposition 3.8. Unfortunately, this is not the case. Although many graphs do indeed admit such coverings, many do not. The Petersen graph, for instance, cannot be covered by two even subgraphs (Exercise 3.5.3a). On the other hand, it may be shown that every graph without cut edges admits a covering by three even subgraphs (Theorem 21.21).

Suppose, now, that every graph without cut edges does indeed have a cycle double cover. It is then natural to ask how few cycles there can be in such a covering; a covering with few cycles may be thought of as an efficient covering, in some sense. Let $\mathcal{C}$ be a cycle double cover of a graph G. As each edge of G is covered exactly twice,

$$\sum_{C \in \mathcal{C}} e(C) = 2m$$

Because $e(C) \leq n$ for all $C \in \mathcal{C}$, we deduce that $|\mathcal{C}| \geq 2m/n$, the average degree of G. In particular, if G is a complete graph K_n, the number of cycles in a cycle double cover of G must be at least $n - 1$. A cycle double cover consisting of no more than this number of cycles is called a *small cycle double cover*. Bondy (1990) conjectures that every simple graph G without cut edges admits such a covering.

Conjecture 3.11   THE SMALL CYCLE DOUBLE COVER CONJECTURE
Every simple graph without cut edges has a small cycle double cover.

Several other strengthenings of the Cycle Double Cover Conjecture have been proposed. One of these is a conjecture put forward by Jaeger (1988).

Conjecture 3.12   THE ORIENTED CYCLE DOUBLE COVER CONJECTURE
Let G be a graph without cut edges. Then the associated digraph $D(G)$ of G admits a decomposition into directed cycles of length at least three.

Further information on these and a number of related conjectures can be found in the book by Zhang (1997).

Exercises

3.5.1 Show that every loopless graph has a double covering by bonds.

3.5.2 Let $\{C_1, C_2, C_3\}$ be a covering of a graph G by three even subgraphs such that $C_1 \cap C_2 \cap C_3 = \emptyset$. Show that $\{C_1 \bigtriangleup C_2,\ C_1 \bigtriangleup C_3\}$ is a covering of G by two even subgraphs.

⋆3.5.3

a) Show that the Petersen graph has no covering by two even subgraphs.

b) Deduce, using Exercise 3.5.2, that this graph has no double cover by four even subgraphs.

c) Find a covering of the Petersen graph by three even subgraphs, and a double cover by five even subgraphs.

3.5.4

a) i) Let $\{C_1, C_2\}$ be a covering of a graph G by two even subgraphs. Show that $\{C_1, C_2, C_1 \bigtriangleup C_2\}$ is a double cover of G by three even subgraphs.

   ii) Deduce that a graph has a covering by two even subgraphs if and only if it has a double cover by three even subgraphs.

b) Let $\{C_1, C_2, C_3\}$ be a covering of a graph G by three even subgraphs. Show that G has a quadruple cover (a covering in which each edge is covered exactly four times) by seven even subgraphs.

(We show in Theorem 21.25 that every graph without cut edges has a covering by three even subgraphs, and hence a quadruple cover by seven even subgraphs.)

3.5.5 Find a small cycle double cover of K_6.

3.5.6 Find a decomposition of $D(K_6)$ into directed cycles of length at least three.

3.5.7 Show that every graph without cut edges has a uniform cycle covering.

3.5.8 Let G be a graph, and let $\mathcal{C}$ be the set of all cycles of G. For $C \in \mathcal{C}$, denote by $\mathbf{f}_C$ the incidence vector of C, and set $\mathbf{F}_\mathcal{C} := \{\mathbf{f}_C : C \in \mathcal{C}\}$.

a) Let $\mathbf{x} \in \mathbb{R}^E$. Show that:

   i) the vector $\mathbf{x}$ lies in the vector space generated by $\mathbf{F}_\mathcal{C}$ if and only if the following two conditions hold:

   ▷ $x(e) = 0$ for every cut edge e,

   ▷ $x(e) = x(f)$ for every edge cut $\{e, f\}$ of cardinality two,

   ii) if $\mathbf{x}$ is a nonnegative linear combination of vectors in $\mathbf{F}_\mathcal{C}$, then for any bond B of G and any edge e of B:

$$x(e) \leq \sum_{f \in B \setminus \{e\}} x(f) \qquad (3.1)$$

(Seymour (1979b) showed that this necessary condition is also sufficient for a nonnegative vector $\mathbf{x}$ to be a nonnegative linear combination of vectors in $\mathbf{F}_\mathcal{C}$.)

iii) if **x** is a nonnegative integer linear combination of vectors in $\mathbf{F}_C$, then for any bond B, in addition to (3.1), **x** must satisfy the condition:

$$\sum_{e \in B} x(e) \equiv 0 \pmod 2 \tag{3.2}$$

b) With the aid of Exercise 2.4.7, give an example showing that conditions (3.1) and (3.2) are not sufficient for a nonnegative integer vector **x** in $\mathbb{R}^E$ to be a nonnegative integer linear combination of vectors in $\mathbf{F}_C$.

(Seymour (1979b) showed, however, that these two conditions are sufficient when G is a planar graph. Furthermore, he conjectured that they are sufficient in any graph if each component of **x** is an even integer. This conjecture clearly implies the Cycle Double Cover Conjecture. For related work, see Alspach et al. (1994).)

3.6 Related Reading

CAGES

Cages were introduced in Exercise 3.1.13. There are many interesting examples of such graphs, the Petersen graph and the Heawood graph being but two. Numerous others are described in the survey by Wong (1982). Two particularly interesting infinite families of examples are those constructed from projective geometries by Benson (1966), namely the $(k, 8)$- and $(k, 12)$-cages, where $k - 1$ is a prime power. For $\ell = 3, 5$, the Benson cages furnish examples of dense graphs (graphs with many edges) containing no 2ℓ-cycles. For $\ell = 2$, examples are provided by polarity graphs of projective planes (see Exercises 12.2.12, 12.2.13, and 12.2.14.) The question as to how many edges a graph on n vertices can have without containing a 2ℓ-cycle is unsolved for other values of ℓ, and in particular for $\ell = 4$; see Appendix A.

The study of *directed cages*, smallest k-diregular digraphs with specified directed girth g, was initiated by Behzad et al. (1970). They conjectured that the directed circulants on $k(g - 1) + 1$ vertices in which each vertex dominates the k vertices succeeding it are directed cages. This conjecture remains open; see Appendix A.

4

Trees

Contents

4.1 Forests and Trees

Recall that an *acyclic* graph is one that contains no cycles. A connected acyclic graph is called a *tree*. The trees on six vertices are shown in Figure 4.1. According to these definitions, each component of an acyclic graph is a tree. For this reason, acyclic graphs are usually called *forests*.

In order for a graph to be connected, there must be at least one path between any two of its vertices. The following proposition, an immediate consequence of Exercise 2.2.12, says that trees are the connected graphs which just meet this requirement.

Proposition 4.1 *In a tree, any two vertices are connected by exactly one path.* □

Following Diestel (2005), we denote the unique path connecting vertices x and y in a tree T by xTy.

By Theorem 2.1, any graph in which all degrees are at least two contains a cycle. Thus, every tree contains a vertex of degree at most one; moreover, if the tree is nontrivial, it must contain a vertex of degree exactly one. Such a vertex is called a *leaf* of the tree. In fact, the following stronger assertion is true (Exercise 2.1.2).

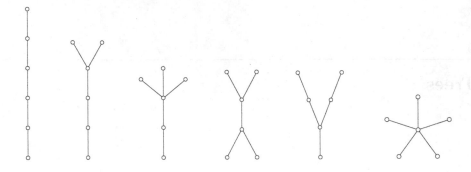

Fig. 4.1. The trees on six vertices

Proposition 4.2 *Every nontrivial tree has at least two leaves.*   □

If x is a leaf of a tree T, the subgraph $T - x$ is a tree with $v(T-x) = v(T)-1$ and $e(T - x) = e(T) - 1$. Because the trivial tree has no edges, we have, by induction on the number of vertices, the following relationship between the numbers of edges and vertices of a tree.

Theorem 4.3 *If T is a tree, then $e(T) = v(T) - 1$.*   □

ROOTED TREES AND BRANCHINGS

A *rooted tree* $T(x)$ is a tree T with a specified vertex x, called the *root* of T. An orientation of a rooted tree in which every vertex but the root has indegree one is called a *branching*. We refer to a rooted tree or branching with root x as an *x-tree* or *x-branching*, respectively.

There is an evident bijection between x-trees and x-branchings. An x-path thus give rise to a simple example of a branching, a directed x-path. Another example of a branching is shown in Figure 4.2.

Observe that the root of this branching is a source. This is always so, because the sum of the indegrees of a digraph is equal to its number of arcs (Exercise 1.5.2)

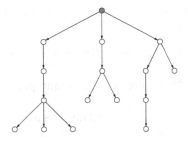

Fig. 4.2. A branching

which, in the case of a branching B, is $v(B) - 1$ by Theorem 4.3. Observe, also, that every vertex of a branching is reachable from its root by a unique directed path. Conversely, in any digraph, reachability from a vertex may be expressed in terms of its branchings. We leave the proof of this fact as an exercise (4.1.6).

Theorem 4.4 *Let x be a vertex of a digraph D, and let X be the set of vertices of D which are reachable from x. Then there is an x-branching in D with vertex set X.* □

Proof Technique: Ordering Vertices

Among the $n!$ linear orderings of the n vertices of a graph, certain ones are especially interesting because they encode particular structural properties. An elementary example is an ordering of the vertices according to their degrees, in decreasing order. More interesting orderings can be obtained by considering the global structure of the graph, rather than just its local structure, as in Exercise 2.2.20. We describe a second example here. Others will be encountered in Chapters 6, 14, and 19, as well as in a number of exercises.

In general, graphs contain copies of many different trees. Indeed, every simple graph with minimum degree k contains a copy of each rooted tree on $k + 1$ vertices, rooted at any given vertex of the graph (Exercise 4.1.9). The analogous question for digraphs (with rooted trees replaced by branchings) is much more difficult. However, in the case of tournaments it can be answered by considering a rather natural ordering of the vertices of the tournament.

A *median order* of a digraph $D = (V, A)$ is a linear order $v_1, v_2, \ldots, v_n$ of its vertex set V such that $|\{(v_i, v_j) : i < j\}|$ (the number of arcs directed from left to right) is as large as possible. In the case of a tournament, such an order can be viewed as a ranking of the players which minimizes the number of upsets (matches won by the lower-ranked player). As we shall see, median orders of tournaments reveal a number of interesting structural properties.

Let us first note two basic properties of median orders of tournaments (Exercise 4.1.10). Let T be a tournament and $v_1, v_2, \ldots, v_n$ a median order of T. Then, for any two indices i, j with $1 \le i < j \le n$:

(M1) the interval $v_i, v_{i+1}, \ldots, v_j$ is a median order of the induced subtournament $T[\{v_i, v_{i+1}, \ldots, v_j\}]$,

(M2) vertex v_i dominates at least half of the vertices $v_{i+1}, v_{i+2}, \ldots, v_j$, and vertex v_j is dominated by at least half of the vertices $v_i, v_{i+1}, \ldots, v_{j-1}$.

In particular, each vertex v_i, $1 \le i < n$, dominates its successor v_{i+1}. The sequence $(v_1, v_2, \ldots, v_n)$ is thus a directed Hamilton path, providing an alternative proof (see Locke (1995)) of Rédei's Theorem (2.3): *every tournament has a directed Hamilton path.*

ORDERING VERTICES (CONTINUED)

By exploiting the properties of median orders, Havet and Thomassé (2000) showed that large tournaments contain all large branchings.

Theorem 4.5 *Any tournament on $2k$ vertices contains a copy of each branching on $k+1$ vertices.*

Proof Let $v_1, v_2, \ldots, v_{2k}$ be a median order of a tournament T on $2k$ vertices, and let B be a branching on $k+1$ vertices. Consider the intervals $v_1, v_2, \ldots, v_i$, $1 \le i \le 2k$. We show, by induction on k, that there is a copy of B in T whose vertex set includes at least half the vertices of any such interval.

This is clearly true for $k = 1$. Suppose, then, that $k \ge 2$. Delete a leaf y of B to obtain a branching B' on k vertices, and set $T' := T - \{v_{2k-1}, v_{2k}\}$. By (M1), $v_1, v_2, \ldots, v_{2k-2}$ is a median order of the tournament T', so there is a copy of B' in T' whose vertex set includes at least half the vertices of any interval $v_1, v_2, \ldots, v_i$, $1 \le i \le 2k - 2$. Let x be the predecessor of y in B. Suppose that x is located at vertex v_i of T'. In T, by (M2), v_i dominates at least half of the vertices $v_{i+1}, v_{i+2}, \ldots, v_{2k}$, thus at least $k - i/2$ of these vertices. On the other hand, B' includes at least $(i-1)/2$ of the vertices $v_1, v_2, \ldots, v_{i-1}$, thus at most $k - (i+1)/2$ of the vertices $v_{i+1}, v_{i+2}, \ldots, v_{2k}$. It follows that, in T, there is an outneighbour v_j of v_i, where $i + 1 \le j \le 2k$, which is not in B'. Locating y at v_j, and adding the vertex y and arc (x, y) to B', we now have a copy of B in T. It is readily checked that this copy of B satisfies the required additional property. □

Three further applications of median orders are described in Exercises 4.1.16, 4.1.17, and 4.1.18.

Rooted trees and branchings turn out to be basic tools in the design of efficient algorithms for solving a variety of problems involving reachability, as we shall show in Chapter 6.

Exercises

4.1.1

a) Show that every tree with maximum degree k has at least k leaves.
b) Which such trees have exactly k leaves?

4.1.2 Show that the following three statements are equivalent.

a) G is connected and has $n - 1$ edges.
b) G is a forest and has $n - 1$ edges.
c) G is a tree.

4.1.3 A *saturated hydrocarbon* is a molecule $C_m H_n$ in which every carbon atom (C) has four bonds, every hydrogen atom (H) has one bond, and no sequence of bonds forms a cycle. Show that, for any positive integer m, the molecule $C_m H_n$ can exist only if $n = 2m + 2$.

4.1.4 Let G be a graph and F a maximal forest of G. Show that $e(F) = v(G) - c(G)$.

4.1.5 Prove Theorem 4.3 by induction on the number of edges of G.

$\star$**4.1.6** Prove Theorem 4.4.

4.1.7 Show that a sequence $(d_1, d_2, \ldots, d_n)$ of positive integers is a degree sequence of a tree if and only if $\sum_{i=1}^{n} d_i = 2(n-1)$.

4.1.8 CENTRE OF A GRAPH
A *centre* of a graph G is a vertex u such that $\max\{d(u, v) : v \in V\}$ is as small as possible.

a) Let T be a tree on at least three vertices, and let T' be the tree obtained from T by deleting all its leaves. Show that T and T' have the same centres.
b) Deduce that every tree has either exactly one centre or two, adjacent, centres.

4.1.9

a) Show that any simple graph with minimum degree k contains a copy of each rooted tree on $k + 1$ vertices, rooted at any given vertex of the graph.
b) Deduce that any simple graph with average degree at least $2(k - 1)$, where $k - 1$ is a positive integer, contains a copy of each tree on $k + 1$ vertices.

(P. Erdős and V. T. Sós (see Erdős (1964)) have conjectured that any simple graph with average degree greater than $k - 1$ contains a copy of each tree on $k + 1$ vertices; see Appendix A.)

4.1.10 Verify the properties (M1) and (M2) of median orders of tournaments.

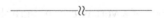

4.1.11 Let G be a simple graph with vertex set $V := \{1, 2, \ldots, n\}$.

a) Show that the set of transpositions $S := \{(i, j) : ij \in E\}$ generates all permutations of V if and only if G is connected.
b) Deduce that S is a minimal set of transpositions that generates all permutations of V if and only if G is a tree.

4.1.12 Let $S := \{x_1, x_2, \ldots, x_n\}$ be an n-set, and let $\mathcal{A} := \{A_1, A_2, \ldots, A_n\}$ be a family of n distinct subsets of S. Construct a graph G with vertex set $\mathcal{A}$, two vertices A_i and A_j being joined by an edge if their symmetric difference $A_i \triangle A_j$ is a singleton. Label the edge $A_i A_j$ by this singleton. By studying this labelled graph, prove that there is an element $x_m \in S$ such that the sets $A_1 \cup \{x_m\}, A_2 \cup \{x_m\}, \ldots, A_n \cup \{x_m\}$ are distinct. (J.A. BONDY)

4.1.13 Give an alternative proof of Exercise 4.1.12 by proceeding as follows. Suppose, by way of contradiction, that there is no such element $x_m \in S$, so that, for all $i \in [1, n]$, there exist distinct indices $j(i)$ and $k(i)$ such that $A_{j(i)} \cup \{x_i\} = A_{k(i)}$.

Let $\mathbf{M}$ be the incidence matrix of the hypergraph $(S, \mathcal{A})$ (so that $m_{ij} = 1$ if $x_i \in A_j$ and $m_{ij} = 0$ otherwise), let $\mathbf{c}_i$ denote the column vector with -1 in position $j(i)$, 1 in position $k(i)$, and 0s elsewhere, let $\mathbf{C}$ denote the $n \times n$ matrix whose ith column is $\mathbf{c}_i$, and let $\mathbf{j}$ be the row vector all of whose entries are 1. Show that $\mathbf{M}\mathbf{C} = \mathbf{I}$ and $\mathbf{j}\mathbf{C} = \mathbf{0}$, and derive a contradiction.   (J. GREENE)

4.1.14 m identical pizzas are to be shared equally amongst n students.

a) Show how this goal can be achieved by dividing the pizzas into a total of $m + n - d$ pieces, where d is the greatest common divisor of m and n.
b) By considering a suitable bipartite graph, show that no division into a smaller number of pieces will achieve the same objective.   (H. BASS)

4.1.15 Rooted trees $T_1(x_1)$ and $T_2(x_2)$ are *isomorphic* if there is an isomorphism from T_1 to T_2 mapping x_1 to x_2. A rooted tree is *uniform* if the degree of a vertex depends only on its distance from the root. Prove that every x-tree on n vertices has exactly n nonisomorphic uniform x-subtrees.

(M.K. GOLDBERG AND I.A. KLIPKER)

4.1.16 Let $v_1, v_2, \ldots, v_n$ be a median order of an even tournament T. Show that $(v_1, v_2, \ldots, v_n, v_1)$ is a directed Hamilton cycle of T.   (S. THOMASSÉ)

4.1.17 A *king* in a tournament is a vertex v from which every vertex is reachable by a directed path of length at most two. Show that every tournament T has a king by proceeding as follows.

Let $v_1, v_2, \ldots, v_n$ be a median order of T.

a) Suppose that v_j dominates v_i, where $i < j$. Show that there is an index k with $i < k < j$ such that v_i dominates v_k and v_k dominates v_j.
b) Deduce that v_1 is a king in T.   (F. HAVET AND S. THOMASSÉ)

4.1.18 A *second outneighbour* of a vertex v in a digraph is a vertex whose distance from v is exactly two. Show that every tournament T has a vertex with at least as many second outneighbours as (first) outneighbours, by proceeding as follows.

Let $v_1, v_2, \ldots, v_n$ be a median order of a tournament T. Colour the outneighbours of v_n red, both v_n and those of its in-neighbours which dominate every red vertex preceding them in the median order black, and the remaining in-neighbours of v_n blue. (Note that every vertex of T is thereby coloured, because T is a tournament.)

a) Show that every blue vertex is a second outneighbour of v_n.
b) Consider the intervals of the median order into which it is subdivided by the black vertices. Using property $(M2)$, show that each such interval includes at least as many blue vertices as red vertices.

c) Deduce that v_n has at least as many second outneighbours as outneighbours.

(F. HAVET AND S. THOMASSÉ)

(P. D. Seymour has conjectured that every oriented graph has a vertex with at least as many second outneighbours as outneighbours; see Appendix A)

4.1.19

a) Show that the cube of a tree on at least three vertices has a Hamilton cycle.

(M. SEKANINA)

b) Find a tree whose square has no Hamilton cycle.

(Fleischner (1974) characterized the graphs whose squares have Hamilton cycles; see also Říha (1991).)

★4.1.20

a) Let T_1 and T_2 be subtrees of a tree T. Show that $T_1 \cap T_2$ and $T_1 \cup T_2$ are subtrees of T if and only if $T_1 \cap T_2 \neq \emptyset$.

b) Let $\mathcal{T}$ be a family of subtrees of a tree T. Deduce, by induction on $|\mathcal{T}|$, that if any two members of $\mathcal{T}$ have a vertex in common, then there is a vertex of T which belongs to all members of $\mathcal{T}$. (In other words, show that the family of subtrees of a tree have the Helly Property (defined in Exercise 1.3.7).)

4.1.21 KÖNIG'S LEMMA
Show that every locally-finite infinite tree contains a one-way infinite path.

(D. KÖNIG)

4.2 Spanning Trees

A *subtree* of a graph is a subgraph which is a tree. If this tree is a spanning subgraph, it is called a *spanning tree* of the graph. Figure 4.3 shows a decomposition of the wheel W_4 into two spanning trees.

Fig. 4.3. Two spanning trees of the wheel W_4

If a graph G has a spanning tree T, then G is connected because any two vertices of G are connected by a path in T, and hence in G. On the other hand, if G is a connected graph which is not a tree, and e is an edge of a cycle of G, then $G \setminus e$ is a spanning subgraph of G which is also connected because, by Proposition

3.2, e is not a cut edge of G. By repeating this process of deleting edges in cycles until every edge which remains is a cut edge, we obtain a spanning tree of G. Thus we have the following theorem, which provides yet another characterization of connected graphs.

Theorem 4.6 *A graph is connected if and only if it has a spanning tree.* $\square$

It is easy to see that every tree is bipartite. We now use Theorem 4.6 to derive a characterization of bipartite graphs.

Theorem 4.7 *A graph is bipartite if and only if it contains no odd cycle.*

Proof   Clearly, a graph is bipartite if and only if each of its components is bipartite, and contains an odd cycle if and only if one of its components contains an odd cycle. Thus, it suffices to prove the theorem for connected graphs.

Let $G[X, Y]$ be a connected bipartite graph. Then the vertices of any path in G belong alternately to X and to Y. Thus, all paths connecting vertices in different parts are of odd length and all paths connecting vertices in the same part are of even length. Because, by definition, each edge of G has one end in X and one end in Y, it follows that every cycle of G is of even length.

Conversely, suppose that G is a connected graph without odd cycles. By Theorem 4.6, G has a spanning tree T. Let x be a vertex of T. By Proposition 4.1, any vertex v of T is connected to x by a unique path in T. Let X denote the set of vertices v for which this path is of even length, and set $Y := V \setminus X$. Then (X, Y) is a bipartition of T. We claim that (X, Y) is also a bipartition of G.

To see this, consider an edge $e = uv$ of $E(G) \setminus E(T)$, and let $P := uTv$ be the unique uv-path in T. By hypothesis, the cycle $P + e$ is even, so P is odd. Therefore the ends of P, and hence the ends of e, belong to distinct parts. It follows that (X, Y) is indeed a bipartition of G. $\square$

According to Theorem 4.7, either a graph is bipartite, or it contains an odd cycle, but not both. An efficient algorithm which finds, in a given graph, either a bipartition or an odd cycle is presented in Chapter 6.

CAYLEY'S FORMULA

There is a remarkably simple formula for the number of labelled trees on n vertices (or, equivalently, for the number of spanning trees in the complete graph K_n). This formula was discovered by Cayley (1889), who was interested in representing certain hydrocarbons by graphs and, in particular, by trees (see Exercise 4.1.3). A wide variety of proofs have since been found for Cayley's Formula (see Moon (1967)). We present here a particularly elegant one, due to Pitman (1999). It makes use of the concept of a *branching forest*, that is, a digraph each of whose components is a branching.

Theorem 4.8   CAYLEY'S FORMULA
The number of labelled trees on n vertices is n^{n-2}.

Proof We show, by counting in two ways, that the number of labelled branchings on n vertices is n^{n-1}. Cayley's Formula then follows directly, because each labelled tree on n vertices gives rise to n labelled branchings, one for each choice of the root vertex.

Consider, first, the number of ways in which a labelled branching on n vertices can be built up, one edge at a time, starting with the empty graph on n vertices. In order to end up with a branching, the subgraph constructed at each stage must be a branching forest. Initially, this branching forest has n components, each consisting of an isolated vertex. At each stage, the number of components decreases by one. If there are k components, the number of choices for the new edge (u, v) is $n(k-1)$: any one of the n vertices may play the role of u, whereas v must be the root of one of the $k-1$ components which do not contain u. The total number of ways of constructing a branching on n vertices in this way is thus

$$\prod_{i=1}^{n-1} n(n-i) = n^{n-1}(n-1)!$$

On the other hand, any individual branching on n vertices is constructed exactly $(n-1)!$ times by this procedure, once for each of the orders in which its $n-1$ edges are selected. It follows that the number of labelled branchings on n vertices is n^{n-1}. □

Another proof of Cayley's Formula is outlined in Exercise 4.2.11.

We denote the number of spanning trees in an arbitrary graph G by $t(G)$. Cayley's Formula says that $t(K_n) = n^{n-2}$. There is a simple recursive formula relating the number of spanning trees of a graph G to the numbers of spanning trees in the two graphs $G \backslash e$ and G / e obtained from G by deleting and contracting a link e (Exercise 4.2.1).

Proposition 4.9 *Let G be a graph and e a link of G. Then*

$$t(G) = t(G \backslash e) + t(G / e)$$ □

Exercises

⋆**4.2.1** Let G be a connected graph and e a link of G.

a) Describe a one-to-one correspondence between the set of spanning trees of G that contain e and the set of spanning trees of G / e.
b) Deduce Proposition 4.9.

4.2.2

a) Let G be a graph with no loops or cut edges. Show that $t(G) \geq e(G)$.

b) For which such graphs does equality hold?

4.2.3 Let G be a connected graph and let x be a specified vertex of G. A spanning x-tree T of G is called a *distance tree* of G with root x if $d_T(x, v) = d_G(x, v)$ for all $v \in V$.

a) Show that G has a distance tree with root x.
b) Deduce that a connected graph of diameter d has a spanning tree of diameter at most $2d$.

$\star$**4.2.4** Show that the incidence matrix of a graph is totally unimodular (defined in Exercise 1.5.7) if and only if the graph is bipartite.

4.2.5 A *fan* is the join $P \vee K_1$ of a path P and a single vertex. Determine the numbers of spanning trees in:

a) the fan F_n on n vertices, $n \geq 2$,
b) the wheel W_n with n spokes, $n \geq 3$.

4.2.6 Let G be an edge-transitive graph.

a) Show that each edge of G lies in exactly $(n-1)t(G)/m$ spanning trees of G.
b) Deduce that $t(G \setminus e) = (m - n + 1)t(G)/m$ and $t(G / e) = (n-1)t(G)/m$.
c) Deduce that $t(K_n)$ is divisible by n, if $n \geq 3$, and that $t(K_{n,n})$ is divisible by n^2.
d) Without appealing to Cayley's Formula (Theorem 4.8), determine $t(K_4)$, $t(K_5)$, and $t(K_{3,3})$.

4.2.7

a) Let G be a simple graph on n vertices, and let H be the graph obtained from G by replacing each edge of G by k multiple edges. Show that $t(H) = k^{n-1}t(G)$.
b) Let G be a graph on n vertices and m edges, and let H be the graph obtained from G by subdividing each edge of G $k-1$ times. Show that $t(H) = k^{m-n+1}t(G)$.

$\star$**4.2.8** Using Theorem 4.7 and Exercise 3.4.11b, show that a digraph contains a directed odd cycle if and only if some strong component is not bipartite.

$\star$**4.2.9** A branching in a digraph is a *spanning branching* if it includes all vertices of the digraph.

a) Show that a digraph D has a spanning x-branching if and only if $\partial^+(X) \neq \emptyset$ for every proper subset X of V that includes x.
b) Deduce that a digraph is strongly connected if and only if it has a spanning v-branching for every vertex v.

4.2.10 NONRECONSTRUCTIBLE INFINITE GRAPHS
Let $T := T_\infty$ denote the infinite tree in which each vertex is of countably infinite degree, and let $F := 2T_\infty$ denote the forest consisting of two disjoint copies of T_∞. Show that (T, F) is a nonreconstructible pair.

—————————————♮—————————————

4.2.11 PRÜFER CODE

Let K_n be the labelled complete graph with vertex set $\{1, 2, \ldots, n\}$, where $n \geq 3$. With each spanning tree T of K_n one can associate a unique sequence $(t_1, t_2, \ldots, t_{n-2})$, known as the *Prüfer code* of T, as follows. Let s_1 denote the first vertex (in the the the ordered set $(1, 2, \ldots, n)$) which is a leaf of T, and let t_1 be the neighbour of s_1 in T. Now let s_2 denote the first vertex which is a leaf of $T - s_1$, and t_2 the neighbour of s_2 in $T - s_1$. Repeat this operation until t_{n-2} is defined and a tree with just two vertices remains. (If $n \leq 2$, the Prüfer code of T is taken to be the empty sequence.)

a) List all the spanning trees of K_4 and their Prüfer codes.
b) Show that every sequence $(t_1, t_2, \ldots, t_{n-2})$ of integers from the set $\{1, 2, \ldots, n\}$ is the Prüfer code of a unique spanning tree of K_n.
c) Deduce Cayley's Formula (see Theorem 4.8).   (H. PRÜFER)

4.2.12

a) For a sequence $d_1, d_2, \ldots, d_n$ of n positive integers whose sum is equal to $2n-2$, let $t(n; d_1, d_2, \ldots, d_n)$ denote the number of trees on n vertices $v_1, v_2, \ldots, v_n$ such that $d(v_i) = d_i$, $1 \leq i \leq n$. Show that

$$t(n; d_1, d_2, \ldots, d_n) = \binom{n-2}{d_1 - 1, d_2 - 1, \ldots, d_n - 1}$$

b) Apply the Multinomial Theorem to deduce Cayley's Formula.

4.2.13 By counting the number of branchings whose root lies in the m-set of $K_{m,n}$, show that $t(K_{m,n}) = m^{n-1}n^{m-1}$.

4.2.14 Show that the Petersen graph has 2000 spanning trees.

4.2.15 Let T be a tree with vertex set V, and let $f : V \rightarrow V$ be a mapping with no fixed point. For $v \in V$, denote by v^+ the successor of v on the path $vTf(v)$, and by D_f the digraph with vertex set V and arc set $\{(v, v^+) : v \in V\}$.

a) Show that each component of D_f contains a unique directed 2-cycle.
b) The *centroid* of T is the set of all vertices v for which the largest component of $T - v$ has as few vertices as possible. For $v \in V$, let $f(v)$ be a vertex of a largest component of $T - v$, and let (x, y, x) be a directed 2-cycle of D_f. Show that the centroid of T is contained in the set $\{x, y\}$, and hence consists either of one vertex or of two adjacent vertices.   (C. JORDAN)
c) An *endomorphism* of a simple graph G is a mapping $f : V \rightarrow V$ such that, for every $xy \in E$, either $f(x) = f(y)$ or $f(x)f(y) \in E$. Let f be an endomorphism of T, and let (x, y, x) be a directed 2-cycle of D_f.
   i) Show that $f(x) = y$ and $f(y) = x$.
   ii) Deduce that every endomorphism of a tree T fixes either a vertex or an edge of T.   (L. LOVÁSZ)

d) Let T be a spanning tree of the n-cube Q_n, let $f(v)$ be the antipodal vertex of vertex v in Q_n (that is, the unique vertex whose distance from v is n), and let (x, y, x) be a directed 2-cycle of D_f.

   i) Show that $d_T(f(x), f(y)) \geq 2n - 1$.

   ii) Deduce that every spanning tree of Q_n has a fundamental cycle of length at least $2n$.   (R.L. GRAHAM)

4.2.16 Let G be a connected simple graph and T a spanning tree of G. Consider the mapping $\phi : \binom{V}{2} \setminus T \to \binom{T}{2}$ (where T is regarded as a subset of E) defined by $\phi(xy) := \{e, f\}$, where e and f are the first and last edges of the path xTy.

a) Show that the mapping ϕ is a bijection.

b) Deduce that $\binom{n}{2} - |T| = \binom{|T|}{2}$.

c) Deduce Theorem 4.3.   (N. GRAHAM, R.C. ENTRINGER, AND L. SZÉKELY)

4.3 Fundamental Cycles and Bonds

The spanning trees of a connected graph, its even subgraphs, and its edge cuts are intimately related. We describe these relationships here. Recall that, in the context of even subgraphs, when we speak of a *cycle* we typically mean its edge set. Likewise, by a *spanning tree*, we understand in this context the edge set of the tree. Throughout this section, G denotes a connected graph and T a spanning tree of G.

COTREES

The complement $E \setminus T$ of a spanning tree T is called a *cotree*, and is denoted $\overline{T}$. Consider, for example, the wheel W_4 shown in Figure 4.4a, and the spanning tree $T := \{1, 2, 4, 5\}$ indicated by solid lines. The cotree $\overline{T}$ is simply the set of light edges, namely $\{3, 6, 7, 8\}$.

By Proposition 4.1, for every edge $e := xy$ of a cotree $\overline{T}$ of a graph G, there is a unique xy-path in T connecting its ends, namely $P := xTy$. Thus $T + e$ contains a unique cycle. This cycle is called the *fundamental cycle* of G with respect to T and e. For brevity, we denote it by C_e, the role of the tree T being implicit. Figure 4.4b shows the fundamental cycles of W_4 with respect to the spanning tree $\{1, 2, 4, 5\}$, namely $C_3 = \{1, 2, 3, 4\}$, $C_6 = \{1, 5, 6\}$, $C_7 = \{1, 2, 5, 7\}$, and $C_8 = \{4, 5, 8\}$.

One can draw interesting conclusions about the structure of a graph from the properties of its fundamental cycles with respect to a spanning tree. For example, if all the fundamental cycles are even, then every cycle of the graph is even and hence, by Theorem 4.7, the graph is bipartite. (This is the idea behind the proof of Theorem 4.7.) The following theorem and its corollaries show why fundamental cycles are important.

Theorem 4.10 *Let T be a spanning tree of a connected graph G, and let S be a subset of its cotree $\overline{T}$. Then $C := \triangle\{C_e : e \in S\}$ is an even subgraph of G. Moreover, $C \cap \overline{T} = S$, and C is the only even subgraph of G with this property.*

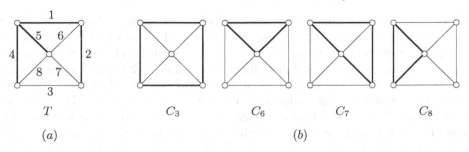

Fig. 4.4. (a) A spanning tree T of the wheel W_4, and (b) the fundamental cycles with respect to T

Proof As each fundamental cycle C_e is an even subgraph, it follows from Corollary 2.16 that C is an even subgraph, too. Furthermore, $C \cap \overline{T} = S$, because each edge of S appears in exactly one member of the family $\{C_e : e \in S\}$.

Let C' be any even subgraph of G such that $C' \cap \overline{T} = S$. Then

$$(C \triangle C') \cap \overline{T} = (C \cap \overline{T}) \triangle (C' \cap \overline{T}) = S \triangle S = \emptyset$$

Therefore the even subgraph $C \triangle C'$ is contained in T. Because the only even subgraph contained in a tree is the empty even subgraph, we deduce that $C' = C$. $\square$

Corollary 4.11 *Let T be a spanning tree of a connected graph G. Every even subgraph of G can be expressed uniquely as a symmetric difference of fundamental cycles with respect to T.*

Proof Let C be an even subgraph of G and let $S := C \cap \overline{T}$. It follows from Theorem 4.10 that $C = \triangle\{C_e : e \in S\}$ and that this is the only way of expressing C as a symmetric difference of fundamental cycles with respect to T. $\square$

The next corollary, which follows from Theorem 4.10 by taking $S := \overline{T}$, has several interesting applications (see, for example, Exercises 4.3.9 and 4.3.10).

Corollary 4.12 *Every cotree of a connected graph is contained in a unique even subgraph of the graph.* $\square$

We now discuss the relationship between spanning trees and edge cuts. We show that, for each of the above statements concerning even subgraphs, there is an analogous statement concerning edge cuts. As before, let G be a connected graph and let T be a spanning tree of G. Note that, because T is connected and spanning, every nonempty edge cut of G contains at least one edge of T. Thus the only edge cut contained in the cotree $\overline{T}$ is the empty edge cut (just as the only even subgraph contained in T is the empty even subgraph).

In order to be able to state the cut-analogue of Theorem 4.10, we need the notion of a fundamental bond. Let $e := xy$ be an edge of T. Then $T \setminus e$ has exactly two components, one containing x and the other containing y. Let X denote the vertex set of the component containing x. The bond $B_e := \partial(X)$ is contained in $\overline{T} \cup \{e\}$ and includes e. Moreover, it is the only such bond. For, let B be any bond contained in $\overline{T} \cup \{e\}$ and including e. By Corollary 2.12, $B \triangle B_e$ is an edge cut. Moreover, this edge cut is contained in $\overline{T}$. But, as remarked above, the only such edge cut is the empty edge cut. This shows that $B = B_e$. The bond B_e is called the *fundamental bond* of G with respect to T and e. For instance, the fundamental bonds of the wheel W_4 with respect to the spanning tree $\{1, 2, 4, 5\}$ (indicated in Figure 4.5a) are $B_1 = \{1, 3, 6, 7\}$, $B_2 = \{2, 3, 7\}$, $B_4 = \{3, 4, 8\}$, and $B_5 = \{5, 6, 7, 8\}$ (see Figure 4.5b).

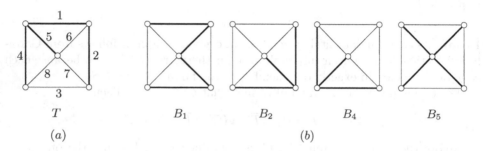

$$T \qquad\qquad B_1 \qquad\qquad B_2 \qquad\qquad B_4 \qquad\qquad B_5$$

$$(a) \qquad\qquad\qquad\qquad\qquad (b)$$

Fig. 4.5. (a) A spanning tree T of the wheel W_4, and (b) the fundamental bonds with respect to T

The proofs of the following theorem and its corollaries are similar to those of Theorem 4.10 and its corollaries, and are left as an exercise (Exercise 4.3.5).

Theorem 4.13 *Let T be a spanning tree of a connected graph G, and let S be a subset of T. Set $B := \triangle\{B_e : e \in S\}$. Then B is an edge cut of G. Moreover $B \cap T = S$, and B is the only edge cut of G with this property.* ☐

Corollary 4.14 *Let T be a spanning tree of a connected graph G. Every edge cut of G can be expressed uniquely as a symmetric difference of fundamental bonds with respect to T.* ☐

Corollary 4.15 *Every spanning tree of a connected graph is contained in a unique edge cut of the graph.* ☐

Corollaries 4.11 and 4.14 imply that the fundamental cycles and fundamental bonds with respect to a spanning tree of a connected graph constitute bases of its cycle and bond spaces, respectively, as defined in Section 2.6 (Exercise 4.3.6). The dimension of the cycle space of a graph is referred to as its *cyclomatic number*.

In this section, we have defined and discussed the properties of fundamental cycles and bonds with respect to spanning trees in connected graphs. All the above theorems are valid for disconnected graphs too, with maximal forests playing the role of spanning trees.

Exercises

4.3.1 Determine the fundamental cycles and fundamental bonds of W_4 with respect to the spanning tree shown in Figure 4.3 (using the edge labelling of Figure 4.4).

4.3.2 TREE EXCHANGE PROPERTY
Let G be a connected graph, let T_1 and T_2 be (the edge sets of) two spanning trees of G, and let $e \in T_1 \setminus T_2$. Show that:

a) there exists $f \in T_2 \setminus T_1$ such that $(T_1 \setminus \{e\}) \cup \{f\}$ is a spanning tree of G,
b) there exists $f \in T_2 \setminus T_1$ such that $(T_2 \setminus \{f\}) \cup \{e\}$ is a spanning tree of G.

(Each of these two facts is referred to as a *Tree Exchange Property*.)

4.3.3 Let G be a connected graph and let S be a set of edges of G. Show that the following statements are equivalent.

a) S is a spanning tree of G.
b) S contains no cycle of G, and is maximal with respect to this property.
c) S meets every bond of G, and is minimal with respect to this property.

4.3.4 Let G be a connected graph and let S be a set of edges of G. Show that the following statements are equivalent.

a) S is a cotree of G.
b) S contains no bond of G, and is maximal with respect to this property.
c) S meets every cycle of G, and is minimal with respect to this property.

4.3.5

a) Prove Theorem 4.13.
b) Deduce Corollaries 4.14 and 4.15.

4.3.6

a) Let T be a spanning tree of a connected graph G. Show that:
    i) the fundamental cycles of G with respect to T form a basis of its cycle space,
    ii) the fundamental bonds of G with respect to T form a basis of its bond space.
b) Determine the dimensions of these two spaces.

(The cycle and bond spaces were defined in Section 2.6.)

4.3.7 Let G be a connected graph, and let $\mathbf{M}$ be its incidence matrix.

a) Show that the columns of $\mathbf{M}$ corresponding to a subset S of E are linearly independent over $GF(2)$ if and only if $G[S]$ is acyclic.
b) Deduce that there is a one-to-one correspondence between the bases of the column space of $\mathbf{M}$ over $GF(2)$ and the spanning trees of G.

(The above statements are special cases of more general results, to be discussed in Section 20.2.)

4.3.8 ALGEBRAIC DUALS

An *algebraic dual* of a graph G is a graph H for which there is a bijection θ : $E(G) \to E(H)$ mapping each cycle of G to a bond of H and each bond of G to a cycle of H.

a) Show that:
    i) the octahedron and the cube are algebraic duals,
    ii) $K_{3,3}$ has no algebraic dual.
b) Let G be a connected graph and H an algebraic dual of G, with bijection θ.
    i) Show that T is a spanning tree of G if and only if $\theta(T)$ is a cotree of H.
    ii) Deduce that $t(G) = t(H)$.

4.3.9 Show that any graph which contains a Hamilton cycle has a covering by two even subgraphs.

★4.3.10 Show that any graph which contains two edge-disjoint spanning trees has:

a) an eulerian spanning subgraph,
b) a covering by two even subgraphs.

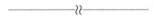

4.4 Related Reading

MATROIDS

One of the characteristic properties of spanning trees of a connected graph is the Tree Exchange Property noted in Exercise 4.3.2a. Because the spanning trees of G correspond to bases of the incidence matrix $\mathbf{M}$ of G (Exercise 4.3.7), the Tree Exchange Property may be seen as a special case of the appropriate exchange property of bases of a vector space. Whitney (1935) observed that many essential properties of spanning trees, such as the ones described in Section 4.3, and more generally of bases of a vector space, may be deduced from that exchange property. Motivated by this observation, he introduced the notion of a matroid.

A *matroid* is an ordered pair $(E, \mathcal{B})$, consisting of a finite set E of *elements* and a nonempty family $\mathcal{B}$ of subsets of E, called *bases*, which satisfy the following *Basis Exchange Property*.

> If $B_1, B_2 \in \mathcal{B}$ and $e \in B_1 \setminus B_2$ then there exists $f \in B_2 \setminus B_1$ such that $(B_1 \setminus \{e\}) \cup \{f\} \in \mathcal{B}$

Let $\mathbf{M}$ be a matrix over a field $\mathbb{F}$, let E denote the set of columns of $\mathbf{M}$, and let $\mathcal{B}$ be the family of subsets of E which are bases of the column space of $\mathbf{M}$. Then $(E, \mathcal{B})$ is a matroid. Matroids which arise in this manner are called *linear matroids.* Various linear matroids may be associated with graphs, one example being the matroid on the edge set of a connected graph in which the bases are the edge sets of spanning trees. (In the matroidal context, statements concerning connected graphs extend easily to all graphs, the role of spanning trees being played by maximal forests when the graph is not connected.)

Much of matroid-theoretic terminology is suggested by the two examples mentioned above. For instance, subsets of bases are called *independent sets*, and minimal dependent sets are called *circuits*. In the matroid whose bases are the spanning trees of a connected graph G, the independent sets of the matroid are the forests of G and its circuits are the cycles of G. For this reason, this matroid is called the *cycle matroid* of G, denoted $M(G)$.

The *dual* of a matroid $M = (E, \mathcal{B})$ is the matroid $M^* = (E, \mathcal{B}^*)$, where $\mathcal{B}^* := \{E \setminus B : B \in \mathcal{B}\}$. When M is the linear matroid associated with a matrix $\mathbf{M}$, the bases of M^* are those subsets of E which are bases of the orthogonal complement of the column space of $\mathbf{M}$. When M is the cycle matroid of a connected graph G, the bases of M^* are the cotrees of G, and its circuits are the bonds of G. For this reason, the dual of the cycle matroid of a graph G is called the *bond matroid* of G, denoted $M^*(G)$. Many manifestations of this cycle–bond duality crop up throughout the book. The reader is referred to Oxley (1992) for a thorough account of the theory of matroids.

5

Nonseparable Graphs

Contents

5.1 Cut Vertices

In Chapter 3, we introduced the notion of a cut edge and discussed various properties of connected graphs without cut edges. Here, we consider the analogous notion for vertices. There are, in fact, two closely related notions, that of a cut vertex and that of a separating vertex.

A *cut vertex* of a graph G is a vertex v such that $c(G-v) > c(G)$. In particular, a cut vertex of a connected graph is a vertex whose deletion results in a disconnected graph. This notion is illustrated in Figure 5.1, the cut vertices being indicated by solid dots.

By Exercise 3.1.3, a graph is connected if any two of its vertices are connected by a path. Connected graphs without cut vertices have a stronger property, described in the theorem below. Two distinct paths are *internally disjoint* if they have no internal vertices in common.

Theorem 5.1 *A connected graph on three or more vertices has no cut vertices if and only if any two distinct vertices are connected by two internally disjoint paths.*

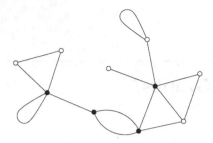

Fig. 5.1. The cut vertices of a graph

Proof Let G be a connected graph, and let v be a vertex of G. If any two vertices of G are connected by two internally disjoint paths, any two vertices of $G - v$ are certainly connected by at least one path, so $G - v$ is connected and v is not a cut vertex of G. This being so for each vertex v, the graph G has no cut vertices.

Conversely, let G be a connected graph on at least three vertices, with no cut vertices. Consider any two vertices u and v of G. We prove, by induction on the distance $d(u, v)$ between u and v, that these vertices are connected by two internally disjoint paths.

Suppose, first, that u and v are adjacent, and let e be an edge joining them. Because neither u nor v is a cut vertex, e is not a cut edge (Exercise 5.1.2) and therefore, by Proposition 3.2, lies in a cycle C of G. It follows that u and v are connected by the internally disjoint paths uev and $C \setminus e$.

Suppose, now, that the theorem holds for any two vertices at distance less than k, where $k \geq 2$, and let $d(u, v) = k$. Consider a uv-path of length k, and let v' be the immediate predecessor of v on this path. Then $d(u, v') = k - 1$. According to the induction hypothesis, u and v' are connected by two internally disjoint paths, P' and Q' (see Figure 5.2).

Because G has no cut vertices, $G - v'$ is connected and therefore contains a uv-path R'. The path R' meets $P' \cup Q'$ at u. Let x be the last vertex of R' at which R' meets $P' \cup Q'$; without loss of generality, we may suppose that x lies on P'. Define $P := uP'xR'v$ and $Q := uQ'v'v$. Then P and Q are internally disjoint uv-paths in G. $\qquad\square$

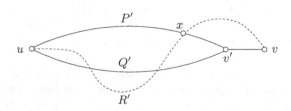

Fig. 5.2. Proof of Theorem 5.1

Generalizations and variants of Theorem 5.1 are discussed in Chapter 7 and Chapter 9.

Exercises

5.1.1 Show that every nontrivial graph has at least two vertices that are not cut vertices.

⋆**5.1.2** Let G be a connected graph on at least three vertices, and let $e = uv$ be a cut edge of G. Show that either u or v is a cut vertex of G.

5.1.3 Let G be a nontrivial connected graph without cut vertices, and let H be obtained from G by adding a new vertex and joining it to two vertices of G. Show that H has no cut vertices.

5.1.4 Let G be a nontrivial connected graph without cut vertices, and let X and Y be two (not necessarily disjoint) sets of vertices of G, each of cardinality at least two. Show that there are two disjoint (X, Y)-paths in G.

5.1.5 Show that any two longest cycles in a loopless connected graph without cut vertices have at least two vertices in common.

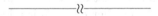

5.2 Separations and Blocks

Whilst the notion of a cut vertex, as defined in Section 5.1, is the most natural analogue for vertices of the notion of cut edge, a slightly more general concept is needed for graphs which may have loops.

A *separation* of a connected graph is a decomposition of the graph into two nonempty connected subgraphs which have just one vertex in common. This common vertex is called a *separating vertex* of the graph. The separating vertices of a disconnected graph are defined to be those of its components. A cut vertex is clearly a separating vertex, but not conversely: a vertex incident with a loop and at least one other edge is a separating vertex but not necessarily a cut vertex. However, in a loopless graph, every separating vertex is indeed a cut vertex, so in this case the two concepts are identical. Whereas the graph shown in Figure 5.1 has four cut vertices, it has five separating vertices, as indicated in Figure 5.3.

NONSEPARABLE GRAPHS

A graph is *nonseparable* if it is connected and has no separating vertices; otherwise, it is *separable*. Up to isomorphism, there are just two nonseparable graphs on one vertex, namely K_1, and K_1 with a loop attached. All nonseparable graphs on two

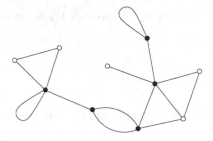

Fig. 5.3. The separating vertices of a graph

or more vertices are loopless. Multiple edges play no role here: a loopless graph is nonseparable if and only if its underlying simple graph is nonseparable. Apart from K_1 and K_2, the most basic nonseparable graphs are the cycles. Whitney (1932c) showed that nonseparable connected graphs may be characterized in terms of their cycles, as follows.

Theorem 5.2 *A connected graph is nonseparable if and only if any two of its edges lie on a common cycle.*

Proof If G is separable, it may be decomposed into two nonempty connected subgraphs, G_1 and G_2, which have just one vertex v in common. Let e_i be an edge of G_i incident with v, $i = 1, 2$. If either e_1 or e_2 is a loop, there is clearly no cycle including both e_1 and e_2. If not, v is a cut vertex of G. Let v_i be the other end of e_i, $i = 1, 2$. Then there is no $v_1 v_2$-path in $G - v$, hence no cycle in G through both e_1 and e_2.

Conversely, suppose that G is nonseparable. Let e_1 and e_2 be two edges of G. Subdivide e_i by a new vertex v_i, $i = 1, 2$. The resulting graph H is also nonseparable (Exercise 5.2.1). By Theorem 5.1, there is a cycle in H through v_1 and v_2, hence a cycle in G through e_1 and e_2. $\square$

BLOCKS

A *block* of a graph is a subgraph which is nonseparable and is maximal with respect to this property. A nonseparable graph therefore has just one block, namely the graph itself. The blocks of a nontrivial tree are the copies of K_2 induced by its edges; and, in general, the blocks of a connected graph fit together in a treelike structure, as illustrated in Figure 5.4. In order to prove this assertion, we note first a number of basic facts about blocks.

Proposition 5.3 *Let G be a graph. Then:*

 a) *any two blocks of G have at most one vertex in common,*
 b) *the blocks of G form a decomposition of G,*
 c) *each cycle of G is contained in a block of G.*

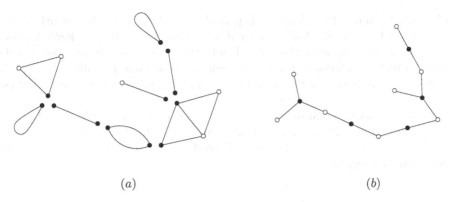

$$(a) \qquad\qquad\qquad\qquad (b)$$

Fig. 5.4. (a) The blocks of the graph of Figure 5.3, and (b) its block tree

Proof (a) We establish the claim by contradiction. Suppose that there are distinct blocks B_1 and B_2 with at least two common vertices. Note that B_1 and B_2 are necessarily loopless. Because they are maximal nonseparable subgraphs of G, neither one contains the other, so $B := B_1 \cup B_2$ properly contains both of them. Let $v \in V(B)$. Then $B - v = (B_1 - v) \cup (B_2 - v)$ is connected, because $B_1 - v$ and $B_2 - v$ are both connected and have at least one common vertex. Thus B has no cut vertices, and so, being loopless, is nonseparable. But this contradicts the maximality of B_1 and B_2.

(b) Each edge of G induces a nonseparable subgraph (on one or two vertices), hence is contained in a maximal nonseparable subgraph, or block, of G. On the other hand, no edge lies in two blocks, by (a). The blocks therefore constitute a decomposition of G.

(c) As noted above, a cycle of G is a nonseparable subgraph, so is contained in a block of G. $\qquad\qquad\qquad\qquad\qquad\qquad\qquad\qquad\qquad\qquad\qquad\qquad\qquad\quad\square$

We may associate with any graph G a bipartite graph $B(G)$ with bipartition $(\mathcal{B}, S)$, where $\mathcal{B}$ is the set of blocks of G and S the set of separating vertices of G, a block B and a separating vertex v being adjacent in $B(G)$ if and only if B contains v. Each path in G connecting vertices in distinct blocks gives rise to a unique path in $B(G)$ connecting these same blocks. It follows that if G is connected, so is $B(G)$. Furthermore, $B(G)$ is acyclic, because a cycle in $B(G)$ would correspond to a cycle in G passing through two or more blocks, contradicting Proposition 5.3c. The graph $B(G)$ is therefore a tree, called the *block tree* of G (see Figure 5.4b). If G is separable, the blocks of G which correspond to leaves of its block tree are referred to as its *end blocks*. An *internal vertex* of a block of a graph G is a vertex which is not a separating vertex of G.

By using this tree structure, one can deduce most properties of connected graphs from the properties of their blocks, just as one can deduce most properties of graphs from those of their components. In other words, one can usually reduce the study of all graphs to the study of their blocks. Examples are given in

Exercises 5.2.5 and 5.2.8b. Another is provided by Proposition 5.3, which implies that a graph has a cycle double cover if and only if each of its blocks has one. It therefore suffices to prove the Cycle Double Cover Conjecture for nonseparable graphs. In fact, it suffices to prove the conjecture for nonseparable cubic graphs. This reduction is based on the operation of splitting off edges from a vertex (see inset).

In the next section, we describe how any nonseparable graph other than K_1 and K_2 can be built up in a very simple way by starting with a cycle and adding paths. We then make use of this structure to deduce several important properties of nonseparable graphs.

PROOF TECHNIQUE: SPLITTING OFF EDGES

Let v be a vertex of a graph G, and let $e_1 := vv_1$ and $e_2 := vv_2$ be two edges of G incident to v. The operation of *splitting off* the edges e_1 and e_2 from v consists of deleting e_1 and e_2 and then adding a new edge e joining v_1 and v_2. This operation is illustrated in Figure 5.5b. (Note that if $v_1 = v_2$, then splitting off e_1 and e_2 from v amounts to replacing these edges by a loop at $v_1 = v_2$.) The following theorem, due to Fleischner (1992), shows that under certain conditions it can be performed without creating cut edges.

Theorem 5.4   THE SPLITTING LEMMA
Let G be a nonseparable graph and let v be a vertex of G of degree at least four with at least two distinct neighbours. Then some two nonparallel edges incident to v can be split off so that the resulting graph is connected and has no cut edges.

Proof   There are two graphs on three vertices and five edges which satisfy the hypotheses of the theorem, and it may be readily checked that the theorem holds for these two graphs. We proceed by induction on m. Let f be an edge of G not incident to v, and set $H := G \setminus f$. If v is an internal vertex of some block B of H, the theorem follows by induction applied to B and v. So we may assume that v is a cut vertex of H. Because G is nonseparable, the block tree of H is a path (Exercise 5.2.11), and the edge f links internal vertices of the two endblocks of H, as illustrated in Figure 5.5a.

Let e_1 and e_2 be two edges incident with v and lying in distinct blocks of H. Consider the graph G' derived from G by splitting off e_1 and e_2. It may be checked that G' is connected and that each edge of G' lies in a cycle (Exercise 5.2.9). By Proposition 3.2, G' has no cut edges. $\square$

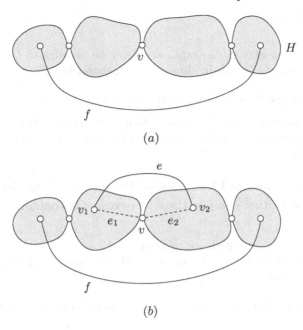

Fig. 5.5. Proof of Theorem 5.4: (a) The block path H and the edge f, (b) the graph G'

SPLITTING OFF EDGES (CONTINUED)

Here is the promised application of the Splitting Lemma to cycle double covers.

Theorem 5.5 *The Cycle Double Conjecture is true if and only if it is true for all nonseparable cubic graphs.*

Proof We have already noted that it suffices to prove the Cycle Double Cover Conjecture for nonseparable graphs. Consider such a graph G. By Veblen's Theorem, we may assume that G has at least one vertex of odd degree. If G has a vertex v of degree two, with neighbours u and w, let G' be the nonseparable graph obtained from $G - v$ on adding a new edge joining u and w. If G has a vertex v of degree four or more, let G' be a nonseparable graph obtained from G by splitting off two edges incident to v. In both cases, it is easy to see that if G' has a cycle double cover, then so has G. Applying these two operations recursively results in a nonseparable cubic graph H, and if H has a cycle double cover, then so has G. □

For another application of the Splitting Lemma, see Exercise 5.2.12.

Exercises

★**5.2.1** Let G be a nonseparable graph, and let e be an edge of G. Show that the graph obtained from G by subdividing e is nonseparable.

★**5.2.2** Let G be a graph, and let e be an edge of G. Show that:

a) if $G \setminus e$ is nonseparable and e is not a loop of G, then G is nonseparable,
b) if G / e is nonseparable and e is neither a loop nor a cut edge of G, then G is nonseparable.

5.2.3 Let G be a graph, and let $\overset{C}{\sim}$ denote the binary relation on E defined by $e \overset{C}{\sim} f$ if and only if either $e = f$ or there is a cycle of G containing both e and f. Show that:

a) the relation $\overset{C}{\sim}$ is an equivalence relation on E,
b) the subgraphs of G induced by the equivalence classes under this relation are its nontrivial blocks.

5.2.4 Show that a connected separable graph has at least two end blocks.

5.2.5 Show that:

a) a graph is even if and only if each of its blocks is even,
b) a graph is bipartite if and only if each of its blocks is bipartite.

5.2.6 We denote a graph G with two distinguished vertices x and y by $G(x,y)$. Prove the following edge analogue of Theorem 5.1.
Let $G(x,y)$ be a connected graph without cut edges. Then there exist two edge-disjoint xy-paths in G.

5.2.7

a) Let $G(x,y)$ be a nonseparable graph. Show that all xy-paths in G have the same parity if and only if G is bipartite.
b) Deduce that each edge of a nonseparable nonbipartite graph lies in an odd cycle.

★**5.2.8**

a) Let B be a block of a graph G, and let P be a path in G connecting two vertices of B. Show that P is contained in B.
b) Deduce that a spanning subgraph T of a connected graph G is a spanning tree of G if and only if $T \cap B$ is a spanning tree of B for every block B of G.

★**5.2.9** Consider the graph G' arising in the proof of Theorem 5.4. Show that:

a) G' is connected,
b) each edge of G' lies in a cycle.

5.2.10 Construct a nonseparable graph each vertex of which has degree at least four and at least two distinct neighbours, and in which splitting off any two adjacent edges results in a separable graph.

⋆**5.2.11** Let G be a nonseparable graph, and let e be an edge of G such that $G \setminus e$ is separable. Show that the block tree of $G \setminus e$ is a path.

(G.A. DIRAC; M.D. PLUMMER)

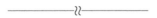

5.2.12

a) By employing the splitting-off operation, show that every even graph has an odd number of cycle decompositions.
b) Deduce that each edge of an even graph lies in an odd number of cycles.

(S. TOIDA)

5.3 Ear Decompositions

Apart from K_1 and K_2, every nonseparable graph contains a cycle. We describe here a simple recursive procedure for generating any such graph starting with an arbitrary cycle of the graph.

Let F be a subgraph of a graph G. An *ear* of F in G is a nontrivial path in G whose ends lie in F but whose internal vertices do not.

Proposition 5.6 *Let F be a nontrivial proper subgraph of a nonseparable graph G. Then F has an ear in G.*

Proof If F is a spanning subgraph of G, the set $E(G) \setminus E(F)$ is nonempty because, by hypothesis, F is a proper subgraph of G. Any edge in $E(G) \setminus E(F)$ is then an ear of F in G. We may suppose, therefore, that F is not spanning.

Since G is connected, there is an edge xy of G with $x \in V(F)$ and $y \in V(G) \setminus V(F)$. Because G is nonseparable, $G - x$ is connected, so there is a $(y, F - x)$-path Q in $G - x$. The path $P := xyQ$ is an ear of F. □

The proofs of the following proposition is left to the reader (Exercise 5.3.1).

Proposition 5.7 *Let F be a nonseparable proper subgraph of a graph G, and let P be an ear of F. Then $F \cup P$ is nonseparable.* □

A *nested sequence* of graphs is a sequence $(G_0, G_1, \ldots, G_k)$ of graphs such that $G_i \subset G_{i+1}, 0 \le i < k$. An *ear decomposition* of a nonseparable graph G is a nested sequence $(G_0, G_1, \ldots, G_k)$ of nonseparable subgraphs of G such that:

▷ G_0 is a cycle,
▷ $G_{i+1} = G_i \cup P_i$, where P_i is an ear of G_i in G, $0 \le i < k$,
▷ $G_k = G$.

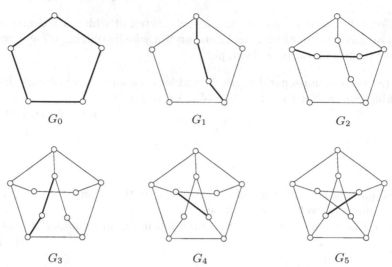

G_0 G_1 G_2

G_3 G_4 G_5

Fig. 5.6. An ear decomposition of the Petersen graph

An ear decomposition of the Petersen graph is shown in Figure 5.6, the initial cycle and the ear added at each stage being indicated by heavy lines.

Using the fact that every nonseparable graph other than K_1 and K_2 has a cycle, we may deduce the following theorem from Propositions 5.6 and 5.7.

Theorem 5.8 *Every nonseparable graph other than K_1 and K_2 has an ear decomposition.*   □

This recursive description of nonseparable graphs can be used to establish many of their properties by induction. We describe below an interesting application of ear decompositions to a problem of traffic flow. Further applications may be found in the exercises at the end of this section.

STRONG ORIENTATIONS

A road network in a city is to be converted into a one-way system, in order that traffic may flow as smoothly as possible. How can this be achieved in a satisfactory manner? This problem clearly involves finding a suitable orientation of the graph representing the road network. Consider, first, the graph shown in Figure 5.7a. No matter how this graph is oriented, the resulting digraph will not be strongly connected, so traffic will not be able to flow freely through the system, certain locations not being accessible from certain others. On the other hand, the graph of Figure 5.7b has the strong orientation shown in Figure 5.7c (one, moreover, in which each vertex is reachable from each other vertex in at most two steps).

Clearly, a necessary condition for a graph to have a strong orientation is that it be free of cut edges. Robbins (1939) showed that this condition is also sufficient. The proof makes use of the following easy proposition (Exercise 5.3.9).

 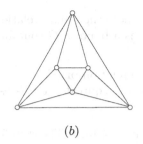 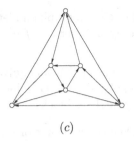

(a)   (b)   (c)

Fig. 5.7. (a) A graph with no strong orientation, and (b) a graph with (c) a strong orientation

Proposition 5.9 *A connected digraph is strongly connected if and only if each of its blocks is strongly connected.* □

Theorem 5.10 *Every connected graph without cut edges has a strong orientation.*

Proof Let G be a connected graph without cut edges. By Proposition 5.9, it suffices to show that each block B of G has a strong orientation. We may assume that $B \neq K_1$. Moreover, because G has no cut edges, $B \neq K_2$. Thus B contains a cycle and, by Theorem 5.8, has an ear decomposition $(G_0, G_1, \ldots, G_k)$. Consider the orientation of B obtained by orienting G_0 as a directed cycle and each ear as a directed path. It can be verified easily, by induction on i, that the resulting orientation of G_i is strong for each i, $0 \leq i \leq k$. In particular, the orientation assigned to $B = G_k$ is strong. □

Exercises

⋆**5.3.1** Deduce Proposition 5.7 from Theorem 5.2.

⋆**5.3.2** An edge e of a nonseparable graph G is called *deletable* if $G \setminus e$ is non-separable, and *contractible* if G / e is nonseparable. Show that every edge of a nonseparable graph is either deletable or contractible.

5.3.3 Show that if G has no even cycles, then each block of G is either an odd cycle or a copy of K_1 or K_2.

5.3.4 Let G be a nonseparable graph, and let x and y be two vertices of G. Show that there is a linear ordering $v_1, v_2, \ldots, v_n$ of the vertices of G such that $v_1 = x$, $v_n = y$, and each vertex v_j, $2 \leq j \leq n - 1$, is joined to some vertex v_i with $i < j$ and some vertex v_k with $k > j$.

5.3.5 Prove the following dual version of Theorem 5.2: *A connected graph is nonseparable if and only if any two of its edges are contained in a common bond.*

5.3.6 Let G be a graph and let $\overset{B}{\sim}$ denote the binary relation on E defined by $e \overset{B}{\sim} f$ if and only if either $e = f$ or there is a bond of G containing both e and f. Show that:

a) the relation $\overset{B}{\sim}$ is an equivalence relation on E,
b) the subgraphs of G induced by the equivalence classes under this relation are its nontrivial blocks.

5.3.7 Deduce the result of Exercise 5.1.4 from Theorem 5.8.

5.3.8 Let G be a nonseparable graph different from K_1 and K_2, and let $(G_0, G_1, \ldots, G_k)$ be an ear decomposition of G.

a) Show that $k = m - n$.
b) Suppose that $G_{i+1} = G_i \cup P_i$, where P_i is an ear of G_i in G_{i+1}, $0 \le i < k$. Set $C_0 := G_0$ and, for $1 \le i \le k$, let C_i be a cycle in G_i containing the ear P_{i-1}. Show that $(C_0, C_1, \ldots, C_k)$ is a basis for $\mathcal{C}(G)$, the cycle space of G.

★5.3.9 Prove Proposition 5.9.

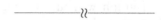

5.3.10 Let G be a nonseparable nonbipartite graph.

a) Show that the cycle space $\mathcal{C}(G)$ of G has a basis consisting of $m - n$ even cycles and one odd cycle.
b) Deduce that the dimension of the subspace of $\mathcal{C}(G)$ generated by the even cycles of G is $m - n$.   (M.A. HENNING AND C.H.C. LITTLE)

5.3.11 Call a family of subgraphs of a graph *linearly independent* if the incidence vectors of their edge sets are linearly independent over $GF(2)$. Let G be a nonseparable graph on at least two vertices.

a) If x and y are two vertices of G, show that there are $m - n + 2$ linearly independent xy-paths in G, and that this number is the greatest possible.
b) Let e be an edge of G. Deduce that the cycle space $\mathcal{C}(G)$ of G has a basis consisting entirely of cycles containing the edge e.
c) Deduce that G has at least $\binom{m-n+2}{2}$ cycles.
d) Which nonseparable graphs G have exactly $\binom{m-n+2}{2}$ cycles?

5.3.12 VINE
A *vine* on a path xPy in a graph G is a sequence $(x_i Q_i y_i : 1 \le i \le r)$ of internally disjoint ears of P in G such that:

$$x = x_1 \prec x_2 \prec y_1 \preceq x_3 \prec y_2 \preceq x_4 \prec \cdots \preceq x_r \prec y_{r-1} \prec y_r = y$$

where $\prec$ is the precedence relation on P (see Figure 5.8).
Let xPy be a path in a nonseparable graph G.

a) Show that there is vine $(x_i Q_i y_i : 1 \le i \le r)$ on P.

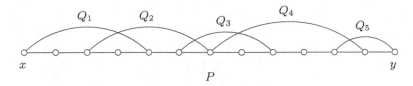

Fig. 5.8. A vine on a path

b) Set $P_i := x_i P y_i$ and $C_i := P_i \cup Q_i$, $1 \leq i \leq r$. Show that $C_{jk} := \triangle\{C_i : j \leq i \leq k\}$ is a cycle of G, $1 \leq j \leq k \leq r$.

c) Suppose that $r = 2t - 1$ is odd.

   i) Show that the t^2 cycles C_{jk}, $1 \leq j \leq t \leq k \leq r$, together cover the path P at least t times and each ear Q_i $\min\{i, 2t - i\}\, t$ times.

   ii) Deduce that if P has length ℓ, then one of these cycles has length at least $(\ell/t) + t$, and hence length at least $2\sqrt{\ell}$.   (G.A. DIRAC)

   iii) Perform a similar computation in the case where r is even.

5.4 Directed Ear Decompositions

There is an analogous theory of ear decompositions for nonseparable strong digraphs. Every strong digraph other than K_1 contains a directed cycle (Exercise 2.5.6). This is the starting point of the ear decomposition which we now describe.

Let F be a subdigraph of a digraph D. A *directed ear* of F in D is a directed path in D whose ends lie in F but whose internal vertices do not.

Proposition 5.11 *Let F be a nontrivial proper nonseparable strong subdigraph of a nonseparable strong digraph D. Then F has a directed ear in D.*

Proof Because D is nonseparable, F has an ear in D, by Proposition 5.6. Among all such ears, we choose one in which the number of reverse arcs (those directed towards its initial vertex) is as small as possible. We show that this path xPy is in fact a directed ear.

Assume the contrary, and let (u, v) be a reverse arc of P (see Figure 5.9a). Because D is strong, there exist in D a directed (F, u)-path Q and a directed (v, F)-path R (one of which might be of length zero). The tail of Q and the head of R must be one and the same vertex, for otherwise the directed walk $QuvR$ would contain a directed ear of F, contradicting the choice of P and our assumption that P is not a directed ear. Let this common vertex be z (see Figure 5.9b). We may suppose that $z \neq x$ (the case $z \neq y$ being analogous). Then the xz-walk $xPvRz$ contains an xz-path that contradicts the choice of P (see Figure 5.9c). Thus P is indeed a directed ear of F.   ☐

The proof of the following proposition is similar to the proof of Proposition 5.7, and is left to the reader (Exercise 5.4.1).

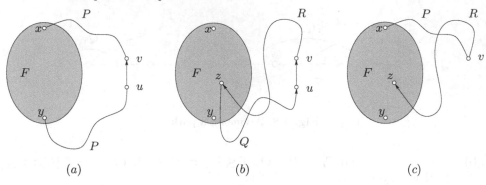

Fig. 5.9. Proof of Proposition 5.11: (a) The ear P of F, (b) the directed paths zQu and vRz, and (c) the xz-walk $xPvRz$

Proposition 5.12 *Let C be a strong subdigraph of a digraph D, and let P be a directed ear of C in D. Then $C \cup P$ is strong.*   □

A *directed ear decomposition* of a nonseparable strong digraph D is a nested sequence $(D_0, D_1, \ldots, D_k)$ of nonseparable strong subdigraphs of D such that:

▷ D_0 is a directed cycle,
▷ $D_{i+1} = D_i \cup P_i$, where P_i is a directed ear of D_i in D, $0 \le i < k$,
▷ $D_k = D$.

A directed ear decomposition of a strong digraph D is shown in Figure 5.10, the initial directed cycle and the directed ear added at each stage being indicated by heavy lines.

Propositions 5.11 and 5.12 imply the following theorem.

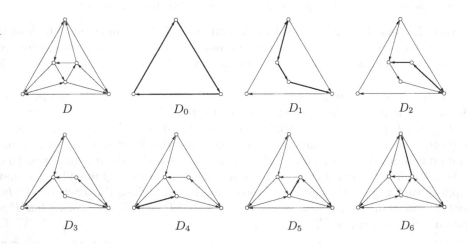

Fig. 5.10. A directed ear decomposition of a strong digraph

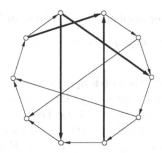

Fig. 5.11. A coherent feedback arc set of a digraph

Theorem 5.13 *Every nonseparable strong digraph other than K_1 has a directed ear decomposition.* □

Recall that a *feedback arc set* of a digraph D is a set S of arcs such that $D \setminus S$ is acyclic (see Exercise 2.5.8). Knuth (1974) proved that, when the digraph D is strongly connected, it has a feedback arc set with an important additional property.

Consider a minimal feedback arc set S of a digraph D. Because S is minimal, for any arc a in S the subdigraph $(D \setminus S) + a$ contains at least one directed cycle. Each such cycle includes the arc a, but no other arc of S. Let us call the directed cycles arising in this way the *fundamental cycles* of D with respect to S. We shall say that S is *coherent* if each arc of D belongs to some fundamental cycle. An example of a coherent feedback arc set is shown in Figure 5.11.

Observe that in order for a digraph to admit a coherent feedback arc set, every component of the digraph must be strong, because each arc should belong to a directed cycle. Knuth (1974) showed that, conversely, every strong digraph has a coherent feedback arc set. The proof makes use of directed ear decompositions.

Theorem 5.14 *Every strong digraph admits a coherent feedback arc set.*

Proof By induction on the number of arcs. Let D be a strong digraph. If D is a directed cycle, the statement is obviously true. If not, then by Theorem 5.13 there exists a proper strong subdigraph D' and a directed ear yPx of D' such that $D = D' \cup P$. By induction, D' has a coherent feedback arc set, and therefore a coherent feedback arc set S' such that $D' \setminus S'$ contains a spanning x-branching (Exercise 5.4.6). The set $S := S' \cup \{a\}$, where a is an arbitrary arc of P, is clearly a feedback arc set of D. Because $D' \setminus S'$ contains a spanning x-branching, there is a directed path xQy in $D' \setminus S'$. Observe that $yPxQy$ is a fundamental cycle with respect to S in D. Because S' is a coherent feedback arc set of D', it now follows that S is a coherent feedback arc set of D. □

Theorem 5.14 was discovered independently by Bessy and Thomassé (2004), and in Chapter 19 we shall see an interesting application of this theorem obtained by them.

We conclude with another application of Theorem 5.13.

Theorem 5.15 *Every strong digraph D has a strong spanning subdigraph with at most $2n - 2$ arcs.*

Proof We may assume that D has no loops, deleting them if necessary. If $D = K_1$, the assertion is trivial. If not, we apply Theorem 5.13 to each block B of D. Consider a directed ear decomposition of B. Delete from B the arcs in directed ears of length one, thereby obtaining a strong spanning subdigraph F of B and a directed ear decomposition $(D_0, D_1, \ldots, D_k)$ of F in which each ear P_i is of length at least two. Thus $k \leq v(F) - v(D_0) \leq v(F) - 2$. Since $e(D_0) = v(D_0)$ and $e(P_i) = v(P_i) - 1$, $1 \leq i \leq k$, we have:

$$e(F) = e(D_0) + \sum_{i=1}^{k} e(P_i) = v(D_0) + \sum_{i=1}^{k}(v(P_i) - 1) = v(F) + k \leq 2v(F) - 2$$

By Proposition 5.9, the union of the strong subdigraphs F (one in each block of D) is a strong spanning subdigraph of D. Because each of the subdigraphs F has at most $2v(F) - 2$ arcs, this spanning subdigraph of D has at most $2n - 2$ arcs. $\square$

Exercises

$\star$**5.4.1** Prove Proposition 5.12.

5.4.2 Which strong digraphs D have no strong spanning subdigraphs with fewer than $2n - 2$ arcs?

5.4.3 Let G be a strong digraph. Show that:

a) G has at least $m - n + 1$ directed cycles,
b) G contains a spanning tree each of whose fundamental cycles is a directed cycle if and only if G has exactly $m - n + 1$ directed cycles.

5.4.4 The *cycle space* of a digraph is the cycle space of its underlying graph. Show that the cycle space of a strong digraph has a basis consisting of directed cycles.

5.4.5 By considering the digraph of Figure 5.11, show that a minimal feedback arc set need not be coherent.

5.4.6 Let D be a strong digraph, and let x be a vertex of D. Suppose that D has a coherent feedback arc set S. Choose S so that the set X of vertices of D reachable from x in $D \setminus S$ is as large as possible.

a) Suppose that $X \neq V$, and set $T := (S \setminus \partial^+(X)) \cup \partial^-(X)$. Show that:
   i) T is a coherent feedback arc set of D,
   ii) the set of vertices of D reachable from x in $D \setminus T$ properly contains X.
b) Deduce that $D \setminus S$ contains a spanning x-branching.   (D.E. KNUTH)

5.5 Related Reading

EVEN CYCLE DECOMPOSITIONS

Veblen's Theorem (2.7) gives a necessary and sufficient condition for a graph to admit a cycle decomposition. If one would like all the constituent cycles of the decomposition to be of even length, not only must the graph be even, but each block must be of even size (number of edges). However, this requirement is still not sufficient: K_5 meets all these conditions, but admits no decomposition into even cycles. On the other hand, Seymour (1981a) showed that a nonseparable even graph of even size does admit an even cycle decomposition if it is planar. Extending the example of K_5, Rizzi (2001) described an infinite class of 4-connected even graphs of even size which do not have even cycle decompositions, and he conjectured that every simple 5-connected even graph of even size admits such a decomposition. (The notion of a k-connected graph is defined in Chapter 9.) For a survey on the topic, we refer the reader to Jackson (1993a), or to the books by Fleischner (1990, 1991).

MATROIDS AND NONSEPARABILITY

Although there is no matroidal analogue of a connected graph, the notion of non-separability extends naturally to matroids. Let M be a matroid on a set E. A partition of E into two nonempty subsets E_1 and E_2 is called a *separation* of M if every basis of M is the union of a basis of E_1 and a basis of E_2, where by a *basis* of E_i we mean a maximal independent subset of E_i. A matroid is *nonseparable* if it has no separation. Whitney (1935) showed that a matroid is nonseparable if and only if any two of its elements belong to a common circuit. This result, when applied to cycle matroids of graphs without isolated vertices, yields Theorem 5.2. Whitney also showed that a matroid is nonseparable if and only if its dual is non-separable. In particular, the cycle matroid of a graph is nonseparable if and only if its bond matroid is nonseparable. Thus, from the point of view of matroids, the statements in Exercises 5.2.3 and 5.3.6 are formally equivalent.

6

Tree-Search Algorithms

Contents

6.1 Tree-Search

We have seen that connectedness is a basic property of graphs. But how does one determine whether a graph is connected? In the case of small graphs, it is a routine matter to do so by inspection, searching for paths between all pairs of vertices. However, in large graphs, such an approach could be time-consuming because the number of paths to examine might be prohibitive. It is therefore desirable to have a systematic procedure, or *algorithm*, which is both efficient and applicable to all graphs. The following property of the trees of a graph provides the basis for such a procedure. For a subgraph F of a graph G, we simply write $\partial(F)$ for $\partial(V(F))$, and refer to this set as the *edge cut* associated with F.

Let T be a tree in a graph G. If $V(T) = V(G)$, then T is a spanning tree of G and we may conclude, by Theorem 4.6, that G is connected. But if $V(T) \subset V(G)$, two possibilities arise: either $\partial(T) = \emptyset$, in which case G is disconnected, or $\partial(T) \neq \emptyset$. In the latter case, for any edge $xy \in \partial(T)$, where $x \in V(T)$ and $y \in V(G) \setminus V(T)$,

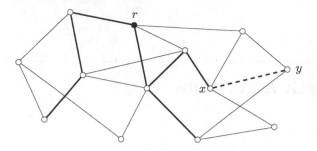

Fig. 6.1. Growing a tree in a graph

the subgraph of G obtained by adding the vertex y and the edge xy to T is again a tree in G (see Figure 6.1).

Using the above idea, one may generate a sequence of rooted trees in G, starting with the trivial tree consisting of a single root vertex r, and terminating either with a spanning tree of the graph or with a nonspanning tree whose associated edge cut is empty. (In practice, this involves scanning the adjacency lists of the vertices already in the tree, one by one, to determine which vertex and edge to add to the tree.) We refer to such a procedure as a *tree-search* and the resulting tree as a *search tree*.

If our objective is just to determine whether a graph is connected, any tree-search will do. In other words, the order in which the adjacency lists are considered is immaterial. However, tree-searches in which specific criteria are used to determine this order can provide additional information on the structure of the graph. For example, a tree-search known as *breadth-first search* may be used to find the distances in a graph, and another, *depth-first search*, to find the cut vertices of a graph.

The following terminology is useful in describing the properties of search trees. Recall that an *r-tree* is a tree with root r. Let T be such a tree. The *level* of a vertex v in T is the length of the path rTv. Each edge of T joins vertices on consecutive levels, and it is convenient to think of these edges as being oriented from the lower to the higher level, so as to form a branching. Several other terms customarily used in the study of rooted trees are borrowed from genealogy. For instance, each vertex on the path rTv, including the vertex v itself, is called an *ancestor* of v, and each vertex of which v is an ancestor is a *descendant* of v. An ancestor or descendant of a vertex is *proper* if it is not the vertex itself. Two vertices are *related* in T if one is an ancestor of the other. The immediate proper ancestor of a vertex v other than the root is its *predecessor* or *parent*, denoted $p(v)$, and the vertices whose predecessor is v are its *successors* or *children*. Note that the (oriented) edge set of a rooted tree $T := (V(T), E(T))$ is determined by its predecessor function p, and conversely

$$E(T) = \{(p(v), v) : v \in V(T) \setminus \{r\}\}$$

where r is the root of T. We often find it convenient to describe a rooted tree by specifying its vertex set and predecessor function.

For the sake of simplicity, we assume throughout this chapter that our graphs and digraphs are connected. This assumption results in no real loss of generality. We may suppose that the components have already been found by means of a tree-search. Each component may then be treated individually. We also assume that our graphs and digraphs are free of loops, which play an insignificant role here.

Breadth-First Search and Shortest Paths

In most types of tree-search, the criterion for selecting a vertex to be added to the tree depends on the order in which the vertices already in the tree T were added. A tree-search in which the adjacency lists of the vertices of T are considered on a first-come first-served basis, that is, in increasing order of their time of incorporation into T, is known as *breadth-first search*. In order to implement this algorithm efficiently, vertices in the tree are kept in a *queue*; this is just a list Q which is updated either by adding a new element to one end (the *tail* of Q) or removing an element from the other end (the *head* of Q). At any moment, the queue Q comprises all vertices from which the current tree could potentially be grown.

Initially, at time $t = 0$, the queue Q is empty. Whenever a new vertex is added to the tree, it joins Q. At each stage, the adjacency list of the vertex at the head of Q is scanned for a neighbour to add to the tree. If every neighbour is already in the tree, this vertex is removed from Q. The algorithm terminates when Q is once more empty. It returns not only the tree (given by its predecessor function p), but also a function $\ell : V \to \mathbb{N}$, which records the level of each vertex in the tree and, more importantly, their distances from r in G. It also returns a function $t : V \to \mathbb{N}$ which records the time of incorporation of each vertex into the tree T. We keep track of the vertices in T by colouring them black. The notation $G(x)$ signifies a graph G with a specified vertex (or *root*) x. Recall that an *x-tree* is a tree rooted at vertex x.

Algorithm 6.1 Breadth-First Search (BFS)

INPUT: a connected graph $G(r)$

OUTPUT: an r-tree T in G with predecessor function p, a level function ℓ such that $\ell(v) = d_G(r, v)$ for all $v \in V$, and a time function t

1: *set $i := 0$ and $Q := \emptyset$*
2: *increment i by 1*
3: *colour r black*
4: *set $\ell(r) := 0$ and $t(r) := i$*
5: *append r to Q*
6: **while** *Q is nonempty* **do**
7:    *consider the head x of Q*
8:    **if** *x has an uncoloured neighbour y* **then**
9:       *increment i by 1*

```
10:        colour y black
11:        set p(y) := x, ℓ(y) := ℓ(x) + 1 and t(y) := i
12:        append y to Q
13:    else
14:        remove x from Q
15:    end if
16: end while
17: return (p, ℓ, t)
```

The spanning tree T returned by BFS is called a *breadth-first search tree*, or *BFS-tree*, of G. An example of a BFS-tree in a connected graph is shown in Figure 6.2. The labels of the vertices in Figure 6.2a indicate the times at which they were added to the tree. The distance function ℓ is shown in Figure 6.2b. The evolution of the queue Q is as follows, the vertices being indicated by their times.

$$\emptyset \to 1 \to 1\,2 \to 1\,2\,3 \to 1\,2\,3\,4 \to 1\,2\,3\,4\,5 \to 2\,3\,4\,5 \to 2\,3\,4\,5\,6$$
$$\to 2\,3\,4\,5\,6\,7 \to 3\,4\,5\,6\,7 \to 3\,4\,5\,6\,7\,8 \to 3\,4\,5\,6\,7\,8\,9 \to 4\,5\,6\,7\,8\,9$$
$$\to 4\,5\,6\,7\,8\,9\,10 \to 5\,6\,7\,8\,9\,10 \to 5\,6\,7\,8\,9\,10\,11 \to 6\,7\,8\,9\,10\,11$$
$$\to 6\,7\,8\,9\,10\,11\,12 \to 7\,8\,9\,10\,11\,12 \to 8\,9\,10\,11\,12 \to 9\,10\,11\,12$$
$$\to 9\,10\,11\,12\,13 \to 10\,11\,12\,13 \to 11\,12\,13 \to 12\,13 \to 13 \to \emptyset$$

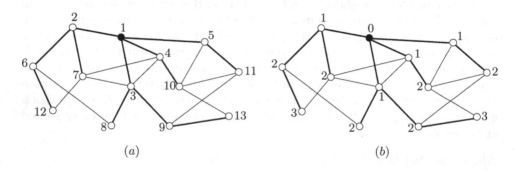

Fig. 6.2. A breadth-first search tree in a connected graph: (a) the time function t, and (b) the level function ℓ

BFS-trees have two basic properties, the first of which justifies our referring to ℓ as a level function.

Theorem 6.2 *Let T be a BFS-tree of a connected graph G, with root r. Then:*

a) for every vertex v of G, $\ell(v) = d_T(r, v)$, the level of v in T,
b) every edge of G joins vertices on the same or consecutive levels of T; that is,

$$|\ell(u) - \ell(v)| \leq 1, \quad \text{for all } uv \in E$$

Proof The proof of (a) is left to the reader (Exercise 6.1.1). To establish (b), it suffices to prove that if $uv \in E$ and $\ell(u) < \ell(v)$, then $\ell(u) = \ell(v) - 1$.

We first establish, by induction on $\ell(u)$, that if u and v are any two vertices such that $\ell(u) < \ell(v)$, then u joined Q before v. This is evident if $\ell(u) = 0$, because u is then the root of T. Suppose that the assertion is true whenever $\ell(u) < k$, and consider the case $\ell(u) = k$, where $k > 0$. Setting $x := p(u)$ and $y := p(v)$, it follows from line 11 of BFS (Algorithm 6.1) that $\ell(x) = \ell(u) - 1 < \ell(v) - 1 = \ell(y)$. By induction, x joined Q before y. Therefore u, being a neighbour of x, joined Q before v.

Now suppose that $uv \in E$ and $\ell(u) < \ell(v)$. If $u = p(v)$, then $\ell(u) = \ell(v) - 1$, again by line 11 of the algorithm. If not, set $y := p(v)$. Because v was added to T by the edge yv, and not by the edge uv, the vertex y joined Q before u, hence $\ell(y) \le \ell(u)$ by the claim established above. Therefore $\ell(v) - 1 = \ell(y) \le \ell(u) \le \ell(v) - 1$, which implies that $\ell(u) = \ell(v) - 1$. □

The following theorem shows that BFS runs correctly.

Theorem 6.3 *Let G be a connected graph. Then the values of the level function ℓ returned by BFS are the distances in G from the root r:*

$$\ell(v) = d_G(r, v), \quad \text{for all } v \in V$$

Proof By Theorem 6.2a, $\ell(v) = d_T(r, v)$. Moreover, $d_T(r, v) \ge d_G(r, v)$ because T is a subgraph of G. Thus $\ell(v) \ge d_G(r, v)$. We establish the opposite inequality by induction on the length of a shortest (r, v)-path.

Let P be a shortest (r, v)-path in G, where $v \ne r$, and let u be the predecessor of v on P. Then rPu is a shortest (r, u)-path, and $d_G(r, u) = d_G(r, v) - 1$. By induction, $\ell(u) \le d_G(r, u)$, and by Theorem 6.2b, $\ell(v) - \ell(u) \le 1$. Therefore

$$\ell(v) \le \ell(u) + 1 \le d_G(r, u) + 1 = d_G(r, v) \qquad □$$

Alternative proofs of Theorems 6.2 and 6.3 are outlined in Exercise 6.1.2.

DEPTH-FIRST SEARCH

Depth-first search is a tree-search in which the vertex added to the tree T at each stage is one which is a neighbour of as recent an addition to T as possible. In other words, we first scan the adjacency list of the most recently added vertex x for a neighbour not in T. If there is such a neighbour, we add it to T. If not, we backtrack to the vertex which was added to T just before x and examine its neighbours, and so on. The resulting spanning tree is called a *depth-first search tree* or *DFS-tree*.

This algorithm may be implemented efficiently by maintaining the vertices of T whose adjacency lists have yet to be fully scanned, not in a queue as we did for breadth-first search, but in a stack. A *stack* is simply a list, one end of which is identified as its *top*; it may be updated either by adding a new element as its top or else by removing its top element. In depth-first search, the stack S is initially empty. Whenever a new vertex is added to the tree T, it is added to S. At each stage, the adjacency list of the top vertex is scanned for a neighbour to add to T. If all of its neighbours are found to be already in T, this vertex is removed from S. The algorithm terminates when S is once again empty. As in breadth-first search, we keep track of the vertices in T by colouring them black.

Associated with each vertex v of G are two times: the time $f(v)$ when v is incorporated into T (that is, added to the stack S), and the time $l(v)$ when all the neighbours of v are found to be already in T, the vertex v is removed from S, and the algorithm backtracks to $p(v)$, the predecessor of v in T. (The time function $l(v)$ is not to be confused with the level function $\ell(v)$ of BFS.) The time increments by one with each change in the stack S. In particular, $f(r) = 1$, $l(v) = f(v) + 1$ for every leaf v of T, and $l(r) = 2n$.

Algorithm 6.4 DEPTH-FIRST SEARCH (DFS)

INPUT: a connected graph G

OUTPUT: a rooted spanning tree of G with predecessor function p, and two time functions f and l

```
 1: set i := 0 and S := ∅
 2: choose any vertex r (as root)
 3: increment i by 1
 4: colour r black
 5: set f(r) := i
 6: add r to S
 7: while S is nonempty do
 8:     consider the top vertex x of S
 9:     increment i by 1
10:     if x has an uncoloured neighbour y then
11:         colour y black
12:         set p(y) := x and f(y) := i
13:         add y to the top of S
14:     else
15:         set l(x) := i
16:         remove x from S
17:     end if
18: end while
19: return (p, f, l)
```

A DFS-tree of a connected graph is shown in Figure 6.3; the tree is indicated by solid lines and each vertex v of the tree is labelled by the pair $(f(v), l(v))$. The evolution of the stack S is as follows, the vertices being indicated by their times of incorporation into T.

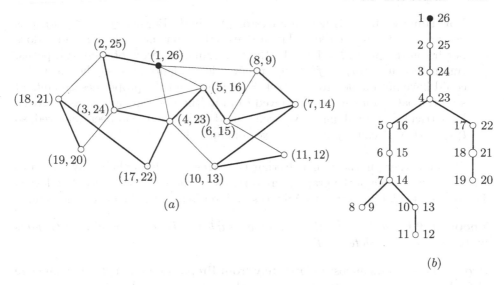

Fig. 6.3. (a) A depth-first search tree of a connected graph, and (b) another drawing of this tree

$$\emptyset \to 1 \to 12 \to 123 \to 1234 \to 12345 \to 123456 \to 1234567$$
$$\to 12345678 \to 1234567 \to 123456710 \to 123456710\,11$$
$$\to 123456710 \to 1234567 \to 123456 \to 12345 \to 1234 \to 123417$$
$$\to 12341718 \to 1234171819 \to 12341718 \to 123417 \to 1234 \to 123$$
$$\to 12 \to 1 \to \emptyset$$

The following proposition provides a link between the input graph G, its DFS-tree T, and the two time functions f and l returned by DFS.

Proposition 6.5 *Let u and v be two vertices of G, with $f(u) < f(v)$.*

a) If u and v are adjacent in G, then $l(v) < l(u)$.
b) u is an ancestor of v in T if and only if $l(v) < l(u)$.

Proof

a) According to lines 8–12 of DFS, the vertex u is removed from the stack S only after all potential children (uncoloured neighbours) have been considered for addition to S. One of these neighbours is v, because $f(u) < f(v)$. Thus v is added to the stack S while u is still in S, and u cannot be removed from S before v is removed. It follows that $l(v) < l(u)$.

b) Suppose that u is an ancestor of v in T. By lines 9 and 12 of DFS, the values of f increase along the path uTv. Applying (a) to each edge of this path yields the inequality $l(v) < l(u)$.

Now suppose that u is not an ancestor of v in T. Because $f(u) < f(v)$, v is not an ancestor of u either. Thus u does not lie on the path rTv and v does not lie on the path rTu. Let s be the last common vertex of these two paths. Again, because $f(u) < f(v)$, the proper descendants of s on the path rTv could have been added to the stack S only after all the proper descendants of s on the path rTu had been removed from it (thereby leaving s as top vertex). In particular, v could only have been added to S after u had been removed, so $l(u) < f(v)$. Because $f(v) < l(v)$, we conclude that $l(u) < l(v)$. $\square$

We saw earlier (in Theorem 6.2b) that BFS-trees are characterized by the property that every edge of the graph joins vertices on the same or consecutive levels. The quintessential property of DFS-trees is described in the following theorem.

Theorem 6.6 *Let T be a DFS-tree of a graph G. Then every edge of G joins vertices which are related in T.*

Proof This follows almost immediately from Proposition 6.5. Let uv be an edge of G. Without loss of generality, suppose that $f(u) < f(v)$. By Proposition 6.5a, $l(v) < l(u)$. Now Proposition 6.5b implies that u is an ancestor of v, so u and v are related in T. $\square$

Finding the Cut Vertices and Blocks of a Graph

In a graph which represents a communications network, the cut vertices of the graph correspond to centres whose breakdown would disrupt communications. It is thus important to identify these sites, so that precautions may be taken to reduce the vulnerability of the network. Tarjan (1972) showed how this problem can be solved efficiently by means of depth-first search.

While performing a depth-first search of a graph G, it is convenient to orient the edges of G with respect to the DFS-tree T. We orient each tree edge from parent to child, and each nontree edge (whose ends are related in T, by Theorem 6.6) from descendant to ancestor. The latter edges are called *back edges*. The following characterization of cut vertices is an immediate consequence of Theorem 6.6.

Theorem 6.7 *Let T be a DFS-tree of a connected graph G. The root of T is a cut vertex of G if and only if it has at least two children. Any other vertex of T is a cut vertex of G if and only if it has a child no descendant of which dominates (by a back edge) a proper ancestor of the vertex.* $\square$

Let us see how depth-first search may be used to find the cut vertices and blocks of a (connected) graph in linear time; that is, in time proportional to the number of edges of the graph.

Let T be a DFS-tree of a connected graph G, and let B be a block of G. Then $T \cap B$ is a tree in G (Exercise 5.2.8b). Moreover, because T is a rooted tree, we may associate with B a unique vertex, the root of the tree $T \cap B$. We call this vertex the *root* of B with respect to T. It is the first vertex of B to be incorporated into T. Note that the cut vertices of G are just the roots of blocks (with the exception of r, if it happens to be the root of a single block). Thus, in order to determine the cut vertices and blocks of G, it suffices to identify these roots. It turns out that one can do so during the execution of depth-first search.

To this end, we consider the function $f^* : V \to \mathbb{N}$ defined as follows. If some proper ancestor of v can be reached from v by means of a directed path consisting of tree edges (possibly none) followed by one back edge, $f^*(v)$ is defined to be the least f-value of such an ancestor; if not, we set $f^*(v) := f(v)$. Observe, now, that a vertex v is the root of a block if and only if it has a child w such that $f^*(w) \geq f(v)$.

The function f^* can be computed while executing depth-first search (see Exercise 6.1.12), and the criterion for roots of blocks may be checked at the same time. Thus the roots of the blocks of G, as well as the blocks themselves, can be determined in linear time.

The roots of the blocks of a graph with respect to a DFS-tree are shown in Figure 6.4. The pair $(f(v), l(v))$ is given for each vertex v. We leave it to the reader to orient the edges of G as described above, and to compute the function f^*.

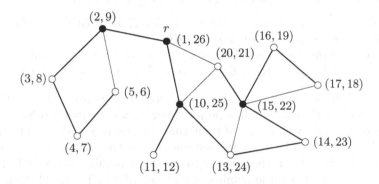

Fig. 6.4. Finding the cut vertices and blocks of a graph by depth-first search

Exercises

⋆**6.1.1** Let T be a BFS-tree of a connected graph G. Show that $\ell(v) = d_T(r, v)$, for all $v \in V$.

6.1.2

a) Let T be a BFS-tree of a connected graph G and let z denote the last vertex to enter T. Show that $T - z$ is a BFS-tree of $G - z$.

b) Using (a), give inductive proofs of Theorems 6.2 and 6.3.

6.1.3 Refine Algorithm 6.1 (breadth-first search) so that it returns either a bipartition of the graph (if the graph is bipartite) or an odd cycle (if it is not).

6.1.4 Describe an algorithm based on breadth-first search for finding a shortest odd cycle in a graph.

6.1.5 Let G be a Moore graph (defined in Exercise 3.1.12). Show that all BFS-trees of G are isomorphic.

6.1.6 Let T be a DFS-tree of a nontrivial connected simple graph G, and let v be the root of a block B of G. Show that the degree of v in $T \cap B$ is one.

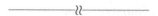

6.1.7 Let G be a connected graph, let x be a vertex of G, and let T be a spanning tree of G which maximizes the function $\sum\{d_T(x,v) : v \in V\}$. Show that T is a DFS-tree of G. (Zs. Tuza)

6.1.8 Solve Exercise 2.2.19 by considering a DFS-tree in G.

6.1.9 For a connected graph G, define $\sigma(G) := \sum\{d(u,v) : u,v \in V\}$.

a) Let G be a connected graph. For $v \in V$, let T_v be a BFS-tree of G rooted at v. Show that $\sum_{v \in V} \sigma(T_v) \le 2(n-1)\sigma(G)$.

b) Deduce that every connected graph G has a spanning tree T such that $\sigma(T) \le 2(1 - \frac{1}{n})\sigma(G)$. (R.C. Entringer, D.J. Kleitman and L. Székely)

6.1.10 Let T be a rooted tree. Two breadth-first searches of T (starting at its root) are distinct if their time functions t differ. Likewise, two depth-first searches of T are distinct if at least one of their time functions f and l differ. Show that the number of distinct breadth-first searches of the tree T is equal to the number of distinct depth-first searches of T, and that this number is precisely $\prod\{n(v)! : v \in V(T)\}$, where $n(v)$ is the number of children of v in T (and $0! = 1$).

$\star$**6.1.11** Let G be a connected graph in which every DFS-tree is a Hamilton path (rooted at one end). Show that G is a cycle, a complete graph, or a complete bipartite graph in which both parts have the same number of vertices.

(G. Chartrand and H.V. Kronk)

$\star$**6.1.12**

a) Let G be a connected graph and T a DFS-tree of G, where the edges of T are oriented from parent to child, and the back edges from descendant to ancestor. For $v \in V$, set:

$$g(v) := \min\{f(w) : (v,w) \in E(G) \setminus E(T)\}$$
$$h(v) := \min\{f^*(w) : (v,w) \in E(T)\}$$

Show that:

i) the function f^* may be computed recursively by the formula

$$f^*(v) = \min\{f(v), g(v), h(v)\}$$

ii) a nonroot vertex v of T is a cut vertex of G if and only if $f(v) \leq h(v)$.

b) Refine Algorithm 6.4 (Depth-First Search) so that it returns the cut vertices and the blocks of a connected graph.   (R.E. TARJAN)

6.1.13 Let G be a simple connected graph, and let $w : V \to \mathbb{Z}$ be a weight function on V such that $\sum_{v \in V} w(v) \geq m - n + 1$. For $X \subset V$, the *move* M_X consists of distributing a unit weight from each vertex of X to each of its neighbours in $V \setminus X$ (so that the weight of a vertex v of $V \setminus X$ increases by $d_X(v)$).

a) Show that the weight can be made nonnegative at each vertex by means of a sequence of moves.

b) Show that this is no longer necessarily true if $\sum_{v \in V} w(v) \leq m - n$.
                                (M. BAKER AND S. NORINE)

6.2 Minimum-Weight Spanning Trees

An electric grid is to be set up in China, linking the cities of Beijing, Chongqing, Guangdong, Nanjing, Shanghai, Tianjin, and Wuhan to the Three Gorges generating station situated at Yichang. The locations of these cities and the distances (in kilometres) between them are given in Figure 6.5. How should the grid be constructed so that the total connection distance is as small as possible?

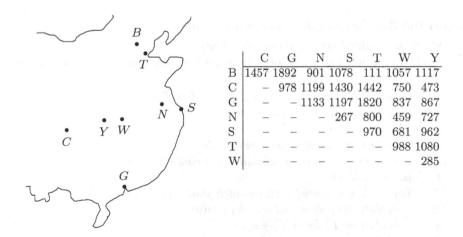

	C	G	N	S	T	W	Y
B	1457	1892	901	1078	111	1057	1117
C	–	978	1199	1430	1442	750	473
G	–	–	1133	1197	1820	837	867
N	–	–	–	267	800	459	727
S	–	–	–	–	970	681	962
T	–	–	–	–	–	988	1080
W	–	–	–	–	–	–	285

Fig. 6.5. The China hydro-electric grid problem

The table in Figure 6.5 determines a weighted complete graph with vertices B, C, G, N, S, T, W, and Y. Our problem amounts to finding, in this graph, a connected spanning subgraph of minimum weight. Because the weights are positive, this subgraph will be a spanning tree.

More generally, we may consider the following problem.

Problem 6.8 MINIMUM-WEIGHT SPANNING TREE
  GIVEN: *a weighted connected graph G,*
  FIND: *a minimum-weight spanning tree T in G.*

For convenience, we refer to a minimum-weight spanning tree as an *optimal tree*.

THE JARNÍK–PRIM ALGORITHM

The Minimum-Weight Spanning Tree Problem (6.8) can be solved by means of a tree-search due to Jarník (1930) and Prim (1957). In this algorithm, which we call the *Jarník–Prim Algorithm*, an arbitrary vertex r is selected as the root of T, and at each stage the edge added to the current tree T is any edge of least weight in the edge cut associated with T.

As in breadth-first and depth-first search, the vertices of T are coloured black. Also, in order to implement the above tree-search efficiently, each uncoloured vertex v is assigned a provisional cost $c(v)$. This is the least weight of an edge linking v to some black vertex u, if there is such an edge, in which case we assign u as provisional predecessor of v, denoted $p(v)$. Initially, each vertex has infinite cost and no predecessor. These two provisional labels are updated at each stage of the algorithm.

Algorithm 6.9 THE JARNÍK–PRIM ALGORITHM
  INPUT: a weighted connected graph (G, w)
  OUTPUT: an optimal tree T of G with predecessor function p, and its weight $w(T)$
    1: *set $p(v) := \emptyset$ and $c(v) := \infty$, $v \in V$, and $w(T) := 0$*
    2: *choose any vertex r (as root)*
    3: *replace $c(r)$ by 0*
    4: **while** *there is an uncoloured vertex* **do**
    5:     *choose such a vertex u of minimum cost $c(u)$*
    6:     *colour u black*
    7:     **for** *each uncoloured vertex v such that $w(uv) < c(v)$* **do**
    8:       *replace $p(v)$ by u and $c(v)$ by $w(uv)$*
    9:       *replace $w(T)$ by $w(T) + c(u)$*
    10:     **end for**
    11: **end while**
    12: *return $(p, w(T))$*

In practice, the set of uncoloured vertices and their costs are kept in a structure called a *priority queue*. Although this is not strictly a queue as defined earlier, the vertex of minimum cost is always located at the head of the queue (hence the 'priority') and can therefore be accessed immediately. Furthermore, the 'queue' is structured so that it can be updated rather quickly when this vertex is removed (coloured black), or when the costs are modified (as in line 9 of the Jarník–Prim Algorithm). As to how this can be achieved is outlined in Section 6.4.

We call a rooted spanning tree output by the Jarník–Prim Algorithm a *Jarník–Prim tree*. The construction of such a tree in the electric grid graph is illustrated (not to scale) in Figure 6.6, the edges being numbered according to the order in which they are added.

In step 1, Yichang (Y) is chosen as the root. No vertex has yet been coloured. Because $c(Y) = 0$, and $c(v) = \infty$ for every other vertex v, vertex Y is chosen as u in step 2, and coloured black. All uncoloured vertices are assigned Y as predecessor, and their costs are reduced to:

$$c(B) = 1117, \quad c(C) = 473, \quad c(G) = 867$$

$$c(N) = 727, \quad c(S) = 962, \quad c(T) = 1080, \quad c(W) = 285$$

The weight of the tree T remains zero.

In the second iteration of step 2, W is selected as the vertex u and coloured black. The predecessors of the uncoloured vertices, and their costs, become:

$$p(B) = W, \quad p(C) = Y, \quad p(G) = W, \quad p(N) = W, \quad p(S) = W, \quad p(T) = W$$

$$c(B) = 1057, \quad c(C) = 473, \quad c(G) = 837, \quad c(N) = 459, \quad c(S) = 681, \quad c(T) = 988$$

and $w(T)$ is increased to 285.

In the third iteration of step 2, N is selected as the vertex u and coloured black. The predecessors of the uncoloured vertices, and their costs, become:

$$p(B) = N, \quad p(C) = Y, \quad p(G) = W, \quad p(S) = N, \quad p(T) = N$$

$$c(B) = 901, \quad c(C) = 473, \quad c(G) = 837, \quad c(S) = 267, \quad c(T) = 800$$

and $w(T)$ is increased to $285 + 459 = 744$.

This procedure continues until all the vertices are coloured black. The total length of the grid thereby constructed is 3232 kilometres.

The following theorem shows that the algorithm runs correctly.

Theorem 6.10 *Every Jarník–Prim tree is an optimal tree.*

Proof Let T be a Jarník–Prim tree with root r. We prove, by induction on $v(T)$, that T is an optimal tree. The first edge added to T is an edge e of least weight in the edge cut associated with $\{r\}$; in other words, $w(e) \leq w(f)$ for all edges f incident with r. To begin with, we show that some optimal tree includes this edge e. Let T^* be an optimal tree. We may assume that $e \notin E(T^*)$. Thus $T^* + e$

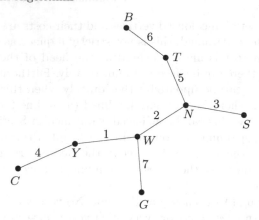

Fig. 6.6. An optimal tree returned by the Jarník–Prim Algorithm

contains a unique cycle C. Let f be the other edge of C incident with r. Then $T^{**} := (T^* + e) \setminus f$ is a spanning tree of G. Moreover, because $w(e) \le w(f)$,

$$w(T^{**}) = w(T^*) + w(e) - w(f) \le w(T^*)$$

As T^* is an optimal tree, equality must hold, so T^{**} is also an optimal tree. Moreover, T^{**} contains e.

Now consider the graph $G' := G / e$, and denote by r' the vertex resulting from the contraction of e. There is a one-to-one correspondence between the set of spanning trees of G that contain e and the set of all spanning trees of G' (Exercise 4.2.1a). Thus, to show that the final tree T is an optimal tree of G, it suffices to show that $T' := T / e$ is an optimal tree of G'. We claim that T' is a Jarník–Prim tree of G' rooted at r'.

Consider the *current* tree T at some stage of the Jarník–Prim Algorithm. We assume that T is not simply the root vertex r, and thus includes the edge e. Let $T' := T / e$. Then $\partial(T) = \partial(T')$, so an edge of minimum weight in $\partial(T)$ is also an edge of minimum weight in $\partial(T')$. Because the *final* tree T is a Jarník–Prim tree of G, we deduce that the final tree T' is a Jarník–Prim tree of G'. As G' has fewer vertices than G, it follows by induction that T' is an optimal tree of G'. We conclude that T is an optimal tree of G. $\qquad\square$

The history of the Jarník–Prim Algorithm is described by Korte and Nešetřil (2001). A second algorithm for solving Problem 6.8, based on another approach, is presented in Section 8.5.

Exercises

★**6.2.1** Let (G, w) be a weighted connected graph whose edges have distinct weights. Show that G has a unique optimal tree.

6.2.2 Let (G, w) be a weighted connected graph. Show that a spanning tree T of G is optimal if and only if, for each edge $e \in E \setminus T$ and each edge $f \in C_e$ (the fundamental cycle of G with respect to T), $w(e) \geq w(f)$.

6.2.3 Let (G, w) be a weighted connected graph. Show that a spanning tree T of G is optimal if and only if, for each edge $e \in T$ and each edge $f \in B_e$ (the fundamental bond of G with respect to T), $w(e) \leq w(f)$.

6.2.4 Let T be an optimal spanning tree in a weighted connected graph (G, w) (with positive weights), and let x and y be two adjacent vertices of T. Show that the path $xTy = xy$ is an xy-path of minimum weight in G.

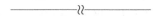

6.2.5 Let (G, w) be a weighted connected graph (with positive weights). Describe an algorithm for finding a spanning tree the product of whose weights is minimum.

6.2.6 Let T be an optimal spanning tree in a weighted connected graph (G, w).

a) Show that T is a spanning tree whose largest edge-weight is minimum.
b) Give an example of a weighted connected graph (G, w) and a spanning tree T of G whose largest edge-weight is minimum, but which is not an optimal spanning tree of G.

6.3 Branching-Search

One can explore directed graphs in much the same way as undirected graphs, but by growing branchings rather than rooted trees. Starting with the branching consisting of a single vertex r, its *root*, one adds one arc at a time, together with its head, the arc being selected from the outcut associated with the current branching. The procedure terminates either with a spanning branching of the digraph or with a nonspanning branching whose associated outcut is empty. Note that the latter outcome may well arise even if the digraph is connected. Indeed, the vertex set of the final branching is precisely the set of vertices of the digraph that are reachable by directed paths from r. We call the above procedure *branching-search*.

As with tree-search, branching-search may be refined by restricting the choice of the arc to be added at each stage. In this way, we obtain directed versions of breadth-first search and depth-first search. We discuss two important applications of branching-search. The first is an extension of directed BFS to weighted directed graphs, the second an application of directed DFS.

FINDING SHORTEST PATHS IN WEIGHTED DIGRAPHS

We have seen how breadth-first search can be used to determine shortest paths in graphs. In practice, one is usually faced with problems of a more complex nature. Given a one-way road system in a city, for instance, one might wish to determine a shortest route between two specified locations in the city. This amounts to finding a directed path of minimum weight connecting two specified vertices in the weighted directed graph whose vertices are the road junctions and whose arcs are the roads linking these junctions.

Problem 6.11   SHORTEST PATH
  GIVEN: *a weighted directed graph* (D, w) *with two specified vertices* x *and* y,
  FIND: *a minimum-weight directed* (x, y)-*path in* D.

For clarity of exposition, we refer to the weight of a directed path in a weighted digraph as its *length*. In the same vein, by a *shortest* directed (x, y)-path we mean one of minimum weight, and this weight is the *distance* from x to y, denoted $d(x, y)$. For example, the path indicated in the graph of Figure 6.7 is a shortest directed (x, y)-path (Exercise 6.3.1) and $d(x, y) = 3 + 1 + 2 + 1 + 2 + 1 + 2 + 4 = 16$. When all the weights are equal to one, these definitions coincide with the usual notions of length and distance.

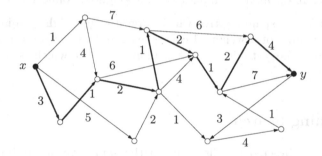

Fig. 6.7. A shortest directed (x, y)-path in a weighted digraph

It clearly suffices to deal with the shortest path problem for strict digraphs, so we assume that this is the case here. We also assume that all weights are positive. Arcs of weight zero can always be contracted. However, the presence of negative weights could well lead to complications. If the digraph should contain directed cycles of negative weight, there might exist (x, y)-walks which are shorter than any (x, y)-path — indeed, ones of arbitrarily small (negative) length — and this eventuality renders shortest path algorithms based on branching-search, such as the one described below, totally ineffective (see Exercise 6.3.3). On the other hand, when all weights are positive, the shortest path problem can be solved efficiently by means of a branching-search due to Dijkstra (1959).

Although similar in spirit to directed breadth-first search, *Dijkstra's Algorithm* bears a resemblance to the Jarník–Prim Algorithm in that provisional labels are assigned to vertices. At each stage, every vertex v of the current branching B is labelled by its predecessor in B, $p(v)$, and its distance from r in B, $\ell(v) := d_B(r, v)$. In addition, each vertex v which is not in B but is an outneighbour of some vertex in B, is labelled with a provisional predecessor $p(v)$ and a provisional distance $\ell(v)$, namely, the length of a shortest directed (r, v)-path in D all of whose internal vertices belong to B. The rule for selecting the next vertex and edge to add to the branching depends only on these provisional distances.

Algorithm 6.12 DIJKSTRA'S ALGORITHM

INPUT: a positively weighted digraph (D, w) with a specified vertex r

OUTPUT: an r-branching in D with predecessor function p, and a function $\ell : V \to \mathbb{R}^+$ such that $\ell(v) = d_D(r, v)$ for all $v \in V$

1: *set $p(v) := \emptyset$, $v \in V$, $\ell(r) := 0$, and $\ell(v) := \infty$, $v \in V \setminus \{r\}$*
2: **while** *there is an uncoloured vertex u with $\ell(u) < \infty$* **do**
3:     *choose such a vertex u for which $\ell(u)$ is minimum*
4:     *colour u black*
5:     **for** *each uncoloured outneighbour v of u with $\ell(v) > \ell(u) + w(u, v)$* **do**
6:         *replace $p(v)$ by u and $\ell(v)$ by $\ell(u) + w(u, v)$*
7:     **end for**
8: **end while**
9: *return (p, ℓ)*

Dijkstra's Algorithm, like the Jarník–Prim Algorithm, may be implemented by maintaining the uncoloured vertices and their distances in a priority queue. We leave it to the reader to verify that the algorithm runs correctly (Exercise 6.3.2).

DIRECTED DEPTH-FIRST SEARCH

Directed BFS (the unweighted version of Dijkstra's Algorithm) is a straightforward analogue of BFS; the labelling procedure is identical, and the branching-search terminates once all vertices reachable from the root have been found. Directed DFS, on the other hand, involves a slight twist: whenever the branching-search comes to a halt, an uncoloured vertex is selected and the search is continued afresh with this vertex as root. The end result is a spanning branching forest of the digraph, which we call a *DFS-branching forest*.

Algorithm 6.13 DIRECTED DEPTH-FIRST SEARCH (DIRECTED DFS)

INPUT: a digraph D

OUTPUT: a spanning branching forest of D with predecessor function p, and two time functions f and l

1: *set $i := 0$ and $S := \emptyset$*
2: **while** *there is an uncoloured vertex* **do**
3:     *choose any uncoloured vertex r (as root)*
4:     *increment i by 1*
5:     *colour r black*
6:     *set $f(r) := i$*
7:     *add r to S*
8:     **while** *S is nonempty* **do**
9:         *consider the top vertex x of S*

```
10:       increment i by 1
11:       if x has an uncoloured outneighbour y then
12:           colour y black
13:           set p(y) := x and f(y) := i
14:           add y as the top vertex of S
15:       else
16:           set l(x) := i
17:           remove x from S
18:       end if
19:   end while
20: end while
21: return (p, f, l)
```

Directed DFS has many applications. One is described below, and several others are outlined in exercises (6.3.6, 6.3.7, 6.3.8, 6.3.13). In these applications, it is convenient to distinguish three types of arcs of D, apart from those in the DFS-branching-forest F.

An arc $(u, v) \in A(D) \setminus A(F)$ is a *forward arc* if u is an ancestor of v in F, a *back arc* if u is a descendant of v in F, and a *cross arc* if u and v are unrelated in F and u was discovered after v. In terms of the time functions f and l:

▷ (u, v) is a *forward arc* if $f(u) < f(v)$ and $l(v) < l(u)$,
▷ (u, v) is a *back arc* if $f(v) < f(u)$ and $l(u) < l(v)$,
▷ (u, v) is a *cross arc* if $l(v) < f(u)$.

The directed analogue of Theorem 6.6, whose proof is left as an exercise (6.3.4), says that these arcs partition $A(D) \setminus A(F)$.

Theorem 6.14 *Let F be a DFS-branching forest of a digraph D. Then each arc of $A(D) \setminus A(F)$ is a forward arc, a back arc, or a cross arc.*   □

Finding the Strong Components of a Digraph

The strong components of a digraph can be found in linear time by using directed DFS. The basic idea is similar to the one employed for finding the blocks of an undirected graph, but is slightly more complicated.

The following proposition shows how the vertices of the strong components of D are disposed in F. Observe that forward arcs play no role with respect to reachability in D because any forward arc can be replaced by the directed path in F connecting its ends. We may therefore assume that there are no such arcs in D.

Proposition 6.15 *Let D be a directed graph, C a strong component of D, and F a DFS-branching forest in D. Then $F \cap C$ is a branching.*

Proof Each component of $F \cap C$ is contained in F, and thus is a branching. Furthermore, vertices of C which are related in F necessarily belong to the same component of $F \cap C$, because the directed path in F connecting them is contained in C also (Exercise 3.4.3).

Suppose that $F \cap C$ has two distinct components, with roots x and y. As remarked above, x and y are not related in F. We may suppose that $f(x) < f(y)$. Because x and y belong to the same strong component C of D, there is a directed (x, y)-path P in C, and because $f(x) < f(y)$, there must be an arc (u, v) of P with $f(u) < f(y)$ and $f(v) \geq f(y)$. This arc can be neither a cross arc nor a back arc, since $f(u) < f(v)$. It must therefore be an arc of F, because we have assumed that there are no forward arcs. Therefore $l(v) < l(u)$. If u and y were unrelated, we would have $l(u) < f(y)$. But this would imply that $f(v) < l(v) < l(u) < f(y)$, contradicting the fact that $f(v) \geq f(y)$. We conclude that u is a proper ancestor of y, and belongs to the same component of $F \cap C$ as y. But this contradicts our assumption that y is the root of this component. $\square$

By virtue of Proposition 6.15, we may associate with each strong component C of D a unique vertex, the root of the branching $F \cap C$. As with blocks, it suffices to identify these roots in order to determine the strong components of D. This can be achieved by means of a supplementary branching-search. We leave the details as an exercise (Exercise 6.3.12).

Exercises

6.3.1 By applying Dijkstra's Algorithm, show that the path indicated in Figure 6.7 is a shortest directed (x, y)-path.

$\star$**6.3.2** Prove that Dijkstra's Algorithm runs correctly.

$\star$**6.3.3** Apply Dijkstra's Algorithm to the directed graph with negative weights shown in Figure 6.8. Does the algorithm determine shortest directed (r, v)-paths for all vertices v?

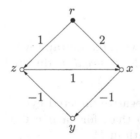

Fig. 6.8. Apply Dijkstra's Algorithm to this weighted directed graph (Exercise 6.3.3)

$\star$**6.3.4** Prove Theorem 6.14.

6.3.5 Describe an algorithm based on directed breadth-first search for finding a shortest directed odd cycle in a digraph.

6.3.6 Describe an algorithm based on directed depth-first search which accepts as input a directed graph D and returns a maximal (but not necessarily maximum) acyclic spanning subdigraph of D.

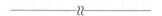

6.3.7 Describe an algorithm based on directed depth-first search which accepts as input a tournament T and returns a directed Hamilton path of T.

6.3.8 Describe an algorithm based on directed depth-first search which accepts as input a directed graph D and returns either a directed cycle in D or a topological sort of D (defined in Exercise 2.1.11).

6.3.9 BELLMAN'S ALGORITHM
Prove the validity of the following algorithm, which accepts as input a topological sort Q of a weighted acyclic digraph (D, w), with first vertex r, and returns a function $\ell : V \rightarrow \mathbb{R}$ such that $\ell(v) = d_D(r, v)$ for all $v \in V$, and a branching B (given by a predecessor function p) such that rBv is a shortest directed (r, v)-path in D for all $v \in V$ such that $d_D(r, v) < \infty$.   (R. BELLMAN)

1: set $\ell(v) := \infty$, $p(v) := \emptyset$, $v \in V$
2: remove r from Q
3: set $\ell(r) := 0$
4: **while** Q is nonempty **do**
5:     remove the first element y from Q
6:     **for all** $x \in N^-(y)$ **do**
7:         **if** $\ell(x) + w(x, y) < \ell(y)$ **then**
8:             replace $\ell(y)$ by $\ell(x) + w(x, y)$ and $p(y)$ by x
9:         **end if**
10:     **end for**
11: **end while**
12: return (ℓ, p).

6.3.10 Let $D := (D, w)$ be a weighted digraph with a specified root r from which all other vertices are reachable. A *negative* directed cycle is one whose weight is negative.

a) Show that if D has no negative directed cycles, then there exists a spanning r-branching B in D such that, for each $v \in V$, the directed path rBv is a shortest directed (r, v)-path in D.
b) Give an example to show that this conclusion need not hold if D has negative directed cycles.

6.3.11 BELLMAN–FORD ALGORITHM
Let $D := (D, w)$ be a weighted digraph with a specified root r from which all other

vertices of D are reachable. For each nonnegative integer k, let $d_k(v)$ denote the weight of a shortest directed (r, v)-walk using at most k arcs, with the convention that $d_k(v) = \infty$ if there is no such walk. (Thus $d_0(r) = 0$ and $d_0(v) = \infty$ for all $v \in V \setminus \{r\}$.)

a) Show that the $d_k(v)$ satisfy the following recursion.

$$d_k(v) = \min\{d_{k-1}(v), \min\{d_{k-1}(u) + w(u, v) : u \in N^-(v)\}\}$$

b) For each of the weighted digraphs shown in Figure 6.9, compute $\mathbf{d}_k := (d_k(v) : v \in V)$ for $k = 0, 1, \ldots, 6$.

c) Show that:
   i) if $\mathbf{d}_k \neq \mathbf{d}_{k-1}$ for all k, $1 \leq k \leq n$, then D contains a negative directed cycle,
   ii) if $\mathbf{d}_k = \mathbf{d}_{k-1}$ for some k, $1 \leq k \leq n$, then D contains no negative directed cycle, and $d_k(v)$ is the distance from r to v, for all $v \in V$.

d) In the latter case, describe how to find a spanning r-branching B of D such that, for each $v \in V$, the directed (r, v)-path in B is a shortest directed (r, v)-path in D.   (R. BELLMAN; L.R. FORD; E.F. MOORE; A. SHIMBEL)

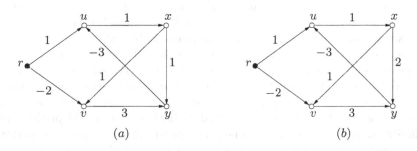

Fig. 6.9. Examples for the Bellman–Ford Algorithm (Exercise 6.3.11)

★**6.3.12** FINDING THE STRONG COMPONENTS OF A DIGRAPH.
Let D be a digraph, and let F be a DFS-branching forest of D. Denote by D' the converse of the digraph obtained from D by deleting all cross edges. By Proposition 6.15, it suffices to consider each component of D' separately, so we assume that D' has just one component.

a) Show that the set of vertices reachable from the root r in D' induces a strong component of D.

b) Apply this idea iteratively to obtain all strong components of D (taking care to select each new root appropriately).

c) Implement this procedure by employing branching-search.

6.3.13 The *diameter* of a directed graph is the maximum distance between any two vertices of the graph. (Thus a directed graph is of finite diameter if and only if it is strong.) Let G be a 2-edge-connected graph and P a longest path in G. By Robbins' Theorem (5.10), G has a strong orientation. Show that:

   a) no strong orientation of G has diameter exceeding the length of P,
   b) some strong orientation of G has diameter equal to the length of P.

(G. GUTIN)

6.4 Related Reading

DATA STRUCTURES

We have discussed in this chapter algorithms for resolving various problems expeditiously. The efficiency of these algorithms can be further enhanced by storing and managing the data involved in an appropriate structure. For example, a data structure known as a *heap* is commonly used for storing elements and their associated values, called *keys* (such as edges and their weights). A heap is a rooted binary tree T whose vertices are in one-to-one correspondence with the elements in question (in our case, vertices or edges). The defining property of a heap is that the key of the element located at vertex v of T is required to be at least as large as the keys of the elements located at vertices of the subtree of T rooted at v. This condition implies, in particular, that the key of the element at the root of T is one of greatest value; that element can thus be accessed instantly. Moreover, heaps can be reconstituted rapidly following small modifications such as the addition of an element, the removal of an element, or a change in the value of a key. A *priority queue* (the data structure used in both Dijkstra's Algorithm and the Jarník–Prim Algorithm) is simply a heap equipped with procedures for performing such readjustments rapidly. Heaps were conceived by Williams (1964).

It should be evident that data structures play a vital role in the efficiency of algorithms. For further information on this topic, we refer the reader to Knuth (1969), Aho et al. (1983), Tarjan (1983), or Cormen et al. (2001).

7

Flows in Networks

Contents

7.1 Transportation Networks

Transportation networks that are used to ship commodities from their production centres to their markets can be most effectively analysed when viewed as digraphs that possess additional structure. The resulting theory has a wide range of interesting applications and ramifications. We present here the basic elements of this important topic.

A *network* $N := N(x, y)$ is a digraph D (the *underlying digraph* of N) with two distinguished vertices, a *source* x and a *sink* y, together with a nonnegative real-valued function c defined on its arc set A. The vertex x corresponds to a production centre, and the vertex y to a market. The remaining vertices are called *intermediate vertices*, and the set of these vertices is denoted by I. The function c is the *capacity function* of N and its value on an arc a the *capacity* of a. The capacity of an arc may be thought of as representing the maximum rate at which a commodity can be transported along it. It is convenient to allow arcs of infinite capacity, along which commodities can be transported at any desired rate. Of course, in practice, one is likely to encounter transportation networks with several production centres and markets, rather than just one. However, this more general situation can be reduced to the case of networks that have just one source and one sink by means of a simple device (see Exercise 7.1.3).

We find the following notation useful. If f is a real-valued function defined on a set A, and if $S \subseteq A$, we denote the sum $\sum_{a \in S} f(a)$ by $f(S)$. Furthermore, when A is the arc set of a digraph D, and $X \subseteq V$, we set

$$f^+(X) := f(\partial^+(X)) \quad \text{and} \quad f^-(X) := f(\partial^-(X))$$

FLOWS

An (x, y)-*flow* (or simply a *flow*) in N is a real-valued function f defined on A satisfying the condition:

$$f^+(v) = f^-(v) \quad \text{for all} \quad v \in I \tag{7.1}$$

The value $f(a)$ of f on an arc a can be likened to the rate at which material is transported along a by the flow f. Condition (7.1) requires that, for any intermediate vertex v, the rate at which material is transported into v is equal to the rate at which it is transported out of v. For this reason, it is known as the *conservation condition*.

A flow f is *feasible* if it satisfies, in addition, the *capacity constraint*:

$$0 \le f(a) \le c(a) \quad \text{for all} \quad a \in A \tag{7.2}$$

The upper bound in condition (7.2) imposes the natural restriction that the rate of flow along an arc cannot exceed the capacity of the arc. Throughout this chapter, the term flow always refers to one that is feasible.

Every network has at least one flow, because the function f defined by $f(a) := 0$, for all $a \in A$, clearly satisfies both (7.1) and (7.2); it is called the *zero flow*. A less trivial example of a flow is given in Figure 7.1. The flow along each arc is indicated in bold face, along with the capacity of the arc.

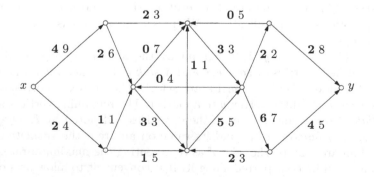

Fig. 7.1. A flow in a network

If X is a set of vertices in a network N and f is a flow in N, then $f^+(X) - f^-(X)$ is called the *net flow* out of X, and $f^-(X) - f^+(X)$ the *net flow* into X, relative to f. The conservation condition (7.1) requires that the net flow $f^+(v) - f^-(v)$ out of any intermediate vertex be zero, thus it is intuitively clear and not difficult to show that, relative to any (x, y)-flow f, the net flow $f^+(x) - f^-(x)$ out of x is equal to the net flow $f^-(y) - f^+(y)$ into y (Exercise 7.1.1b). This common quantity is called the *value* of f, denoted val (f). For example, the value of the flow indicated in Figure 7.1 is $2 + 4 = 6$. The value of a flow f may, in fact, be expressed as the net flow out of any subset X of V such that $x \in X$ and $y \in V \setminus X$, as we now show.

Proposition 7.1 *For any flow f in a network $N(x, y)$ and any subset X of V such that $x \in X$ and $y \in V \setminus X$,*

$$\operatorname{val}(f) = f^+(X) - f^-(X) \tag{7.3}$$

Proof From the definition of a flow and its value, we have

$$f^+(v) - f^-(v) = \begin{cases} \operatorname{val}(f) & \text{if } v = x \\ 0 & \text{if } v \in X \setminus \{x\} \end{cases}$$

Summing these equations over X and simplifying (Exercise 7.1.2), we obtain

$$\operatorname{val}(f) = \sum_{v \in X} (f^+(v) - f^-(v)) = f^+(X) - f^-(X) \qquad \square$$

A flow in a network N is a *maximum flow* if there is no flow in N of greater value. Maximum flows are of obvious importance in the context of transportation networks. A network $N(x, y)$ which has a directed (x, y)-path all of whose arcs are of infinite capacity evidently admits flows of arbitrarily large value. However, such networks do not arise in practice, and we assume that all the networks discussed here have maximum flows. We study the problem of finding such flows efficiently.

Problem 7.2 MAXIMUM FLOW
GIVEN: *a network $N(x, y)$,*
FIND: *a maximum flow from x to y in N.*

CUTS

It is convenient to denote a digraph D with two distinguished vertices x and y by $D(x, y)$. An (x, y)-*cut* in a digraph $D(x, y)$ is an outcut $\partial^+(X)$ such that $x \in X$ and $y \in V \setminus X$, and a *cut* in a network $N(x, y)$ is an (x, y)-cut in its underlying digraph. We also say that such a cut *separates y from x*. In the network of Figure 7.2, the heavy lines indicate a cut $\partial^+(X)$, where X is the set of solid vertices. The *capacity* of a cut $K := \partial^+(X)$ is the sum of the capacities of its arcs, $c^+(X)$. We denote the capacity of K by cap (K). The cut indicated in Figure 7.2 has capacity $3 + 7 + 1 + 5 = 16$.

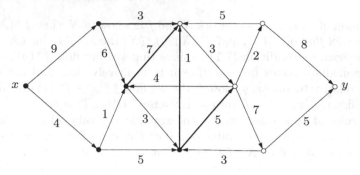

Fig. 7.2. A cut in a network

Flows and cuts are related in a simple fashion: the value of any (x, y)-flow is bounded above by the capacity of any cut separating y from x. In proving this inequality, it is convenient to call an arc a f-*zero* if $f(a) = 0$, f-*positive* if $f(a) > 0$, f-*unsaturated* if $f(a) < c(a)$, and f-*saturated* if $f(a) = c(a)$.

Theorem 7.3 *For any flow f and any cut $K := \partial^+(X)$ in a network N,*

$$\mathrm{val}\,(f) \le \mathrm{cap}\,(K)$$

Furthermore, equality holds in this inequality if and only if each arc in $\partial^+(X)$ is f-saturated and each arc in $\partial^-(X)$ is f-zero.

Proof By (7.2),
$$f^+(X) \le c^+(X) \quad \text{and} \quad f^-(X) \ge 0 \tag{7.4}$$
Thus, applying Proposition 7.1,

$$\mathrm{val}\,(f) = f^+(X) - f^-(X) \le c^+(X) = \mathrm{cap}\,(K)$$

We have $\mathrm{val}\,(f) = \mathrm{cap}\,(K)$ if and only if equality holds in (7.4), that is, if and only if each arc of $\partial^+(X)$ is f-saturated and each arc of $\partial^-(X)$ is f-zero. $\square$

A cut K in a network N is a *minimum cut* if no cut in N has a smaller capacity.

Corollary 7.4 *Let f be a flow and K a cut. If $\mathrm{val}\,(f) = \mathrm{cap}\,(K)$, then f is a maximum flow and K is a minimum cut.*

Proof Let f^* be a maximum flow and K^* a minimum cut. By Theorem 7.3,

$$\mathrm{val}\,(f) \le \mathrm{val}\,(f^*) \le \mathrm{cap}\,(K^*) \le \mathrm{cap}\,(K)$$

But, by hypothesis, $\mathrm{val}\,(f) = \mathrm{cap}\,(K)$. It follows that $\mathrm{val}\,(f) = \mathrm{val}\,(f^*)$ and $\mathrm{cap}\,(K^*) = \mathrm{cap}\,(K)$. Thus f is a maximum flow and K is a minimum cut. $\square$

Exercises

$\star$**7.1.1** Let $D = (V, A)$ be a digraph and f a real-valued function on A. Show that:

a) $\sum\{f^+(v) : v \in V\} = \sum\{f^-(v) : v \in V\}$,
b) if f is an (x, y)-flow, the net flow $f^+(x) - f^-(x)$ out of x is equal to the net flow $f^-(y) - f^+(y)$ into y.

$\star$**7.1.2**

a) Show that, for any flow f in a network N and any set $X \subseteq V$,

$$\sum_{v \in X}(f^+(v) - f^-(v)) = f^+(X) - f^-(X)$$

b) Give an example of a flow f in a network such that $\sum_{v \in X} f^+(v) \neq f^+(X)$ and $\sum_{v \in X} f^-(v) \neq f^-(X)$.

$\star$**7.1.3** Let $N := N(X, Y)$ be a network with source set X and sink set Y. Construct a new network $N' := N'(x, y)$ as follows.

▷   Adjoin two new vertices x and y.
▷   Join x to each source by an arc of infinite capacity.
▷   Join each sink to y by an arc of infinite capacity.

For any flow f in N, consider the function f' defined on the arc set of N' by:

$$f'(a) := \begin{cases} f(a) & \text{if } a \text{ is an arc of } N \\ f^+(v) & \text{if } a = (x, v) \\ f^-(v) & \text{if } a = (v, y) \end{cases}$$

a) Show that f' is a flow in N' with the same value as f.
b) Show, conversely, that the restriction of a flow in N' to the arc set of N is a flow in N of the same value.

7.1.4 Let $N(x, y)$ be a network which contains no directed (x, y)-path. Show that the value of a maximum flow and the capacity of a minimum cut in N are both zero.

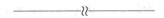

7.2 The Max-Flow Min-Cut Theorem

We establish here the converse of Corollary 7.4, namely that the value of a maximum flow is always equal to the capacity of a minimum cut.

Let f be a flow in a network $N := N(x, y)$. With each x-path P in N (not necessarily a directed path), we associate a nonnegative integer $\epsilon(P)$ defined by:

$$\epsilon(P) := \min\{\epsilon(a) : a \in A(P)\}$$

where

$$\epsilon(a) := \begin{cases} c(a) - f(a) & \text{if } a \text{ is a forward arc of } P \\ f(a) & \text{if } a \text{ is a reverse arc of } P \end{cases}$$

As we now explain, $\epsilon(P)$ is the largest amount by which the flow f can be increased along P without violating the constraints (7.2). The path P is said to be f-*saturated* if $\epsilon(P) = 0$ and f-*unsaturated* if $\epsilon(P) > 0$ (that is, if each forward arc of P is f-unsaturated and each reverse arc of P is f-positive). Put simply, an f-unsaturated path is one that is not being used to its full capacity. An f-*incrementing* path is an f-unsaturated (x, y)-path. For example, in the network of Figure 7.3a, the path $P := xv_1v_2v_3y$ is such a path. The forward arcs of P are (x, v_1) and (v_3, y), and $\epsilon(P) = \min\{5, 2, 5, 4\} = 2$.

The existence of an f-incrementing path P is significant because it implies that f is not a maximum flow. By sending an additional flow of $\epsilon(P)$ along P, one obtains a new flow f' of greater value. More precisely, define $f' : A \to \mathbb{R}$ by:

$$f'(a) := \begin{cases} f(a) + \epsilon(P) & \text{if } a \text{ is a forward arc of } P \\ f(a) - \epsilon(P) & \text{if } a \text{ is a reverse arc of } P \\ f(a) & \text{otherwise} \end{cases} \tag{7.5}$$

We then have the following proposition, whose proof is left as an exercise (7.2.1).

Proposition 7.5 *Let f be a flow in a network N. If there is an f-incrementing path P, then f is not a maximum flow. More precisely, the function f' defined by (7.5) is a flow in N of value $\mathrm{val}\,(f') = \mathrm{val}\,(f) + \epsilon(P)$.* ☐

We refer to the flow f' defined by (7.5) as the *incremented flow* based on P. Figure 7.3b shows the incremented flow in the network of Figure 7.3a based on the f-incrementing path $xv_1v_2v_3y$.

What if there is no f-incrementing path? The following proposition addresses this eventuality.

Proposition 7.6 *Let f be a flow in a network $N := N(x, y)$. Suppose that there is no f-incrementing path in N. Let X be the set of all vertices reachable from x by f-unsaturated paths, and set $K := \partial^+(X)$. Then f is a maximum flow in N and K is a minimum cut.*

Proof   Clearly $x \in X$. Also, $y \in V \setminus X$ because there is no f-incrementing path. Therefore K is a cut in N.

Consider an arc $a \in \partial^+(X)$, with tail u and head v. Because $u \in X$, there exists an f-unsaturated (x, u)-path Q. If a were f-unsaturated, Q could be extended by the arc a to yield an f-unsaturated (x, v)-path. But $v \in V \setminus X$, so there is no such path. Therefore a must be f-saturated. Similar reasoning shows that if $a \in \partial^-(X)$,

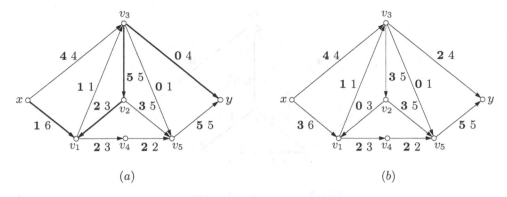

Fig. 7.3. (a) An f-incrementing path P, and (b) the incremented flow based on P

then a must be f-zero. By Theorem 7.3, we have $\mathrm{val}\,(f) = \mathrm{cap}\,(K)$. Corollary 7.4 now implies that f is a maximum flow in N and that K is a minimum cut.   □

A far-reaching consequence of Propositions 7.5 and 7.6 is the following theorem, due independently to Elias et al. (1956) and Ford and Fulkerson (1956).

Theorem 7.7 THE MAX-FLOW MIN-CUT THEOREM
In any network, the value of a maximum flow is equal to the capacity of a minimum cut.

Proof Let f be a maximum flow. By Proposition 7.5, there can be no f-incrementing path. The theorem now follows from Proposition 7.6.   □

The Max-Flow Min-Cut Theorem (7.7) shows that one can always demonstrate the optimality of a maximum flow simply by exhibiting a cut whose capacity is equal to the value of the flow. Many results in graph theory are straightforward consequences of this theorem, as applied to suitably chosen networks. Among these are two fundamental theorems due to K. Menger, discussed at the end of this chapter (Theorems 7.16 and 7.17). Other important applications of network flows are given in Chapter 16.

THE FORD–FULKERSON ALGORITHM

The following theorem is a direct consequence of Propositions 7.5 and 7.6.

Theorem 7.8 *A flow f in a network is a maximum flow if and only if there is no f-incrementing path.*   □

This theorem is the basis of an algorithm for finding a maximum flow in a network. Starting with a known flow f, for instance the zero flow, we search for an f-incrementing path by means of a tree-search algorithm. An x-tree T is f-*unsaturated* if, for every vertex v of T, the path xTv is f-unsaturated. An example

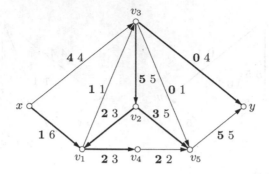

Fig. 7.4. An f-unsaturated tree

is shown in the network of Figure 7.4. It is a tree T of this type that we grow in searching for an f-incrementing path.

Initially, the tree T consists of just the source x. At any stage, there are two ways in which the tree may be grown. If there exists an f-unsaturated arc a in $\partial^+(X)$, where $X = V(T)$, both a and its head are adjoined to T. Similarly, if there exists an f-positive arc a in $\partial^-(X)$, both a and its tail are adjoined to T. If the tree T reaches the sink y, the path xTy is an f-incrementing path, and we replace f by the flow f' defined in (7.5). If T fails to reach the sink, and is a maximal f-unsaturated tree, each arc in $\partial^+(X)$ is f-saturated and each arc in $\partial^-(X)$ is f-zero. We may then conclude, by virtue of Theorem 7.3, that the flow f is a maximum flow and the cut $\partial^+(X)$ a minimum cut. We refer to this tree-search algorithm as *Incrementing Path Search* (IPS) and to a maximal f-unsaturated tree which does not include the sink as an *IPS-tree*.

Algorithm 7.9 MAX-FLOW MIN-CUT (MFMC)

INPUT: a network $N := N(x, y)$ and a feasible flow f in N
OUTPUT: a maximum flow f and a minimum cut $\partial^+(X)$ in N

1: set $X := \{x\}$, $p(v) := \emptyset$, $v \in V$
*2: **while** there is either an f-unsaturated arc $a := (u, v)$ or an f-positive arc $a := (v, u)$ with $u \in X$ and $v \in V \setminus X$ **do***
3:   replace X by $X \cup \{v\}$
4:   replace $p(v)$ by u
*5: **end while***
*6: **if** $y \in X$ **then***
7:   compute $\epsilon(P) := \min\{\epsilon(a) : a \in A(P)\}$, where P is the xy-path in the tree whose predecessor function is p
8:   for each forward arc a of P, replace $f(a)$ by $f(a) + \epsilon(P)$
9:   for each reverse arc a of P, replace $f(a)$ by $f(a) - \epsilon(P)$
10:   return to 1
*11: **end if***
12: return $(f, \partial^+(X))$

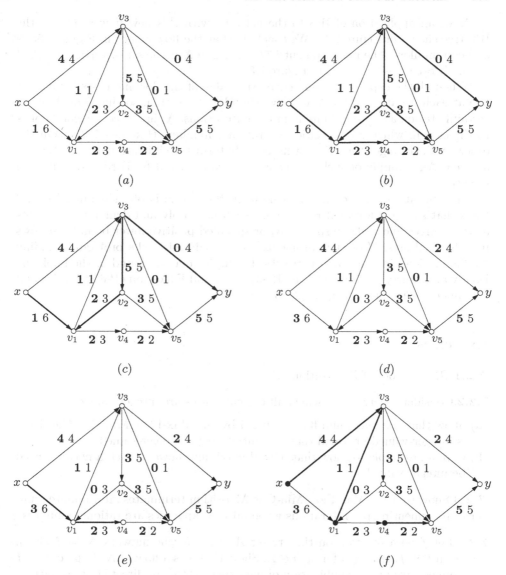

Fig. 7.5. (a) A flow f, (b) an f-unsaturated tree, (c) the f-incrementing path, (d) the f-incremented flow, (e) an IPS-tree, and (f) a minimum cut

As an example, consider the network shown in Figure 7.5a, with the indicated flow. Applying IPS, we obtain the f-unsaturated tree shown in Figure 7.5b. Because this tree includes the sink y, the xy-path contained in it, namely $xv_1v_2v_3y$, is an f-incrementing path (see Figure 7.5c). By incrementing f along this path, we obtain the incremented flow shown in Figure 7.5d.

Now, an application of IPS to the network with this new flow results in the IPS-tree shown in Figure 7.5e. We conclude that the flow shown in Figure 7.5d is a maximum flow. The minimum cut $\partial^+(X)$, where X is the set of vertices reached by the IPS-tree, is indicated in Figure 7.5f.

When all the capacities are integers, the value of the flow increases by at least one at each iteration of the Max-Flow Min-Cut Algorithm, so the algorithm will certainly terminate after a finite number of iterations. A similar conclusion applies to the case in which all capacities are rational numbers (Exercise 7.2.3). On the other hand, the algorithm will not necessarily terminate if irrational capacities are allowed. An example of such a network was constructed by Ford and Fulkerson (1962).

In applications of the theory of network flows, one is often required to find flows that satisfy additional restrictions, such as supply and demand constraints at the sources and sinks, respectively, or specified positive lower bounds on flows in individual arcs. Most such problems can be reduced to the problem of finding maximum flows in associated networks. Examples may be found in the books by Bondy and Murty (1976), Chvátal (1983), Ford and Fulkerson (1962), Lovász and Plummer (1986), and Schrijver (2003).

Exercises

★**7.2.1** Give a proof of Proposition 7.5.

7.2.2 Consider a network in which all the capacities are integer-valued.

a) Show that the maximum flow returned by the Max-Flow Min-Cut Algorithm is integer-valued provided that the initial flow is integer-valued.
b) Give an example to show that not all maximum flows in such a network need be integer-valued.

★**7.2.3** Show that the Max-Flow Min-Cut Algorithm terminates after a finite number of incrementing path iterations when all the capacities are rational numbers.

7.2.4 Let f be a function on the arc set A of an acyclic network $N := N(x,y)$, such that $0 \le f(a) \le c(a)$ for all $a \in A$. Show that f is a flow in N if and only if f is a nonnegative linear combination of incidence vectors of directed (x,y)-paths.

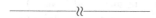

7.2.5 DEGREE SEQUENCES OF BIPARTITE GRAPHS
Let $\mathbf{p} := (p_1, p_2, \ldots, p_m)$ and $\mathbf{q} := (q_1, q_2, \ldots, q_n)$ be two sequences of nonnegative integers. The pair $(\mathbf{p}, \mathbf{q})$ is said to be *realizable* by a simple bipartite graph if there exists a simple bipartite graph G with bipartition $(\{x_1, x_2, \ldots, x_m\}, \{y_1, y_2, \ldots, y_n\})$, such that $d(x_i) = p_i$, for $1 \le i \le m$, and $d(y_j) = q_j$, for $1 \le j \le n$.

a) Formulate as a network flow problem the problem of determining whether a given pair $(\mathbf{p}, \mathbf{q})$ is realizable by a simple bipartite graph.

b) Suppose that $q_1 \geq q_2 \geq \cdots \geq q_n$. Deduce from the Max-Flow Min-Cut Theorem that $(\mathbf{p}, \mathbf{q})$ is realizable by a simple bipartite graph if and only if:

$$\sum_{i=1}^{m} p_i = \sum_{j=1}^{n} q_j \quad \text{and} \quad \sum_{i=1}^{m} \min\{p_i, k\} \geq \sum_{j=1}^{k} q_j \quad \text{for} \ \ 1 \leq k \leq n$$

(D. GALE AND H.J. RYSER)

7.2.6 DEGREE SEQUENCES OF DIRECTED GRAPHS

Let D be a strict digraph and let p and q be two nonnegative integer-valued functions on V.

a) Consider the problem of determining whether D has a spanning subdigraph H such that:

$$d_H^-(v) = p(v) \quad \text{and} \quad d_H^+(v) = q(v) \quad \text{for all} \ \ v \in V$$

Formulate this as a network flow problem.

b) Deduce from the Max-Flow Min-Cut Theorem that D has a subdigraph H satisfying the condition in (a) if and only if:

i) $\displaystyle\sum_{v \in V} p(v) = \sum_{v \in V} q(v)$,

ii) $\displaystyle\sum_{v \in S} q(v) \leq \sum_{v \in T} p(v) + a(S, V \setminus T)$ for all $S, T \subseteq V$.

c) Taking D to be the complete directed graph on n vertices and applying (b), find necessary and sufficient conditions for two sequences $\mathbf{p} := (p_1, p_2, \ldots, p_n)$ and $\mathbf{q} := (q_1, q_2, \ldots, q_n)$ to be realizable as the in- and outdegree sequences of a strict digraph on n vertices.

7.3 Arc-Disjoint Directed Paths

A communications network N with one-way communication links may be modelled by a directed graph D whose vertices correspond to the stations of N and whose arcs correspond to its links. In order to be able to relay information in N from station x to station y, the digraph D must clearly contain a directed (x, y)-path. In practice, however, the possible failure of communication links (either by accident or by sabotage) must also be taken into account. For example, if all the directed (x, y)-paths in D should happen to contain one particular arc, and if the communication link corresponding to that arc should fail or be destroyed, it would no longer be possible to relay information from x to y. This situation would not arise if D contained two arc-disjoint directed (x, y)-paths. More generally, if D had k arc-disjoint directed (x, y)-paths, x would still be able to send messages to y even if $k - 1$ links should fail. The maximum number of arc-disjoint directed (x, y)-paths is therefore a relevant parameter in this context, and we are led to the following problem.

Problem 7.10 ARC-DISJOINT DIRECTED PATHS (ADDP)
  GIVEN: *a digraph* $D := D(x, y)$,
  FIND: *a maximum family of arc-disjoint directed* (x, y)-*paths in* D.

Let us now look at the network from the viewpoint of a saboteur who wishes to disrupt communications from x to y. The saboteur will seek to eliminate all directed (x, y)-paths in D by destroying arcs, preferably as few as possible. Now, a minimal set of arcs whose deletion destroys all directed (x, y)-paths is nothing but an (x, y)-cut. The saboteur's problem can thus be stated as follows.

Problem 7.11 MINIMUM ARC CUT
  GIVEN: *a digraph* $D := D(x, y)$,
  FIND: *a minimum* (x, y)-*cut in* D.

As the reader might have guessed, these problems can be solved by applying network flow theory. The concept of a circulation provides the essential link.

CIRCULATIONS

A *circulation* in a digraph D is a function $f : A \to \mathbb{R}$ which satisfies the conservation condition at every vertex:

$$f^+(v) = f^-(v), \quad \text{for all } v \in V \tag{7.6}$$

Figure 7.6a shows a circulation in a digraph.

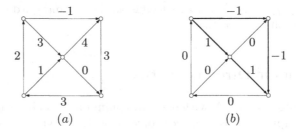

(a)   (b)

Fig. 7.6. (a) A circulation in a digraph, and (b) a circulation associated with a cycle

Circulations in a digraph D can be expressed very simply in terms of the incidence matrix of D. Recall that this is the matrix $\mathbf{M} = (m_{va})$ whose rows and columns are indexed by the vertices and arcs of D, respectively, where, for a vertex v and arc a,

$$m_{va} := \begin{cases} 1 & \text{if } a \text{ is a link and } v \text{ is the tail of } a \\ -1 & \text{if } a \text{ is a link and } v \text{ is the head of } a \\ 0 & \text{otherwise} \end{cases}$$

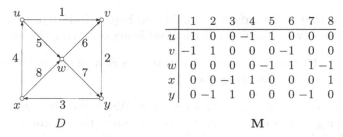

Fig. 7.7. A digraph and its incidence matrix

The incidence matrix of a digraph is shown in Figure 7.7.

We frequently identify a real-valued function f defined on a set S with the vector $\mathbf{f} := (f(a) : a \in S)$. With this convention, the conservation condition (7.6) for a function f to be a circulation in D may be expressed in matrix notation as:

$$\mathbf{Mf} = \mathbf{0} \qquad (7.7)$$

where $\mathbf{M}$ is the $n \times m$ incidence matrix of D and $\mathbf{0}$ the $n \times 1$ zero-vector.

Circulations and flows can be readily transformed into one another. If f is a circulation in a digraph $D := (V, A)$, and if $a = (y, x)$ is an arc of D, the restriction f' of f to $A \setminus a$ is an (x, y)-flow of value $f(a)$ in the digraph $D' := D \setminus a$ (Exercise 7.3.2). Conversely, if f is an (x, y)-flow in a digraph $D := (V, A)$, and if D' is the digraph obtained from D by adding a new arc a' from y to x, the extension f' of f to $A \cup \{a'\}$ defined by $f'(a') := \mathrm{val}\,(f)$ is a circulation in D'. By virtue of these transformations, results on flows and circulations go hand in hand. Often, it is more convenient to study circulations rather than flows because the conservation condition (7.6) is then satisfied uniformly, at all vertices.

The *support* of a real-valued function is the set of elements at which its value is nonzero.

Lemma 7.12 *Let f be a nonzero circulation in a digraph. Then the support of f contains a cycle. Moreover, if f is nonnegative, then the support of f contains a directed cycle.*

Proof The first assertion follows directly from Theorem 2.1, because the support of a nonzero circulation can contain no vertex of degree less than two. Likewise, the second assertion follows from Exercise 2.1.11a. □

Certain circulations are of particular interest, namely those associated with cycles. Let C be a cycle, together with a given sense of traversal. An arc of C is a *forward arc* if its direction agrees with the sense of traversal of C, and a *reverse arc* otherwise. We denote the sets of forward and reverse arcs of C by C^+ and C^-, respectively, and associate with C the circulation f_C defined by:

$$f_C(a) := \begin{cases} 1 & \text{if } a \in C^+ \\ -1 & \text{if } a \in C^- \\ 0 & \text{if } a \notin C \end{cases}$$

It can be seen that f_C is indeed a circulation. Figure 7.6b depicts a circulation associated with a cycle (the sense of traversal being counterclockwise).

Proposition 7.13 *Every circulation in a digraph is a linear combination of the circulations associated with its cycles.*

Proof Let f be a circulation, with support S. We proceed by induction on $|S|$. There is nothing to prove if $S = \emptyset$. If S is nonempty, then S contains a cycle C by Lemma 7.12. Let a be any arc of C, and choose the sense of traversal of C so that $f_C(a) = 1$. Then $f' := f - f(a)f_C$ is a circulation whose support is a proper subset of S. By induction, f' is a linear combination of circulations associated with cycles, so $f = f' + f(a)f_C$ is too.   □

There is an analogous statement to Proposition 7.13 in the case where the circulation is nonnegative. The proof is essentially the same (Exercise 7.3.4).

Proposition 7.14 *Every nonnegative circulation in a digraph is a nonnegative linear combination of the circulations associated with its directed cycles. Moreover, if the circulation is integer-valued, the coefficients of the linear combination may be chosen to be nonnegative integers.*   □

The relationship between circulations and flows described above implies the following corollary.

Corollary 7.15 *Let $N := N(x, y)$ be a network in which each arc is of unit capacity. Then N has an (x, y)-flow of value k if and only if its underlying digraph $D(x, y)$ has k arc-disjoint directed (x, y)-paths.*   □

MENGER'S THEOREM

In view of Corollary 7.15, Problems 7.10 and 7.11 can both be solved by the Max-Flow Min-Cut Algorithm. Moreover, the Max-Flow Min-Cut Theorem in this special context becomes a fundamental min–max theorem on digraphs, due to Menger (1927).

Theorem 7.16   MENGER'S THEOREM (ARC VERSION)
In any digraph $D(x, y)$, the maximum number of pairwise arc-disjoint directed (x, y)-paths is equal to the minimum number of arcs in an (x, y)-cut.   □

There is a corresponding version of Menger's Theorem for undirected graphs. As with networks and digraphs, it is convenient to adopt the notation $G(x, y)$ to signify a graph G with two distinguished vertices x and y. By an *xy-cut* in a graph $G(x, y)$, we mean an edge cut $\partial(X)$ such that $x \in X$ and $y \in V \setminus X$. We say that such an edge cut *separates x and y*.

Theorem 7.17   MENGER'S THEOREM (EDGE VERSION)
In any graph $G(x, y)$, the maximum number of pairwise edge-disjoint xy-paths is equal to the minimum number of edges in an xy-cut.   □

Theorem 7.17 can be derived quite easily from Theorem 7.16. Likewise, the undirected version of Problem 7.10 can be solved by applying the Max-Flow Min-Cut Algorithm to an appropriate network (Exercise 7.3.5). In Chapter 8, we explain how vertex versions of Menger's Theorems (7.16 and 7.17) can be derived from Theorem 7.16. These theorems play a central role in graph theory, as is shown in Chapter 9.

Exercises

⋆**7.3.1**

a) Let $D = (V, A)$ be a digraph, and let f be a real-valued function on A. Show that f is a circulation in D if and only if $f^+(X) = f^-(X)$ for all $X \subseteq V$.
b) Let f be a circulation in a digraph D, with support S. Deduce that:
  i) $D[S]$ has no cut edges,
  ii) if f is nonnegative, then $D[S]$ has no directed bonds.

⋆**7.3.2** Let f be a circulation in a digraph $D := (V, A)$, and let $a = (y, x)$ be an arc of D. Show that the restriction f' of f to $A' := A \setminus a$ is an (x, y)-flow in $D' := (V, A')$ of value $f(a)$.

7.3.3 Let f and f' be two flows of equal value in a network N. Show that $f - f'$ is a circulation in N.

⋆**7.3.4** Prove Proposition 7.14.

⋆**7.3.5**

a) Deduce Theorem 7.17 from Theorem 7.16.
b) The undirected version of Problem 7.10 may be expressed as follows.

  Problem 7.18   EDGE-DISJOINT PATHS (EDP)
   GIVEN: *a graph $G := G(x, y)$,*
   FIND: *a maximum family of edge-disjoint xy-paths in G.*

Explain how this problem can be solved by applying the Max-Flow Min-Cut Algorithm to an appropriate network.

————————⟨⟨————————

7.4 Related Reading

MULTICOMMODITY FLOWS

In this chapter, we have dealt with the problem of transporting a single commodity along the arcs of a network. In practice, transportation networks are generally shared by many users, each wishing to transport a different commodity from one location to another. This gives rise to the notion of a *multicommodity flow*. Let N be a network with k source-sink pairs (x_i, y_i), $1 \le i \le k$, and let d_i denote the demand at y_i for commodity i, $1 \le i \le k$. The k-*commodity flow problem* consists of finding functions $f_i : A \to \mathbb{R}$, $1 \le i \le k$, such that:

(i) f_i is a flow in N of value d_i from x_i to y_i, $1 \le i \le k$,
(ii) for each arc a of D, $\sum_{i=1}^{k} f_i(a) \le c(a)$.

For a subset X of V, let $d(X)$ denote the quantity $\sum\{d_i : x_i \in X, y_i \in V \setminus X\}$. If there is a solution to the k-commodity flow problem, the inequality $d(X) \le c^+(X)$, known as the *cut condition*, must hold for all subsets X of V. For $k = 1$, this cut condition is equivalent to the condition $\mathrm{val}(f) \le \mathrm{cap}(K)$ of Theorem 7.3. By the Max-Flow Min-Cut Theorem (7.7), this condition is sufficient for the existence of a flow of value d_1. However, even for $k = 2$, the cut condition is not sufficient for the 2-commodity flow problem to have a solution, as is shown by the network with unit capacities and demands depicted in Figure 7.8a.

There is another noteworthy distinction between the single commodity and the multicommodity flow problems. Suppose that all capacities and demands are integers and that there is a k-commodity flow meeting all the requirements. When $k = 1$, this implies the existence of such a flow which is integer-valued (Exercise 7.2.2). The same is not true for $k \ge 2$. Consider, for example, the network in Figure 7.8b, again with unit capacities and demands. This network has the 2-commodity flow (f_1, f_2), where $f_1(a) = f_2(a) = 1/2$ for all $a \in A$, but it has no 2-commodity flow which takes on only integer values.

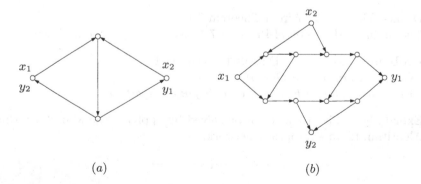

(a)   (b)

Fig. 7.8. Examples of networks: (a) satisfies the cut condition but has no 2-commodity flow, (b) has a fractional 2-commodity flow, but not one which is integral

8

Complexity of Algorithms

Contents

8.1 Computational Complexity

In this chapter, we see how problems may be classified according to their level of difficulty.

Most problems that we consider in this book are of a general character, applying to all members of some family of graphs or digraphs. By an *instance* of a problem, we mean the problem as applied to one specific member of the family. For example, an instance of the Minimum-Weight Spanning Tree Problem is the problem of finding an optimal tree in a particular weighted connected graph.

An *algorithm* for solving a problem is a well-defined computational procedure which accepts any instance of the problem as *input* and returns a solution to the problem as *output*. For example, the Jarník–Prim Algorithm (6.9) accepts as input a weighted connected graph G and returns as output an optimal tree.

As we have seen, many problems of practical importance can be formulated in terms of graphs. Designing computationally efficient algorithms for solving these problems is one of the main concerns of graph theorists and computer scientists. The two aspects of theoretical interest in this regard are, firstly, to verify that a proposed algorithm does indeed perform correctly and, secondly, to analyse how efficient a procedure it is. We have already encountered algorithms for solving a number of basic problems. In each case, we have established their validity. Here, we discuss the efficiency of these and other algorithms.

By the *computational complexity* (or, for short, *complexity*) of an algorithm, we mean the number of basic computational steps (such as arithmetical operations and comparisons) required for its execution. This number clearly depends on the size and nature of the input. In the case of graphs, the complexity is a function of the number of bits required to encode the adjacency list of the input graph G, a function of n and m. (The number of bits required to encode an integer k is $\lceil \log_2 k \rceil$.) Naturally, when the input includes additional information, such as weights on the vertices or edges of the graph, this too must be taken into account in calculating the complexity. If the complexity is bounded above by a polynomial in the input size, the algorithm is called a *polynomial-time algorithm*. Such an algorithm is further qualified as *linear-time* if the polynomial is a linear function, *quadratic-time* if it is a quadratic function, and so on.

The Class $\mathcal{P}$

The significance of polynomial-time algorithms is that they are usually found to be computationally feasible, even for large input graphs. By contrast, algorithms whose complexity is exponential in the size of the input have running times which render them unusable even on inputs of moderate size. For example, an algorithm which checks whether two graphs on n vertices are isomorphic by considering all $n!$ bijections between their vertex sets is feasible only for small values of n (certainly no greater than 20), even on the fastest currently available computers. The class of problems solvable by polynomial-time algorithms is denoted by $\mathcal{P}$.

The tree-search algorithms discussed in Chapter 6 are instances of polynomial-time algorithms. In breadth-first search, each edge is examined for possible inclusion in the tree just twice, when the adjacency lists of its two ends are scanned. The same is true of depth-first search. Therefore both of these algorithms are linear in m, the number of edges. The Jarník–Prim Algorithm involves, in addition, comparing weights of edges, but it is easily seen that the number of comparisons is also bounded by a polynomial in m.

Unlike the other algorithms described in Chapter 6, the Max-Flow Min-Cut Algorithm is not a polynomial-time algorithm even when all the capacities are integers; the example in Exercise 8.1.1 shows that, in the worst case, the algorithm may perform an arbitrarily large number of iterations before returning a maximum flow. Fortunately, this eventuality can be avoided by modifying the way in which IPS is implemented, as was shown by Edmonds and Karp (1970) and Dinic (1970). Among all the arcs in $\partial(T)$ that qualify for inclusion in T, preference is given to those which are incident to the vertex that entered T the earliest, just as in breadth-first search, resulting in a shortest incrementing path. It can be shown that, with this refinement, the number of iterations of IPS is bounded by a polynomial in n and thus yields a polynomial-time algorithm.

Although our analysis of these algorithms is admittedly cursory, and leaves out many pertinent details, it should be clear that they do indeed run in polynomial time. A thorough analysis of these and other graph algorithms can be found in the books by Aho et al. (1975) and Papadimitriou (1994). On the other hand, there are many basic problems for which polynomial-time algorithms have yet to be found, and indeed might well not exist. Determining which problems are solvable in polynomial time and which are not is evidently a fundamental question. In this connection, a class of problems denoted by $\mathcal{NP}$ (standing for *nondeterministic polynomial-time*) plays an important role. We give here an informal definition of this class; a precise treatment can be found in Chapter 29 of the *Handbook of Combinatorics* (Graham et al. (1995)), or in the book by Garey and Johnson (1979).

THE CLASSES $\mathcal{NP}$ AND CO-$\mathcal{NP}$

A *decision problem* is a question whose answer is either 'yes' or 'no'. Such a problem belongs to the class $\mathcal{P}$ if there is a polynomial-time algorithm that solves any instance of the problem in polynomial time. It belongs to the class $\mathcal{NP}$ if, given any instance of the problem whose answer is 'yes', there is a certificate validating this fact which can be checked in polynomial time; such a certificate is said to be *succinct*. Analogously, a decision problem belongs to the class co-$\mathcal{NP}$ if, given any instance of the problem whose answer is 'no', there is a succinct certificate which confirms that this is so. It is immediate from these definitions that $\mathcal{P} \subseteq \mathcal{NP}$, inasmuch as a polynomial-time algorithm constitutes, in itself, a succinct certificate. Likewise, $\mathcal{P} \subseteq$ co-$\mathcal{NP}$. Thus

$$\mathcal{P} \subseteq \mathcal{NP} \cap \text{co-}\mathcal{NP}$$

Consider, for example, the problem of determining whether a graph is bipartite. This decision problem belongs to $\mathcal{NP}$, because a bipartition is a succinct certificate: given a bipartition (X, Y) of a bipartite graph G, it suffices to check that each edge of G has one end in X and one end in Y. The problem also belongs to co-$\mathcal{NP}$ because, by Theorem 4.7, every nonbipartite graph contains an odd cycle, and any such cycle constitutes a succinct certificate of the graph's nonbipartite character.

It thus belongs to $\mathcal{NP} \cap$ co-$\mathcal{NP}$. In fact, as indicated in Exercise 6.1.3, it belongs to $\mathcal{P}$.

As a second example, consider the problem of deciding whether a graph $G(x, y)$ has k edge-disjoint xy-paths. This problem is clearly in $\mathcal{NP}$, because a family of k edge-disjoint xy-paths is a succinct certificate: given such a family of paths, one may check in polynomial-time that it indeed has the required properties. The problem is also in co-$\mathcal{NP}$ because, by Theorem 7.17, a graph that does not have k edge-disjoint xy-paths has an xy-edge cut of size less than k. Such an edge cut serves as a succinct certificate for the nonexistence of k edge-disjoint xy-paths. Finally, because the maximum number of edge-disjoint xy-paths can be found in polynomial time by applying the Max-Flow Min-Cut Algorithm (7.9) (see Exercise 7.3.5) this problem belongs to $\mathcal{P}$, too.

Consider, now, the problem of deciding whether a graph has a Hamilton cycle.

Problem 8.1 HAMILTON CYCLE
  GIVEN: *a graph G,*
  DECIDE: *Does G have a Hamilton cycle?*

If the answer is 'yes', then any Hamilton cycle would serve as a succinct certificate. However, should the answer be 'no', what would constitute a succinct certificate confirming this fact? In contrast to the two problems described above, no such certificate is known! In other words, notwithstanding that HAMILTON CYCLE is clearly a member of the class $\mathcal{NP}$, it has not yet been shown to belong to co-$\mathcal{NP}$, and might very well not belong to this class. The same is true of the decision problem for Hamilton paths. These two problems are discussed in detail in Chapter 18.

Many problems that arise in practice, such as the Shortest Path Problem (6.11), are optimization problems rather than decision problems. Nonetheless, each such problem implicitly includes an infinitude of decision problems. For example, the Shortest Path Problem includes, for each real number ℓ, the following decision problem. Given a weighted directed graph (D, w) with two specified vertices x and y, is there a directed (x, y)-path in D of length at most ℓ?

We have noted three relations of inclusion among the classes $\mathcal{P}$, $\mathcal{NP}$, and co-$\mathcal{NP}$, and it is natural to ask whether these inclusions are proper. Because $\mathcal{P} = \mathcal{NP}$ if and only if $\mathcal{P} = $ co-$\mathcal{NP}$, two basic questions arise, both of which have been posed as conjectures.

THE COOK–EDMONDS–LEVIN CONJECTURE

Conjecture 8.2 $\mathcal{P} \neq \mathcal{NP}$

EDMONDS' CONJECTURE

Conjecture 8.3 $\mathcal{P} = \mathcal{NP} \cap co\text{-}\mathcal{NP}$

Conjecture 8.2 is one of the most fundamental open questions in all of mathematics. (A prize of one million dollars has been offered for its resolution.) It is widely (but not universally) believed that the conjecture is true, that there are problems in $\mathcal{NP}$ for which no polynomial-time algorithm exists. One such problem would be HAMILTON CYCLE. As we show in Section 8.3, this problem, and its directed analogue DIRECTED HAMILTON CYCLE, are at least as hard to solve as any problem in the class $\mathcal{NP}$; more precisely, if a polynomial-time algorithm for either of these problems should be found, it could be adapted to solve any problem in $\mathcal{NP}$ in polynomial time by means of a suitable transformation. Conjecture 8.2 was, in essence, put forward by J. Edmonds in the mid-1960s, when he asserted that there could exist no 'good' (that is, polynomial-time) algorithm for the Travelling Salesman Problem (Problem 2.6). The conjecture thus predates the formal definition of the class $\mathcal{NP}$ by Cook (1971) and Levin (1973).

Conjecture 8.3, also proposed by Edmonds (1965c), is strongly supported by empirical evidence. Most decision problems which are known to belong to $\mathcal{NP} \cap$ co-$\mathcal{NP}$ are also known to belong to $\mathcal{P}$. A case in point is the problem of deciding whether a given integer is prime. Although it had been known for some time that this problem belongs to both $\mathcal{NP}$ and co-$\mathcal{NP}$, a polynomial-time algorithm for testing primality was discovered only much more recently, by Agrawal et al. (2004).

Exercises

$\star$**8.1.1**

a) Show that, starting with the zero flow, an application of the Max-Flow Min-Cut Algorithm (7.9) to the network N in Figure 8.1 might execute $2M + 1$ incrementing path iterations before finding a maximum flow.
b) Deduce that this algorithm is not a polynomial-time algorithm.

8.1.2 Show that Fleury's Algorithm (3.3) is a polynomial-time algorithm.

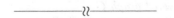

8.1.3 Given a graph $G(x, y)$, consider the problem of deciding whether G has an xy-path of odd (respectively, even) length.

a) Show that this problem:
   i) belongs to $\mathcal{NP}$,

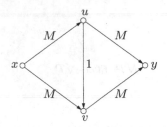

Fig. 8.1. A network on which Algorithm 7.9 might require many iterations

ii) belongs to co-$\mathcal{NP}$.
b) Describe a polynomial-time algorithm for solving the problem.

8.1.4 Describe a polynomial-time algorithm for deciding whether two trees are isomorphic.

8.2 Polynomial Reductions

A common approach to problem-solving is to transform the given problem into one whose solution is already known, and then convert that solution into a solution of the original problem. Of course, this approach is feasible only if the transformation can be made rapidly. The concept of polynomial reduction captures this requirement.

A *polynomial reduction* of a problem P to a problem Q is a pair of polynomial-time algorithms, one which transforms each instance I of P to an instance J of Q, and the other which transforms a solution for the instance J to a solution for the instance I. If such a reduction exists, we say that P is *polynomially reducible* to Q, and write $P \preceq Q$; this relation is clearly both reflexive and transitive. The significance of polynomial reducibility is that if $P \preceq Q$, and if there is a polynomial-time algorithm for solving Q, then this algorithm can be converted into a polynomial-time algorithm for solving P. In symbols:

$$P \preceq Q \ \text{ and } \ Q \in \mathcal{P} \Rightarrow P \in \mathcal{P} \tag{8.1}$$

A very simple example of the above paradigm is the polynomial reduction to the Minimum-Weight Spanning Tree Problem (6.8) of the following problem.

Problem 8.4   MAXIMUM-WEIGHT SPANNING TREE
GIVEN: *a weighted connected graph G,*
FIND: *a maximum-weight spanning tree in G.*

In order to solve an instance of this problem, it suffices to replace each weight by its negative and apply the Jarník–Prim Algorithm (6.9) to find an optimal tree in the resulting weighted graph. The very same tree will be one of maximum weight in the original weighted graph. (We remark that one can similarly reduce the problem of finding a longest xy-path in a graph to the Shortest Path Problem (6.11). However no polynomial-time algorithm is known for solving the latter problem when there are negative edge weights.)

Not all reductions are quite as straightforward as this one. Recall that two directed (x, y)-paths are *internally disjoint* if they have no internal vertices in common. Consider the following problem, the analogue for internally disjoint paths of Problem 7.10, the Arc-Disjoint Directed Paths Problem (ADDP).

Problem 8.5   INTERNALLY DISJOINT DIRECTED PATHS (IDDP)
GIVEN: *a digraph* $D := D(x, y)$,
FIND: *a maximum family of internally disjoint directed (x, y)-paths in D.*

A polynomial reduction of IDDP to ADDP can be obtained by constructing a new digraph $D' := D'(x, y)$ from D as follows.

▷   Split each vertex $v \in V \setminus \{x, y\}$ into two new vertices v^- and v^+, joined by a new arc (v^-, v^+).
▷   For each arc (u, v) of D, replace its tail u by u^+ (unless $u = x$ or $u = y$) and its head v by v^- (unless $v = x$ or $v = y$).

This construction is illustrated in Figure 8.2.

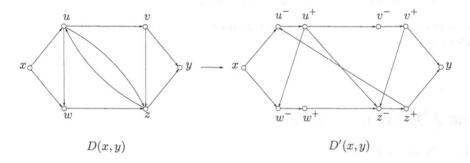

$$D(x, y) \qquad\qquad D'(x, y)$$

Fig. 8.2. Reduction of $IDDP$ to $ADDP$

It can be seen that there is a bijection between families of internally disjoint directed (x, y)-paths in D and families of arc-disjoint directed (x, y)-paths in D'. Thus, finding a maximum family of internally disjoint directed (x, y)-paths in $D(x, y)$ amounts to finding a maximum family of arc-disjoint directed (x, y)-paths in $D'(x, y)$. This transformation of the instance $D(x, y)$ of IDDP to the instance $D'(x, y)$ of ADDP is a polynomial reduction because $v(D') = 2v(D) - 2$ and $a(D') = a(D) + v(D) - 2$. Hence $IDDP \preceq ADDP$.

The Max-Flow Min-Cut Algorithm (7.9) is a polynomial-time algorithm for solving ADDP. Therefore $ADDP \in \mathcal{P}$. Because $IDDP \preceq ADDP$, we may conclude that $IDDP \in \mathcal{P}$, also.

Most problems concerning paths in undirected graphs can be reduced to their analogues in directed graphs by the simple artifice of considering the associated digraph. As an example, let $G := G(x, y)$ be an undirected graph and let $D :=$ $D(G)$ be its associated digraph. There is an evident bijection between families of internally disjoint xy-paths in G and families of internally disjoint directed (x, y)-paths in D. Thus $IDP \preceq IDDP$, where IDP is the problem of finding a maximum family of internally disjoint xy-paths in a given graph $G(x, y)$. We showed above that $IDDP \in \mathcal{P}$. It now follows from the transitivity of the relation $\preceq$ that $IDP \in \mathcal{P}$.

Exercises

8.2.1 Consider a network in which a nonnegative integer $m(v)$ is associated with each intermediate vertex v. Show how a maximum flow f satisfying the constraint $f^-(v) \leq m(v)$, for all $v \in I$, can be found by applying the Max-Flow Min-Cut Algorithm to a suitably modified network.

8.2.2 Consider the following problem.

Problem 8.6 DISJOINT PATHS
   GIVEN: *a graph G, a positive integer k, and two k-subsets X and Y of V,*
   DECIDE: *Does G have k disjoint (X, Y)-paths?*

Describe a polynomial reduction of this problem to IDP (INTERNALLY DISJOINT PATHS).

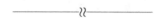

8.3 $\mathcal{NP}$-Complete Problems

THE CLASS $\mathcal{NPC}$

We have just seen how polynomial reductions may be used to produce new polynomial-time algorithms from existing ones. By the same token, polynomial reductions may also be used to link 'hard' problems, ones for which no polynomial-time algorithm exists, as can be seen by writing (8.1) in a different form:

$$P \preceq Q \text{ and } P \notin \mathcal{P} \Rightarrow Q \notin \mathcal{P}$$

This viewpoint led Cook (1971) and Levin (1973) to define a special class of seemingly intractable decision problems, the class of $\mathcal{NP}$-complete problems. Informally, these are the problems in the class $\mathcal{NP}$ which are 'at least as hard to solve' as any problem in $\mathcal{NP}$.

Formally, a problem P in $\mathcal{NP}$ is $\mathcal{NP}$-*complete* if $P' \preceq P$ for every problem P' in $\mathcal{NP}$. The class of $\mathcal{NP}$-complete problems is denoted by $\mathcal{NPC}$. It is by no means obvious that $\mathcal{NP}$-complete problems should exist at all. On the other hand, once one such problem has been found, the $\mathcal{NP}$-completeness of other problems may be established by means of polynomial reductions, as follows.

In order to prove that a problem Q in $\mathcal{NP}$ is $\mathcal{NP}$-complete, it suffices to find a polynomial reduction to Q of some known $\mathcal{NP}$-complete problem P. Why is this so? Suppose that P is $\mathcal{NP}$-complete. Then $P' \preceq P$ for all $P' \in \mathcal{NP}$. If $P \preceq Q$, then $P' \preceq Q$ for all $P' \in \mathcal{NP}$, by the transitivity of the relation $\preceq$. In other words, Q is $\mathcal{NP}$-complete. In symbols:

$$P \preceq Q \text{ and } P \in \mathcal{NPC} \Rightarrow Q \in \mathcal{NPC}$$

Cook (1971) and Levin (1973) made a fundamental breakthrough by showing that there do indeed exist $\mathcal{NP}$-complete problems. More precisely, they proved that the satisfiability problem for boolean formulae is $\mathcal{NP}$-complete. We now describe this problem, and examine the theoretical and practical implications of their discovery.

BOOLEAN FORMULAE

A *boolean variable* is a variable which takes on one of two values, 0 ('false') or 1 ('true'). Boolean variables can be combined into *boolean formulae*, which may be defined recursively as follows.

▷ Every boolean variable is a boolean formula.
▷ If f is a boolean formula, then so too is $(\neg f)$, the *negation* of f.
▷ If f and g are boolean formulae, then so too are:
- $(f \vee g)$, the *disjunction* of f and g,
- $(f \wedge g)$, the *conjunction* of f and g.

These three operations may be thought of informally as 'not f', 'f or g', and 'f and g', respectively. The negation of a boolean variable x is often written as $\overline{x}$. Thus the expression

$$(\neg(x_1 \vee \overline{x_2}) \vee x_3) \wedge (x_2 \vee \overline{x_3}) \tag{8.2}$$

is a boolean formula in the variables x_1, x_2, x_3. Note that the parentheses are needed here to avoid ambiguity as to the order of execution of the various operations. (For ease of reading, we omit the outer pair of parentheses.)

An assignment of values to the variables of a boolean formula is called a *truth assignment*. Given a truth assignment, the value of the formula may be computed according to the following rules.

$\neg$	
0	1
1	0

$\vee$	0	1
0	0	1
1	1	1

$\wedge$	0	1
0	0	0
1	0	1

For instance, if $x_1 = 1$, $x_2 = 0$, and $x_3 = 1$, the value of formula (8.2) is:

$$(\neg(1 \vee \overline{0}) \vee 1) \wedge (0 \vee \overline{1}) = (\neg(1 \vee 1) \vee 1) \wedge (0 \vee 0) = (\overline{1} \vee 1) \wedge 0 = (0 \vee 1) \wedge 0 = 1 \wedge 0 = 0$$

Two boolean formulae are *equivalent* (written $\equiv$) if they take the same value for each truth assignment of the variables involved. It follows easily from the above rules that negation is an *involution*:

$$\neg(\neg f) \equiv f$$

and that disjunction and conjunction are *commutative*, *associative*, and *idempotent*:

$$f \vee g \equiv g \vee f, \qquad f \wedge g \equiv g \wedge f$$
$$f \vee (g \vee h) \equiv (f \vee g) \vee h, \qquad f \wedge (g \wedge h) \equiv (f \wedge g) \wedge h$$
$$f \vee f \equiv f, \qquad f \wedge f \equiv f.$$

Furthermore, disjunction and conjunction together satisfy the *distributive laws*:

$$f \vee (g \wedge h) \equiv (f \vee g) \wedge (f \vee h), \qquad f \wedge (g \vee h) \equiv (f \wedge g) \vee (f \wedge h)$$

and interact with negation according to *de Morgan's laws*:

$$\neg(f \vee g) \equiv (\neg f) \wedge (\neg g), \qquad \neg(f \wedge g) \equiv (\neg f) \vee (\neg g)$$

Finally, there are the *tautologies*:

$$f \vee \neg f = 1, \qquad f \wedge \neg f = 0.$$

Boolean formulae may be transformed into equivalent ones by applying these laws. For instance:

$$
\begin{aligned}
(\neg(x_1 \vee \overline{x_2}) \vee x_3) \wedge (x_2 \vee \overline{x_3}) &\equiv ((\overline{x_1} \wedge x_2) \vee x_3) \wedge (x_2 \vee \overline{x_3}) \\
&\equiv ((\overline{x_1} \vee x_3) \wedge (x_2 \vee x_3)) \wedge (x_2 \vee \overline{x_3}) \\
&\equiv (\overline{x_1} \vee x_3) \wedge ((x_2 \vee x_3) \wedge (x_2 \vee \overline{x_3})) \\
&\equiv (\overline{x_1} \vee x_3) \wedge (x_2 \vee (x_3 \wedge \overline{x_3})) \\
&\equiv (\overline{x_1} \vee x_3) \wedge x_2
\end{aligned}
$$

SATISFIABILITY OF BOOLEAN FORMULAE

A boolean formula is *satisfiable* if there is a truth assignment of its variables for which the value of the formula is 1. In this case, we say that the formula is *satisfied* by the assignment. It can be seen that formula (8.2) is satisfiable, for instance by the truth assignment $x_1 = 0$, $x_2 = 1$, $x_3 = 0$. But not all boolean formulae are satisfiable ($x \wedge \overline{x}$ being a trivial example). This poses the general problem:

Problem 8.7 BOOLEAN SATISFIABILITY (SAT)
GIVEN: *a boolean formula f,*
DECIDE: *Is f satisfiable?*

Observe that SAT belongs to $\mathcal{NP}$: given appropriate values of the variables, it can be checked in polynomial time that the value of the formula is indeed 1. These values of the variables therefore constitute a succinct certificate. Cook (1971) and Levin (1973) proved, independently, that SAT is an example of an $\mathcal{NP}$-complete problem.

Theorem 8.8   THE COOK–LEVIN THEOREM
The problem SAT *is* $\mathcal{NP}$-*complete.*   □

The proof of the Cook–Levin Theorem involves the notion of a Turing machine, and is beyond the scope of this book. A proof may be found in Garey and Johnson (1979) or Sipser (2005).

By applying Theorem 8.8, Karp (1972) showed that many combinatorial problems are $\mathcal{NP}$-complete. One of these is DIRECTED HAMILTON CYCLE. In order to explain the ideas underlying his approach, we need a few more definitions.

A variable x, or its negation $\overline{x}$, is a *literal*, and a disjunction or conjunction of literals is a *disjunctive* or *conjunctive clause*. Because the operations of disjunction and conjunction are associative, parentheses may be dispensed with within clauses. There is no ambiguity, for example, in the following formula, a conjunction of three disjunctive clauses, each consisting of three literals.

$$f := (x_1 \vee \overline{x_2} \vee x_3) \wedge (x_2 \vee \overline{x_3} \vee x_4) \wedge (\overline{x_1} \vee \overline{x_3} \vee \overline{x_4})$$

Any conjunction of disjunctive clauses such as this one is referred to as a formula in *conjunctive normal form*. It can be shown that every boolean formula is equivalent, via a polynomial reduction, to one in conjunctive normal form (Exercise 8.3.1). Furthermore, as we explain below in the proof of Theorem 8.10, every boolean formula in conjunctive normal form is equivalent, again via a polynomial reduction, to one in conjunctive normal form with exactly three literals per clause. The decision problem for such boolean formulae is known as 3-SAT.

Problem 8.9   BOOLEAN 3-SATISFIABILITY (3-SAT)
GIVEN: *a boolean formula f in conjunctive normal form with three literals per clause,*
DECIDE: *Is f satisfiable?*

Theorem 8.10 *The problem* 3-SAT *is* $\mathcal{NP}$-*complete.*

Proof By the Cook–Levin Theorem (8.8), it suffices to prove that SAT $\preceq$ 3-SAT. Let f be a boolean formula in conjunctive normal form. We show how to construct, in polynomial time, a boolean formula f' in conjunctive normal form such that:

i) each clause in f' has three literals,
ii) f is satisfiable if and only if f' is satisfiable.

Such a formula f' may be obtained by the addition of new variables and clauses, as follows.

Suppose that some clause of f has just two literals, for instance the clause $(x_1 \lor x_2)$. In this case, we simply replace this clause by two clauses with three literals, $(x_1 \lor x_2 \lor x)$ and $(\overline{x} \lor x_1 \lor x_2)$, where x is a new variable. Clearly,

$$(x_1 \lor x_2) \equiv (x_1 \lor x_2 \lor x) \land (\overline{x} \lor x_1 \lor x_2)$$

Clauses with single literals may be dealt with in a similar manner (Exercise 8.3.2).

Now suppose that some clause $(x_1 \lor x_2 \lor \cdots \lor x_k)$ of f has k literals, where $k \geq 4$. In this case, we add $k - 3$ new variables $y_1, y_2, \ldots, y_{k-3}$ and form the following $k - 2$ clauses, each with three literals.

$$(x_1 \lor x_2 \lor y_1), \; (\overline{y}_1 \lor x_3 \lor y_2), \; (\overline{y}_2 \lor x_4 \lor y_3), \; \cdots \; (\overline{y}_{k-4} \lor x_{k-2} \lor y_{k-3}), \; (\overline{y}_{k-3} \lor x_{k-1} \lor x_k)$$

One may verify that $(x_1 \lor x_2 \lor \cdots \lor x_k)$ is equivalent to the conjunction of these $k - 2$ clauses. We leave the details as an exercise (8.3.3). $\qquad \square$

Theorem 8.10 may be used to establish the $\mathcal{NP}$-completeness of decision problems in graph theory such as DIRECTED HAMILTON CYCLE by means of polynomial reductions.

As we have observed, in order to show that a decision problem Q in $\mathcal{NP}$ is $\mathcal{NP}$-complete, it suffices to find a polynomial reduction to Q of a known $\mathcal{NP}$-complete problem P. This is generally easier said than done. What is needed is to first decide on an appropriate $\mathcal{NP}$-complete problem P and then come up with a suitable polynomial reduction. In the case of graphs, the latter step is often achieved by means of a construction whereby certain special subgraphs, referred to as 'gadgets', are inserted into the instance of P so as to obtain an instance of Q with the required properties. An illustration of this technique is described in the inset overleaf, where we show how 3-SAT may be reduced to DIRECTED HAMILTON CYCLE via an intermediate problem, EXACT COVER.

Almost all of the decision problems that we come across in this book are known to belong either to the class $\mathcal{P}$ or to the class $\mathcal{NPC}$. One notable exception is the isomorphism problem:

Problem 8.11   GRAPH ISOMORPHISM
GIVEN: *two graphs G and H,*
DECIDE: *Are G and H isomorphic?*

The complexity status of this problem remains a mystery. Whilst the problem clearly belongs to $\mathcal{NP}$, whether it belongs to $\mathcal{P}$, to co-$\mathcal{NP}$, or to $\mathcal{NPC}$ is not known. Polynomial-time isomorphism-testing algorithms have been found for certain classes of graphs, including planar graphs (Hopcroft and Wong (1974)) and graphs of bounded degree (Luks (1982)), but these algorithms are not valid for all graphs. GRAPH ISOMORPHISM might, conceivably, be a counterexample to Conjecture 8.3.

PROOF TECHNIQUE: POLYNOMIAL REDUCTION

We establish the NP-completeness of DIRECTED HAMILTON CYCLE by reducing 3-SAT to it via an intermediate problem, EXACT COVER, which we now describe.

Let $\mathcal{A}$ be a family of subsets of a finite set X. An *exact cover* of X by $\mathcal{A}$ is a partition of X, each member of which belongs to $\mathcal{A}$. For instance, if $X := \{x_1, x_2, x_3\}$ and $\mathcal{A} := \{\{x_1\}, \{x_1, x_2\}, \{x_2, x_3\}\}$, then $\{\{x_1\}, \{x_2, x_3\}\}$ is an exact cover of X by $\mathcal{A}$. This notion gives rise to the following decision problem.

Problem 8.12   EXACT COVER
   GIVEN: *a set X and a family $\mathcal{A}$ of subsets of X,*
   DECIDE: *Is there an exact cover of X by $\mathcal{A}$?*

We first describe a polynomial reduction of 3-SAT to EXACT COVER, and then a polynomial reduction of EXACT COVER to DIRECTED HAMILTON CYCLE. The chain of reductions:

$$\text{SAT} \preceq 3 - \text{SAT} \preceq \text{EXACT COVER} \preceq \text{DIRECTED HAMILTON CYCLE}$$

will then imply that DIRECTED HAMILTON CYCLE is $\mathcal{NP}$-complete, by virtue of the Cook–Levin Theorem (8.8).

Theorem 8.13 3-SAT $\preceq$ EXACT COVER.

Proof   Let f be an instance of 3-SAT, with clauses $f_1, \ldots, f_n$ and variables $x_1, \ldots, x_m$. The first step is to construct a graph G from f, by setting:

$$V(G) := \{x_i : 1 \le i \le m\} \cup \{\overline{x_i} : 1 \le i \le m\} \cup \{f_j : 1 \le j \le n\}$$
$$E(G) := \{x_i \overline{x_i} : 1 \le i \le m\} \cup \{x_i f_j : x_i \in f_j\} \cup \{\overline{x_i} f_j : \overline{x_i} \in f_j\}$$

where the notation $x_i \in f_j$ ($\overline{x_i} \in f_j$) signifies that x_i ($\overline{x_i}$) is a literal of the clause f_j. The next step is to obtain an instance $(X, \mathcal{A})$ of EXACT COVER from this graph G. We do so by setting:

$$X := \{f_j : 1 \le j \le n\} \cup E(G)$$
$$\mathcal{A} := \{\partial(x_i) : 1 \le i \le m\} \cup \{\partial(\overline{x_i}) : 1 \le i \le m\}$$
$$\cup \{\{f_j\} \cup F_j : F_j \subset \partial(f_j), 1 \le j \le n\}$$

It can be verified that the formula f is satisfiable if and only if the set X has an exact cover by the family $\mathcal{A}$ (Exercise 8.3.4). $\square$

POLYNOMIAL REDUCTION (CONTINUED)

For instance, if $f := (x_1 \vee \overline{x_2} \vee x_3) \wedge (x_2 \vee \overline{x_3} \vee x_4) \wedge (\overline{x_1} \vee \overline{x_3} \vee \overline{x_4})$, the graph G obtained by this construction is:

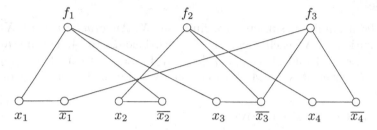

In this example, the given formula is satisfied by the truth assignment $x_1 = 1$, $x_2 = 1$, $x_3 = 0$, $x_4 = 0$, and this truth assignment corresponds to the exact cover:

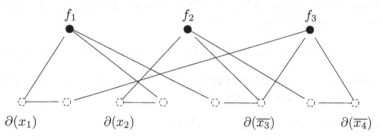

To round off the proof that DIRECTED HAMILTON CYCLE is an $\mathcal{NP}$-complete problem, we describe a polynomial reduction of EXACT COVER to DIRECTED HAMILTON CYCLE.

Theorem 8.14 EXACT COVER $\preceq$ DIRECTED HAMILTON CYCLE.

Proof Let $(X, \mathcal{A})$ be an instance of EXACT SET COVER, where $X := \{x_i : 1 \leq i \leq m\}$ and $\mathcal{A} := \{A_j : 1 \leq j \leq n\}$. We construct a directed graph G from $(X, \mathcal{A})$ as follows. Let P be a directed path whose arcs are labelled by the elements of X, Q a directed path whose arcs are labelled by the elements of $\mathcal{A}$ and, for $1 \leq j \leq n$, R_j a directed path whose vertices are labelled by the elements of A_j. The paths P, Q, and R_j, $1 \leq j \leq n$, are assumed to be pairwise disjoint. We add an arc from the initial vertex of P to the initial vertex of Q, and from the terminal vertex of Q to the terminal vertex of P.

POLYNOMIAL REDUCTION (CONTINUED)

For $1 \leq j \leq n$, we also add an arc from the initial vertex of the arc A_j of Q to the initial vertex of R_j, and from the terminal vertex of R_j to the terminal vertex of A_j:

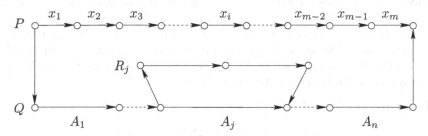

For $1 \leq j \leq n$, we now transform the directed path R_j into a digraph D_j by replacing each vertex x_i of R_j by a 'path' P_{ij} of length two whose edges are pairs of oppositely oriented arcs. Moreover, for every such 'path' P_{ij}, we add an arc from the initial vertex of P_{ij} to the initial vertex of the arc x_i of P, and one from the terminal vertex of x_i to the terminal vertex of P_{ij}:

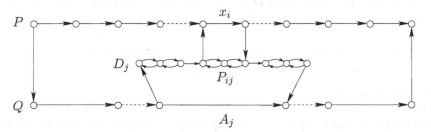

We denote the resulting digraph by D. This construction, with $X :=\{x_1, x_2, x_3\}$ and $\mathcal{A} := \{\{x_1\}, \{x_1, x_2\}, \{x_2, x_3\}\}$, is illustrated in the following figure.

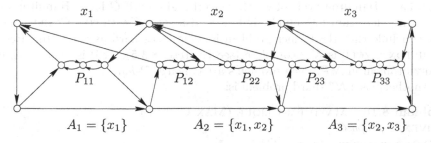

POLYNOMIAL REDUCTION (CONTINUED)

Observe, now, that the digraph D has a directed Hamilton cycle C if and only if the set X has an exact cover by the family of subsets $\mathcal{A}$. If C does not use the arc A_j, it is obliged to traverse D_j from its initial to its terminal vertex. Conversely, if C uses the arc A_j, it is obliged to include each one of the paths P_{ij} of D_j in its route from the terminal vertex of P to the initial vertex of P. Moreover, C traces exactly one of the paths P_{ij} ($x_i \in A_j$) in travelling from the head of the arc x_i to its tail. The arcs A_j of Q which are included in C therefore form a partition of X. Conversely, to every partition of X there corresponds a directed Hamilton cycle of D.

Finally, the numbers of vertices and arcs of D are given by:

$$v(D) = |X| + |\mathcal{A}| + 3\sum_{j=1}^{n} |A_j| + 2$$
$$a(D) = |X| + 2|\mathcal{A}| + 7\sum_{j=1}^{n} |A_j| + 2$$

Because both of these parameters are bounded above by linear functions of the size of the instance $(X, \mathcal{A})$, the above reduction is indeed polynomial. $\square$

Corollary 8.15 *The problem* DIRECTED HAMILTON CYCLE *is $\mathcal{NP}$-complete.*

$\square$

$\mathcal{NP}$-HARD PROBLEMS

We now turn to the computational complexity of optimization problems such as the Travelling Salesman Problem (TSP) (Problem 2.6). This problem contains HAMILTON CYCLE as a special case. To see this, associate with a given graph G the weighted complete graph on $V(G)$ in which the weight attached to an edge uv is zero if $uv \in E(G)$, and one otherwise. The resulting weighted complete graph has a Hamilton cycle of weight zero if and only if G has a Hamilton cycle. Thus, any algorithm for solving TSP will also solve HAMILTON CYCLE, and we may conclude that the former problem is at least as hard as the latter. Because HAMILTON CYCLE is $\mathcal{NP}$-complete (see Exercise 8.3.5), TSP is at least as hard as any problem in $\mathcal{NP}$. Such problems are called $\mathcal{NP}$-hard .

Another basic $\mathcal{NP}$-hard problem is:

Problem 8.16   MAXIMUM CLIQUE (MAX CLIQUE)
  GIVEN: *a graph G,*
  FIND: *a maximum clique in G.*

In order to solve this problem, one needs to know, for a given value of k, whether G has a k-clique. The largest such k is called the *clique number* of G, denoted $\omega(G)$.

If k is a fixed integer not depending on n, the existence of a k-clique can be decided in polynomial time, simply by means of an exhaustive search, because the number of k-subsets of V is bounded above by n^k. However, if k depends on n, this is no longer true. Indeed, the problem of deciding whether a graph G has a k-clique, where k depends on n, is $\mathcal{NP}$-complete (Exercise 8.3.9).

The complementary notion of a clique is a *stable set*, a set of vertices no two of which are adjacent. A stable set in a graph is *maximum* if the graph contains no larger stable set. The cardinality of a maximum stable set in a graph G is called the *stability number* of G, denoted $\alpha(G)$. Clearly, a subset S of V is a stable set in G if and only if S is a clique in $\overline{G}$, the complement of G. Consequently, the following problem is polynomially equivalent to MAX CLIQUE, and thus is $\mathcal{NP}$-hard also.

Problem 8.17 MAXIMUM STABLE SET (MAX STABLE SET)
GIVEN: *a graph G,*
FIND: *a maximum stable set in G.*

Exercises

$\star$**8.3.1** Let $f := f_1 \wedge f_2 \wedge \cdots \wedge f_k$ and $g := g_1 \wedge g_2 \wedge \cdots \wedge g_\ell$ be two boolean formulae in conjunctive normal form (where f_i, $1 \leq i \leq k$, and g_j, $1 \leq j \leq \ell$, are disjunctive clauses).

a) Show that:
   i) $f \wedge g$ is in conjunctive normal form,
   ii) $f \vee g$ is equivalent to a boolean formula in conjunctive normal form,
   iii) $\neg f$ is in disjunctive normal form, and is equivalent to a boolean formula in conjunctive normal form.
b) Deduce that every boolean formula is equivalent to a boolean formula in conjunctive normal form.

$\star$**8.3.2** Show that every clause consisting of just one literal is equivalent to a boolean formula in conjunctive normal form with exactly three literals per clause.

$\star$**8.3.3** Let $(x_1 \vee x_2 \vee \cdots \vee x_k)$ be a disjunctive clause with k literals, where $k \geq 4$, and let $y_1, y_2, \ldots, y_{k-2}$ be boolean variables. Show that:

$$(x_1 \vee x_2 \vee \cdots \vee x_k) \equiv (x_1 \vee x_2 \vee y_1) \wedge (\overline{y}_1 \vee x_3 \vee y_2) \wedge (\overline{y}_2 \vee x_4 \vee y_3) \wedge \cdots$$
$$\cdots \wedge (\overline{y}_{k-4} \vee x_{k-2} \vee y_{k-3}) \wedge (\overline{y}_{k-3} \vee x_{k-1} \vee x_k)$$

$\star$**8.3.4** Let $f := f_1 \wedge f_2 \wedge \cdots \wedge f_n$ be an instance of 3-SAT, with variables $x_1, x_2, \ldots, x_m$. Form a graph G from f, and an instance $(X, \mathcal{A})$ of EXACT COVER from G, as described in the proof of Theorem 8.13.

a) Show that the formula f is satisfiable if and only if the set X has an exact cover by the family $\mathcal{A}$.

b) Show also that the pair $(X, \mathcal{A})$ can be constructed from f in polynomial time (in the parameters m and n).

c) Deduce that EXACT COVER $\in \mathcal{NPC}$.

d) Explain why constructing a graph in the same way, but from an instance of SAT rather than 3-SAT, does not provide a polynomial reduction of SAT to EXACT COVER.

⋆8.3.5

a) Describe a polynomial reduction of DIRECTED HAMILTON CYCLE to HAMILTON CYCLE.

b) Deduce that HAMILTON CYCLE $\in \mathcal{NPC}$.

8.3.6 Let HAMILTON PATH denote the problem of deciding whether a given graph has a Hamilton path.

a) Describe a polynomial reduction of HAMILTON CYCLE to HAMILTON PATH.

b) Deduce that HAMILTON PATH $\in \mathcal{NPC}$.

8.3.7 Two problems P and Q are *polynomially equivalent*, written $P \equiv Q$, if $P \preceq Q$ and $Q \preceq P$.

a) Show that:

$$\text{HAMILTON PATH} \equiv \text{HAMILTON CYCLE} \equiv \text{DIRECTED HAMILTON CYCLE}$$

b) Let MAX PATH denote the problem of finding the length of a longest path in a given graph. Show that MAX PATH $\equiv$ HAMILTON PATH.

8.3.8

a) Let k be a fixed positive integer. Describe a polynomial-time algorithm for deciding whether a given graph has a path of length k.

b) The length of a longest path in a graph G can be determined by checking, for each k, $1 \leq k \leq n$, whether G has a path of length k. Does your algorithm for the problem in part (a) lead to a polynomial-time algorithm for MAX PATH?

⋆8.3.9

a) Let $f = f_1 \wedge f_2 \wedge \cdots \wedge f_k$ be an instance of 3-SAT (where the f_i, $1 \leq i \leq k$, are disjunctive clauses, each containing three literals). Construct a k-partite graph G on $7k$ vertices (seven vertices in each part) such that f is satisfiable if and only if G has a k-clique.

b) Deduce that 3-SAT and MAX CLIQUE are polynomially equivalent.

————————————— ⸮⸮ —————————————

8.3.10 Let k be a positive integer. The following problem is a generalization of 3-SAT.

Problem 8.18 BOOLEAN k-SATISFIABILITY *(k-SAT)*
GIVEN: *a boolean formula f in conjunctive normal form with k literals per clause,*
DECIDE: *Is f satisfiable?*

Show that:

a) 2-SAT $\in \mathcal{P}$,
b) k-SAT $\in \mathcal{NPC}$ for $k \geq 3$.

8.4 Approximation Algorithms

For $\mathcal{NP}$-hard optimization problems of practical interest, such as the Travelling Salesman Problem, the best that one can reasonably expect of a polynomial-time algorithm is that it should always return a feasible solution which is not too far from optimality.

Given a real number $t \geq 1$, a *t-approximation algorithm* for a minimization problem is an algorithm that accepts any instance of the problem as input and returns a feasible solution whose value is no more than t times the optimal value; the smaller the value of t, the better the approximation. Naturally, the running time of the algorithm is an equally important factor. We give two examples.

Problem 8.19 MAXIMUM CUT (MAX CUT)
GIVEN: *a weighted graph (G, w),*
FIND: *a maximum-weight spanning bipartite subgraph F of G.*

This problem admits a polynomial-time 2-approximation algorithm, based on the ideas for the unweighted case presented in Chapter 2 (Exercise 2.2.2). We leave the details as an exercise (8.4.1).

A somewhat less simple approximation algorithm was obtained by Rosenkrantz et al. (1974), who considered the special case of the Travelling Salesman Problem in which the weights satisfy the *triangle inequality*:

$$w(xy) + w(yz) \geq w(xz), \quad \text{for any three vertices } x, y, z. \tag{8.3}$$

Problem 8.20 METRIC TRAVELLING SALESMAN PROBLEM (METRIC TSP)
GIVEN: *a weighted complete graph G whose weights satisfy inequality (8.3),*
FIND: *a minimum-weight Hamilton cycle C of G.*

Theorem 8.21 METRIC TSP *admits a polynomial-time 2-approximation algorithm.*

Proof Applying the Jarnìk–Prim Algorithm (6.9), we first find a minimum-weight spanning tree T of G. Suppose that C is a minimum-weight Hamilton cycle of G. By deleting any edge of C, we obtain a Hamilton path P of G. Because P is a spanning tree of G and T is a spanning tree of minimum weight,

$$w(T) \leq w(P) \leq w(C)$$

We now duplicate each edge of T, thereby obtaining a connected even graph H with $V(H) = V(G)$ and $w(H) = 2w(T)$. Note that this graph H is not even a subgraph of G, let alone a Hamilton cycle. The idea is to transform H into a Hamilton cycle of G, and to do so without increasing its weight. More precisely, we construct a sequence $H_0, H_1, \ldots, H_{n-2}$ of connected even graphs, each with vertex set $V(G)$, such that $H_0 = H$, H_{n-2} is a Hamilton cycle of G, and $w(H_{i+1}) \leq w(H_i)$, $0 \leq i \leq n-3$. We do so by reducing the number of edges, one at a time, as follows.

Let C_i be an Euler tour of H_i, where $i < n-2$. The graph H_i has $2(n-2)-i > n$ edges, and thus has a vertex v of degree at least four. Let $x e_1 v e_2 y$ be a segment of the tour C_i; it will follow by induction that $x \neq y$. We replace the edges e_1 and e_2 of C_i by a new edge e of weight $w(xy)$ linking x and y, thereby bypassing v and modifying C_i to an Euler tour C_{i+1} of $H_{i+1} := (H_i \setminus \{e_1, e_2\}) + e$. By the triangle inequality (8.3),

$$w(H_{i+1}) = w(H_i) - w(e_1) - w(e_2) + w(e) \leq w(H_i)$$

The final graph H_{n-2}, being a connected even graph on n vertices and n edges, is a Hamilton cycle of G. Furthermore,

$$w(H_{n-2}) \leq w(H_0) = 2w(T) \leq 2w(C) \qquad \square$$

The relevance of minimum-weight spanning trees to the Travelling Salesman Problem was first observed by Kruskal (1956). A $\frac{3}{2}$-approximation algorithm for METRIC TSP was found by Christofides (1976). This algorithm makes use of a polynomial-time algorithm for weighted matchings (discussed in Chapter 16; see Exercise 16.5.15). For other approaches to the Travelling Salesman Problem, see Jünger et al. (1995).

The situation with respect to the general Travelling Salesman Problem, in which the weights are not subject to the triangle inequality, is dramatically different: for any integer $t \geq 2$, there cannot exist a polynomial-time t-approximation algorithm for solving TSP unless $\mathcal{P} = \mathcal{NP}$ (Exercise 8.4.4). The book by Vazirani (2001) treats the topic of approximation algorithms in general. For the state of the art regarding computational aspects of TSP, we refer the reader to Applegate et al. (2007).

Exercises

$\star$**8.4.1** Describe a polynomial-time 2-approximation algorithm for MAX CUT (Problem 8.19).

8.4.2 EUCLIDEAN TSP
The *Euclidean Travelling Salesman Problem* is the special case of METRIC TSP in which the vertices of the graph are points in the plane, the edges are straight-line

segments linking these points, and the weight of an edge is its length. Show that, in any such graph, the minimum-weight Hamilton cycles are crossing-free (that is, no two of their edges cross).

8.4.3 Show that METRIC TSP is $\mathcal{NP}$-hard.

$\star$**8.4.4**

a) Let G be a simple graph with $n \geq 3$, and let t be a positive integer. Consider the weighted complete graph (K, w), where $K := G \cup \overline{G}$, in which $w(e) := 1$ if $e \in E(G)$ and $w(e) := (t-1)n + 2$ if $e \in E(\overline{G})$. Show that:

   i) (K, w) has a Hamilton cycle of weight n if and only if G has a Hamilton cycle,

   ii) any Hamilton cycle of (K, w) of weight greater than n has weight at least $tn + 1$.

b) Deduce that, unless $\mathcal{P} = \mathcal{NP}$, there cannot exist a polynomial-time t-approximation algorithm for solving TSP.

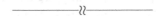

8.5 Greedy Heuristics

A *heuristic* is a computational procedure, generally based on some simple rule, which intuition tells one should usually yield a good approximate solution to the problem at hand.

One particularly simple and natural class of heuristics is the class of greedy heuristics. Informally, a *greedy heuristic* is a procedure which selects the best current option at each stage, without regard to future consequences. As can be imagined, such an approach rarely leads to an optimal solution in each instance. However, there are cases in which the greedy approach does indeed work. In such cases, we call the procedure a *greedy algorithm*. The following is a prototypical example of such an algorithm.

THE BORŮVKA–KRUSKAL ALGORITHM

The Jarník–Prim algorithm for the Minimum-Weight Spanning Tree Problem, described in Section 6.2, starts with the root and determines a nested sequence of trees, terminating with a minimum-weight spanning tree. Another algorithm for this problem, due to Borůvka (1926a,b) and, independently, Kruskal (1956), starts with the empty spanning subgraph and finds a nested sequence of forests, terminating with an optimal tree. This sequence is constructed by adding edges, one at a time, in such a way that the edge added at each stage is one of minimum weight, subject to the condition that the resulting subgraph is still a forest.

Algorithm 8.22 THE BORŮVKA–KRUSKAL ALGORITHM

    INPUT: a weighted connected graph $G = (G, w)$
    OUTPUT: an optimal tree $T = (V, F)$ of G, and its weight $w(F)$
    1: set $F := \emptyset$, $w(F) := 0$ *(F denotes the edge set of the current forest)*
    2: **while** *there is an edge $e \in E \setminus F$ such that $F \cup \{e\}$ is the edge set of a forest* **do**
    3:   *choose such an edge e of minimum weight*
    4:   *replace F by $F \cup \{e\}$ and $w(F)$ by $w(F) + w(e)$*
    5: **end while**
    6: *return* $((V, F), w(F))$

Because the graph G is assumed to be connected, the forest (V, F) returned by the Borůvka–Kruskal Algorithm is a spanning tree of G. We call it a *Borůvka–Kruskal tree*. The construction of such a tree in the electric grid graph of Section 6.2 is illustrated in Figure 8.3. As before, the edges are numbered according to the order in which they are added. Observe that this tree is identical to the one returned by the Jarník–Prim Algorithm (even though its edges are selected in a different order). This is because all the edge weights in the electric grid graph happen to be distinct (see Exercise 6.2.1).

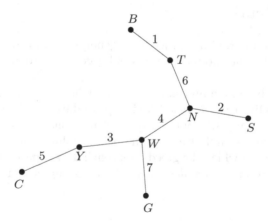

Fig. 8.3. An optimal tree returned by the Borůvka–Kruskal Algorithm

In order to implement the Borůvka–Kruskal Algorithm efficiently, one needs to be able to check easily whether a candidate edge links vertices in different components of the forest. This can be achieved by colouring vertices in the same component by the same colour and vertices in different components by distinct colours. It then suffices to check that the ends of the edge have different colours. Once the edge has been added to the forest, all the vertices in one of the two merged components are recoloured with the colour of the other component. We leave the details as an exercise (Exercise 8.5.1).

The following theorem shows that the Borůvka–Kruskal Algorithm runs correctly. Its proof resembles that of Theorem 6.10, and we leave it to the reader (Exercise 8.5.2).

Theorem 8.23 *Every Borůvka–Kruskal tree is an optimal tree.* ☐

The problem of finding a maximum-weight spanning tree of a connected graph can be solved by the same approach; at each stage, instead of picking an edge of minimum weight subject to the condition that the resulting subgraph remains a forest, we pick one of maximum weight subject to the same condition (see Exercise 8.5.3). The origins of the Borůvka–Kruskal Algorithm are recounted in Nešetřil et al. (2001) and Kruskal (1997).

INDEPENDENCE SYSTEMS

One can define a natural family of greedy heuristics which includes the Borůvka–Kruskal Algorithm in the framework of set systems.

A set system $(V, \mathcal{F})$ is called an *independence system* on V if $\mathcal{F}$ is nonempty and, for any member F of $\mathcal{F}$, all subsets of F also belong to $\mathcal{F}$. The members of $\mathcal{F}$ are then referred to as *independent sets* and their maximal elements as *bases*. (The independent sets of a matroid, defined in Section 4.4, form an independence system.)

Many independence systems can be defined on graphs. For example, if $G = (V, E)$ is a connected graph, we may define an independence system on V by taking as independent sets the cliques of G (including the empty set). Likewise, we may define an independence system on E by taking the edge sets of the forests of G as independent sets; the bases of this independence system are the edge sets of spanning trees. (This latter independence system is the cycle matroid of the graph, defined in Section 4.4.)

Consider, now, an arbitrary independence system $(V, \mathcal{F})$. Suppose that, with each element x of V, there is an associated nonnegative weight $w(x)$, and that we wish to find an independent set of maximum weight, where the *weight* of a set is defined to be the sum of the weights of its elements. A naive approach to this problem would be to proceed as follows.

Heuristic 8.24 GREEDY HEURISTIC (FOR INDEPENDENCE SYSTEMS)

INPUT: an independence system $(V, \mathcal{F})$ with weight function $w : V \to \mathbb{R}^+$
OUTPUT: a maximal independent set F of $(V, \mathcal{F})$, and its weight $w(F)$

1: *set* $F := \emptyset$, $w(F) := 0$
2: **while** *there is an element* $x \in V \setminus F$ *such that* $F \cup \{x\}$ *is independent* **do**
3: *choose such an element* x *of minimum weight* $w(x)$
4: *replace* F *by* $F \cup \{x\}$ *and* $w(F)$ *by* $w(F) + w(x)$
5: **end while**
6: *return* $(F, w(X))$

As we have seen, the GREEDY HEURISTIC always returns an optimal solution when the independence system consists of the edge sets of the forests of a graph, no matter what the edge weights. (More generally, as Rado (1957) observed, the GREEDY HEURISTIC performs optimally whenever $\mathcal{F}$ is the family of independent sets of a matroid, regardless of the weight function w.) By contrast, when the independent sets are the cliques of a graph, the GREEDY HEURISTIC rarely returns an optimal solution, even if all the weights are 1, because most graphs have maximal cliques which are not maximum cliques. (If the GREEDY HEURISTIC unfailingly returns a maximum weight independent set, no matter what the weight function of the independence system, then the system must, it turns out, be a matroid (see, for example, Oxley (1992)).)

We remark that greedy heuristics are by no means limited to the framework of independence systems. For example, if one is looking for a longest x-path in a graph, an obvious greedy heuristic is to start with the trivial x-path consisting of just the vertex x and iteratively extend the current x-path by any available edge. This amounts to applying depth-first search from x, stopping when one is forced to backtrack. The path so found will certainly be a maximal x-path, but not necessarily a longest x-path. Even so, this simple-minded greedy heuristic proves to be effective when combined with other ideas, as we show in Chapter 18.

Exercises

⋆**8.5.1** Refine the Borůvka–Kruskal Algorithm in such a way that, at each stage, vertices in the same component of the forest F are assigned the same colour and vertices in different components are assigned distinct colours.

⋆**8.5.2** Prove Theorem 8.23.

⋆**8.5.3** Show that the problem of finding a maximum-weight spanning tree of a connected graph can be solved by choosing, at each stage, an edge of maximum weight subject to the condition that the resulting subgraph is still a forest.

8.5.4 Give an example of a weighted independence system, all of whose bases have the same number of elements, but for which the GREEDY HEURISTIC (8.24) fails to return an optimal solution.

8.5.5 Consider the following set of real vectors.

$$V := \{(1,0,0,1),(1,1,1,1),(1,0,1,2),(0,1,0,1),(0,2,-1,1),(1,-1,0,0)\}$$

Find a linearly independent subset of V whose total number of zeros is maximum by applying the GREEDY HEURISTIC (8.24).

————————〰————————

8.6 Linear and Integer Programming

A *linear program* (*LP*) is a problem of maximizing or minimizing a linear function of real variables that are subject to linear equality or inequality constraints. By means of simple substitutions, such as replacing an equation by two inequalities, any LP may be transformed into one of the following form.

$$\text{maximize } \mathbf{cx} \text{ subject to } \mathbf{Ax} \leq \mathbf{b} \text{ and } \mathbf{x} \geq \mathbf{0} \qquad (8.4)$$

where $\mathbf{A} = (a_{ij})$ is an $m \times n$ matrix, $\mathbf{c} = (c_1, c_2, \ldots, c_n)$ a $1 \times n$ row vector, and $\mathbf{b} = (b_1, b_2, \ldots, b_m)$ an $m \times 1$ column vector. The m inequalities $\sum_{j=1}^{n} a_{ij} x_j \leq b_i$, $1 \leq i \leq m$, and the n nonnegativity conditions $x_j \geq 0$, $1 \leq j \leq n$, are known as the *constraints* of the problem. The function $\mathbf{cx}$ to be maximized is called the *objective function*.

A column vector $\mathbf{x} = (x_1, x_2, \ldots, x_n)$ in $\mathbb{R}^n$ is a *feasible solution* to (8.4) if it satisfies all $m + n$ constraints, and a feasible solution at which the objective function $\mathbf{cx}$ attains its maximum is an *optimal solution*. This maximum is the *optimal value* of the LP.

Associated with every LP, there is another LP, called its dual. The *dual* of the LP (8.4) is the LP:

$$\text{minimize } \mathbf{yb} \text{ subject to } \mathbf{yA} \geq \mathbf{c} \text{ and } \mathbf{y} \geq \mathbf{0} \qquad (8.5)$$

With reference to this dual LP, the original LP (8.4) is called the *primal* LP.

Not every LP has a feasible solution. Moreover, even if it does have one, it need not have an optimal solution: the objective function might be unbounded over the set of feasible solutions, and thus not achieve a maximum (or minimum). Such an LP is said to be *unbounded*.

The following proposition implies that if both the primal and the dual have feasible solutions, neither is unbounded.

Proposition 8.25   WEAK DUALITY THEOREM
Let $\mathbf{x}$ be a feasible solution to (8.4) and $\mathbf{y}$ a feasible solution to its dual (8.5). Then

$$\mathbf{cx} \leq \mathbf{yb} \qquad (8.6)$$

Proof Because $\mathbf{c} \leq \mathbf{yA}$ and $\mathbf{x} \geq \mathbf{0}$, we have $\mathbf{cx} \leq \mathbf{yAx}$. Likewise $\mathbf{yAx} \leq \mathbf{yb}$. Inequality (8.6) follows. $\square$

Corollary 8.26 *Let $\mathbf{x}$ be a feasible solution to (8.4) and $\mathbf{y}$ a feasible solution to its dual (8.5). Suppose that $\mathbf{cx} = \mathbf{yb}$. Then $\mathbf{x}$ is an optimal solution to (8.4) and $\mathbf{y}$ is an optimal solution to (8.5).* $\square$

The significance of this corollary is that if equality holds in (8.6), the primal solution $\mathbf{x}$ serves as a succinct certificate for the optimality of the dual solution $\mathbf{y}$, and *vice versa*. A remarkable and fundamental theorem due to von Neumann (1928) guarantees that one can always certify optimality in this manner.

Theorem 8.27   DUALITY THEOREM
*If an LP has an optimal solution, then its dual also has an optimal solution, and
the optimal values of these two LPs are equal.*   □

A wide variety of graph-theoretical problems may be formulated as LPs, albeit
with additional *integrality constraints*, requiring that the variables take on only
integer values. In some cases, these additional constraints may be ignored without
affecting the essential nature of the problem, because the LP under consideration
can be shown to always have optimal solutions that are integral. In such cases,
the duals generally have natural interpretations in terms of graphs, and interesting
results may be obtained by applying the Duality Theorem. Such results are referred
to as *min–max theorems*. As a simple example, let us consider the problem of
finding a maximum stable set in a graph.

It clearly suffices to consider graphs without isolated vertices. Let G be such
a graph. With any stable set S of G, we may associate its $(0, 1)$-valued incidence
vector $\mathbf{x} := (x_v : v \in V)$, where $x_v := 1$ if $v \in S$, and $x_v := 0$ otherwise. Because
no stable set can include more than one of the two ends of any edge, every such
vector $\mathbf{x}$ satisfies the constraint $x_u + x_v \leq 1$, for each $uv \in E$. Thus, the problem
MAX STABLE SET is equivalent to the following LP (where $\mathbf{M}$ is the incidence
matrix of G).

$$\text{maximize } \mathbf{1x} \text{ subject to } \mathbf{M}^t \mathbf{x} \leq \mathbf{1} \text{ and } \mathbf{x} \geq \mathbf{0} \tag{8.7}$$

together with the requirement that $\mathbf{x}$ be integral. The dual of (8.7) is the following
LP, in which there is a variable y_e for each edge e of G.

$$\text{minimize } \mathbf{y1} \text{ subject to } \mathbf{yM}^t \geq \mathbf{1} \text{ and } \mathbf{y} \geq \mathbf{0} \tag{8.8}$$

Consider an integer-valued feasible solution $\mathbf{y}$ to this dual LP. The support of $\mathbf{y}$
is a set of edges of G that together meet every vertex of G. Such a set of edges is
called an *edge covering* of G. The number of edges in a minimum edge covering of
a graph G without isolated vertices is denoted by $\beta'(G)$.

Conversely, the incidence vector of any edge covering of G is a feasible solution
to (8.8). Thus the optimal value of (8.8) is a lower bound on $\beta'(G)$. Likewise, the
optimal value of (8.7) is an upper bound on $\alpha(G)$. By the Weak Duality Theorem,
it follows that, for any graph G without isolated vertices, $\alpha(G) \leq \beta'(G)$. In general,
these two quantities are not equal (consider, for example, K_3). They are, however,
always equal for bipartite graphs (see inset).

A linear program in which the variables are constrained to take on only integer
values is called an *integer linear program* (*ILP*). Any ILP may be transformed to
one of the following form.

$$\text{maximize } \mathbf{cx} \text{ subject to } \mathbf{Ax} \leq \mathbf{b}, \ \mathbf{x} \geq \mathbf{0}, \text{ and } \mathbf{x} \in \mathbb{Z} \tag{8.9}$$

As already mentioned, MAX STABLE SET can be formulated as an ILP. Because
MAX STABLE SET is $\mathcal{NP}$-hard, so is ILP. On the other hand, there do exist
polynomial-time algorithms for solving linear programs, so LP is in $\mathcal{P}$.

PROOF TECHNIQUE: TOTAL UNIMODULARITY

Recall that a matrix $\mathbf{A}$ is *totally unimodular* if the determinant of each of its square submatrices is equal to 0, $+1$, or -1. The following theorem provides a sufficient condition for an LP to have an integer-valued optimal solution.

Theorem 8.28 *Suppose that $\mathbf{A}$ is a totally unimodular matrix and that $\mathbf{b}$ is an integer vector. If (8.4) has an optimal solution, then it has an integer optimal solution.*

Proof The set of points in $\mathbb{R}^n$ at which any single constraint holds with equality is a hyperplane in $\mathbb{R}^n$. Thus each constraint is satisfied by the points of a closed half-space of $\mathbb{R}^n$, and the set of feasible solutions is the intersection of all these half-spaces, a convex polyhedron P.

Because the objective function is linear, its level sets are hyperplanes. Thus, if the maximum value of $\mathbf{cx}$ over P is z^*, the hyperplane $\mathbf{cx} = z^*$ is a supporting hyperplane of P. Hence $\mathbf{cx} = z^*$ contains an extreme point (a corner) of P. It follows that the objective function attains its maximum at one of the extreme points of P.

Every extreme point of P is at the intersection of n or more hyperplanes determined by the constraints. It is thus a solution to a subsystem of $\mathbf{Ax} = \mathbf{b}$. Using the hypotheses of the theorem and applying Cramér's rule, we now conclude that each extreme point of P is an integer vector, and hence that (8.4) has an integer optimal solution.   □

Because the incidence matrix of a bipartite graph is totally unimodular (Exercise 4.2.4), as a consequence of the above theorem, we have:

Theorem 8.29 *Let G be a bipartite graph with incidence matrix $\mathbf{M}$. Then the LPs*

$$\text{maximize } \mathbf{1x} \quad \text{subject to } \mathbf{M}^t\mathbf{x} \le \mathbf{1} \quad \text{and} \quad \mathbf{x} \ge \mathbf{0} \qquad (8.10)$$

$$\text{minimize } \mathbf{y1} \quad \text{subject to } \mathbf{yM}^t \ge \mathbf{1} \quad \text{and} \quad \mathbf{y} \ge \mathbf{0} \qquad (8.11)$$

both have $(0,1)$-valued optimal solutions.   □

This theorem, in conjunction with the Duality Theorem, now implies the following min–max equality, due independently to D. König and R. Rado (see Schrijver (2003)).

Theorem 8.30 THE KÖNIG–RADO THEOREM
In any bipartite graph without isolated vertices, the number of vertices in a maximum stable set is equal to the number of edges in a minimum edge covering.   □

TOTAL UNIMODULARITY (CONTINUED)

The König–Rado Theorem (8.30) implies that the problem of deciding whether a bipartite graph has a stable set of cardinality k is in co-$\mathcal{NP}$; when the answer is 'no', an edge cover of size less than k provides a succinct certificate of this fact. In fact, as shown in Chapter 16, there is a polynomial algorithm for finding a maximum stable set in any bipartite graph.

A second application of this proof technique is given below, and another is presented in Section 19.2.

One approach to the problem of determining the value of a graph-theoretic parameter such as α is to express the problem as an ILP of the form (8.9) and then solve its *LP relaxation*, that is, the LP (8.4) obtained by dropping the integrality constraint $\mathbf{x} \in \mathbb{Z}$. If the optimal solution found happens to be integral, as in Theorem 8.29, it will also be an optimal solution to the ILP, and thus determine the exact value of the parameter. In any event, the value returned by the LP will be an upper bound on the value of the parameter. This upper bound is referred to as the *fractional version* of the parameter. For example, the LP (8.10) returns the *fractional stability number*, denoted α^*.

The fractional stability number of a graph may be computed in polynomial time. However, in general, α can be very much smaller than α^*. For example, $\alpha(K_n) = 1$, whereas $\alpha^*(K_n) = n/2$ for $n \geq 2$. Taking into cognisance the fact that no stable set of a graph can include more than one vertex of any clique of the graph, one may obtain a LP associated with MAX STABLE SET with tighter constraints than (8.7) (see Exercise 8.6.3).

MATCHINGS AND COVERINGS IN BIPARTITE GRAPHS

We now describe a second application of total unimodularity, to matchings in bipartite graphs. A *matching* in a graph is a set of pairwise nonadjacent links. With any matching M of a graph G, we may associate its $(0, 1)$-valued incidence vector. Since no matching has more than one edge incident with any vertex, every such vector $\mathbf{x}$ satisfies the constraint $\sum \{x_e : e \in \partial(v)\} \leq 1$, for all $v \in V$. Thus the problem of finding a largest matching in a graph is equivalent to the following ILP.

$$\text{maximize } \mathbf{1x} \text{ subject to } \mathbf{Mx} \leq \mathbf{1} \text{ and } \mathbf{x} \geq \mathbf{0} \qquad (8.12)$$

(where $\mathbf{M}$ is the incidence matrix of G), together with the requirement that $\mathbf{x}$ be integral. The dual of (8.12) is the following LP.

$$\text{minimize } \mathbf{y1} \text{ subject to } \mathbf{yM} \geq \mathbf{1} \text{ and } \mathbf{y} \geq \mathbf{0} \qquad (8.13)$$

Because the incidence matrix of a bipartite graph is totally unimodular (Exercise 4.2.4), as a consequence of the above theorem, we now have:

Theorem 8.31 *When G is bipartite, (8.12) and (8.13) have $(0,1)$-valued optimal solutions.*   □

If $\mathbf{x}$ is a $(0,1)$-valued feasible solution to (8.12), then no two edges of the set $M := \{e \in E : x_e = 1\}$ have an end in common; that is, M is a matching of G. Analogously, if $\mathbf{y}$ is a $(0,1)$-valued feasible solution to (8.13), then each edge of G has at least one end in the set $K := \{v \in V : y_v = 1\}$; such a set is called a *covering* of G. These two observations, together with the Duality Theorem, now imply the following fundamental min–max theorem, due independently to König (1931) and Egerváry (1931).

Theorem 8.32   THE KÖNIG–EGERVÁRY THEOREM
In any bipartite graph, the number of edges in a maximum matching is equal to the number of vertices in a minimum covering.   □

Just as the König–Rado Theorem (8.30) shows that the problem of deciding whether a bipartite graph $G[X,Y]$ has a stable set of k vertices is in co-$\mathcal{NP}$, the König–Egerváry Theorem shows that the problem of deciding whether such a graph has a matching of k edges is in co-$\mathcal{NP}$. When the answer is 'no', a covering of cardinality less than k provides a succinct certificate of this fact. The König–Egerváry Theorem can also be derived from the arc version of Menger's Theorem (7.16) (see Exercise 8.6.7). The maximum number of edges in a matching of a graph G is called the *matching number* of G and denoted $\alpha'(G)$.

If G is nonbipartite, (8.12) may have optimal solutions that are not integral. For example, when G is a triangle, it can be seen that $(1/2, 1/2, 1/2)$ is an optimal solution. However, Edmonds (1965b) showed that one may introduce additional constraints that are satisfied by all the incidence vectors of matchings in a graph so that the resulting linear program has integer optimal solutions (see Exercise 8.6.8). This was the basis for his solution to the optimal matching problem. Matchings are discussed in detail in Chapter 16.

Using the fact that the incidence matrix of a directed graph is totally unimodular (Exercise 1.5.7a), Menger's Theorem (7.16) may be derived from the Duality Theorem. Further examples of min–max theorems are presented in Chapters 16 and 19. For additional information on these and other applications of linear programming, see Chvátal (1983), Lovász and Plummer (1986), and Schrijver (2003).

Exercises

8.6.1 COMPLEMENTARY SLACKNESS
Let $\mathbf{x}$ and $\mathbf{y}$ be feasible solutions to the LP (8.4) and its dual (8.5), respectively. Show that these solutions are optimal if and only if:

$$\sum_{j=1}^{n} a_{ij}x_j < b_i \Rightarrow y_i = 0, \quad 1 \le i \le m, \quad \text{and}$$
$$\sum_{i=1}^{m} a_{ij}y_i > c_j \Rightarrow x_j = 0, \quad 1 \le j \le n$$

(The above conditions are known as the *complementary slackness conditions* for optimality. They play an important role in the solution of optimization problems involving weighted graphs.)

⋆**8.6.2** CLIQUE COVERING

A *clique covering* of a graph is a set of cliques whose union is the entire vertex set of the graph.

a) Show that the stability number of a graph is bounded above by the minimum number of cliques in a clique covering.
b) Give an example of a graph in which these two quantities are unequal.

8.6.3 Let $\mathcal{K}$ denote the set of all cliques of a graph G and let $\mathbf{K}$ denote the incidence matrix of the hypergraph $(V, \mathcal{K})$. Consider the LP:

$$\text{maximize } \mathbf{1x} \text{ subject to } \mathbf{K}^t \mathbf{x} \le \mathbf{1} \text{ and } \mathbf{x} \ge \mathbf{0} \qquad (8.14)$$

and its dual:

$$\text{minimize } \mathbf{y1} \text{ subject to } \mathbf{yK}^t \ge \mathbf{1} \text{ and } \mathbf{y} \ge \mathbf{0} \qquad (8.15)$$

Show that:

a) an integer-valued vector $\mathbf{x}$ in $\mathbb{R}^V$ is a feasible solution to (8.14) if and only if it is the incidence vector of a stable set of G,
b) a $(0, 1)$-valued vector $\mathbf{y}$ in $\mathbb{R}^{\mathcal{K}}$ is a feasible solution to (8.15) if and only if it is the incidence vector of a clique covering.

8.6.4 Show that the set of feasible solutions to (8.14) is a subset of the set of feasible solutions to (8.7), with equality when G is triangle-free.

8.6.5 Let $\alpha^{**}(G)$ denote the optimal value of (8.14).

a) Show that, for any graph G, $\alpha \le \alpha^{**} \le \alpha^*$.
b) Give examples of graphs for which these inequalities are strict.

8.6.6 Let G be a simple graph with $n \ge 3$, and let $\mathbf{x} := (x_e : e \in E) \in \mathbb{R}^E$. Consider the following system of linear inequalities.

$$\sum_{e \in \partial(X)} x_e \ge 2 \text{ if } \emptyset \subset X \subset V$$

$$\sum_{e \in \partial(v)} x_e = 2 \text{ for all } v \in V$$

$$x_e \le 1 \text{ for all } e \in E$$

$$x_e \ge 0 \text{ for all } e \in E$$

a) Show that the integer-valued feasible solutions to this system are precisely the incidence vectors of the Hamilton cycles of G.

b) Let $\mathbf{c} \in \mathbb{R}^E$ be a weight function on G. Deduce from (a) that an optimal solution to the TSP for this weighted graph is provided by an optimal solution to the ILP that consists of maximizing the objective function $\mathbf{cx}$ subject to the above constraints, together with the integrality constraint $\mathbf{x} \in \mathbb{Z}$.
(Grötschel et al. (1988) have given a polynomial-time algorithm for solving the LP relaxation of this ILP.)

$\star$**8.6.7**

a) Transform the problem of finding a maximum matching in a bipartite graph $G[X, Y]$ into the problem of finding a maximum collection of arc-disjoint directed (x, y)-paths in a related digraph $D(x, y)$.

b) Deduce the König–Egerváry Theorem from Menger's Theorem (7.16).

8.6.8 Show that every integer feasible solution $\mathbf{x}$ to the LP (8.12) satisfies the inequality

$$\sum_{e \in E(X)} x_e \leq \frac{1}{2}(|X| - 1)$$

for any odd subset X of V of cardinality three or more.
(Edmonds (1965b) showed that, by adding these inequalities to the set of constraints in (8.12), one obtains an LP every optimal solution of which is $(0,1)$-valued.)

$\star$**8.6.9** FARKAS' LEMMA
The following two LPs are duals of each other.

$$\text{maximize} \quad \mathbf{0x} \quad \text{subject to} \quad \mathbf{Ax} = \mathbf{0} \quad \text{and} \quad \mathbf{x} \geq \mathbf{b}$$
$$\text{minimize} \quad -\mathbf{zb} \quad \text{subject to} \quad \mathbf{yA} - \mathbf{z} = \mathbf{0} \quad \text{and} \quad \mathbf{z} \geq \mathbf{0}$$

Farkas' Lemma (see Section 20.1) says that exactly one of the two linear systems:

$$\mathbf{Ax} = \mathbf{0}, \quad \mathbf{x} \geq \mathbf{b} \quad \text{and} \quad \mathbf{yA} \geq \mathbf{0}, \quad \mathbf{yAb} > 0$$

has a solution. Deduce Farkas' Lemma from the Duality Theorem (8.27).

$\star$**8.6.10** The following two LPs are duals of each other.

$$\text{minimize} \quad \mathbf{y0} \quad \text{subject to} \quad \mathbf{yA} \geq \mathbf{c}$$
$$\text{maximize} \quad \mathbf{cx} \quad \text{subject to} \quad \mathbf{Ax} = \mathbf{0} \quad \text{and} \quad \mathbf{x} \geq \mathbf{0}$$

A variant of Farkas' Lemma says that exactly one of the two linear systems:

$$\mathbf{yA} \geq \mathbf{c} \quad \text{and} \quad \mathbf{Ax} = \mathbf{0}, \quad \mathbf{x} \geq \mathbf{0}, \quad \mathbf{cx} > 0$$

has a solution. Deduce this variant of Farkas' Lemma from the Duality Theorem (8.27).

8.6.11 Prove the inequality $\alpha(G) \leq \beta'(G)$ directly, without appealing to the Weak Duality Theorem (8.25).

8.7 Related Reading

ISOMORPHISM-COMPLETENESS

As mentioned earlier, the complexity status of GRAPH ISOMORPHISM is unknown. There is strong theoretical evidence to support the belief that the problem is not $\mathcal{NP}$-complete (see, for example, Babai (1995)), and its rather unique status has led to the notion of isomorphism-completeness: a problem is said to be *isomorphism-complete* if it is polynomially equivalent to GRAPH ISOMORPHISM. The Legitimate Deck Problem, mentioned in Section 2.8, is one such problem (see Harary et al. (1982) and Mansfield (1982)). The problem of finding the orbits of a graph is 'isomorphism-hard'. For these and other examples, we refer the reader to Babai (1995).

9

***

Connectivity

Contents

9.1 Vertex Connectivity

In Section 3.1, we discussed the concept of connection in graphs. Consider, now, the four connected graphs in Figure 9.1.

G_1 is a tree, a minimal connected graph; deleting any edge disconnects it. G_2 cannot be disconnected by the deletion of a single edge, but can be disconnected

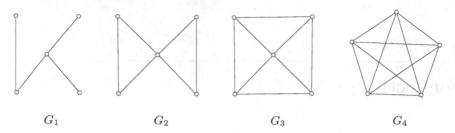

G_1 G_2 G_3 G_4

Fig. 9.1. Four connected graphs

by the deletion of one vertex, its cut vertex. There are no cut edges or cut vertices in G_3, but even so G_3 is clearly not as well connected as G_4, the complete graph on five vertices. Thus, intuitively, each successive graph is better connected than the previous one. We now introduce two parameters of a graph, its connectivity and edge connectivity, which measure the extent to which it is connected. We first define these parameters in terms of numbers of disjoint paths connecting pairs of vertices, and then relate those definitions to sizes of vertex and edge cuts, as suggested by the above examples.

CONNECTIVITY AND LOCAL CONNECTIVITY

We begin by discussing the notion of vertex connectivity, commonly referred to simply as connectivity. Recall that xy-paths P and Q in G are *internally disjoint* if they have no internal vertices in common, that is, if $V(P) \cap V(Q) = \{x, y\}$. The *local connectivity* between distinct vertices x and y is the maximum number of pairwise internally disjoint xy-paths, denoted $p(x, y)$; the local connectivity is undefined when $x = y$. The matrix in Figure 9.2b displays the local connectivities between all pairs of vertices of the graph shown in Figure 9.2a. (The function shown in Figure 9.2c will be defined shortly.)

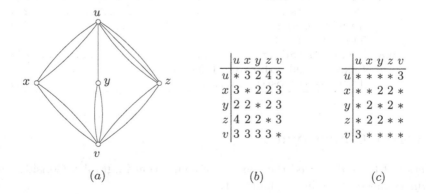

	u	x	y	z	v
u	*	3	2	4	3
x	3	*	2	2	3
y	2	2	*	2	3
z	4	2	2	*	3
v	3	3	3	3	*

	u	x	y	z	v
u	*	*	*	*	3
x	*	*	2	2	*
y	*	2	*	2	*
z	*	2	2	*	*
v	3	*	*	*	*

(a)   (b)   (c)

Fig. 9.2. (a) A graph, (b) its local connectivity function, and (c) its local cut function

A nontrivial graph G is *k-connected* if $p(u, v) \geq k$ for any two distinct vertices u and v. By convention, a trivial graph is 0-connected and 1-connected, but is not k-connected for any $k > 1$. The *connectivity* $\kappa(G)$ of G is the maximum value of k for which G is k-connected. Thus, for a nontrivial graph G,

$$\kappa(G) := \min\{p(u, v) :\ u, v \in V,\ u \neq v\} \tag{9.1}$$

A graph is 1-connected if and only if it is connected; equivalently, a graph has connectivity zero if and only if it is disconnected. Nonseparable graphs on at least three vertices are 2-connected; conversely, every 2-connected loopless graph is nonseparable. For the four graphs shown in Figure 9.1, $\kappa(G_1) = 1$, $\kappa(G_2) = 1$, $\kappa(G_3) = 3$, and $\kappa(G_4) = 4$. Thus, of these four graphs, the only graph that is 4-connected is G_4. Graphs G_3 and G_4 are 2-connected and 3-connected, whereas G_1 and G_2 are not. And, because all four graphs are connected, they are all 1-connected. The graph in Figure 9.2a is 1-connected and 2-connected, but is not 3-connected. Thus, the connectivity of this graph is two.

VERTEX CUTS AND MENGER'S THEOREM

We now rephrase the definition of connectivity in terms of 'vertex cuts'. This is not totally straightforward because complete graphs (and, more generally, graphs in which any two vertices are adjacent) have no such cuts. For this reason, we first determine the connectivities of these graphs.

Distinct vertices x and y of K_n are connected by one path of length one and $n - 2$ internally disjoint paths of length two. It follows that $p(x, y) = n - 1$ and that $\kappa(K_n) = n - 1$ for $n \geq 2$. More generally, if the underlying simple graph of a graph G is complete, and x and y are joined by $\mu(x, y)$ links, there are $\mu(x, y)$ paths of length one, and $n - 2$ internally disjoint paths of length two connecting x and y; thus $p(x, y) = n - 2 + \mu(x, y)$. Hence the connectivity of a nontrivial graph G in which any two vertices are adjacent is $n - 2 + \mu$, where μ is the minimum edge multiplicity in the graph. On the other hand, if x and y are nonadjacent, there are at most $n - 2$ internally disjoint paths connecting x and y. Thus, if the underlying simple graph of a graph G is not complete, its connectivity κ cannot exceed $n - 2$. For such a graph, the connectivity is equal to the minimum number of vertices whose deletion results in a disconnected graph, as we now explain.

Let x and y be distinct nonadjacent vertices of G. An *xy-vertex-cut* is a subset S of $V \setminus \{x, y\}$ such that x and y belong to different components of $G - S$. We also say that such a subset S *separates* x *and* y. The minimum size of a vertex cut separating x and y is denoted by $c(x, y)$. This function, the *local cut function* of G, is not defined if either $x = y$ or x and y are adjacent. The matrix displayed in Figure 9.2c gives the values of the local cut function of the graph shown in Figure 9.2a.

A vertex cut separating some pair of nonadjacent vertices of G is a *vertex cut* of G, and one with k elements is a *k-vertex cut*. A complete graph has no vertex cuts; moreover, the only graphs which do not have vertex cuts are those whose underlying simple graphs are complete. We now show that, if G has at least one pair of nonadjacent vertices, the size of a minimum vertex cut of G is equal to the connectivity of G. The main ingredient required is a version of Menger's Theorem which relates the two functions p and c.

Finding the maximum number of internally disjoint xy-paths in a graph $G := G(x, y)$ amounts to finding the maximum number of internally disjoint directed (x, y)-paths in the associated digraph $D(x, y) := D(G)$. In turn, as noted in Section 8.3, the latter problem may be reduced to one of finding the maximum number of arc-disjoint directed (x, y)-paths in a related digraph $D'(x, y)$ (of order $2n - 2$), and this number can be determined by the Max-Flow Min-Cut Algorithm (7.9). Thus the Max-Flow Min-Cut Algorithm may be adapted to find, in polynomial time, the maximum number of internally disjoint xy-paths in G. The same algorithm will also return an xy-vertex-cut whose cardinality is equal to the maximum number of internally disjoint xy-paths, implying the validity of the following fundamental theorem of Menger (1927). We include here a simple inductive proof of this theorem due to Göring (2000), as an alternative to the above-mentioned constructive proof.

For this purpose, we need the operation of shrinking a set of vertices in a graph. Let G be a graph and let X be a proper subset of V. To *shrink* X is to delete all edges between vertices of X and then identify the vertices of X into a single vertex. We denote the resulting graph by G / X.

Theorem 9.1   MENGER'S THEOREM (UNDIRECTED VERTEX VERSION)
In any graph $G(x, y)$, where x and y are nonadjacent, the maximum number of pairwise internally disjoint xy-paths is equal to the minimum number of vertices in an xy-vertex-cut, that is,

$$p(x, y) = c(x, y)$$

Proof   By induction on $e(G)$. For convenience, let us set $k := c_G(x, y)$. Note that $p_G(x, y) \leq k$, because any family $\mathcal{P}$ of internally disjoint xy-paths meets any xy-vertex-cut in at least $|\mathcal{P}|$ distinct vertices. Thus it suffices to show that $p_G(x, y) \geq k$. We may assume that there is an edge $e = uv$ incident neither with x nor with y; otherwise, every xy-path is of length two, and the conclusion follows easily.

Set $H := G \setminus e$. Because H is a subgraph of G, $p_G(x, y) \geq p_H(x, y)$. Also, by induction, $p_H(x, y) = c_H(x, y)$. Furthermore, $c_G(x, y) \leq c_H(x, y) + 1$ because any xy-vertex-cut in H, together with either end of e, is an xy-vertex-cut in G. We therefore have:

$$p_G(x, y) \geq p_H(x, y) = c_H(x, y) \geq c_G(x, y) - 1 = k - 1$$

We may assume that equality holds throughout; if not, $p_G(x, y) \geq k$ and there is nothing more to prove. Thus, in particular, $c_H(x, y) = k - 1$. Let $S := \{v_1, \ldots, v_{k-1}\}$ be a minimum xy-vertex-cut in H, let X be the set of vertices reachable from x in $H - S$, and let Y be the set of vertices reachable from y in $H - S$. Because $|S| = k - 1$, the set S is not an xy-vertex-cut of G, so there is an

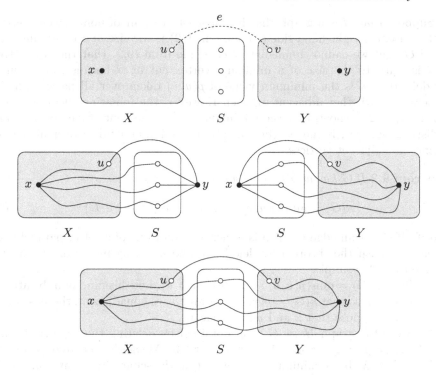

Fig. 9.3. Proof of Menger's Theorem (9.1)

xy-path in $G - S$. This path necessarily includes the edge e. We may thus assume, without loss of generality, that $u \in X$ and $v \in Y$.

Now consider the graph G / Y obtained from G by shrinking Y to a single vertex y. Every xy-vertex-cut T in G / Y is necessarily an xy-vertex-cut in G, because if P were an xy-path in G avoiding T, the subgraph P / Y of G / Y would contain an xy-path in G / Y avoiding T, too. Therefore $c_{G/Y}(x, y) \geq k$. On the other hand, $c_{G/Y}(x, y) \leq k$ because $S \cup \{u\}$ is an xy-vertex-cut of G / Y. It follows that $S \cup \{u\}$ is a minimum xy-vertex-cut of G / Y. By induction, there are k internally disjoint xy-paths $P_1, \ldots, P_k$ in G / Y, and each vertex of $S \cup \{u\}$ lies on one of them. Without loss of generality, we may assume that $v_i \in V(P_i)$, $1 \leq i \leq k - 1$, and $u \in V(P_k)$. Likewise, there are k internally disjoint xy-paths $Q_1, \ldots, Q_k$ in the graph G / X obtained by shrinking X to x with $v_i \in V(Q_i)$, $1 \leq i \leq k - 1$, and $v \in Q_k$. It follows that there are k internally disjoint xy-paths in G, namely $xP_iv_iQ_iy$, $1 \leq i \leq k - 1$, and xP_kuvQ_ky (see Figure 9.3, where the vertices not in $X \cup S \cup Y$ are omitted, as they play no role in the proof.). $\square$

As a consequence of Theorem 9.1 we have:

$$\min\{p(u, v) : u, v \in V, u \neq v, uv \notin E\} = \min\{c(u, v) : u, v \in V, u \neq v, uv \notin E\} \tag{9.2}$$

Suppose that G is a graph that has at least one pair of nonadjacent vertices. In this case, the right-hand side of equation (9.2) is the size of a minimum vertex cut of G. But we cannot immediately conclude from (9.2) that the connectivity of G is equal to the size of a minimum vertex cut of G because, according to our definition, κ is the minimum value of $p(u,v)$ taken over all pairs of distinct vertices u, v (whether adjacent or not). However, the following theorem, due to Whitney (1932a), shows that the minimum local connectivity taken over all pairs of distinct vertices, is indeed the same as the minimum taken over all pairs of distinct nonadjacent vertices.

Theorem 9.2 *If G has at least one pair of nonadjacent vertices,*

$$\kappa(G) = \min\{p(u,v) : \ u,v \in V, \ u \neq v, uv \notin E\} \tag{9.3}$$

Proof If G has an edge e which is either a loop or one of a set of parallel edges, we can establish the theorem by deleting e and applying induction. So we may assume that G is simple.

By (9.1), $\kappa(G) = \min\{p(u,v) : \ u,v \in V, \ u \neq v\}$. Let this minimum be attained for the pair x, y, so that $\kappa(G) = p(x,y)$. If x and y are nonadjacent, there is nothing to prove. So suppose that x and y are adjacent.

Consider the graph $H := G \setminus xy$, obtained by deleting the edge xy from G. Clearly, $p_G(x,y) = p_H(x,y) + 1$. Furthermore, by Menger's Theorem, $p_H(x,y) = c_H(x,y)$. Let X be a minimum vertex cut in H separating x and y, so that $p_H(x,y) = c_H(x,y) = |X|$, and $p_G(x,y) = |X| + 1$. If $V \setminus X = \{x,y\}$, then

$$\kappa(G) = p_G(x,y) = |X| + 1 = (n-2) + 1 = n - 1$$

But this implies that G is complete, which is contrary to the hypothesis. So we may assume that $V \setminus X$ has at least three vertices, x, y, z. We may also assume, interchanging the roles of x and y if necessary, that x and z belong to different components of $H - X$. Then x and z are nonadjacent in G and $X \cup \{y\}$ is a vertex cut of G separating x and z. Therefore,

$$c(x,z) \leq |X \cup \{y\}| = p(x,y)$$

On the other hand, by Menger's Theorem, $p(x,z) = c(x,z)$. Hence $p(x,z) \leq p(x,y)$. By the choice of $\{x,y\}$, we have $p(x,z) = p(x,y) = \kappa(G)$. Because x and z are nonadjacent,

$$\kappa(G) = p(x,z) = \min\{p(u,v) : u,v \in V, \ u \neq v, uv \notin E\} \qquad \square$$

It follows from Theorems 9.1 and 9.2 that the connectivity of a graph G which has at least one pair of nonadjacent vertices is equal to the size of a minimum vertex cut of G. In symbols,

$$\kappa(G) = \min\{c(u,v) : \ u,v \in V, \ u \neq v, uv \notin E\} \tag{9.4}$$

The vertex cuts of a graph are the same as those of its underlying simple graph, thus (9.4) implies that the connectivity of a graph which has at least one pair of nonadjacent vertices is the same as the connectivity of its underlying simple graph.

As noted in Section 8.2, for every nonadjacent pair x, y of vertices of G, the value of $c(x, y)$ may be computed by running the Max-Flow Min-Cut Algorithm (7.9) on an auxiliary digraph of order $2n - 2$ with unit capacities. It follows that the connectivity of any graph may be computed in polynomial time.

Exercises

9.1.1 Consider the vertices $x = (0, 0, \ldots, 0)$ and $y = (1, 1, \ldots, 1)$ of the n-cube Q_n. Describe a maximum collection of edge-disjoint xy-paths in Q_n and a minimum vertex cut of Q_n separating x and y.

9.1.2 Let G and H be simple graphs. Show that $\kappa(G \vee H) = \min\{v(G) + \kappa(H), v(H) + \kappa(G)\}$.

9.1.3

a) Show that if G is simple and $\delta \geq n - 2$, then $\kappa = \delta$.
b) For each $n \geq 4$, find a simple graph G with $\delta = n - 3$ and $\kappa < \delta$.

9.1.4 Show that if G is simple, with $n \geq k + 1$ and $\delta \geq (n + k - 2)/2$, then G is k-connected.

$\star$**9.1.5** Let G be a 2-connected graph and let e be an edge of G such that $G \,/\, e$ is not 2-connected. Prove that $G \,/\, e$ has exactly one cut vertex, namely the vertex resulting from the contraction of the edge e.

9.1.6 An edge e of a 2-connected graph G is called *contractible* if $G \,/\, e$ is 2-connected also. (The analogous concept, for nonseparable graphs, was defined in Exercise 5.3.2.) Show that every 2-connected graph on three or more vertices has a contractible edge.

9.1.7 An edge of a graph G is *deletable* (with respect to connectivity) if $\kappa(G \setminus e) = \kappa(G)$. Show that each edge of a 2-connected graph on at least four vertices is either deletable or contractible.

9.1.8 Let G be a connected graph which is not complete. Show that G is k-connected if and only if any two vertices at distance two are connected by k internally disjoint paths.

9.1.9 Consider the following statement, which resembles Menger's Theorem. *Let $G(x, y)$ be a graph of diameter d, where x and y are nonadjacent vertices. Then the maximum number of internally disjoint xy-paths of length d or less is equal to the minimum number of vertices whose deletion destroys all xy-paths of length d or less.*

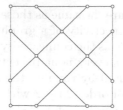

Fig. 9.4. A counterexample to Menger's Theorem for short paths (Exercise 9.1.9)

a) Prove this statement for $d = 2$.
b) Verify that the graph shown in Figure 9.4 is a counterexample to the statement in general. (J.A. BONDY AND P. HELL)

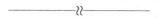

9.1.10 A k-connected graph G is *minimally k-connected* if the graph $G \setminus e$ is not k-connected for any edge e (that is, if no edge is deletable).

a) Let G be a minimally 2-connected graph. Show that:
   i) $\delta = 2$,
   ii) if $n \geq 4$, then $m \leq 2n - 4$.
b) For all $n \geq 4$, find a minimally 2-connected graph with n vertices and $2n - 4$ edges.

9.1.11

a) Show that if G is a k-connected graph and e is any edge of G, then G / e is $(k-1)$-connected.
b) For each $k \geq 4$, find a k-connected graph $G \neq K_{k+1}$ such that $\kappa(G / e) = k - 1$, for every edge e of G.

9.1.12

a) Let $D := D(X, Y)$ be a directed graph, where X and Y are disjoint subsets of V. Obtain an undirected graph G from D as follows.
   ▷ For each vertex v of D, replace v by two adjacent vertices, v^- and v^+.
   ▷ For each arc (u, v) of D, join u^+ and v^- by an edge.
   ▷ Delete the set of vertices $\{x^- : x \in X\} \cup \{y^+ : y \in Y\}$.
   Observe that G is a bipartite graph with bipartition

   $$(\{v^- : v \in V(D)\} \setminus \{x^- : x \in X\}, \{v^+ : v \in V(D)\} \setminus \{y^+ : y \in Y\})$$

   Show that:
   i) $\alpha'(G) = |V(D)| - |X \cup Y| + p_D(X, Y)$, where $p_D(X, Y)$ denotes the maximum number of disjoint directed (X, Y)-paths in D,
   ii) $\beta(G) = |V(D)| - |X \cup Y| + c_D(X, Y)$, where $c_D(X, Y)$ denotes the minimum number of vertices whose deletion destroys all directed (X, Y)-paths in D.

(A. Schrijver)

b) Derive Menger's Theorem (9.8) from the König–Egerváry Theorem (8.32).

9.1.13 Let xPy be a path in a graph G. Two vines on P (defined in Exercise 5.3.12) are *disjoint* if:

▷ their constituent paths are internally disjoint,
▷ x is the only common initial vertex of two paths in these vines,
▷ y is the only common terminal vertex of two paths in these vines.

If G is k-connected, where $k \geq 2$, show that there are $k-1$ pairwise disjoint vines on P.   (S.C. Locke)

9.1.14 Let P be a path in a 3-connected cubic graph G.

a) Consider two disjoint vines on P. Denote by F the union of P and the constituent paths of these two vines. Show that F admits a double cover by three cycles.
b) Deduce that if P is of length l, then G has a cycle of length greater than $2l/3$. (Compare Exercise 5.3.12.)   (J.A. Bondy and S.C. Locke)

9.1.15 Let G be a 3-connected graph, and let e and f be two edges of G. Show that:

a) the subspace generated by the cycles through e and f has dimension $\dim(\mathcal{C})-1$,
                                                                        (C. Thomassen)
b) G has an odd cycle through e and f unless $G \setminus \{e, f\}$ is bipartite.
                                                    (W.D. McCuaig and M. Rosenfeld)

9.2 The Fan Lemma

One can deduce many properties of a graph merely from a knowledge of its connectivity. In this context, Menger's Theorem, or a derivative of it, invariably plays a principal role. We describe here a very useful consequence of Menger's Theorem known as the *Fan Lemma*, and apply it to deduce a theorem of Dirac (1952b) about cycles in k-connected graphs.

The following lemma establishes a simple but important property of k-connected graphs.

Lemma 9.3 *Let G be a k-connected graph and let H be a graph obtained from G by adding a new vertex y and joining it to at least k vertices of G. Then H is also k-connected.*

Proof The conclusion clearly holds if any two vertices of H are adjacent, because $v(H) \geq k+1$. Let S be a subset of $V(H)$ with $|S| = k-1$. To complete the proof, it suffices to show that $H - S$ is connected.

Suppose first that $y \in S$. Then $H - S = G - (S \setminus \{y\})$. By hypothesis, G is k-connected and $|S \setminus \{y\}| = k-2$. We deduce that $H - S$ is connected.

Now suppose that $y \notin S$. Since, by hypothesis, y has at least k neighbours in $V(G)$ and $|S| = k - 1$, there is a neighbour z of y which does not belong to S. Because G is k-connected, $G - S$ is connected. Furthermore, z is a vertex of $G - S$, and hence yz is an edge of $H - S$. It follows that $(G - S) + yz$ is a spanning connected subgraph of $H - S$. Hence $H - S$ is connected. $\square$

The following useful property of k-connected graphs can be deduced from Lemma 9.3.

Proposition 9.4 *Let G be a k-connected graph, and let X and Y be subsets of V of cardinality at least k. Then there exists in G a family of k pairwise disjoint (X, Y)-paths.*

Proof Obtain a new graph H from G by adding vertices x and y and joining x to each vertex of X and y to each vertex of Y. By Lemma 9.3, H is k-connected. Therefore, by Menger's Theorem, there exist k internally disjoint xy-paths in H. Deleting x and y from each of these paths, we obtain k disjoint paths $Q_1, Q_2, \ldots, Q_k$ in G, each of which has its initial vertex in X and its terminal vertex in Y. Every path Q_i necessarily contains a segment P_i with initial vertex in X, terminal vertex in Y, and no internal vertex in $X \cup Y$, that is, an (X, Y)-path. The paths $P_1, P_2, \ldots, P_k$ are pairwise disjoint (X, Y)-paths. $\square$

A family of k internally disjoint (x, Y)-paths whose terminal vertices are distinct is referred to as a k-*fan* from x to Y. The following assertion is another very useful consequence of Menger's Theorem. Its proof is similar to the proof of Proposition 9.4 (Exercise 9.2.1).

Proposition 9.5   THE FAN LEMMA
Let G be a k-connected graph, let x be a vertex of G, and let $Y \subseteq V \setminus \{x\}$ be a set of at least k vertices of G. Then there exists a k-fan in G from x to Y. $\square$

We now give the promised application of the Fan Lemma. By Theorem 5.1, in a 2-connected graph any two vertices are connected by two internally disjoint paths; equivalently, any two vertices in a 2-connected graph lie on a common cycle. Dirac (1952b) generalized this latter statement to k-connected graphs.

Theorem 9.6 *Let S be a set of k vertices in a k-connected graph G, where $k \geq 2$. Then there is a cycle in G which includes all the vertices of S.*

Proof By induction on k. We have already observed that the assertion holds for $k = 2$, so assume that $k \geq 3$. Let $x \in S$, and set $T := S \setminus x$. Because G is k-connected, it is $(k - 1)$-connected. Therefore, by the induction hypothesis, there is a cycle C in G which includes T. Set $Y := V(C)$. If $x \in Y$, then C includes all the vertices of S. Thus we may assume that $x \notin Y$. If $|Y| \geq k$, the Fan Lemma (Proposition 9.5) ensures the existence of a k-fan in G from x to Y. Because $|T| = k - 1$, the set T divides C into $k - 1$ edge-disjoint segments. By the Pigeonhole Principle, some two paths of the fan, P and Q, end in the same

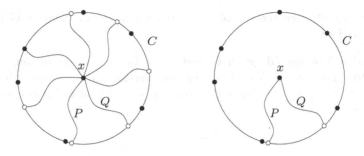

Fig. 9.5. Proof of Theorem 9.6

segment. The subgraph $C \cup P \cup Q$ contains three cycles, one of which includes $S = T \cup \{x\}$ (see Figure 9.5). If $|Y| = k - 1$, the Fan Lemma yields a $(k-1)$-fan from x to Y in which each vertex of Y is the terminus of one path, and we conclude as before. $\qquad\square$

It should be pointed out that the order in which the vertices of S occur on the cycle whose existence is established in Theorem 9.6 cannot be specified in advance. For example, the 4-connected graph shown in Figure 9.6 has no cycle including the four vertices x_1, y_1, x_2, y_2 in this exact order, because every x_1y_1-path intersects every x_2y_2-path.

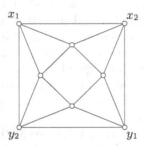

Fig. 9.6. No cycle includes x_1, y_1, x_2, y_2 in this order

Exercises

⋆**9.2.1** Give a proof of the Fan Lemma (Proposition 9.5).

9.2.2 Show that a 3-connected nonbipartite graph contains at least four odd cycles.

⋆**9.2.3** Let C be a cycle of length at least three in a nonseparable graph G, and let S be a set of three vertices of C. Suppose that some component H of $G - V(C)$

is adjacent to all three vertices of S. Show that there is a 3-fan in G from some vertex v of H to S.

9.2.4 Find a 5-connected graph G and a set $\{x_1, y_1, x_2, y_2\}$ of four vertices in G, such that no cycle of G contains all four vertices in the given order. (In a 6-connected graph, it can be shown that there is a cycle containing any four vertices in any prescribed order.)

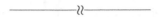

9.2.5 Let G be a graph, x a vertex of G, and Y and Z subsets of $V \setminus \{x\}$, where $|Y| < |Z|$. Suppose that there are fans from x to Y and from x to Z. Show that there is a fan from x to $Y \cup \{z\}$ for some $z \in Z \setminus Y$.   (H. PERFECT)

9.3 Edge Connectivity

We now turn to the notion of edge connectivity. The *local edge connectivity* between distinct vertices x and y is the maximum number of pairwise edge-disjoint xy-paths, denoted $p'(x, y)$; the local edge connectivity is undefined when $x = y$. A nontrivial graph G is *k-edge-connected* if $p'(u, v) \geq k$ for any two distinct vertices u and v of G. By convention, a trivial graph is both 0-edge-connected and 1-edge-connected, but is not k-edge-connected for any $k > 1$. The *edge connectivity* $\kappa'(G)$ of a graph G is the maximum value of k for which G is k-edge-connected.

A graph is 1-edge-connected if and only if it is connected; equivalently, the edge connectivity of a graph is zero if and only if it is disconnected. For the four graphs shown in Figure 9.1, $\kappa'(G_1) = 1$, $\kappa'(G_2) = 2$, $\kappa'(G_3) = 3$, and $\kappa'(G_4) = 4$. Thus, of these four graphs, the only graph that is 4-edge-connected is G_4. Graphs G_3 and G_4 are 3-edge-connected, but G_1 and G_2 are not. Graphs G_2, G_3, and G_4 are 2-edge-connected, but G_1 is not. And, because all four graphs are connected, they are all 1-edge-connected.

For distinct vertices x and y of a graph G, recall that an edge cut $\partial(X)$ *separates* x and y if $x \in X$ and $y \in V \setminus X$. We denote by $c'(x, y)$ the minimum cardinality of such an edge cut. With this notation, we may now restate the edge version of Menger's Theorem (7.17).

Theorem 9.7 MENGER'S THEOREM (EDGE VERSION)
For any graph $G(x, y)$,
$$p'(x, y) = c'(x, y)$$

This theorem was proved in Chapter 7 using flows. It may also be deduced from Theorem 9.1 by considering a suitable line graph (see Exercise 9.3.12).

A *k-edge cut* is an edge cut $\partial(X)$, where $\emptyset \subset X \subset V$ and $|\partial(X)| = k$, that is, an edge cut of k elements which separates some pair of vertices. Because every nontrivial graph has such edge cuts, it follows from Theorem 9.7 that the edge connectivity $\kappa'(G)$ of a nontrivial graph G is equal to the least integer k for which

G has a k-edge cut. For any particular pair x, y of vertices of G, the value of $c'(x, y)$ can be determined by an application of the Max-Flow Min-Cut Algorithm (7.9). Therefore the parameter κ' can obviously be determined by $\binom{n}{2}$ applications of that algorithm. However, the function c' takes at most $n - 1$ distinct values (Exercise 9.3.14b). Moreover, Gomory and Hu (1961) have shown that κ' can be computed by just $n - 1$ applications of the Max-Flow Min-Cut Algorithm. A description of their approach is given in Section 9.6.

ESSENTIAL EDGE CONNECTIVITY

The vertex and edge connectivities of a graph G, and its minimum degree, are related by the following basic inequalities (Exercise 9.3.2).

$$\kappa \leq \kappa' \leq \delta$$

Thus, for 3-regular graphs, the connectivity and edge connectivity do not exceed three. They are, moreover, always equal for such graphs (Exercise 9.3.5). These two measures of connectivity therefore fail to distinguish between the triangular prism $K_3 \,\square\, K_2$ and the complete bipartite graph $K_{3,3}$, both of which are 3-regular graphs with connectivity and edge connectivity equal to three. Nonetheless, one has the distinct impression that $K_{3,3}$ is better connected than $K_3 \,\square\, K_2$. Indeed, $K_3 \,\square\, K_2$ has a 3-edge cut which separates the graph into two nontrivial subgraphs, whereas $K_{3,3}$ has no such cut.

Recall that a *trivial* edge cut is one associated with a single vertex. A k-edge-connected graph is termed *essentially* $(k+1)$-*edge-connected* if all of its k-edge cuts are trivial. For example, $K_{3,3}$ is essentially 4-edge-connected whereas $K_3 \,\square\, K_2$ is not. If a k-edge-connected graph has a k-edge cut $\partial(X)$, the graphs $G \,/\, X$ and $G \,/\, \overline{X}$ (obtained by shrinking X to a single vertex x and $\overline{X} := V \setminus X$ to a single vertex $\overline{x}$, respectively) are also k-edge-connected (Exercise 9.3.8). By iterating this shrinking procedure, any k-edge-connected graph with $k \geq 1$, can be 'decomposed' into a set of essentially $(k + 1)$-edge-connected graphs. For many problems, it is enough to treat each of these 'components' separately. (When $k = 0$ — that is, when the graph is disconnected — this procedure corresponds to considering each of its components individually.)

The notion of essential edge connectivity is particularly useful for 3-regular graphs. For instance, to show that a 3-connected 3-regular graph has a cycle double cover, it suffices to verify that each of its essentially 4-edge-connected components has one; the individual cycle double covers can then be spliced together to yield a cycle double cover of the entire graph (Exercise 9.3.9).

CONNECTIVITY IN DIGRAPHS

The definitions of connectivity and edge connectivity have straightforward extensions to digraphs. It suffices to replace 'path' by 'directed path' throughout. We have already seen three versions of Menger's Theorem, namely the arc version

(Theorem 7.16), and the edge and vertex versions for undirected graphs (Theorems 7.17 and 9.1). Not surprisingly, there is also a vertex version for directed graphs. It can be deduced easily from the reduction of IDDP to ADDP described in Section 8.3. An (x, y)-*vertex-cut* is a subset S of $V \setminus \{x, y\}$ whose deletion destroys all directed (x, y)-paths.

Theorem 9.8   MENGER'S THEOREM (DIRECTED VERTEX VERSION)
In any digraph $D(x, y)$, where $(x, y) \notin A(D)$, the maximum number of pairwise internally disjoint directed (x, y)-paths is equal to the minimum number of vertices in an (x, y)-vertex-cut. □

As already noted, of the four versions of Menger's Theorem, Theorem 7.16 implies the other three. Also, Theorem 9.8 clearly implies Theorem 9.1. Although less obvious, the converse implication holds too (see Exercise 9.1.12). By using a suitable line graph, Theorem 9.7 may be derived from Theorem 9.1 (see Exercise 9.3.12).

Exercises

9.3.1 Determine the connectivity and the edge connectivity of the Kneser graph $KG_{m,n}$.

★9.3.2

a) Show that every graph G satisfies the inequalities $\kappa \leq \kappa' \leq \delta$.
b) Find a graph G with $\kappa = 3$, $\kappa' = 4$, and $\delta = 5$.

9.3.3 Let G be a simple graph of diameter two. Show that $\kappa' = \delta$.   (J. PLESNÍK)

9.3.4

a) Show that if G is simple and $\delta \geq (n - 1)/2$, then $\kappa' = \delta$.
b) For each even $n \geq 2$, find a simple graph G with $\delta = (n/2) - 1$ and $\kappa' < \delta$.

★9.3.5 Show that if G is cubic, then $\kappa = \kappa'$.

9.3.6

a) Show that if G is k-edge-connected, where $k > 0$, and if S is a set of k edges of G, then $c(G \setminus S) \leq 2$.
b) For $k > 0$, find a k-connected graph G and a set S of k vertices of G such that $c(G - S) > 2$.

9.3.7 Show that if G is a k-edge-connected graph on at least three vertices, and e is any edge of G, then G / e is k-edge-connected.

★9.3.8 Show that if $\partial(X)$ is a k-edge cut of a k-edge-connected graph G, the graphs G / X and $G / \overline{X}$ are also k-edge-connected, where $\overline{X} := V \setminus X$.

⋆**9.3.9** Let $\partial(X)$ be a 3-edge cut of a cubic graph G. Show that G has a cycle double cover if and only if both $G\,/\,X$ and $G\,/\,\overline{X}$ have cycle double covers, where $\overline{X} := V \setminus X$.

9.3.10 Show that in a nontrivial connected graph, any minimal edge cut separating two of its vertices is a bond.

9.3.11 Let $G = (V, E)$ be a $(2k-1)$-edge-connected graph and let F be a spanning bipartite subgraph of G with as many edges as possible. Show that F is k-edge-connected.   (C. THOMASSEN)

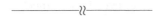

9.3.12 Deduce Theorem 9.7 from Theorem 9.1.   (F. HARARY)

9.3.13 Let S be a set of three pairwise-nonadjacent edges in a simple 3-connected graph G. Show that there is a cycle in G containing all three edges of S unless S is an edge cut of G.   (L. LOVÁSZ, N. ROBERTSON)

⋆**9.3.14**

a) Show that, for any three vertices x, y, and z of a graph G:

$$c'(x, z) \geq \min\{c'(x, y), c'(y, z)\}$$

  and that at least two of the values $c'(x, y)$, $c'(x, z)$, and $c'(y, z)$ are equal.
b) Deduce from (a) that:
    i) the function c' takes on at most $n - 1$ distinct values,
    ii) for any sequence $(v_1, v_2, \ldots, v_k)$ of vertices of a graph G,

$$c'(v_1, v_k) \geq \min\{c'(v_1, v_2), c'(v_2, v_3), \ldots, c'(v_{k-1}, v_k)\}$$

9.3.15 A k-edge-connected graph G is *minimally k-edge-connected* if, for any edge e of G, the graph $G \setminus e$ is not k-edge-connected.

a) Let G be a minimally k-edge-connected graph. Prove that:
    i) every edge e of G is contained in a k-edge cut of G,
    ii) G has a vertex of degree k,
    iii) $m \leq k(n - 1)$.
b) Deduce that every k-edge-connected graph G contains a spanning k-edge-connected subgraph with at most $k(n - 1)$ edges.   (W. MADER)

(Halin (1969) and Mader (1971b) found analogues of the above statements for vertex connectivity.)

9.4 Three-Connected Graphs

As we observed in Chapter 5, in most instances it is possible to draw conclusions about a graph by examining each of its blocks individually. For example, a graph has a cycle double cover if and only if each of its blocks has a cycle double cover. Because blocks on more than two vertices are 2-connected, the question of the existence of a cycle double cover can therefore be restricted, or 'reduced', to the study of 2-connected graphs. A similar reduction applies to the problem of deciding whether a given graph is planar, as we show in Chapter 10.

In many cases, further reductions can be applied, allowing one to restrict the analysis to 3-connected graphs, or even to 3-connected essentially 4-edge-connected graphs. The basic idea is to decompose a 2-connected graph which has a 2-vertex cut into smaller 2-connected graphs. Loops do not play a significant role in this context. For clarity, we therefore assume that all graphs considered in this section are loopless.

Let G be a connected graph which is not complete, let S be a vertex cut of G, and let X be the vertex set of a component of $G - S$. The subgraph H of G induced by $S \cup X$ is called an S-*component* of G. In the case where G is 2-connected and $S := \{x, y\}$ is a 2-vertex cut of G, we find it convenient to modify each S-component by adding a new edge between x and y. We refer to this edge as a *marker edge* and the modified S-components as *marked S-components*. The set of marked S-components constitutes the *marked S-decomposition* of G. The graph G can be recovered from its marked S-decomposition by taking the union of its marked S-components and deleting the marker edge. This procedure is illustrated in Figure 9.7, the cut S and the marker edge being indicated by solid dots and lines.

Theorem 9.9 *Let G be a 2-connected graph and let S be a 2-vertex cut of G. Then the marked S-components of G are also 2-connected.*

Proof Let H be a marked S-component of G, with vertex set $S \cup X$. Then $|V(H)| = |S| + |X| \geq 3$. Thus if H is complete, it is 2-connected. On the other hand, if H is not complete, every vertex cut of H is also a vertex cut of G, hence of cardinality at least two.   □

DECOMPOSITION TREES

By Theorem 9.9, a 2-connected graph G with a 2-vertex cut S has a marked S-decomposition into 2-connected graphs. If any one of these marked S-components itself has a 2-vertex cut, it in turn can be decomposed into still smaller marked 2-connected graphs. This decomposition process may be iterated until G has been decomposed into 2-connected graphs without 2-vertex cuts. The marked S-components which arise during the entire procedure form the vertices of a *decomposition tree* of G, as illustrated in Figure 9.8.

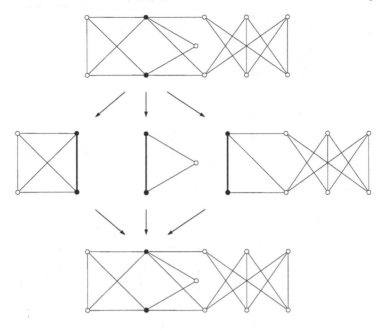

Fig. 9.7. A marked decomposition of a 2-connected graph and its recomposition

The root of this decomposition tree is G, and its leaves are either 3-connected graphs or else 2-connected graphs whose underlying graphs are complete (and which therefore have at most three vertices). We refer to the 3-connected graphs in such a decomposition as the *3-connected components* of G. The 3-connected components of the root graph in Figure 9.8 are K_3 (both with and without multiple edges), K_4, and $K_{3,3}$.

At any stage, there may be a choice of cuts along which to decompose a graph. Consequently, two separate applications of this decomposition procedure may well result in different sets of graphs (Exercise 9.4.1). However, it was shown by Cunningham and Edmonds (1980) that any two applications of the procedure always result in the same set of 3-connected components (possibly with different edge multiplicities).

To observe the relevance of the above decomposition to cycle double covers, let G be a 2-connected graph with a 2-vertex cut S. If each marked S-component of G has a cycle double cover, one can show that G also has a cycle double cover (Exercise 9.4.2). Because 2-connected graphs on at most three vertices clearly have cycle double covers, we conclude that if the Cycle Double Cover Conjecture is true for all 3-connected graphs, it is true for all 2-connected graphs. This fact may be expressed more strikingly in terms of potential counterexamples to the conjecture: if the Cycle Double Cover Conjecture is false, a smallest counterexample to it (that is, one with the minimum possible number of vertices) must be 3-connected. Jaeger (1976) and Kilpatrick (1975) proved that every 4-edge-connected graph has a cycle

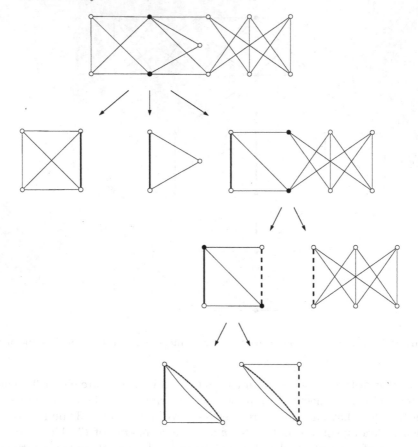

Fig. 9.8. A decomposition tree of a 2-connected graph

double cover (see Theorem 21.24). Thus, if the Cycle Double Cover Conjecture happens to be false, a minimum counterexample must have connectivity precisely three.

CONTRACTIONS OF THREE-CONNECTED GRAPHS

The relevance of 3-connectivity to the study of planar graphs is discussed in Section 10.5. In this context, the following property of 3-connected graphs, established by Thomassen (1981), plays an extremely useful role.

Theorem 9.10 *Let G be a 3-connected graph on at least five vertices. Then G contains an edge e such that $G \,/\, e$ is 3-connected.*

The proof of Theorem 9.10 requires the following lemma.

Lemma 9.11 *Let G be a 3-connected graph on at least five vertices, and let $e = xy$ be an edge of G such that G / e is not 3-connected. Then there exists a vertex z such that $\{x, y, z\}$ is a 3-vertex cut of G.*

Proof Let $\{z, w\}$ be a 2-vertex cut of G / e. At least one of these two vertices, say z, is not the vertex resulting from the contraction of e. Set $F := G - z$. Because G is 3-connected, F is certainly 2-connected. However $F / e = (G - z) / e = (G / e) - z$ has a cut vertex, namely w. Hence w must be the vertex resulting from the contraction of e (Exercise 9.1.5). Therefore $G - \{x, y, z\} = (G / e) - \{z, w\}$ is disconnected, in other words, $\{x, y, z\}$ is a 3-vertex cut of G. $\square$

Proof of Theorem 9.10. Suppose that the theorem is false. Then, for any edge $e = xy$ of G, the contraction G / e is not 3-connected. By Lemma 9.11, there exists a vertex z such that $\{x, y, z\}$ is a 3-vertex cut of G (see Figure 9.9).

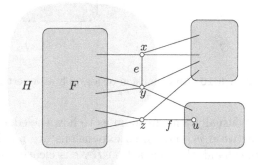

Fig. 9.9. Proof of Theorem 9.10

Choose the edge e and the vertex z in such a way that $G - \{x, y, z\}$ has a component F with as many vertices as possible. Consider the graph $G - z$. Because G is 3-connected, $G - z$ is 2-connected. Moreover $G - z$ has the 2-vertex cut $\{x, y\}$. It follows that the $\{x, y\}$-component $H := G[V(F) \cup \{x, y\}]$ is 2-connected.

Let u be a neighbour of z in a component of $G - \{x, y, z\}$ different from F. Since $f := zu$ is an edge of G, and G is a counterexample to Theorem 9.10, there is a vertex v such that $\{z, u, v\}$ is a 3-vertex cut of G, too. (The vertex v is not shown in Figure 9.9; it might or might not lie in H.) Moreover, because H is 2-connected, $H - v$ is connected (where, if $v \notin V(H)$, we set $H - v := H$), and thus is contained in a component of $G - \{z, u, v\}$. But this component has more vertices than F (because H has two more vertices than F), contradicting the choice of the edge e and the vertex v. $\square$

Although the proof of Theorem 9.10 proceeds by way of contradiction, the underlying idea can be used to devise a polynomial-time algorithm for finding an edge e in a 3-connected graph G such that G / e is 3-connected (Exercise 9.4.3).

EXPANSIONS OF THREE-CONNECTED GRAPHS

We now define an operation on 3-connected graphs which may be thought of as an inverse to contraction. Let G be a 3-connected graph and let v be a vertex of G of degree at least four. Split v into two vertices, v_1 and v_2, add a new edge e between v_1 and v_2, and distribute the edges of G incident to v among v_1 and v_2 in such a way that v_1 and v_2 each have at least three neighbours in the resulting graph H. This graph H is called an *expansion* of G at v (see Figure 9.10).

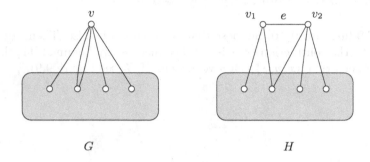

Fig. 9.10. An expansion of a graph at a vertex

Note that there is usually some freedom as to how the edges of G incident with v are distributed between v_1 and v_2, so expansions are not in general uniquely defined. On the other hand, the contraction H / e is clearly isomorphic to G.

The following theorem may be regarded as a kind of converse of Theorem 9.10.

Theorem 9.12 *Let G be a 3-connected graph, let v be a vertex of G of degree at least four, and let H be an expansion of G at v. Then H is 3-connected.*

Proof Because $G - v$ is 2-connected and v_1 and v_2 each have at least two neighbours in $G - v$, the graph $H \setminus e$ is 2-connected, by Lemma 9.3. Using the fact that any two vertices of G are connected by three internally disjoint paths, it is now easily seen that any two vertices of H are also connected by three internally disjoint paths. □

In light of Theorems 9.10 and 9.12, every 3-connected graph G can be obtained from K_4 by means of edge additions and vertex expansions. More precisely, given any 3-connected graph G, there exists a sequence $G_1, G_2, \ldots, G_k$ of graphs such that (i) $G_1 = K_4$, (ii) $G_k = G$, and (iii) for $1 \le i \le k - 1$, G_{i+1} is obtained by adding an edge to G_i or by expanding G_i at a vertex of degree at least four.

It is not possible to obtain a simple 3-connected graph, different from K_4, by means of the above construction if we wish to stay within the realm of simple graphs. However, Tutte (1961b) has shown that, by starting with the class of all wheels, all simple 3-connected graphs may be constructed by means of the two above-mentioned operations without ever creating parallel edges. This result may be deduced from Theorem 9.10 (see Exercise 9.4.7).

Recursive constructions of 3-connected graphs have been used to prove many interesting theorems in graph theory; see, for example, Exercise 9.4.9. For further examples, see Tutte (1966a).

For $k \geq 4$, no recursive procedure for generating all k-connected graphs is known. This is in striking contrast with the situation for k-edge-connected graphs (see Exercise 9.5.5). We refer the reader to Frank (1995) for a survey of recursive procedures for generating k-connected and k-edge-connected graphs.

Exercises

9.4.1 Find a 2-connected graph and two decomposition trees of your graph which result in different collections of leaves.

9.4.2 Let G be a 2-connected graph with a 2-vertex cut $S := \{u, v\}$. Prove that if each marked S-component of G has a cycle double cover then so has G.

9.4.3 Describe a polynomial-time algorithm to find, in a 3-connected graph G on five or more vertices, an edge e such that G / e is 3-connected.

9.4.4

a) Let G be a 4-regular 4-connected graph each edge of which lies in a triangle. Show that no edge-contraction of G is 4-connected.
b) For each integer $k \geq 4$, find a k-connected graph G on at least $k + 2$ vertices, none of whose edge contractions is k-connected.

9.4.5 Show how the Petersen graph can be obtained from the wheel W_6 by means of vertex expansions.

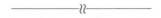

9.4.6 Let G be a graph and let $e = xy$ and $e' = x'y'$ be two distinct (but possibly adjacent) edges of G. The operation which consists of subdividing e by inserting a new vertex v between x and y, subdividing e' by inserting a new vertex v' between x' and y', and joining v and v' by a new edge, is referred to as an *edge-extension* of G. Show that:

a) any edge-extension of a 3-connected cubic graph is also 3-connected and cubic,
b) every 3-connected cubic graph can be obtained from K_4 by means of a sequence of edge-extensions,
c) an edge-extension of an essentially 4-edge-connected cubic graph G is also essentially 4-edge-connected provided that the two edges e and e' of G involved in the extension are nonadjacent in G.

(Wormald (1979) has shown that all essentially 4-edge-connected cubic graphs may be obtained from K_4 and the cube by means of edge-extensions involving nonadjacent pairs of edges.)

9.4.7 Let G be a 3-connected graph with $n \geq 5$. Show that, for any edge e, either G / e is 3-connected or $G \setminus e$ can be obtained from a 3-connected graph by subdividing at most two edges.   (W.T. TUTTE)

9.4.8 Let G be a simple 3-connected graph different from a wheel. Show that, for any edge e, either G / e or $G \setminus e$ is also a 3-connected simple graph.
                                                       (W.T. TUTTE)

$\star$**9.4.9**

  a) Let $\mathcal{G}$ be the family of graphs consisting of K_5, the wheels W_n, $n \geq 3$, and all graphs of the form $H \vee \overline{K}_n$, where H is a spanning subgraph of K_3 and $\overline{K}_n$ is the complement of K_n, $n \geq 3$. Show that a 3-connected simple graph G does not contain two disjoint cycles if and only if $G \in \mathcal{G}$.
                               (W.G. BROWN; L. LOVÁSZ)
  b) Deduce from (a) that any simple graph not containing two disjoint cycles has three vertices whose deletion results in an acyclic graph.

  (The same result holds for directed cycles in digraphs, although the proof, due to McCuaig (1993), is very much harder. For undirected graphs, Erdős and Pósa (1965) showed that there exists a constant c such that any graph either contains k disjoint cycles or has $ck \log k$ vertices whose deletion results in an acyclic graph. This is discussed in Section 19.1.)

9.5 Submodularity

A real-valued function f defined on the set of subsets of a set S is *submodular* if, for any two subsets X and Y of S,

$$f(X \cup Y) + f(X \cap Y) \leq f(X) + f(Y)$$

The degree function d defined on the set of subsets of the vertex set of a graph G by $d(X) := |\partial(X)|$ for all $X \subseteq V$ is a typical example of a submodular function associated with a graph (see Exercise 2.5.4). Another example is described in Exercise 9.5.7.

  Submodular functions play an important role in combinatorial optimization (see Fujishige (2005)). Here, we describe three interesting consequences of the submodularity of the degree function. One of these is Theorem 9.16, which has many applications, including a theorem on orientations of graphs due to Nash-Williams (1960). A second use of submodularity is described below, and a third is given in Section 9.6.

  It is convenient both here and in the next section to denote the complement $V \setminus X$ of a set X by $\overline{X}$.

Edge Connectivity of Vertex-Transitive Graphs

Two subsets X and Y of a set V are said to *cross* if the subsets $X \cap Y$, $X \cap \overline{Y}$, $\overline{X} \cap Y$, and $\overline{X} \cap \overline{Y}$ (shown in the Venn diagram of Figure 9.11) are all nonempty. When

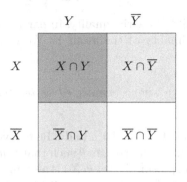

Fig. 9.11. Crossing sets X and Y

V is the vertex set of a graph G, we say that the edge cuts $\partial(X)$ and $\partial(Y)$ *cross* if the sets X and Y cross. In such cases, it is often fruitful to consider the edge cuts $\partial(X \cup Y)$ and $\partial(X \cap Y)$ and invoke the submodularity of the degree function. Here, we apply this idea to show that the edge connectivity of a nontrivial connected vertex-transitive graph is always equal to its degree, a result due independently to Mader (1971a) and Watkins (1970). Its proof relies on the concept of an atom.

An *atom* of a graph G is a minimal subset X of V such that $d(X) = \kappa'$ and $|X| \le n/2$. Thus if $\kappa' = \delta$, then any vertex of minimum degree is a singleton atom. On the other hand, if $\kappa' < \delta$, then G has no singleton atoms.

Proposition 9.13 *The atoms of a graph are pairwise disjoint.*

Proof Let X and Y be two distinct atoms of a graph G. Suppose that $X \cap Y \ne \emptyset$. Because X and Y are atoms, neither is properly contained in the other, so $X \cap \overline{Y}$ and $\overline{X} \cap Y$ are both nonempty. We show that $\overline{X} \cap \overline{Y}$ is nonempty, too, and thus that X and Y cross.

Noting that $\overline{X} \cup Y$ and $X \cap \overline{Y}$ are complementary sets, and that $X \cap \overline{Y}$ is a nonempty proper subset of the atom X, we have

$$d(\overline{X} \cup Y) = d(X \cap \overline{Y}) > d(X) = d(Y)$$

It follows that $\overline{X} \cup Y \ne Y$ or, equivalently, $\overline{X} \cap \overline{Y} \ne \emptyset$. So X and Y do indeed cross.

Because $\partial(X)$ and $\partial(Y)$ are minimum edge cuts,

$$d(X \cup Y) \ge d(X) \quad \text{and} \quad d(X \cap Y) \ge d(Y)$$

Therefore

$$d(X \cup Y) + d(X \cap Y) \geq d(X) + d(Y)$$

On the other hand, because d is a submodular function,

$$d(X \cup Y) + d(X \cap Y) \leq d(X) + d(Y)$$

These inequalities thus all hold with equality. In particular, $d(X \cap Y) = d(Y)$. But this contradicts the minimality of the atom Y. We conclude that X and Y are disjoint.   □

Theorem 9.14 *Let G be a simple connected vertex-transitive graph of positive degree d. Then $\kappa' = d$.*

Proof Let X be an atom of G, and let u and v be two vertices in X. Because G is vertex-transitive, it has an automorphism θ such that $\theta(u) = v$. Being the image of an atom under an automorphism, the set $\theta(X)$ is also an atom of G. As v belongs to both X and $\theta(X)$, it follows from Proposition 9.13 that $\theta(X) = X$, which implies that $\theta|_X$ is an automorphism of the graph $G[X]$, with $\theta|_X(u) = v$. This being so for any two vertices u, v in X, we deduce that $G[X]$ is vertex-transitive.

Suppose that $G[X]$ is k-regular. Because G is simple, $|X| \geq k+1$, and because G is connected, $\partial(X) \neq \emptyset$. Therefore $d \geq k + 1$, and we have:

$$\kappa' = d(X) = |X|(d - k) \geq (k+1)(d-k) = d + k(d-k-1) \geq d$$

Since κ' cannot exceed d, we conclude that $\kappa' = d$.   □

NASH-WILLIAMS' ORIENTATION THEOREM

By Theorem 5.10 every 2-edge-connected graph admits a strongly connected orientation. Nash-Williams (1960) established the following beautiful generalization of this result. (In the remainder of this section, k denotes a positive integer.)

Theorem 9.15 *Every $2k$-edge-connected graph has a k-arc-connected orientation.*

Mader (1978) proved an elegant theorem concerning the splitting off of edges (an operation introduced in Chapter 5) and deduced Theorem 9.15 from it. We present here a special case of Mader's result which is adequate for proving Theorem 9.15. The proof is due to Frank (1992).

Let v be a vertex of a graph G. We say that G is *locally $2k$-edge-connected modulo v* if the local edge connectivity between any two vertices different from v is at least $2k$. Using Menger's theorem and the fact that $d(X) = d(\overline{X})$, it can be seen that a graph G on at least three vertices is locally $2k$-edge-connected modulo v if and only if:

$$d(X) \geq 2k, \quad \text{for all } X, \quad \emptyset \subset X \subset V \setminus \{v\}$$

Theorem 9.16 *Let G be a graph which is locally $2k$-edge-connected modulo v, where v is a vertex of even degree in G. Given any link uv incident with v, there exists a second link vw incident with v such that the graph G' obtained by splitting off uv and vw at v is also locally $2k$-edge-connected modulo v.*

Proof We may assume that $n \geq 3$ as the statement holds trivially when $n = 2$. We may also assume that G is loopless. Consider all nonempty proper subsets X of $V \setminus \{v\}$. Splitting off uv and another link vw incident with v preserves the degree of X if at most one of u and w belongs to X, and reduces it by two if both u and w belong to X. Thus if all such sets either do not contain u or have degree at least $2k + 2$, any link vw may be chosen as the companion of uv. Suppose that this is not the case and that there is a proper subset X of $V \setminus \{v\}$ with $u \in X$ and $d(X) \leq 2k + 1$. Call such a set *tight*. We show that the union of two tight sets X and Y is also tight. We may assume that X and Y cross; otherwise, $X \cup Y$ is equal either to X or to Y. Therefore $X \cap \overline{Y}$ and $\overline{X} \cap Y$ are nonempty subsets of $V \setminus \{v\}$. Note, also, that $uv \in E[X \cap Y, \overline{X} \cap \overline{Y}]$. We thus have (using Exercise 2.5.4):

$$(2k + 1) + (2k + 1) \geq d(X) + d(Y)$$
$$= d(X \cap \overline{Y}) + d(\overline{X} \cap Y) + 2e(X \cap Y, \overline{X} \cap \overline{Y}) \geq 2k + 2k + 2$$

so

$$d(X) = d(Y) = 2k + 1, \quad d(X \cap \overline{Y}) = d(\overline{X} \cap Y) = 2k, \quad \text{and} \quad e(X \cap Y, \overline{X} \cap \overline{Y}) = 1$$

One may now deduce that $e(X \cap \overline{Y}, \overline{X} \cap \overline{Y}) = e(\overline{X} \cap Y, \overline{X} \cap \overline{Y})$ (Exercise 9.5.4). Thus $d(\overline{X} \cap \overline{Y})$ is odd. Because the degree of v is even, by hypothesis, $X \cup Y \neq V \setminus \{v\}$. Therefore $\emptyset \subset X \cup Y \subset V \setminus \{v\}$. Moreover, by submodularity,

$$d(X \cup Y) \leq d(X) + d(Y) - d(X \cap Y) \leq (2k + 1) + (2k + 1) - 2k = 2k + 2$$

Since $d(X \cup Y) = d(\overline{X} \cap \overline{Y})$ is odd, we may conclude that $d(X \cup Y) \leq 2k + 1$. Thus the union of any two tight sets is tight, as claimed. Now let S denote the union of all tight sets and let w be an element of $V \setminus S$ distinct from v. Because w belongs to no tight set in G, the graph G' obtained from G by splitting off uv and vw is locally $2k$-edge-connected modulo v. $\square$

Proof of Theorem 9.15. By induction on the number of edges. Let G be a $2k$-edge-connected graph. Suppose first that G has an edge e such that $G \setminus e$ is also $2k$-edge-connected. Then, by induction, $G \setminus e$ has an orientation such that the resulting digraph is k-arc-connected. That orientation of $G \setminus e$ may be extended to a k-arc-connected orientation of G itself by orienting e arbitrarily. Thus, we may assume that G is minimally $2k$-edge-connected and so has a vertex of degree $2k$ (Exercise 9.3.15). Let v be such a vertex.

By Theorem 9.16, the $2k$ edges incident with v may be divided into k pairs and each of these pairs may be split off, one by one, to obtain k new edges $e_1, e_2, \ldots, e_k$ and a $2k$-edge-connected graph H. By the induction hypothesis, there is an orientation $\overline{H}$ of H which is k-arc-connected. Let $a_1, a_2, \ldots, a_k$, respectively, be the

k arcs of $\overrightarrow{H}$ corresponding to the edges $e_1, e_2, \ldots, e_k$ of H. By subdividing, for $1 \le i \le k$, the arc a_i by a vertex v_i, and then identifying the k vertices $v_1, v_2, \ldots, v_k$ to form vertex v, we obtain an orientation $\overrightarrow{G}$ of G. Using the fact that $\overrightarrow{H}$ is k-arc-connected, one may easily verify that $\overrightarrow{G}$ is also k-arc-connected. We leave the details to the reader as Exercise 9.5.5. □

Nash-Williams (1960) in fact proved a far stronger result than Theorem 9.15. He showed that any graph G admits an orientation $\overrightarrow{G}$ such that, for any two vertices u and v, the size of a minimum outcut in $\overrightarrow{G}$ separating v from u is at least $\lfloor \frac{1}{2} c'(u,v) \rfloor$. We refer the reader to Schrijver (2003) for further details.

Exercises

9.5.1 Let X be an atom of a graph G. Show that the induced subgraph $G[X]$ is connected.

9.5.2 Give an example of a connected cubic vertex-transitive graph that is not 3-edge-connected.
(This shows that Theorem 9.14 is not valid for graphs with multiple edges.)

9.5.3 Give an example of a simple connected vertex-transitive k-regular graph whose connectivity is strictly less than k.
(Watkins (1970) showed that the connectivity of any such graph exceeds $2k/3$.)

⋆**9.5.4** In the proof of Theorem 9.16, show that $e(X \cap \overline{Y}, \overline{X} \cap \overline{Y}) = e(\overline{X} \cap Y, \overline{X} \cap \overline{Y})$.

⋆**9.5.5** Let $\overrightarrow{G}$ and $\overrightarrow{H}$ be the digraphs described in the proof of Theorem 9.15. Deduce that $\overrightarrow{G}$ is k-arc-connected from the fact that $\overrightarrow{H}$ is k-arc-connected.

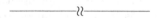

9.5.6 Let G be a $2k$-edge-connected graph with an Euler trail. Show that G has an orientation in which any two vertices u and v are connected by at least k arc-disjoint directed (u,v)-paths.

9.5.7 Let G be a graph. For a subset S of E, denote by $c(S)$ the number of components of the spanning subgraph of G with edge set S.

a) Show that the function $c : 2^E \to \mathbb{N}$ is *supermodular*: for any two subsets X and Y of E,
$$c(X \cup Y) + c(X \cap Y) \ge c(X) + c(Y)$$

b) Deduce that the function $r : 2^E \to \mathbb{N}$ defined by $r(S) := n - c(S)$ for all $S \subseteq E$ is submodular. (This function r is the rank function of a certain matroid associated with G.)

9.5.8 Given any graph G and k distinct edges $e_1, e_2, \ldots, e_k$ (loops or links) of G, the operation of *pinching together* those k edges consists of subdividing, for $1 \leq i \leq k$, the edge e_i by a vertex v_i, and then identifying the k vertices $v_1, v_2, \ldots, v_k$ to form a new vertex of degree $2k$.

a) Show that if G is $2k$-edge-connected, then the graph G' obtained from G by pinching together any k edges of G is also $2k$-edge-connected.

b) Using Theorem 9.16, show that, given any $2k$-edge-connected graph G, there exists a sequence $(G_1, G_2, \ldots, G_r)$ of graphs such that (i) $G_1 = K_1$, (ii) $G_r = G$, and (iii) for $1 \leq i \leq r - 1$, G_{i+1} is obtained from G_i either by adding an edge (a loop or a link) or by pinching together k of its edges.
(Mader (1978) found an analogous construction for $(2k + 1)$-edge-connected graphs.)

9.6 Gomory–Hu Trees

As mentioned earlier, Gomory and Hu (1961) showed that only $n - 1$ applications of the Max-Flow Min-Cut Algorithm (7.9) are needed in order to determine the edge connectivity of a graph G. The following theorem, in which two edge cuts $\partial(X)$ and $\partial(Y)$ that cross are replaced by two, $\partial(X)$ and $\partial(X \cap Y)$, that do not cross, is the basis of their approach. This procedure is referred to as *uncrossing*. We leave the proof of the theorem, which makes use of submodularity, as an exercise (9.6.1).

Theorem 9.17 *Let $\partial(X)$ be a minimum edge cut in a graph G separating two vertices x and y, where $x \in X$, and let $\partial(Y)$ be a minimum edge cut in G separating two vertices u and v of X, where $y \notin Y$. Then $\partial(X \cap Y)$ is a minimum edge cut in G separating u and v.* □

A consequence of Theorem 9.17 is that, given a minimum edge cut $\partial(X)$ in G separating vertices x and y, in order to find a minimum edge cut in G separating u and v, where $\{u, v\} \subset X$, it suffices to consider the graph $G / \overline{X}$ obtained from G by shrinking $\overline{X} := V \setminus X$ to a single vertex. Using this idea, Gomory and Hu showed how to find all the $\binom{n}{2}$ values of the function c' by just $n - 1$ applications of the Max-Flow Min-Cut Algorithm (7.9). They also showed that the $n - 1$ cuts found by their procedure have certain special properties which may be conveniently visualized in terms of an appropriately weighted tree associated with G. We first describe the characteristics of this weighted tree and then explain how to construct it.

Given any tree T with vertex set V, and an edge e of T, there is a unique edge cut $B_e := \partial(X)$ of G associated with e, where X is the vertex set of one component of $T \setminus e$. (This is akin to the notion of a fundamental bond, introduced in Chapter 4, except that here we do not insist on T being a spanning tree of G.) A weighted tree (T, w) on V is a *Gomory–Hu tree* of G if, for each edge $e = xy$ of T,

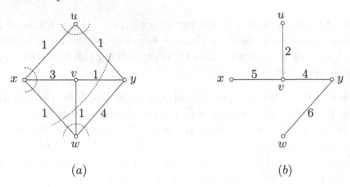

Fig. 9.12. (a) A graph G, and (b) a Gomory–Hu tree T of G

i) $w(e) = c'(x, y)$,
ii) the cut B_e associated with e is a minimum edge cut in G separating x and y.

As an example, consider the graph G on five vertices shown in Figure 9.12a, where the weights indicate edge multiplicities. Figure 9.12b is a Gomory–Hu tree T of G. The four edge cuts of G corresponding to the four edges of T are indicated by dashed lines in Figure 9.12a. Note that this particular tree T is not a spanning tree of G.

The $n-1$ edge cuts associated with a Gomory–Hu tree are pairwise noncrossing. As a consequence of the following proposition, these $n-1$ cuts are sufficient for determining $\kappa'(G)$.

Proposition 9.18 *Let (T, w) be a Gomory–Hu tree of a graph G. For any two vertices x and y of G, $c'(x, y)$ is the minimum of the weights of the edges on the unique xy-path in T.*

Proof Clearly, for every edge e on the xy-path in T, the edge cut B_e associated with e separates x and y. If $v_1, v_2, \ldots, v_k$ is the xy-path in T, where $x = v_1$ and $y = v_k$, it follows that

$$c'(x, y) \leq \min\{c'(v_1, v_2), c'(v_2, v_3), \ldots, c'(v_{k-1}v_k)\}$$

On the other hand, by Exercise 9.3.14b,

$$c'(x, y) \geq \min\{c'(v_1, v_2), c'(v_2, v_3), \ldots, c'(v_{k-1}v_k)\}$$

The required equality now follows.   □

Determining Edge Connectivity

We conclude this section with a brief description of the Gomory–Hu Algorithm. For this purpose, we consider trees whose vertices are the parts in a partition of

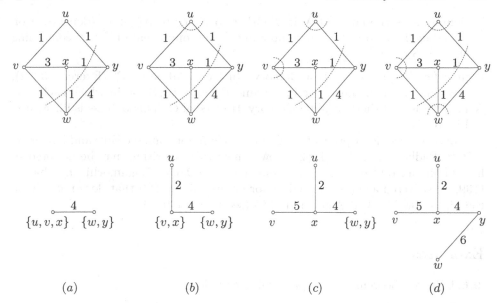

Fig. 9.13. Growing a Gomory–Hu tree

V; every edge of such a tree determines a unique edge cut of G. A weighted tree (T, w) whose vertex set is a partition $\mathcal{P}$ of V is a *Gomory–Hu tree* of G *relative to $\mathcal{P}$* if, for any edge $e := XY$ of T (where $X, Y \in \mathcal{P}$), there is an element x of X and an element y of Y such that $c'(x, y) = w(e)$ and the edge cut B_e associated with e is a minimum edge cut in G separating x and y. For example, if $\partial(X)$ is a minimum edge cut in G separating x and y, the tree consisting of two vertices X and $\overline{X} := V \setminus X$ joined by an edge with weight $c'(x, y) = d(X)$ is the Gomory–Hu tree relative to the partition $\{X, \overline{X}\}$ (see Figure 9.13a).

Suppose that we are given a Gomory–Hu tree (T, w) relative to a certain partition $\mathcal{P}$. If each part is a singleton, then (T, w) is already a Gomory–Hu tree of G. Thus, suppose that there is a vertex X of T (that is, a part X in $\mathcal{P}$) which contains two distinct elements u and v. It may be deduced from Theorem 9.17 that, in order to find a minimum edge cut in G separating u and v, it suffices to consider the graph G' obtained from G by shrinking, for each component of $T - X$, the union of the vertices (parts) in that component to a single vertex. Let $\partial(S)$ be a minimum edge cut separating u and v in G', and suppose that $u \in S$ and $v \in \overline{S}$, where $\overline{S} := V(G') \setminus S$. Now let $X_1 := X \cap S$ and $X_2 := X \cap \overline{S}$ and let $\mathcal{P}'$ be the partition obtained from $\mathcal{P}$ by replacing X by X_1 and X_2 and leaving all other parts as they are. A weighted tree T' with vertex set $\mathcal{P}'$ may now be obtained from T by:

i) splitting the vertex X into X_1 and X_2, and joining them by an edge of weight $c'(u, v) = d(S)$,

ii) joining a neighbour Y of X in T either to X_1 or to X_2 in T' (depending on whether the vertex of G' corresponding to the component of $T - X$ containing Y is in S or in $\overline{S}$).

It may be shown that T' is a Gomory–Hu tree relative to $\mathcal{P}'$ (Exercise 9.6.2). Proceeding in this manner, one may refine $\mathcal{P}$ into a partition in which each part is a singleton and thereby find a Gomory–Hu tree of G. This process is illustrated in Figure 9.13.

For a detailed description of the Gomory–Hu Algorithm, see Ford and Fulkerson (1962). Padberg and Rao (1982) showed how this algorithm may be adapted to find minimum odd cuts in graphs (see Exercise 9.6.3). Nagamochi and Ibaraki (1992) discovered a simple procedure for determining $\kappa'(G)$ that does not rely on the Max-Flow Min-Cut Algorithm (7.9) (see Exercise 9.6.4).

Exercises

$\star$**9.6.1** Prove Theorem 9.17 by proceeding as follows.

a) Show that $\partial(X \cup Y)$ is an edge cut separating x and y, and that $\partial(X \cap Y)$ is an edge cut separating u and v.
b) Deduce that $d(X \cup Y) \geq d(X)$ and $d(X \cap Y) \geq d(Y)$.
c) Apply the submodularity inequality.

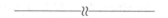

9.6.2 Show that the weighted tree T' obtained from T in the Gomory–Hu Algorithm is a Gomory–Hu tree of G relative to $\mathcal{P}'$.

9.6.3 Let G be a graph with at least two vertices of odd degree.

a) Suppose that $\partial(X)$ is an edge cut of smallest size among those separating pairs of vertices of odd degree in G. Show that:
   i) if $d(X)$ is odd, it is a smallest odd edge cut of G,
   ii) if $d(X)$ is even, a smallest odd edge cut of G is an edge cut of either G / X or $G / \overline{X}$, where $\overline{X} := V \setminus X$.
b) Using (a), show how to find a smallest odd edge cut of a graph by applying the Gomory–Hu Algorithm.   (M.W. PADBERG AND M.R. RAO)

9.6.4 Call an ordering $(v_1, v_2, \ldots, v_n)$ of the vertices of a connected graph G a *cut-greedy order* if, for $2 \leq i \leq n$,

$$d(v_i, \{v_1, v_2, \ldots, v_{i-1}\}) \geq d(v_j, \{v_1, v_2, \ldots, v_{i-1}\}), \text{ for all } j \geq i,$$

a) Show that one can find, starting with any vertex of G, a cut-greedy order of the vertices of G in time $O(m)$.
b) If $(v_1, v_2, \ldots, v_n)$ is a cut-greedy order of the vertices of G, show that

$$c'(v_{n-1}, v_n) = d(v_n)$$

c) Describe a polynomial-time algorithm for finding $\kappa'(G)$ based on part (b).

(H. NAGAMOCHI AND T. IBARAKI)

d) Find the edge connectivity of the graph in Figure 9.12 by applying the above algorithm.

9.6.5 WELL-BALANCED ORIENTATION

An orientation D of a graph G is *well-balanced* if its local arc connectivities $p'_D(u, v)$ satisfy $p'_D(u, v) \geq \lfloor p'_G(u, v)/2 \rfloor$ for all ordered pairs (u, v) of vertices. Show that every well-balanced orientation of an eulerian graph is eulerian.   (Z. SZIGETI)

9.7 Chordal Graphs

A *chordal graph* is a simple graph in which every cycle of length greater than three has a chord. Equivalently, the graph contains no induced cycle of length four or more. Thus every induced subgraph of a chordal graph is chordal. An example of a chordal graph is shown in Figure 9.14.

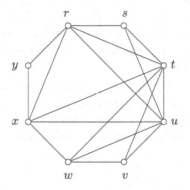

Fig. 9.14. A chordal graph

Complete graphs and trees are simple instances of chordal graphs. Moreover, as we now show, all chordal graphs have a treelike structure composed of complete graphs (just as trees are composed of copies of K_2). In consequence, many $\mathcal{NP}$-hard problems become polynomial when restricted to chordal graphs.

CLIQUE CUTS

A *clique cut* is a vertex cut which is also a clique. In a chordal graph, every minimal vertex cut is a clique cut.

Theorem 9.19 *Let G be a connected chordal graph which is not complete, and let S be a minimal vertex cut of G. Then S is a clique cut of G.*

Proof Suppose that S contains two nonadjacent vertices x and y. Let G_1 and G_2 be two components of $G - S$. Because S is a minimal cut, both x and y are joined to vertices in both G_1 and G_2. Let P_i be a shortest xy-path all of whose internal vertices lie in G_i, $i = 1, 2$. Then $P_1 \cup P_2$ is an induced cycle of length at least four, a contradiction. $\qquad \square$

From Theorem 9.19, one may deduce that every connected chordal graph can be built by pasting together complete graphs in a treelike fashion.

Theorem 9.20 *Let G be a connected chordal graph, and let V_1 be a maximal clique of G. Then the maximal cliques of G can be arranged in a sequence $(V_1, V_2, \ldots, V_k)$ such that $V_j \cap (\cup_{i=1}^{j-1} V_i)$ is a clique of G, $2 \le j \le k$.*

Proof There is nothing to prove if G is complete, so we may assume that G has a minimal vertex cut S. By Theorem 9.19, S is a clique of G. Let H_i, $1 \le i \le p$, be the S-components of G, and let Y_i be a maximal clique of H_i containing S, $1 \le i \le p$. Observe that the maximal cliques of $H_1, H_2 \ldots, H_p$ are also maximal cliques of G, and that every maximal clique of G is a maximal clique of some H_i (Exercise 9.7.1). Without loss of generality, suppose that V_1 is a maximal clique of H_1. By induction, the maximal cliques of H_1 can be arranged in a sequence starting with V_1 and having the stated property. Likewise, for $2 \le i \le p$, the maximal cliques of H_i can be arranged in a suitable sequence starting with Y_i. The concatenation of these sequences is a sequence of the maximal cliques of G satisfying the stated property. $\qquad \square$

A sequence $(V_1, V_2, \ldots, V_k)$ of maximal cliques as described in Theorem 9.20 is called a *simplicial decomposition* of the chordal graph G. The graph in Figure 9.14 has the simplicial decomposition shown in Figure 9.15. Dirac (1961) proved that a graph is chordal if and only if it has such a decomposition (see Exercise 9.7.2).

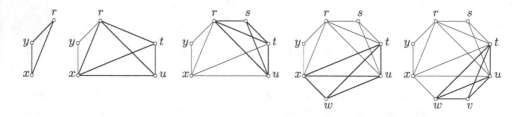

Fig. 9.15. A simplicial decomposition of the chordal graph of Figure 9.14

SIMPLICIAL VERTICES

A *simplicial vertex* of a graph is a vertex whose neighbours induce a clique. Dirac (1961) showed that every noncomplete chordal graph has at least two such vertices (just as every nontrivial tree has at least two vertices of degree one). The graph in Figure 9.14, for example, has three simplicial vertices, namely s, v, and y.

Theorem 9.21 *Every chordal graph which is not complete has two nonadjacent simplicial vertices.*

Proof Let $(V_1, V_2, \ldots, V_k)$ be a simplicial decomposition of a chordal graph, and let $x \in V_k \setminus (\cup_{i=1}^{k-1} V_i)$. Then x is a simplicial vertex. Now consider a simplicial decomposition $(V_{\pi(1)}, V_{\pi(2)}, \ldots, V_{\pi(k)})$, where π is a permutation of $\{1, 2, \ldots, k\}$ such that $\pi(1) = k$. Let $y \in V_{\pi(k)} \setminus (\cup_{i=1}^{k-1} V_{\pi(i)})$. Then y is a simplicial vertex nonadjacent to x. $\square$

A *simplicial order* of a graph G is an enumeration $v_1, v_2, \ldots, v_n$ of its vertices such that v_i is a simplicial vertex of $G[\{v_i, v_{i+1}, \ldots, v_n\}]$, $1 \leq i \leq n$. Because induced subgraphs of chordal graphs are chordal, it follows directly from Theorem 9.21 that every chordal graph has a simplicial order. Conversely, if a graph has a simplicial order, it is necessarily chordal (Exercise 9.7.3).

Corollary 9.22 *A graph is chordal if and only if it has a simplicial order.* $\square$

There is a linear-time algorithm due to Rose et al. (1976), and known as *lexicographic breadth-first search*, for finding a simplicial order of a graph if one exists. A brief description is given in Section 9.8.

TREE REPRESENTATIONS

Besides the characterizations of chordal graphs given above in terms of simplicial decompositions and simplicial orders, chordal graphs may also be viewed as intersection graphs of subtrees of a tree.

Theorem 9.23 *A graph is chordal if and only if it is the intersection graph of a family of subtrees of a tree.*

Proof Let G be a chordal graph. By Theorem 9.20, G has a simplicial decomposition $(V_1, V_2, \ldots, V_k)$. We prove by induction on k that G is the intersection graph of a family of subtrees $\mathcal{T} = \{T_v : v \in V\}$ of a tree T with vertex set $\{x_1, x_2, \ldots, x_k\}$ such that $x_i \in T_v$ for all $v \in V_i$. If $k = 1$, then G is complete and we set $T_v := T$ for all $v \in V$. If $k \geq 2$, let $G' = (V', E')$ be the chordal graph with simplicial decomposition $(V_1, V_2, \ldots, V_{k-1})$. By induction, G' is the intersection graph of a family of subtrees $\mathcal{T}' = \{T'_v : v \in V'\}$ of a tree T' with vertex set $\{x_1, x_2, \ldots, x_{k-1}\}$. Let V_j be a maximal clique of G' such that $V_j \cap V_k \neq \emptyset$. We form the tree T by adding a new vertex x_k adjacent to x_j. For $v \in V_j$, we form the tree T_v by adding x_k to T'_v and joining it to x_j. For $v \in V' \setminus V_j$, we set $T_v := T'_v$. Finally, for $v \in V_k \setminus V'$, we set $T_v := x_k$. It can be checked that G is the intersection graph of $\{T_v : v \in V\}$. We leave the proof of the converse statement as an exercise (9.7.4). $\square$

We refer to the pair $(T, \mathcal{T})$ described in the proof of Theorem 9.23 as a *tree representation* of the chordal graph G.

Exercises

⋆9.7.1 Let G be a connected chordal graph which is not complete, and let S be a clique cut of G. Show that the maximal cliques of the S-components of G are also maximal cliques of G, and that every maximal clique of G is a maximal clique of some S-component of G.

⋆9.7.2 Show that a graph is chordal if it has a simplicial decomposition.

⋆9.7.3 Show that a graph is chordal if it has a simplicial order.

⋆9.7.4

a) Show that the intersection graph of a family of subtrees of a tree is a chordal graph.
b) Represent the chordal graph of Figure 9.14 as the intersection graph of a family of subtrees of a tree.

9.7.5

a) Let G be a chordal graph and v a simplicial vertex of G. Set $X := N(v) \cup \{v\}$ and $G' := G - X$, and let S' be a maximum stable set and $\mathcal{K}'$ a minimum clique covering of G'. Show that:
  i) $S := S' \cup \{v\}$ is a maximum stable set of G,
  ii) $\mathcal{K} := \mathcal{K}' \cup \{X\}$ is a minimum clique covering of G,
  iii) $|S| = |\mathcal{K}|$.
b) Describe a linear-time algorithm which accepts as input a simplicial order of a chordal graph G and returns a maximum stable set and a minimum clique covering of G.

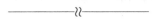

9.8 Related Reading

LEXICOGRAPHIC BREADTH-FIRST SEARCH

By Exercise 9.7.3b, a graph is chordal if and only if it has a simplicial order. Breadth-first search, with a special rule for determining the head of the queue, may be used to find a simplicial order of an input graph, if one exists. The rule, which gives the procedure its name, involves assigning sequences of integers to vertices and comparing them lexicographically to break ties. (Sequences of integers from the set $\{1.2, \ldots, n\}$ may be thought of as words of a language whose alphabet consists of n letters $1, 2, \ldots, n$, the first letter being 1, the second letter 2, and so on. A sequence S is *lexicographically smaller* than another sequence S' if S appears before S' in a dictionary of that language.) If G happens to be chordal, the sequence of vertices generated by this tree-search will be the converse of a simplicial order.

We choose an arbitrary vertex of the input graph G as root, and denote the vertex incorporated into the tree at time t by v_t, the root being v_1. Each vertex v of the graph is assigned a sequence $S(v)$ of integers, initially the empty sequence. When vertex v_t enters the tree, for each $v \in N(v_t) \setminus \{v_1, v_2, \ldots, v_{t-1}\}$, we modify $S(v)$ by appending to it the integer $n - t + 1$. The next vertex selected to enter the tree is any vertex in the queue whose label is lexicographically largest.

Rose et al. (1976), who introduced lexicographic breadth-first search (Lex BFS), showed that it will find a simplicial order of the input graph if there is one. A very readable account of chordal graphs, including a proof of the validity of Lex BFS, can be found in Golumbic (2004). In recent years, Lex BFS has been used extensively in algorithms for recognizing various other classes of graphs (see, for example, Corneil (2004)).

TREE-DECOMPOSITIONS

Due to their rather simple structure, chordal graphs can be recognized in polynomial time, as outlined above. Moreover, many $\mathcal{NP}$-hard problems, such as MAX STABLE SET, can be solved in polynomial time when restricted to chordal graphs (see Exercise 9.7.5). A more general class of graphs for which polynomial-time algorithms exist for such $\mathcal{NP}$-hard problems was introduced by Robertson and Seymour (1986).

Recall that by Theorem 9.23 every chordal graph G has a tree representation, that is, an ordered pair $(T, \mathcal{T})$, where T is a tree and $\mathcal{T} := \{T_v : v \in V\}$ is a family of subtrees of T such that $T_u \cap T_v \neq \emptyset$ if and only if $uv \in E$. For an arbitrary simple graph G, a *tree-decomposition* of G is an ordered pair $(T, \mathcal{T})$, where T is a tree and $\mathcal{T} := \{T_v : v \in V\}$ is a family of subtrees of T such that $T_u \cap T_v \neq \emptyset$ if (but not necessarily only if) $uv \in E$. Equivalently, $(T, \mathcal{T})$ is a tree-decomposition of a simple graph G if and only if G is a spanning subgraph of the chordal graph with tree representation $(T, \mathcal{T})$.

Every simple graph G has the *trivial* tree-decomposition $(T, \mathcal{T})$, where T is an arbitrary tree and $T_v = T$ for all $v \in V$ (the corresponding chordal graph being K_n). For algorithmic purposes, one is interested in finer tree-decompositions, as measured by a parameter called the width of the decomposition. A nontrivial tree-decomposition of $K_{2,3}$ is shown in Figure 9.16.

Let $(T, \{T_v : v \in V\})$ be a tree-decomposition of a graph G, where $V(T) = X$ and $V(T_v) = X_v$, $v \in V$. The dual of the hypergraph $(X, \{X_v : v \in V\})$ is the hypergraph $(V, \{V_x : x \in X\})$, where $V_x := \{v \in V : x \in X_v\}$. For instance, if G is a chordal graph, the sets V_x, $x \in X$, are the cliques in its simplicial decomposition. The greatest cardinality of an edge of this dual hypergraph, $\max\{|V_x| : x \in X\}$, is called the *width* of the decomposition.[1] The tree-decomposition of $K_{2,3}$ shown in

---

[1] WARNING: the value of the width as defined here is *one greater than the standard definition*. This difference has no bearing on qualitative statements about tree-width, many of which are of great significance. On the other hand, as regards quantitative statements, this is certainly the right definition from an aesthetic viewpoint.

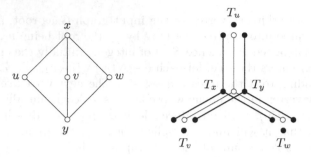

Fig. 9.16. A tree-decomposition of $K_{2,3}$, of width three

Figure 9.16 has width three, the sets V_x, $x \in X$, being $\{u, x, y\}$, $\{v, x, y\}$, $\{w, x, y\}$, and $\{x, y\}$.

As another example, consider the tree-decomposition of the (3×3)-grid $P_3 \square P_3$ with vertex set $\{(i, j) : 1 \le i, j \le 3\}$ shown in Figure 9.17. This tree-decomposition has width four, all six sets V_x (the horizontal sets) being of cardinality four.

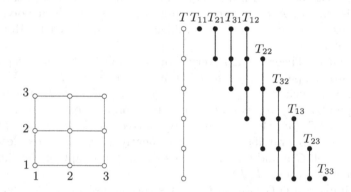

Fig. 9.17. A tree-decomposition of the (3×3)-grid, of width four

In general, a graph may have many different tree-decompositions. The *tree-width* of the graph is the minimum width among all tree-decompositions. Thus the tree-width of a chordal graph is its clique number; in particular, every nontrivial tree has tree-width two. Cycles also have tree-width two. More generally, one can show that every series-parallel graph (defined in Exercise 10.5.11) has tree-width at most three. The $(n \times n)$-grid has tree-width $n + 1$; that this is an upper bound follows from a generalization of the tree-decomposition given in Figure 9.17, but establishing the lower bound is more difficult (see Section 10.7). For graphs in general, Arnborg et al. (1987) showed that computing the tree-width is an $\mathcal{NP}$-hard problem. On the other hand, there exists polynomial-time algorithm for deciding whether a graph has tree-width at most k, where k is a fixed integer (Robertson and Seymour (1986)).

If a graph has a small tree-width, then it has a treelike structure, resembling a 'thickened' tree, and this structure has enabled the development of polynomial-time algorithms for many $\mathcal{NP}$-hard problems (see, for example, Arnborg and Proskurowski (1989)). More significantly, tree-decompositions have proved to be a fundamental tool in the work of Robertson and Seymour on linkages and graph minors (see Section 10.7).

A number of other width parameters have been studied, including the path-width (where the tree T is constrained to be a path), the branch-width, and the cut-width. We refer the reader to one of the many surveys on this topic; for example, Bienstock and Langston (1995), Reed (2003), or Bodlaender (2006).

10

Planar Graphs

Contents

10.1 Plane and Planar Graphs

A graph is said to be *embeddable in the plane*, or *planar*, if it can be drawn in the plane so that its edges intersect only at their ends. Such a drawing is called a *planar embedding* of the graph. A planar embedding $\widetilde{G}$ of a planar graph G can be regarded as a graph isomorphic to G; the vertex set of $\widetilde{G}$ is the set of points representing the vertices of G, the edge set of $\widetilde{G}$ is the set of lines representing the edges of G, and a vertex of $\widetilde{G}$ is incident with all the edges of $\widetilde{G}$ that contain it. For this reason, we often refer to a planar embedding $\widetilde{G}$ of a planar graph G as a *plane graph*, and we refer to its points as vertices and its lines as edges. However, when discussing both a planar graph G and a plane embedding $\widetilde{G}$ of G, in order to distinguish the two graphs, we call the vertices of $\widetilde{G}$ *points* and its edges *lines*; thus, by the *point v* of $\widetilde{G}$ we mean the point of $\widetilde{G}$ that represents the vertex v of G, and by the *line e* of $\widetilde{G}$ we mean the line of $\widetilde{G}$ that represents the edge e of G. Figure 10.1b depicts a planar embedding of the planar graph $K_5 \setminus e$, shown in Figure 10.1a.

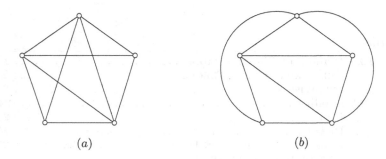

(a)   (b)

Fig. 10.1. (a) The planar graph $K_5 \setminus e$, and (b) a planar embedding of $K_5 \setminus e$

THE JORDAN CURVE THEOREM

It is evident from the above definition that the study of planar graphs necessarily involves the topology of the plane. We do not attempt here to be strictly rigorous in topological matters, however, and are content to adopt a naive point of view toward them. This is done so as not to obscure the combinatorial aspects of the theory, which is our main interest. An elegant and rigorous treatment of the topological aspects can be found in the book by Mohar and Thomassen (2001).

The results of topology that are especially relevant in the study of planar graphs are those which deal with simple curves. By a *curve*, we mean a continuous image of a closed unit line segment. Analogously, a *closed curve* is a continuous image of a circle. A curve or closed curve is *simple* if it does not intersect itself (in other words, if the mapping is one-to-one). Properties of such curves come into play in the study of planar graphs because cycles in plane graphs are simple closed curves.

A subset of the plane is *arcwise-connected* if any two of its points can be connected by a curve lying entirely within the subset. The basic result of topology that we need is the Jordan Curve Theorem.

Theorem 10.1   THE JORDAN CURVE THEOREM
Any simple closed curve C in the plane partitions the rest of the plane into two disjoint arcwise-connected open sets.   □

Although this theorem is intuitively obvious, giving a formal proof of it is quite tricky. The two open sets into which a simple closed curve C partitions the plane are called the *interior* and the *exterior* of C. We denote them by $\text{int}(C)$ and $\text{ext}(C)$, and their closures by $\text{Int}(C)$ and $\text{Ext}(C)$, respectively (thus $\text{Int}(C) \cap \text{Ext}(C) = C$). The Jordan Curve Theorem implies that every arc joining a point of $\text{int}(C)$ to a point of $\text{ext}(C)$ meets C in at least one point (see Figure 10.2).

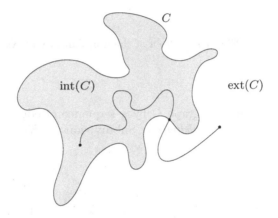

Fig. 10.2. The Jordan Curve Theorem

Figure 10.1b shows that the graph $K_5 \setminus e$ is planar. The graph K_5, on the other hand, is not planar. Let us see how the Jordan Curve Theorem can be used to demonstrate this fact.

Theorem 10.2 K_5 *is nonplanar.*

Proof   By contradiction. Let G be a planar embedding of K_5, with vertices v_1, v_2, v_3, v_4, v_5. Because G is complete, any two of its vertices are joined by an edge. Now the cycle $C := v_1 v_2 v_3 v_1$ is a simple closed curve in the plane, and the vertex v_4 must lie either in $\text{int}(C)$ or in $\text{ext}(C)$. Without loss of generality, we may suppose that $v_4 \in \text{int}(C)$. Then the edges $v_1 v_4, v_2 v_4, v_3 v_4$ all lie entirely in $\text{int}(C)$, too (apart from their ends v_1, v_2, v_3) (see Figure 10.3).

Consider the cycles $C_1 := v_2 v_3 v_4 v_2$, $C_2 := v_3 v_1 v_4 v_3$, and $C_3 := v_1 v_2 v_4 v_1$. Observe that $v_i \in \text{ext}(C_i)$, $i = 1, 2, 3$. Because $v_i v_5 \in E(G)$ and G is a plane graph, it follows from the Jordan Curve Theorem that $v_5 \in \text{ext}(C_i)$, $i = 1, 2, 3$, too. Thus $v_5 \in \text{ext}(C)$. But now the edge $v_4 v_5$ crosses C, again by the Jordan Curve Theorem. This contradicts the planarity of the embedding G.   □

A similar argument can be used to establish that $K_{3,3}$ is nonplanar, too (Exercise 10.1.1b).

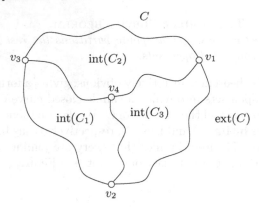

Fig. 10.3. Proof of the nonplanarity of K_5

SUBDIVISIONS

Any graph derived from a graph G by a sequence of edge subdivisions is called a *subdivision* of G or a *G-subdivision*. Subdivisions of K_5 and $K_{3,3}$ are shown in Figure 10.4.

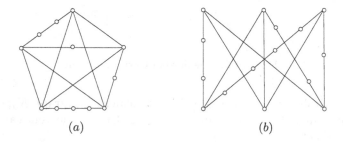

(a) (b)

Fig. 10.4. (a) A subdivision of K_5, (b) a subdivision of $K_{3,3}$

The proof of the following proposition is straightforward (Exercise 10.1.2).

Proposition 10.3 *A graph G is planar if and only if every subdivision of G is planar.* □

Because K_5 and $K_{3,3}$ are nonplanar, Proposition 10.3 implies that no planar graph can contain a subdivision of either K_5 or $K_{3,3}$. A fundamental theorem due to Kuratowski (1930) states that, conversely, every nonplanar graph necessarily contains a copy of a subdivision of one of these two graphs. A proof of Kuratowski's Theorem is given in Section 10.5.

As mentioned in Chapter 1 and illustrated in Chapter 3, one may consider embeddings of graphs on surfaces other than the plane. We show in Section 10.6

that, for every surface S, there exist graphs which are not embeddable on S. Every graph can, however, be embedded in the 3-dimensional euclidean space $\mathbb{R}^3$ (Exercise 10.1.7).

Planar graphs and graphs embeddable on the sphere are one and the same. To see this, we make use of a mapping known as stereographic projection. Consider a sphere S resting on a plane P, and denote by z the point that is diametrically opposite the point of contact of S and P. The mapping $\pi : S \setminus \{z\} \to P$, defined by $\pi(s) = p$ if and only if the points z, s, and p are collinear, is called a *stereographic projection* from z; it is illustrated in Figure 10.5.

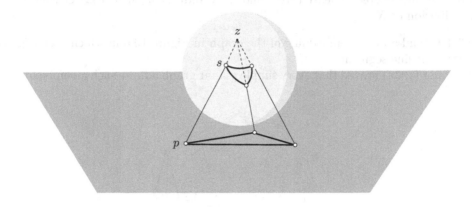

Fig. 10.5. Stereographic projection

Theorem 10.4 *A graph G is embeddable on the plane if and only if it is embeddable on the sphere.*

Proof Suppose that G has an embedding $\widetilde{G}$ on the sphere. Choose a point z of the sphere not in $\widetilde{G}$. Then the image of $\widetilde{G}$ under stereographic projection from z is an embedding of G on the plane. The converse is proved similarly. $\qquad \square$

On many occasions, it is advantageous to consider embeddings of planar graphs on the sphere; one instance is provided by the proof of Proposition 10.5 in the next section.

Exercises

$\star$**10.1.1** Show that:

  a) every proper subgraph of $K_{3,3}$ is planar,
  b) $K_{3,3}$ is nonplanar.

$\star$**10.1.2** Show that a graph is planar if and only if every subdivision of the graph is planar.

⋆**10.1.3**

a) Show that the Petersen graph contains a subdivision of $K_{3,3}$.
b) Deduce that the Petersen graph is nonplanar.

⋆**10.1.4**

a) Let G be a planar graph, and let e be a link of G. Show that G / e is planar.
b) Is the converse true?

⋆**10.1.5** Let G be a simple nontrivial graph in which each vertex, except possibly one, has degree at least three. Show, by induction on n, that G contains a subdivision of K_4.

10.1.6 Find a planar embedding of the graph in Figure 10.6 in which each edge is a straight line segment.
(Wagner (1936) proved that every simple planar graph admits such an embedding.)

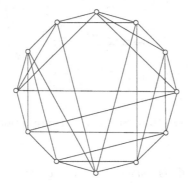

Fig. 10.6. Find a straight-line planar embedding of this graph (Exercise 10.1.6)

10.1.7 A k-*book* is a topological subspace of $\mathbb{R}^3$ consisting of k unit squares, called its *pages*, that have one side in common, called its *spine*, but are pairwise disjoint otherwise. Show that any graph G is embeddable in $\mathbb{R}^3$ by showing that it is embeddable in a k-book, for some k.

10.1.8 Consider a drawing $\widetilde{G}$ of a (not necessarily planar) graph G in the plane. Two edges of $\widetilde{G}$ *cross* if they meet at a point other than a vertex of $\widetilde{G}$. Each such point is called a *crossing* of the two edges. The *crossing number* of G, denoted by $\mathrm{cr}(G)$, is the least number of crossings in a drawing of G in the plane. Show that:

a) $\mathrm{cr}(G) = 0$ if and only if G is planar,
b) $\mathrm{cr}(K_5) = \mathrm{cr}(K_{3,3}) = 1$,
c) $\mathrm{cr}(P_{10}) = 2$, where P_{10} is the Petersen graph,
d) $\mathrm{cr}(K_6) = 3$.

10.1.9 Show that $\operatorname{cr}(K_n)/\binom{n}{4}$ is a monotonically increasing function of n.

10.1.10 A graph G is *crossing-minimal* if $\operatorname{cr}(G \setminus e) < \operatorname{cr}(G)$ for all $e \in E$. Show that every nonplanar edge-transitive graph is crossing-minimal.

10.1.11 A *thrackle* is a graph embedded in the plane in such a way that any two edges intersect exactly once (possibly at an end). Such an embedding is called a *thrackle embedding*. Show that:

  a) every tree has a thrackle embedding,
  b) the 4-cycle has no thrackle embedding,
  c) the triangle and every cycle of length five or more has a thrackle embedding.

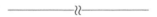

10.1.12 Show that every simple graph can be embedded in $\mathbb{R}^3$ in such a way that:

  a) each vertex lies on the curve $\{(t, t^2, t^3) : t \in \mathbb{R}\}$,
  b) each edge is a straight line segment.   (C. THOMASSEN)

10.2 Duality

FACES

A plane graph G partitions the rest of the plane into a number of arcwise-connected open sets. These sets are called the *faces* of G. Figure 10.7 shows a plane graph with five faces, $f_1, f_2, f_3, f_4,$ and f_5. Each plane graph has exactly one unbounded face, called the *outer face*. In the plane graph of Figure 10.7, the outer face is f_1. We denote by $F(G)$ and $f(G)$, respectively, the set of faces and the number of faces of a plane graph G. The notion of a face applies also to embeddings of graphs on other surfaces.

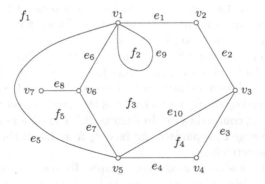

Fig. 10.7. A plane graph with five faces

The *boundary* of a face f is the boundary of the open set f in the usual topological sense. A face is said to be *incident* with the vertices and edges in its boundary, and two faces are *adjacent* if their boundaries have an edge in common. In Figure 10.7, the face f_1 is incident with the vertices v_1, v_2, v_3, v_4, v_5 and the edges e_1, e_2, e_3, e_4, e_5; it is adjacent to the faces f_3, f_4, f_5.

We denote the boundary of a face f by $\partial(f)$. The rationale for this notation becomes apparent shortly, when we discuss duality. The boundary of a face may be regarded as a subgraph. Moreover, when there is no scope for confusion, we use the notation $\partial(f)$ to denote the edge set of this subgraph.

Proposition 10.5 *Let G be a planar graph, and let f be a face in some planar embedding of G. Then G admits a planar embedding whose outer face has the same boundary as f.*

Proof Consider an embedding $\widetilde{G}$ of G on the sphere; such an embedding exists by virtue of Theorem 10.4. Denote by $\widetilde{f}$ the face of $\widetilde{G}$ corresponding to f. Let z be a point in the interior of $\widetilde{f}$, and let $\pi(\widetilde{G})$ be the image of $\widetilde{G}$ under stereographic projection from z. Clearly $\pi(\widetilde{G})$ is a planar embedding of G with the desired property. $\qquad\square$

By the Jordan Curve Theorem, a planar embedding of a cycle has exactly two faces. In the ensuing discussion of plane graphs, we assume, without proof, a number of other intuitively obvious statements concerning their faces. We assume, for example, that a planar embedding of a tree has just one face, and that each face boundary in a connected plane graph is itself connected. Some of these facts rely on another basic result of plane topology, known as the Jordan–Schönfliess Theorem.

Theorem 10.6 THE JORDAN–SCHÖNFLIESS THEOREM
Any homeomorphism of a simple closed curve in the plane onto another simple closed curve can be extended to a homeomorphism of the plane. $\qquad\square$

One implication of this theorem is that any point p on a simple closed curve C can be connected to any point not on C by means of a simple curve which meets C only in p. We refer the reader to Mohar and Thomassen (2001) for further details.

A cut edge in a plane graph has just one incident face, but we may think of the edge as being incident twice with the same face (once from each side); all other edges are incident with two distinct faces. We say that an edge *separates* the faces incident with it. The *degree*, $d(f)$, of a face f is the number of edges in its boundary $\partial(f)$, cut edges being counted twice. In Figure 10.7, the edge e_9 separates the faces f_2 and f_3 and the edge e_8 separates the face f_5 from itself; the degrees of f_3 and f_5 are 6 and 5, respectively.

Suppose that G is a connected plane graph. To *subdivide* a face f of G is to add a new edge e joining two vertices on its boundary in such a way that, apart from its endpoints, e lies entirely in the interior of f. This operation results in a plane graph $G + e$ with exactly one more face than G; all faces of G except f are

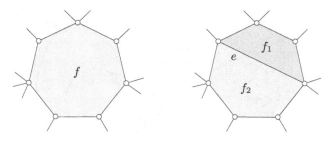

Fig. 10.8. Subdivision of a face f by an edge e

also faces of $G + e$, and the face f is replaced by two new faces, f_1 and f_2, which meet in the edge e, as illustrated in Figure 10.8.

In a connected plane graph the boundary of a face can be regarded as a closed walk in which each cut edge of the graph that lies in the boundary is traversed twice. This is clearly so for plane trees, and can be established in general by induction on the number of faces (Exercise 10.2.2.) In the plane graph of Figure 10.7, for instance,

$$\partial(f_3) = v_1 e_1 v_2 e_2 v_3 e_{10} v_5 e_7 v_6 e_6 v_1 e_9 v_1 \quad \text{and} \quad \partial(f_5) = v_1 e_6 v_6 e_8 v_7 e_8 v_6 e_7 v_5 e_5 v_1$$

Moreover, in the case of nonseparable graphs, these boundary walks are simply cycles, as was shown by Whitney (1932c).

Theorem 10.7 *In a nonseparable plane graph other than K_1 or K_2, each face is bounded by a cycle.*

Proof Let G be a nonseparable plane graph. Consider an ear decomposition $G_0, G_1, \ldots, G_k$ of G, where G_0 is a cycle, $G_k = G$, and, for $0 \le i \le k - 2$, $G_{i+1} := G_i \cup P_i$ is a nonseparable plane subgraph of G, where P_i is an ear of G_i. Since G_0 is a cycle, the two faces of G_0 are clearly bounded by cycles. Assume, inductively, that all faces of G_i are bounded by cycles, where $i \ge 0$. Because G_{i+1} is a plane graph, the ear P_i of G_i is contained in some face f of G_i. (More precisely, G_{i+1} is obtained from G_i by subdividing the face f by an edge joining the ends of P_i and then subdividing that edge by inserting the internal vertices of P_i.) Each face of G_i other than f is a face of G_{i+1} as well, and so, by the induction hypothesis, is bounded by a cycle. On the other hand, the face f of G_i is divided by P_i into two faces of G_{i+1}, and it is easy to see that these, too, are bounded by cycles. $\qquad\square$

One consequence of Theorem 10.7 is that all planar graphs without cut edges have cycle double covers (Exercise 10.2.4). Another is the following.

Corollary 10.8 *In a loopless 3-connected plane graph, the neighbours of any vertex lie on a common cycle.*

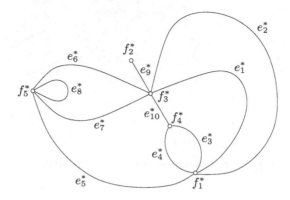

Fig. 10.9. The dual of the plane graph of Figure 10.7

Proof Let G be a loopless 3-connected plane graph and let v be a vertex of G. Then $G - v$ is nonseparable, so each face of $G - v$ is bounded by a cycle, by Theorem 10.7. If f is the face of $G - v$ in which the vertex v was situated, the neighbours of v lie on its bounding cycle $\partial(f)$. $\square$

DUALS

Given a plane graph G, one can define a second graph G^* as follows. Corresponding to each face f of G there is a vertex f^* of G^*, and corresponding to each edge e of G there is an edge e^* of G^*. Two vertices f^* and g^* are joined by the edge e^* in G^* if and only if their corresponding faces f and g are separated by the edge e in G. Observe that if e is a cut edge of G, then $f = g$, so e^* is a loop of G^*; conversely, if e is a loop of G, the edge e^* is a cut edge of G^*. The graph G^* is called the *dual* of G. The dual of the plane graph of Figure 10.7 is drawn in Figure 10.9.

   In the dual G^* of a plane graph G, the edges corresponding to those which lie in the boundary of a face f of G are just the edges incident with the corresponding vertex f^*. When G has no cut edges, G^* has no loops, and this set is precisely the trivial edge cut $\partial(f^*)$; that is,

$$\partial(f^*) = \{e^* : e \in \partial(f)\}$$

It is for this reason that the notation $\partial(f)$ was chosen.

   It is easy to see that the dual G^* of a plane graph G is itself a planar graph; in fact, there is a natural embedding of G^* in the plane. We place each vertex f^* in the corresponding face f of G, and then draw each edge e^* in such a way that it crosses the corresponding edge e of G exactly once (and crosses no other edge of G). This procedure is illustrated in Figure 10.10, where the dual is indicated by heavy lines.

   It is intuitively clear that we can always draw the dual as a plane graph in this way, but we do not prove this fact. We refer to such a drawing of the dual as a *plane dual* of the plane graph G.

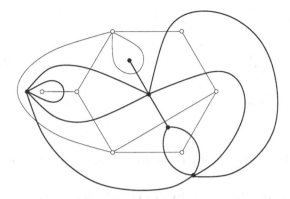

Fig. 10.10. The plane graph of Figure 10.7 and its plane dual

Proposition 10.9 *The dual of any plane graph is connected.*

Proof Let G be a plane graph and G^* a plane dual of G. Consider any two vertices of G^*. There is a curve in the plane connecting them which avoids all vertices of G. The sequence of faces and edges of G traversed by this curve corresponds in G^* to a walk connecting the two vertices. □

Although defined abstractly, it is often convenient to regard the dual G^* of a plane graph G as being itself a plane graph, embedded as described above. One may then consider the dual G^{**} of G^*. When G is connected, it is not difficult to prove that $G^{**} \cong G$ (Exercise 10.2.6); a glance at Figure 10.10 indicates why this is so.

It should be noted that isomorphic plane graphs may well have nonisomorphic duals. For example, although the plane graphs in Figure 10.11 are isomorphic, their duals are not: the plane graph shown in Figure 10.11a has two faces of degree three, whereas that of Figure 10.11b has only one such face. Thus the notion of a dual graph is meaningful only for plane graphs, and not for planar graphs in general. We show, however (in Theorem 10.28) that every simple 3-connected planar graph has a unique planar embedding (in the sense that its face boundaries are uniquely determined) and hence has a unique dual.

The following relations are direct consequences of the definition of the dual G^*.

$$v(G^*) = f(G), \quad e(G^*) = e(G), \quad \text{and} \quad d_{G^*}(f^*) = d_G(f) \quad \text{for all} \quad f \in F(G) \quad (10.1)$$

The next theorem may be regarded as a dual version of Theorem 1.1.

Theorem 10.10 *If G is a plane graph,*

$$\sum_{f \in F} d(f) = 2m$$

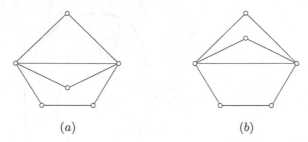

(a) (b)

Fig. 10.11. Isomorphic plane graphs with nonisomorphic duals

Proof Let G^* be the dual of G. By (10.1) and Theorem 1.1,

$$\sum_{f \in F(G)} d(f) = \sum_{f^* \in V(G^*)} d(f^*) = 2e(G^*) = 2e(G) = 2m \qquad \square$$

A simple connected plane graph in which all faces have degree three is called a *plane triangulation* or, for short, a *triangulation*. The tetrahedron, the octahedron, and the icosahedron (depicted in Figure 1.14) are all triangulations. As a consequence of (10.1) we have:

Proposition 10.11 *A simple connected plane graph is a triangulation if and only if its dual is cubic.* $\qquad \square$

It is easy to show that every simple plane graph on three or more vertices is a spanning subgraph of a triangulation (Exercise 10.2.3). On the other hand, as we show in Section 10.3, no simple spanning supergraph of a triangulation is planar. For this reason, triangulations are also known as *maximal planar graphs*. They play an important role in the theory of planar graphs.

DELETION–CONTRACTION DUALITY

Let G be a planar graph and $\widetilde{G}$ a plane embedding of G. For any edge e of G, a plane embedding of $G \setminus e$ can be obtained by simply deleting the line e from $\widetilde{G}$. Thus, the deletion of an edge from a planar graph results in a planar graph. Although less obvious, the contraction of an edge of a planar graph also results in a planar graph (Exercise 10.1.4b). Indeed, given any edge e of a planar graph G and a planar embedding $\widetilde{G}$ of G, the line e of $\widetilde{G}$ can be contracted to a single point (and the lines incident to its ends redrawn) so that the resulting plane graph is a planar embedding of $G \,/\, e$.

The following two propositions show that the operations of contracting and deleting edges in plane graphs are related in a natural way under duality.

Proposition 10.12 *Let G be a connected plane graph, and let e be an edge of G that is not a cut edge. Then*

$$(G \setminus e)^* \cong G^* \,/\, e^*$$

Proof Because e is not a cut edge, the two faces of G incident with e are distinct; denote them by f_1 and f_2. Deleting e from G results in the amalgamation of f_1 and f_2 into a single face f (see Figure 10.12). Any face of G that is adjacent to f_1 or f_2 is adjacent in $G \setminus e$ to f; all other faces and adjacencies between them are unaffected by the deletion of e. Correspondingly, in the dual, the two vertices f_1^* and f_2^* of G^* which correspond to the faces f_1 and f_2 of G are now replaced by a single vertex of $(G \setminus e)^*$, which we may denote by f^*, and all other vertices of G^* are vertices of $(G \setminus e)^*$. Furthermore, any vertex of G^* that is adjacent to f_1^* or f_2^* is adjacent in $(G \setminus e)^*$ to f^*, and adjacencies between vertices of $(G \setminus e)^*$ other than v are the same as in G^*. The assertion follows from these observations. $\square$

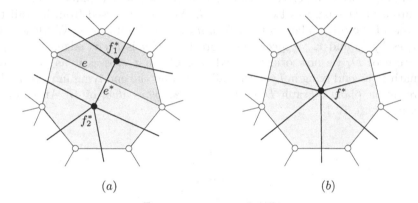

(a)   (b)

Fig. 10.12. (a) G and G^*, (b) $G \setminus e$ and G^* / e^*

Dually, we have:

Proposition 10.13 *Let G be a connected plane graph, and let e be a link of G. Then*

$$(G / e)^* \cong G^* \setminus e^*$$

Proof Because G is connected, $G^{**} \cong G$ (Exercise 10.2.6). Also, because e is not a loop of G, the edge e^* is not a cut edge of G^*, so $G^* \setminus e^*$ is connected. By Proposition 10.12,

$$(G^* \setminus e^*)^* \cong G^{**} / e^{**} \cong G / e$$

The proposition follows on taking duals. $\square$

We now apply Propositions 10.12 and 10.13 to show that nonseparable plane graphs have nonseparable duals. This fact turns out to be very useful.

Theorem 10.14 *The dual of a nonseparable plane graph is nonseparable.*

Proof By induction on the number of edges. Let G be a nonseparable plane graph. The theorem is clearly true if G has at most one edge, so we may assume that G has at least two edges, hence no loops or cut edges. Let e be an edge of G. Then either $G \setminus e$ or G / e is nonseparable (Exercise 5.3.2). If $G \setminus e$ is nonseparable, so is $(G \setminus e)^* \cong G^* / e^*$, by the induction hypothesis and Proposition 10.12. Applying Exercise 5.2.2b, we deduce that G^* is nonseparable. The case where G / e is nonseparable can be established by an analogous argument. □

The dual of any plane graph is connected, and it follows from Theorem 10.14 that the dual of a loopless 2-connected plane graph is 2-connected. Furthermore, one can show that the dual of a simple 3-connected plane graph is 3-connected (Exercise 10.2.9).

The notion of plane duality can be extended to directed graphs. Let D be a plane digraph, with underlying plane graph G. Consider a plane dual G^* of G. Each arc a of D separates two faces of G. As a is traversed from its tail to its head, one of these faces lies to the left of a and one to its right. We denote these two faces by l_a and r_a, respectively; note that if a is a cut edge, $l_a = r_a$. For each arc a of D, we now orient the edge of G^* that crosses it as an arc a^* by designating the end lying in l_a as the tail of a^* and the end lying in r_a as its head. The resulting plane digraph D^* is the *directed plane dual* of D. An example is shown in Figure 10.13.

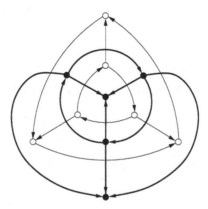

Fig. 10.13. An orientation of the triangular prism, and its directed plane dual

VECTOR SPACES AND DUALITY

We have seen that the cycle and bond spaces are orthogonal complements (Exercise 2.6.4a). In the case of plane graphs, this relationship of orthogonality can also be expressed in terms of duality, as we now explain. As usual in this context, we identify cycles, trees, and cotrees with their edge sets.

We observed earlier that all duals are connected (Proposition 10.9). A similar argument, based on the fact that the interior of a cycle in a plane graph is arcwise-connected, establishes the following proposition.

Proposition 10.15 *Let G be a plane graph, G^* a plane dual of G, C a cycle of G, and X^* the set of vertices of G^* that lie in $\text{int}(C)$. Then $G^*[X^*]$ is connected.*
□

For a subset S of $E(G)$, we denote by S^* the subset $\{e^* : e \in S\}$ of $E(G^*)$.

Theorem 10.16 *Let G be a connected plane graph, and let G^* be a plane dual of G.*

a) If C is a cycle of G, then C^* is a bond of G^*.
b) If B is a bond of G, then B^* is a cycle of G^*.

Proof a) Let C be a cycle of G, and let X^* denote the set of vertices of G^* that lie in $\text{int}(C)$. Then C^* is the edge cut $\partial(X^*)$ in G^*. By Proposition 10.15, the subgraph of G^* induced by X^* is connected. Likewise, the subgraph of G^* induced by $V(G^*) \setminus X^*$ is connected. It follows from Theorem 2.15 that C^* is a bond of G^*. We leave part (b) (the converse of (a)) as an exercise (Exercise 10.2.7).   □

As a straightforward consequence of Theorem 10.16, we have:

Corollary 10.17 *For any plane graph G, the cycle space of G is isomorphic to the bond space of G^*.*
□

The relationship between cycles and bonds expressed in Theorem 10.16 may be refined by taking orientations into account and considering directed duals, as defined above. Let D be a plane digraph and D^* its directed plane dual. For a subset S of $A(D)$, denote by S^* the subset $\{a^* : a \in S\}$ of $A(D^*)$.

Theorem 10.18 *Let D be a connected plane digraph and let D^* be a plane directed dual of D.*

a) Let C be a cycle of D, with a prescribed sense of traversal. Then C^* is a bond $\partial(X^*)$ of D^*. Moreover the set of forward arcs of C corresponds to the outcut $\partial^+(X^*)$ and the set of reverse arcs of C to the incut $\partial^-(X^*)$.
b) Let $B := \partial(X)$ be a bond of D. Then B^* is a cycle of D^*. Moreover the outcut $\partial^+(X)$ corresponds to the set of forward arcs of B^* and the incut $\partial^-(X)$ corresponds to the set of reverse arcs of B^* (with respect to a certain sense of traversal of B^*).
□

The proof of Theorem 10.18 is left to the reader (Exercise 10.2.14).

Exercises

10.2.1

a) Show that a graph is planar if and only if each of its blocks is planar.
b) Deduce that every minimal nonplanar graph is both simple and nonseparable.

⋆**10.2.2** Prove that the boundary of a face of a connected plane graph can be regarded as a closed walk in which each cut edge of the graph lying in the boundary is traversed twice.

⋆**10.2.3** Show that every simple connected plane graph on n vertices, where $n \geq 3$, is a spanning subgraph of a triangulation.

10.2.4 Show that every planar graph without cut edges has a cycle double cover.

10.2.5 Determine the duals of the five platonic graphs (Figure 1.14).

⋆**10.2.6** Let G be a plane graph. Show that $G^{**} \cong G$ if and only if G is connected.

⋆**10.2.7** Let B be a bond of a plane graph G. Show that B^* is a cycle of its plane dual G^*.

10.2.8 Show that the dual of a triangulation on at least four vertices is a simple nonseparable cubic plane graph.

10.2.9 Show that the dual of a simple 3-connected plane graph is both simple and 3-connected.

10.2.10 Show that the dual of an even plane graph is bipartite.

10.2.11 A *Hamilton bond* of a connected graph G is a bond B such that both components of $G \setminus B$ are trees. Let G be a plane graph which contains a Hamilton cycle, and let C be (the edge set of) such a cycle. Show that C^* is a Hamilton bond of G^*.

10.2.12 OUTERPLANAR GRAPH
A graph G is *outerplanar* if it has a planar embedding $\widetilde{G}$ in which all vertices lie on the boundary of its outer face. An outerplanar graph equipped with such an embedding is called an *outerplane graph*. Show that:

a) if G is an outerplane graph, then the subgraph of G^* induced by the vertices corresponding to the interior faces of G is a tree,
b) every simple 2-connected outerplanar graph other than K_2 has a vertex of degree two.

10.2.13 Let T be a spanning tree of a connected plane graph G. Show that $(E \setminus T)^*$ is a spanning tree of G^*.

⋆**10.2.14** Prove Theorem 10.18.

10.2.15 A *Halin graph* is a graph $H := T \cup C$, where T is a plane tree on at least four vertices in which no vertex has degree two, and C is a cycle connecting the leaves of T in the cyclic order determined by the embedding of T. Show that:

a) every Halin graph is minimally 3-connected,
b) every Halin graph has a Hamilton cycle.

10.2.16 The *medial graph* of a plane graph G is the 4-regular graph $M(G)$ with vertex set $E(G)$ in which two vertices are joined by k edges if, in G, they are adjacent edges which are incident to k common faces ($k = 0, 1, 2$). (The medial graph has a natural planar embedding.) Let G be a nonseparable plane graph. Show that:

a) $M(G)$ is a 4-regular planar graph,
b) $M(G) \cong M(G^*)$.

10.3 Euler's Formula

There is a simple formula relating the numbers of vertices, edges, and faces in a connected plane graph. It was first established for polyhedral graphs by Euler (1752), and is known as *Euler's Formula*.

Theorem 10.19   EULER'S FORMULA
For a connected plane graph G,

$$v(G) - e(G) + f(G) = 2 \tag{10.2}$$

Proof By induction on $f(G)$, the number of faces of G. If $f(G) = 1$, each edge of G is a cut edge and so G, being connected, is a tree. In this case $e(G) = v(G) - 1$, by Theorem 4.3, and the assertion holds. Suppose that it is true for all connected plane graphs with fewer than f faces, where $f \geq 2$, and let G be a connected plane graph with f faces. Choose an edge e of G that is not a cut edge. Then $G \setminus e$ is a connected plane graph with $f - 1$ faces, because the two faces of G separated by e coalesce to form one face of $G \setminus e$. By the induction hypothesis,

$$v(G \setminus e) - e(G \setminus e) + f(G \setminus e) = 2$$

Using the relations

$$v(G \setminus e) = v(G), \qquad e(G \setminus e) = e(G) - 1, \quad \text{and} \quad f(G \setminus e) = f(G) - 1$$

we obtain

$$v(G) - e(G) + f(G) = 2$$

The theorem follows by induction.   □

Corollary 10.20 *All planar embeddings of a connected planar graph have the same number of faces.*

Proof Let $\widetilde{G}$ be a planar embedding of a planar graph G. By Euler's Formula (10.2), we have

$$f(\widetilde{G}) = e(\widetilde{G}) - v(\widetilde{G}) + 2 = e(G) - v(G) + 2$$

Thus the number of faces of $\widetilde{G}$ depends only on the graph G, and not on its embedding. □

Corollary 10.21 *Let G be a simple planar graph on at least three vertices. Then $m \leq 3n - 6$. Furthermore, $m = 3n - 6$ if and only if every planar embedding of G is a triangulation.*

Proof It clearly suffices to prove the corollary for connected graphs. Let G be a simple connected planar graph with $n \geq 3$. Consider any planar embedding $\widetilde{G}$ of G. Because G is simple and connected, on at least three vertices, $d(f) \geq 3$ for all $f \in F(\widetilde{G})$. Therefore, by Theorem 10.10 and Euler's Formula (10.2)

$$2m = \sum_{f \in F(\widetilde{G})} d(f) \geq 3f(\widetilde{G}) = 3(m - n + 2) \tag{10.3}$$

or, equivalently,

$$m \leq 3n - 6 \tag{10.4}$$

Equality holds in (10.4) if and only if it holds in (10.3), that is, if and only if $d(f) = 3$ for each $f \in F(\widetilde{G})$. □

Corollary 10.22 *Every simple planar graph has a vertex of degree at most five.*

Proof This is trivial for $n < 3$. If $n \geq 3$, then by Theorem 1.1 and Corollary 10.21,

$$\delta n \leq \sum_{v \in V} d(v) = 2m \leq 6n - 12$$

It follows that $\delta \leq 5$. □

We have already seen that K_5 and $K_{3,3}$ are nonplanar (Theorem 10.2 and Exercise 10.1.1b). Here, we derive these two basic facts from Euler's Formula (10.2).

Corollary 10.23 K_5 *is nonplanar.*

Proof If K_5 were planar, Corollary 10.21 would give

$$10 = e(K_5) \leq 3v(K_5) - 6 = 9$$

Thus K_5 must be nonplanar. □

Corollary 10.24 $K_{3,3}$ *is nonplanar.*

Proof Suppose that $K_{3,3}$ is planar and let G be a planar embedding of $K_{3,3}$. Because $K_{3,3}$ has no cycle of length less than four, every face of G has degree at least four. Therefore, by Theorem 10.10, we have

$$4f(G) \le \sum_{f \in F} d(f) = 2e(G) = 18$$

implying that $f(G) \le 4$. Euler's Formula (10.2) now implies that

$$2 = v(G) - e(G) + f(G) \le 6 - 9 + 4 = 1$$

which is absurd. $\qquad\qquad\qquad\qquad\qquad\qquad\qquad\qquad\qquad\qquad\qquad\square$

Exercises

★**10.3.1** Show that the crossing number satisfies the inequality $\mathrm{cr}(G) \ge m - 3n + 6$, provided that $n \ge 3$.

10.3.2

a) Let G be a connected planar graph with girth k, where $k \ge 3$. Show that $m \le k(n-2)/(k-2)$.
b) Deduce that the Petersen graph is nonplanar.

10.3.3 Deduce Euler's Formula (10.2) from Exercise 10.2.13.

10.3.4

a) Show that the complement of a simple planar graph on at least eleven vertices is nonplanar.
b) Find a simple planar graph on eight vertices whose complement is planar.

10.3.5 A plane graph is *face-regular* if all of its faces have the same degree.

a) Characterize the plane graphs which are both regular and face-regular.
b) Show that exactly five of these graphs are simple and 3-connected. (They are the platonic graphs.)

10.3.6 The *thickness* $\theta(G)$ of a graph G is the minimum number of planar graphs whose union is G. (Thus $\theta(G) = 1$ if and only if G is planar.)

a) Let G be a simple graph. Show that $\theta(G) \ge \lceil m/(3n-6) \rceil$.
b) Deduce that $\theta(K_n) \ge \lfloor (n+1)/6 \rfloor + 1$ and show, using Exercise 10.3.4b, that equality holds for all $n \le 8$.
   (Beineke and Harary (1965) proved that equality holds for all $n \ne 9, 10$; Battle et al. (1962) showed that $\theta(K_9) = 3$.)

c) Express the Turán graph $T_{6,12}$ (defined in Exercise 1.1.11) as the union of two graphs, each isomorphic to the icosahedron.

d) Deduce from (b) and (c) that $\theta(K_{12}) = 3$.

10.3.7

a) Let G be a simple bipartite graph. Show that $\theta(G) \geq \lceil m/(2n - 4) \rceil$.

b) Deduce that $\theta(K_{m,n}) \geq \lceil mn/(2m + 2n - 4) \rceil$.

(Beineke et al. (1964) showed that equality holds if mn is even. It is conjectured that equality holds in all cases.)

10.3.8 A plane graph is *self-dual* if it is isomorphic to its dual.

a) Show that:
   i) if G is self-dual, then $e(G) = 2v(G) - 2$,
   ii) the four plane graphs shown in Figure 10.14 are self-dual.

b) Find four infinite families of self-dual plane graphs of which those four graphs are members.

(Smith and Tutte (1950) proved that every self-dual plane graph belongs to one of four infinite families.)

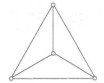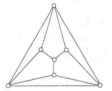

Fig. 10.14. Self-dual plane graphs

10.3.9

a) Let S be a set of n points in the plane, where $n \geq 3$ and the distance between any two points of S is at least one. Show that no more than $3n - 6$ pairs of points of S can be at distance exactly one. (P. ERDŐS)

b) By considering the triangular lattice (shown in Figure 1.27) find, for each positive integer k, a set S of $3k^2 + 3k + 1$ points in the plane such that the distance between any two points of S is at least one, and such that $9k^2 + 3k$ pairs of points of S are at distance exactly one.

10.3.10 THE SYLVESTER–GALLAI THEOREM

a) Let $\mathcal{L}$ be a finite set of lines in the plane, no two of which are parallel and not all of which are concurrent. Using Euler's Formula (10.2), show that some point is the point of intersection of precisely two lines of $\mathcal{L}$.

b) Deduce from (a) the *Sylvester–Gallai Theorem*: if S is a finite set of points in the plane, not all of which are collinear, there is a line that contains precisely two points of S. (E. MELCHIOR)

10.4 Bridges

In the study of planar graphs, certain subgraphs, called bridges, play an important role. We now define these subgraphs and discuss their properties.

Let H be a proper subgraph of a connected graph G. The set $E(G) \setminus E(H)$ may be partitioned into classes as follows.

▷ For each component F of $G - V(H)$, there is a class consisting of the edges of F together with the edges linking F to H.

▷ Each remaining edge e (that is, one which has both ends in $V(H)$) defines a singleton class $\{e\}$.

The subgraphs of G induced by these classes are the *bridges* of H in G. It follows immediately from this definition that bridges of H can intersect only in vertices of H, and that any two vertices of a bridge of H are connected by a path in the bridge that is internally disjoint from H. For a bridge B of H, the elements of $V(B) \cap V(H)$ are called its *vertices of attachment* to H; the remaining vertices of B are its *internal vertices*. A bridge is *trivial* if it has no internal vertices (that is, if it is of the second type). In a connected graph, every bridge has at least one vertex of attachment; moreover, in a nonseparable graph, every bridge has at least two vertices of attachment. A bridge with k vertices of attachment is called a *k-bridge*. Two bridges with the same vertices of attachment are *equivalent* bridges. Figure 10.15 shows a variety of bridges of a cycle in a graph; edges of different bridges are distinguished by different kinds of lines. Bridges B_1 and B_2 are equivalent 3-bridges; B_3 and B_6 are trivial bridges.

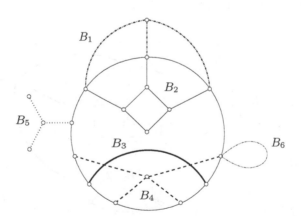

Fig. 10.15. Bridges of a cycle

BRIDGES OF CYCLES

We are concerned here with bridges of cycles, and all bridges are understood to be bridges of a given cycle C. Thus, to avoid repetition, we abbreviate 'bridge of C' to 'bridge' in the coming discussion.

The vertices of attachment of a k-bridge B with $k \geq 2$ effect a partition of C into k edge-disjoint paths, called the *segments* of B. Two bridges *avoid* each other if all the vertices of attachment of one bridge lie in a single segment of the other bridge; otherwise, they *overlap*. In Figure 10.15, B_2 and B_3 avoid each other, whereas B_1 and B_2 overlap, as do B_3 and B_4. Two bridges B and B' are *skew* if there are distinct vertices of attachment u, v of B, and u', v' of B', which occur in the cyclic order u, u', v, v' on C. In Figure 10.15, B_3 and B_4 are skew, whereas B_1 and B_2 are not.

Theorem 10.25 *Overlapping bridges are either skew or else equivalent 3-bridges.*

Proof Suppose that bridges B and B' overlap. Clearly, each must have at least two vertices of attachment. If either B or B' is a 2-bridge, it is easily verified that they must be skew. We may therefore assume that both B and B' have at least three vertices of attachment.

If B and B' are not equivalent bridges, then B' has a vertex u' of attachment between two consecutive vertices of attachment u and v of B. Because B and B' overlap, some vertex of attachment v' of B' does not lie in the segment of B connecting u and v. It follows that B and B' are skew.

If B and B' are equivalent k-bridges, then $k \geq 3$. If $k \geq 4$, B and B' are skew; if $k = 3$, they are equivalent 3-bridges.   □

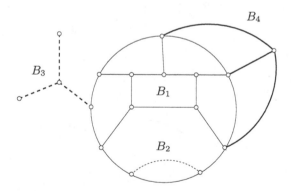

Fig. 10.16. Bridges of a cycle in a plane graph

We now consider bridges of cycles in plane graphs. Suppose that G is a plane graph and that C is a cycle in G. Because C is a simple closed curve in the plane, each bridge of C in G is contained in one of the two regions Int(C) or Ext(C). A bridge contained in $\text{Int}(C)$ is called an *inner bridge*, a bridge contained in $\text{Ext}(C)$ an *outer bridge*. In Figure 10.16, B_1 and B_2 are inner bridges, and B_3 and B_4 are outer bridges.

Theorem 10.26 *Inner (outer) bridges avoid one another.*

Proof Let B and B' be inner bridges of a cycle C in a plane graph G. Suppose that they overlap. By Theorem 10.25, they are either skew or equivalent 3-bridges. In both cases, we obtain contradictions.

Case 1: B and B' are skew. By definition, there exist distinct vertices u, v in B and u', v' in B', appearing in the cyclic order u, u', v, v' on C. Let uPv be a path in B and $u'P'v'$ a path in B', both internally disjoint from C. Consider the subgraph $H := C \cup P \cup P'$ of G (see Figure 10.17a). Because G is plane, so is H. Let K be the plane graph obtained from H by adding a vertex in $\text{ext}(C)$ and joining it to u, u', v, v' (see Figure 10.17b). Then K is a subdivision of K_5. But this is impossible, K_5 being nonplanar.

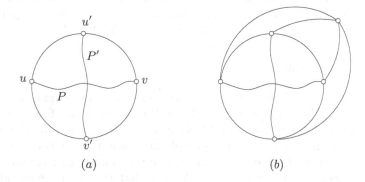

(a) $\qquad\qquad\qquad\qquad\qquad$ (b)

Fig. 10.17. Proof of Theorem 10.26, Case 1: (a) the subgraph H, (b) the subdivision K of K_5

Case 2: B and B' are equivalent 3-bridges. Denote by $S := \{v_1, v_2, v_3\}$ their common set of vertices of attachment. By Exercise 9.2.3, there exists a (v, S)-fan F in B, for some internal vertex v of B; likewise, there exists a (v', S)-fan F' in B', for some internal vertex v' of B'. Consider the subgraph $H := F \cup F'$ of G. Because G is plane, so is H. Let K be the plane graph obtained from H by adding a vertex in $\text{ext}(C)$ and joining it to the three vertices of S. Then K is a subdivision of $K_{3,3}$. But this is impossible, because $K_{3,3}$ is nonplanar (see Figure 10.18).

We conclude that inner bridges avoid one another. Similarly, outer bridges avoid one another. $\qquad\qquad\qquad\qquad\qquad\qquad\qquad\qquad\qquad\qquad\qquad\qquad\qquad$ □

It is convenient to visualize the above theorem in terms of the bridge-overlap graph. Let G be a graph and let C be a cycle of G. The *bridge-overlap graph* of C is the graph whose vertex set is the set of all bridges of C in G, two bridges being

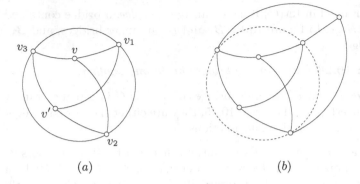

Fig. 10.18. Proof of Theorem 10.26, Case 2: (a) the subgraph H, (b) the subdivision K of $K_{3,3}$

adjacent if they overlap. Theorem 10.26 simply states that the bridge-overlap graph of any cycle of a plane graph is bipartite. Thus, a necessary condition for a graph to be planar is that the bridge-overlap graph of each of its cycles be bipartite. This condition also suffices to guarantee planarity (Exercise 10.5.7).

UNIQUE PLANE EMBEDDINGS

Just as there is no unique way of representing graphs by diagrams, there is no unique way of embedding planar graphs in the plane. Apart from the positions of points representing vertices and the shapes of lines representing the edges, two different planar embeddings of the same planar graph may differ in the incidence relationships between their edge and face sets; they may even have different face-degree sequences, as in Figure 10.11. We say that two planar embeddings of a planar graph G are *equivalent* if their face boundaries (regarded as sets of edges) are identical. A planar graph for which any two planar embeddings are equivalent is said have an *unique embedding* in the plane. Using the theory of bridges developed above, we show that every simple 3-connected planar graph is uniquely embeddable in the plane; note that the graph of Figure 10.11 is not 3-connected. The notion of a nonseparating cycle plays a crucial role in this proof.

A cycle is *nonseparating* if it has no chords and at most one nontrivial bridge. Thus, in a loopless graph G which is not itself a cycle, a cycle C is nonseparating if and only if it is an induced subgraph of G and $G - V(C)$ is connected. In the case of simple 3-connected plane graphs, Tutte (1963) proved that facial and nonseparating cycles are one and the same.

Theorem 10.27 *A cycle in a simple 3-connected plane graph is a facial cycle if and only if it is nonseparating.*

Proof Let G be a simple 3-connected plane graph and let C be a cycle of G. Suppose, first, that C is not a facial cycle of G. Then C has at least one inner bridge and at least one outer bridge. Because G is simple and connected, these bridges are not loops. Thus either they are both nontrivial or at least one of them is a chord. It follows that C is not a nonseparating cycle.

Now suppose that C is a facial cycle of G. By Proposition 10.5, we may assume that C bounds the outer face of G, so all its bridges are inner bridges. By Theorem 10.26, these bridges avoid one another. If C had a chord xy, the set $\{x, y\}$ would be a vertex cut separating the internal vertices of the two xy-segments of C. Likewise, if C had two nontrivial bridges, the vertices of attachment of one of these bridges would all lie on a single xy-segment of the other bridge, and $\{x, y\}$ would be a vertex cut of G separating the internal vertices of the two bridges. In either case, the 3-connectedness of G would be contradicted. Thus C is nonseparating. $\square$

A direct consequence of Theorem 10.27 is the following fundamental theorem, due to Whitney (1933).

Theorem 10.28 *Every simple 3-connected planar graph has a unique planar embedding.*

Proof Let G be a simple 3-connected planar graph. By Theorem 10.27, the facial cycles in any planar embedding of G are precisely its nonseparating cycles. Because the latter are defined solely in terms of the abstract structure of the graph, they are the same for every planar embedding of G. $\square$

The following corollary is immediate.

Corollary 10.29 *Every simple 3-connected planar graph has a unique dual graph.* $\square$

Exercises

$\star$**10.4.1** Let G_1 and G_2 be planar graphs whose intersection is isomorphic to K_2. Show that $G_1 \cup G_2$ is planar.

10.4.2 Let H be a subgraph of a graph G. Consider the binary relation $\sim$ on $E(G) \setminus E(H)$, where $e_1 \sim e_2$ if there exists a walk W in G such that:

▷ the first and the last edges of W are e_1 and e_2, respectively,
▷ W is internally disjoint from H (that is, no internal vertex of W is a vertex of H).

Show that:

a) the relation $\sim$ is an equivalence relation on $E(G) \setminus E(H)$,
b) the subgraphs of $G \setminus E(H)$ induced by the equivalence classes under this equivalence relation are the bridges of H in G.

———————————$\lambda\lambda$———————————

10.4.3 A *3-polytope* is the convex hull of a set of points in $\mathbb{R}^3$ which do not lie on a common plane. Show that the polyhedral graph of such a polytope is simple, planar, and 3-connected.
(Steinitz (1922) proved that, conversely, *every simple 3-connected planar graph is the polyhedral graph of some 3-polytope.*)

10.4.4 Show that any 3-connected cubic plane graph on n vertices, where $n \geq 6$, may be obtained from one on $n - 2$ vertices by subdividing two edges in the boundary of a face and joining the resulting new vertices by an edge subdividing the face.

10.4.5 A *rooting* of a plane graph G is a triple (v, e, f), where v is a vertex, called the *root vertex*, e is an edge of G incident with v, called the *root edge*, and f is a face incident with e, called the *root face*.

  a) Show that the only automorphism of a simple 3-connected plane graph which fixes a given rooting is the identity automorphism.
  b) Let G be a simple 3-connected planar graph. Deduce from (a) that:
    i) $aut(G)$ divides $4m$,
    ii) $aut(G) = 4m$ if and only if G is one of the five platonic graphs.
                       (F. HARARY AND W.T. TUTTE; L. WEINBERG)

10.5 Kuratowski's Theorem

Planarity being such a fundamental property, the problem of deciding whether a given graph is planar is clearly of great importance. A major step towards this goal is provided by the following characterization of planar graphs, due to Kuratowski (1930).

Theorem 10.30 KURATOWSKI'S THEOREM
A graph is planar if and only if it contains no subdivision of either K_5 or $K_{3,3}$.

A subdivision of K_5 or $K_{3,3}$ is consequently called a *Kuratowski subdivision*.
We first present a proof of Kuratowski's Theorem due to Thomassen (1981), and then explain how it gives rise to a polynomial-time decision algorithm for planarity. Before proving the theorem, we reformulate it in terms of minors.

MINORS

A *minor* of a graph G is any graph obtainable from G by means of a sequence of vertex and edge deletions and edge contractions. Alternatively, consider a partition $(V_0, V_1, \ldots, V_k)$ of V such that $G[V_i]$ is connected, $1 \leq i \leq k$, and let H be the graph obtained from G by deleting V_0 and shrinking each induced subgraph $G[V_i]$, $1 \leq i \leq k$, to a single vertex. Then any spanning subgraph F of H is a *minor* of G. For instance, K_5 is a minor of the Petersen graph because it can be obtained by contracting the five 'spoke' edges of the latter graph (see Figure 10.19).

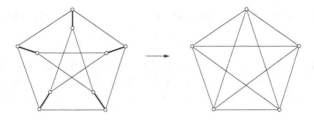

Fig. 10.19. Contracting the Petersen graph to K_5

If F is a minor of G, we write $F \preceq G$. By an F-*minor* of G, where F is an arbitrary graph, we mean a minor of G which is isomorphic to F. It is important to point out that any graph which contains an F-subdivision also has an F-minor: to obtain F as a minor, one simply deletes the vertices and edges not in the subdivision, and then contracts each subdivided edge to a single edge. For example, because the Petersen graph contains a $K_{3,3}$-subdivision (Exercise 10.1.3), it also has a $K_{3,3}$-minor. Conversely, provided that F is a graph of maximum degree three or less, any graph which has an F-minor also contains an F-subdivision (Exercise 10.5.3a).

WAGNER'S THEOREM

As observed in Section 10.2, the deletion or contraction of an edge in a planar graph results again in a planar graph. Thus we have:

Proposition 10.31 *Minors of planar graphs are planar.*   □

A minor which is isomorphic to K_5 or $K_{3,3}$ is called a *Kuratowski minor*. Because K_5 and $K_{3,3}$ are nonplanar, Proposition 10.31 implies that any graph which has a Kuratowski minor is nonplanar. Wagner (1937) proved that the converse is true.

Theorem 10.32   WAGNER'S THEOREM
A graph is planar if and only if it has no Kuratowski minor.

We remarked above that a graph which contains an F-subdivision also has an F-minor. Thus Kuratowski's Theorem implies Wagner's Theorem. On the other hand, because $K_{3,3}$ has maximum degree three, any graph which has a $K_{3,3}$-minor contains a $K_{3,3}$-subdivision (Exercise 10.5.3a). Furthermore, any graph which has a K_5-minor necessarily contains a Kuratowski subdivision (Exercise 10.5.3b). Thus Wagner's Theorem implies Kuratowski's Theorem, and the two are therefore equivalent.

It turns out to be slightly more convenient to prove Wagner's variant of Kuratowski's Theorem. Before doing so, we need to establish two simple lemmas.

Lemma 10.33 *Let G be a graph with a 2-vertex cut $\{x, y\}$. Then each marked $\{x, y\}$-component of G is isomorphic to a minor of G.*   □

Proof Let H be an $\{x, y\}$-component of G, with marker edge e, and let xPy be a path in another $\{x, y\}$-component of G. Then $H \cup P$ is a subgraph of G. But $H \cup P$ is isomorphic to a subdivision of $G + e$, so $G + e$ is isomorphic to a minor of G.   □

Lemma 10.34 *Let G be a graph with a 2-vertex cut $\{x, y\}$. Then G is planar if and only if each of its marked $\{x, y\}$-components is planar.*

Proof Suppose, first, that G is planar. By Lemma 10.33, each marked $\{x, y\}$-component of G is isomorphic to a minor of G, hence is planar by Proposition 10.31.

Conversely, suppose that G has k marked $\{x, y\}$-components each of which is planar. Let e denote their common marker edge. Applying Exercise 10.4.1 and induction on k, it follows that $G + e$ is planar, hence so is G.   □

In view of Lemmas 10.33 and 10.34, it suffices to prove Wagner's Theorem for 3-connected graphs. It remains, therefore, to show that every 3-connected nonplanar graph has either a K_5-minor or a $K_{3,3}$-minor. We present an elegant proof of this statement. It is due to Thomassen (1981), and is based on his Theorem 9.10.

Theorem 10.35 *Every 3-connected nonplanar graph has a Kuratowski minor.*

Proof Let G be a 3-connected nonplanar graph. We may assume that G is simple. Because all graphs on four or fewer vertices are planar, we have $n \geq 5$. We proceed by induction on n. By Theorem 9.10, G contains an edge $e = xy$ such that $H := G / e$ is 3-connected. If H is nonplanar, it has a Kuratowski minor, by induction. Since every minor of H is also a minor of G, we deduce that G too has a Kuratowski minor. So we may assume that H is planar.

Consider a plane embedding $\widetilde{H}$ of H. Denote by z the vertex of H formed by contracting e. Because H is loopless and 3-connected, by Corollary 10.8 the neighbours of z lie on a cycle C, the boundary of some face f of $\widetilde{H} - z$. Denote by B_x and B_y, respectively, the bridges of C in $G \setminus e$ that contain the vertices x and y.

Suppose, first, that B_x and B_y avoid each other. In this case, B_x and B_y can be embedded in the face f of $\widetilde{H} - z$ in such a way that the vertices x and y belong to the same face of the resulting plane graph $(\widetilde{H} - z) \cup \widetilde{B}_x \cup \widetilde{B}_y$ (see Figure 10.20). The edge xy can now be drawn in that face so as to obtain a planar embedding of G itself, contradicting the hypothesis that G is nonplanar.

It follows that B_x and B_y do not avoid each other, that is, they overlap. By Theorem 10.25, they are therefore either skew or else equivalent 3-bridges. In the former case, G has a $K_{3,3}$-minor; in the latter case, G has a K_5-minor (see Figure 10.21).   □

We note that the same proof serves to show that every simple 3-connected planar graph admits a *convex embedding*, that is, a planar embedding all of whose

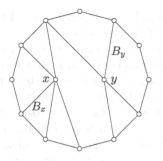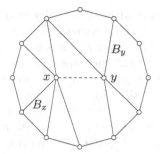

Fig. 10.20. A planar embedding of G (B_x and B_y avoid each other)

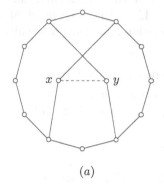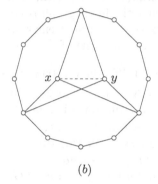

(a)   (b)

Fig. 10.21. (a) A $K_{3,3}$-minor (B_x and B_y skew), (b) a K_5-minor (B_x and B_y equivalent 3-bridges)

faces are bounded by convex polygons. All that is needed is a bit more care in placing the bridges B_x and B_y, and the edge $e = xy$, in the face f (Exercise 10.5.5).

There are several other interesting characterizations of planar graphs, all of which can be deduced from Kuratowski's Theorem (see Exercises 10.5.7, 10.5.8, and 10.5.9).

RECOGNIZING PLANAR GRAPHS

There are many practical situations in which it is important to decide whether a given graph is planar, and if so, to find a planar embedding of the graph. In the layout of printed circuits, for example, one is interested in knowing if a particular electrical network is planar.

It is easy to deduce from Lemma 10.34 that a graph is planar if and only if each of its 3-connected components is planar. Thus the problem of deciding whether a given graph is planar can be solved by considering each 3-connected component separately. The proof of Wagner's Theorem presented above can be transformed without difficulty into a polynomial-time algorithm for determining whether a given 3-connected graph is planar. The idea is as follows.

First, the input graph is contracted, one edge at a time, to a complete graph on four vertices (perhaps with loops and multiple edges) in such a way that all intermediate graphs are 3-connected. This contraction phase can be executed in polynomial time by proceeding as indicated in the proof of Theorem 9.10. The resulting four-vertex graph is then embedded in the plane. The contracted edges are now expanded one by one (in reverse order). At each stage of this expansion phase, one of two eventualities may arise: either the edge can be expanded while preserving planarity, and the algorithm proceeds to the next contracted edge, or else two bridges are found which overlap, yielding a Kuratowski minor. In the second eventuality, the algorithm outputs one of these nonplanar minors, thereby certifying that the input graph is nonplanar. If, on the other hand, all contracted edges are expanded without encountering overlapping bridges, the algorithm outputs a planar embedding of G.

Algorithm 10.36 PLANARITY RECOGNITION AND EMBEDDING

INPUT: a 3-connected graph G on four or more vertices
OUTPUT: a Kuratowski minor of G or a planar embedding of G

1: set $i := 0$ and $G_0 := G$
   CONTRACTION PHASE:
2: **while** $i < n - 4$ **do**
3:    find a link $e_i := x_i y_i$ of G_i such that G_i / e_i is 3-connected
4:    set $G_{i+1} := G_i / e_i$
5:    replace i by $i + 1$
6: **end while**
   EXPANSION PHASE:
7: find a planar embedding $\widetilde{G}_{n-4}$ of the four-vertex graph G_{n-4}
8: set $i := n - 4$
9: **while** $i > 0$ **do**
10:    let C_i be the facial cycle of $\widetilde{G}_i - z_i$ that includes all the neighbours of z_i in $\widetilde{G}_i$, where z_i denotes the vertex of $\widetilde{G}_i$ resulting from the contraction of the edge e_{i-1} of G_{i-1}
11:    let B_i and B_i', respectively, denote the bridges of C_i containing the vertices x_{i-1} and y_{i-1} in the graph obtained from G_{i-1} by deleting e_{i-1} and all other edges linking x_{i-1} and y_{i-1}
12:    **if** B_i and B_i' are skew **then**
13:       find a $K_{3,3}$-minor K of G_{i-1}
14:       return K
15:    **end if**
16:    **if** B_i and B_i' are equivalent 3-bridges **then**
17:       find a K_5-minor K of G_{i-1}
18:       return K
19:    **end if**
20:    **if** B_i and B_i' avoid each other **then**

21:        extend the planar embedding $\widetilde{G}_i$ of G_i to a planar embedding $\widetilde{G}_{i-1}$ of
           G_{i-1}
22:        replace i by $i-1$
23:    **end if**
24: **end while**
25: return $\widetilde{G}_0$

Each step in the contraction phase and each step in the expansion phase can
be executed in polynomial time. It follows that the problem of deciding whether a
graph is planar belongs to $\mathcal{P}$. There is, in fact, a linear-time planarity recognition
algorithm, due to Hopcroft and Tarjan (1974). There also exist efficient planarity
algorithms based on the characterization of planarity in terms of the bridge-overlap
graph given in Exercise 10.5.7; for details, see Bondy and Murty (1976).

Exercises

10.5.1 Show that a simple graph has a K_3-minor if and only if it contains a cycle.

10.5.2 Show that the (3×3)-grid has a K_4-minor.

$\star$**10.5.3**

a) Let F be a graph with maximum degree at most three. Show that a graph has
   an F-minor if and only if it contains an F-subdivision.
b) Show that any graph which has a K_5-minor contains a Kuratowski subdivision.

10.5.4 Consider the two 3-connected graphs shown in Figure 10.22. In each case,
the contraction of the edge 12 results in a graph that is 3-connected and planar.
Obtain a planar embedding of this resulting graph, and apply the Planarity Recognition and Embedding Algorithm (10.36) to either obtain a planar embedding of
the given graph or else find a Kuratowski minor of the graph.

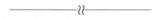

10.5.5 Prove that every simple 3-connected planar graph admits a convex planar
embedding.

10.5.6 Let G be a simple graph. A *straight-line embedding* of G is an embedding
of G in the plane in which each edge is a straight-line segment. The *rectilinear
crossing number* of G, denoted by $\overline{cr}(G)$ is the minimum number of crossings in a
straight-line embedding of G.

a) Show that:
     i) $cr(G) \leq \overline{cr}(G)$,
     ii) if $cr(G) = 1$, then $\overline{cr}(G) = 1$.

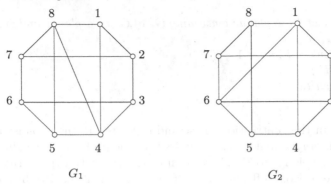

Fig. 10.22. Apply Algorithm 10.36 to these graphs (Exercise 10.5.4)

(Bienstock and Dean (1993) have shown that $\operatorname{cr}(G) = \overline{\operatorname{cr}}(G)$ if G is simple and $\operatorname{cr}(G) \leq 3$. They have also given examples of graphs G with $\operatorname{cr}(G) = 4 < \overline{\operatorname{cr}}(G)$.)

b) Show that $\overline{\operatorname{cr}}(K_{m,n}) \leq \lfloor m/2 \rfloor \lfloor (m-1)/2 \rfloor \lfloor n/2 \rfloor \lfloor (n-1)/2 \rfloor$.
   (It was conjectured by P. Turán that this bound is best possible.)

10.5.7 Using Kuratowski's Theorem (10.30), show that a graph is planar if and only if the bridge-overlap graph of each cycle is bipartite.   (W.T. TUTTE)

10.5.8 A basis of the cycle space of a graph is a 2-*basis* if each member of the basis is a cycle of the graph, and each edge of the graph lies in at most two of these cycles.

a) Show that:
   i) the cycle space of any planar graph has a 2-basis,
   ii) the cycle spaces of K_5 and $K_{3,3}$ do not have 2-bases.
b) A theorem due to MacLane (1937) states that a graph is planar if and only if its cycle space has a 2-basis. Deduce MacLane's Theorem from Kuratowski's Theorem (10.30).

10.5.9 A graph H is called an *algebraic dual* of a graph G if there is a bijection $\phi : E(G) \to E(H)$ such that a subset C of $E(G)$ is a cycle of G if and only if $\phi(C)$ is a bond of H.

a) Show that:
   i) every planar graph has an algebraic dual,
   ii) K_5 and $K_{3,3}$ do not have algebraic duals.
b) A theorem due to Whitney (1932c) states that a graph is planar if and only if it has an algebraic dual. Deduce Whitney's Theorem from Kuratowski's Theorem (10.30).

10.5.10 k-SUM
Let G_1 and G_2 be two graphs whose intersection $G_1 \cap G_2$ is a complete graph on

k vertices. The graph obtained from their union $G_1 \cup G_2$ by deleting the edges of $G_1 \cap G_2$ is called the k-*sum* of G_1 and G_2.

a) Show that if G_1 and G_2 are planar and $k = 0, 1$, or 2, then the k-sum of G_1 and G_2 is also planar.
b) Express the nonplanar graph $K_{3,3}$ as a 3-sum of two planar graphs.

10.5.11 SERIES-PARALLEL GRAPH
A *series extension* of a graph is the subdivision of a link of the graph; a *parallel extension* is the addition of a new link joining two adjacent vertices. A *series-parallel* graph is one that can be obtained from K_2 by a sequence of series and parallel extensions.

a) Show that a series-parallel graph has no K_4-minor.
b) By applying Exercise 10.1.5, deduce that a graph has no K_4-minor if and only if it can be obtained from K_1, the loop graph L_1 (a loop incident with a single vertex), and the family of series-parallel graphs, by means of 0-sums, 1-sums, and 2-sums.   (G.A. DIRAC)

10.5.12 Show that a graph is outerplanar if and only if it has neither a K_4-minor nor a $K_{2,3}$-minor.

10.5.13 EXCLUDED $K_{3,3}$-MINOR
Show that:

a) every 3-connected nonplanar graph on six or more vertices has a $K_{3,3}$-minor,
b) any graph with no $K_{3,3}$-minor can be obtained from the family of planar graphs and K_5 by means of 0-sums, 1-sums, and 2-sums. (D.W. HALL; K. WAGNER)

10.5.14 EXCLUDED K_5-MINOR
Show that:

a) the *Wagner graph*, depicted in Figure 10.23, has no K_5-minor,
b) if G_1 and G_2 are two graphs, each of which is either a planar graph or the Wagner graph, then no 0-sum, 1-sum, 2-sum, or 3-sum of G_1 and G_2 has a K_5-minor.

(Wagner (1936) showed that every 4-connected nonplanar graph has a K_5-minor and deduced the converse of (b), namely that any graph with no K_5-minor can be obtained from the family of planar graphs and the Wagner graph by means of 0-sums, 1-sums, 2-sums, and 3-sums.)

10.6 Surface Embeddings of Graphs

During the nineteenth century, in their attempts to discover generalizations of Euler's Formula (10.2) and the Four-Colour Conjecture (discussed in the next chapter), graph theorists were led to the study of embeddings of graphs on surfaces

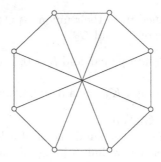

Fig. 10.23. The Wagner graph

other than the plane and the sphere. In recent years, embeddings have been used to investigate a wide variety of problems in graph theory, and have proved to be an essential tool in the study of an important graph-theoretic parameter, the tree-width, whose theory was developed in an extensive series of papers by N. Robertson and P. D. Seymour (see Sections 9.8 and 10.7). The books by Bonnington and Little (1995), Fréchet and Fan (2003), Gross and Tucker (1987), and Mohar and Thomassen (2001) have excellent accounts of the theory of embeddings of graphs on surfaces. We present here a brief account of some of the basic notions and results of the subject, without proofs, and without making any attempt to be rigorous.

ORIENTABLE AND NONORIENTABLE SURFACES

A *surface* is a connected 2-dimensional manifold. Apart from the plane and the sphere, examples of surfaces include the cylinder, the Möbius band, and the torus. The *cylinder* may be obtained by gluing together two opposite sides of a rectangle, the *Möbius band* by gluing together two opposite sides of a rectangle after making one half-twist, and the *torus* by gluing together the two open ends of a cylinder. The Möbius band and the torus are depicted in Figure 10.24. (Drawings from Crossley (2005), courtesy Martin Crossley.)

There are two basic types of surface: those which are orientable and those which are nonorientable. To motivate the distinction between these two types, let us consider the Möbius band. First note that, unlike what the physical model suggests, the Möbius band has no 'thickness'. Moreover, unlike the cylinder, it is 'one-sided'. Now consider a line running along the middle of a Möbius band, and imagine an ant crawling on the surface along it. After one complete revolution, the ant would return to where it started from. However, it would have the curious experience of finding its 'left' and 'right' reversed; those points of the surface which were to the left of the ant at the beginning would now be to its right: it is not possible to 'globally' distinguish left from right on the Möbius band. Surfaces which have this property are said to be *nonorientable*; all other surfaces are *orientable*. The plane, the cylinder, the sphere, and the torus are examples of orientable surfaces.

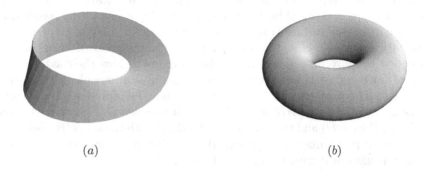

Fig. 10.24. (a) The Möbius band, and (b) the torus

A surface is *closed* if it is bounded but has no boundary. The Möbius band has a boundary which is *homeomorphic* (that is, continuously deformable) to a circle and, hence, is not a closed surface. The plane is clearly not bounded, hence is not a closed surface either. The simplest closed surface is the sphere. Other closed surfaces are sometimes referred to as *higher surfaces*. Starting with the sphere, all higher surfaces can be constructed by means of two operations.

Let S be a sphere, let D_1 and D_2 be two disjoint discs of equal radius on S, and let H be a cylinder of the same radius as D_1 and D_2. The operation of *adding a handle* to S at D_1 and D_2 consists of cutting out D_1 and D_2 from S and then bending and attaching H to S in such a way that the rim of one of the ends of H coincides with the boundary of D_1 and the rim of the other end of H coincides with the boundary of D_2. Any number of disjoint handles may be added to S by selecting disjoint pairs of discs on S and adding a handle at each of those pairs of discs. A *sphere with k handles* is the surface obtained from a sphere by adding k handles; it is denoted by S_k, and the index k is its *genus*. The torus is homeomorphic to a sphere with one handle, S_1. More generally, every orientable surface is homeomorphic to a sphere with k handles for some $k \geq 0$.

As mentioned above (see also Section 3.5), given any rectangle $ABCD$, one may obtain a torus by identifying the side AB with the side DC and the side AD with the side BC. More generally, any orientable surface may be constructed from a suitable polygon by identifying its sides in a specified manner. For example, the surface S_2, also known as the *double torus*, may be obtained by a suitable identification of the sides of an octagon (see Exercise 10.6.2).

We now turn to nonorientable surfaces. Let S be a sphere, let D be a disc on S, and let B be a Möbius band whose boundary has the same length as the circumference of D. The operation of *adding a cross-cap to S at D* consists of attaching B to S so that the boundaries of D and B coincide. Equivalently, this operation consists of 'sewing' or 'identifying' every point on the boundary of D to the point of D that is antipodal to it. Just as with handles, we may attach any number of cross-caps to a sphere. The surface obtained from the sphere by attaching one cross-cap is known as the *projective plane* and is the simplest nonorientable surface. A sphere with k cross-caps is denoted by N_k, the index k being its *cross-cap number*. Every closed nonorientable surface is homeomorphic to N_k for some $k \geq 1$.

As with orientable closed surfaces, all nonorientable closed surfaces may be represented by polygons, along with indications as to how their sides are to be identified (although it is not possible to obtain physical models of these surfaces in this way). The projective plane, for example, may be represented by a rectangle $ABCD$ in which the side AB is identified with the side CD (so that A coincides with C and B with D) and the side AD is identified with the side CB. Equivalently, the projective plane may be represented by a disc in which every point on the boundary is identified with its antipodal point.

An important theorem of the topology of surfaces, known as the *classification theorem for surfaces*, states that every closed surface is homeomorphic to either S_k or N_k, for a suitable value of k. One may, of course, obtain surfaces by adding both handles and cross-caps to spheres. However, one does not produce any new surfaces in this way. It turns out that the surface obtained from the sphere by adding $k > 0$ handles and $\ell > 0$ cross-caps is homeomorphic to $N_{2k+\ell}$.

In Chapter 3, we presented embeddings of K_7 and the Petersen graph on the torus (see Figure 3.9). Embeddings of K_6 and the Petersen graph on the projective plane are shown in Figure 10.25.

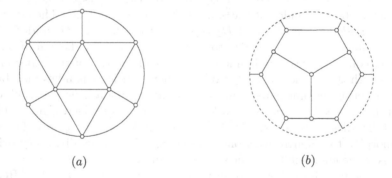

(a)   (b)

Fig. 10.25. Embeddings on the projective plane of (a) K_6, and (b) the Petersen graph

Polygonal representations of surfaces are convenient for displaying embeddings of graphs on surfaces of low genus or cross-cap number. However, for more complicated surfaces, such representations are unwieldy. Convenient algebraic and combinatorial schemes exist for describing embeddings on arbitrary surfaces.

THE EULER CHARACTERISTIC

An embedding $\widetilde{G}$ of a graph G on a surface Σ is a *cellular embedding* if each of the arcwise-connected regions of $\Sigma \setminus \widetilde{G}$ is homeomorphic to the open disc. These regions are the *faces* of $\widetilde{G}$, and their number is denoted by $f(\widetilde{G})$.

Consider, for example, the two embeddings of K_4 on the torus shown in Figure 10.26. The first embedding is cellular: it has two faces, bounded by the closed walks 12341 and 124134231, respectively. The second embedding is not cellular, because one of its faces is homeomorphic to a cylinder, bounded by the cycles 1231 and 1431.

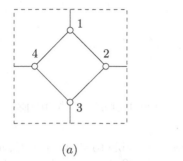   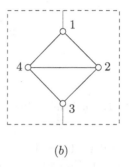

(a)   (b)

Fig. 10.26. Two embeddings of K_4 on the torus: (a) a cellular embedding, and (b) a noncellular embedding

Most theorems of interest about embeddings are valid only for cellular embeddings. For this reason, all the embeddings that we discuss are assumed to be cellular.

The *Euler characteristic* of a surface Σ, denoted $c(\Sigma)$, is defined by:

$$c(\Sigma) := \begin{cases} 2 - 2k & \text{if } \Sigma \text{ is homeomorphic to } S_k \\ 2 - k & \text{if } \Sigma \text{ is homeomorphic to } N_k \end{cases}$$

Thus the Euler characteristics of the sphere, the projective plane, and the torus are 2, 1, and 0, respectively. The following theorem is a generalization of Euler's Formula (10.2) for graphs embedded on surfaces.

Theorem 10.37 *Let $\widetilde{G}$ be an embedding of a connected graph G on a surface Σ. Then:*

$$v(\widetilde{G}) - e(\widetilde{G}) + f(\widetilde{G}) = c(\Sigma) \qquad \square$$

The following easy corollaries of Theorem 10.37 generalize Corollaries 10.20 and 10.21 to higher surfaces (Exercise 10.6.3).

Corollary 10.38 *All embeddings of a connected graph on a given surface have the same number of faces.* $\qquad \square$

Corollary 10.39 *Let G be a simple connected graph that is embeddable on a surface Σ. Then*

$$m \leq 3(n - c(\Sigma)) \qquad \square$$

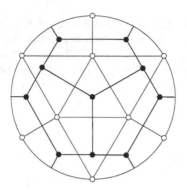

Fig. 10.27. Dual embeddings of K_6 and the Petersen graph on the projective plane

Using Euler's Formula for the sphere, we were able to show that K_5 and $K_{3,3}$ are not planar. Similarly, by using Corollary 10.39, one can show that for any surface there are graphs that are not embeddable on that surface. For example, K_7 is not embeddable on the projective plane and K_8 is not embeddable on the torus (Exercise 10.6.4). On the other hand, K_6 is embeddable on the projective plane (see Figure 10.25a), as is K_7 on the torus (see Figure 3.9a).

Duals of graphs embedded on surfaces may be defined in the same way as duals of plane graphs. It can be seen from Figure 10.27 that the dual of the embedding of K_6 shown in Figure 10.25a is the Petersen graph, embedded as shown in Figure 10.25b. Likewise, the dual of the embedding of K_7 shown in Figure 3.9a is the Heawood graph (Exercise 10.6.1).

We proved in Section 9.2 that all faces of a loopless 2-connected plane graph are bounded by cycles. The analogous statement for loopless 2-connected graphs embedded on other surfaces is not true, as can be seen from the embeddings of K_4 on the torus shown in Figure 10.26. An embedding $\widetilde{G}$ of a graph G on a surface Σ is a *circular embedding* if all the faces of $\widetilde{G}$ are bounded by cycles.

The following conjecture is due to Jaeger (1988). It refines at the same time the Circular Embedding Conjecture (3.10) and the Oriented Cycle Double Cover Conjecture (3.12).

THE ORIENTABLE EMBEDDING CONJECTURE

Conjecture 10.40 *Every loopless 2-connected graph has a circular embedding on some orientable surface.*

Exercises

10.6.1 Show that the dual of the embedding of K_7 shown in Figure 3.9a is the Heawood graph (depicted in Figure 1.16).

10.6.2 Show that the surface obtained by identifying the identically labelled edges of the octagon shown in Figure 10.28 along the directions indicated is the *double torus*.

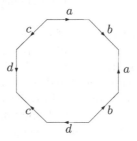

Fig. 10.28. A representation of the double torus

$\star$**10.6.3** Prove Corollaries 10.38 and 10.39.

10.6.4 Show that:

a) K_7 is not embeddable on the projective plane,
b) K_8 is not embeddable on the torus.

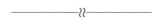

10.6.5 The (orientable) *genus* $\gamma(G)$ of a graph G is the minimum value of k such that G is embeddable in S_k. (Thus the genus of a planar graph is zero and the genus of a nonplanar graph that is embeddable in the torus is one.)

a) Show that $\gamma(K_{m,n}) \geq (m-2)(n-2)/4$.
b) By finding an embedding of $K_{4,4}$ on the torus, deduce that $\gamma(K_{4,4}) = 1$.

(G. Ringel has shown that $\gamma(K_{m,n}) = \lceil (m-2)(n-2)/4 \rceil$ for all m, n; see Hartsfield and Ringel (1994).)

10.6.6 Show that $\gamma(K_n) \geq (n-3)(n-4)/12$.
(A major result in the theory of graph embeddings, due to G. Ringel and J.W.T. Youngs, and known as the *Map Colour Theorem*, states that $\gamma(K_n) = \lceil (n-3)(n-4)/12 \rceil$ for all n; see Ringel (1974).)

10.7 Related Reading

GRAPH MINORS

In a long and impressive series of papers entitled *Graph Minors*, N. Robertson and P. D. Seymour proved a conjecture of K. Wagner which states that any infinite sequence $G_1, G_2, \ldots$ of (finite) graphs includes two graphs G_i and G_j, with $i < j$, such that G_i is a minor of G_j. In doing so, they introduced and employed a wealth of new concepts and ways of viewing graphical structure that are destined to play a major role in future developments of graph theory (see Robertson and Seymour (2004)).

A class $\mathcal{G}$ of graphs is *minor-closed* if every minor of a member of $\mathcal{G}$ is also a member of $\mathcal{G}$. For example, the class of all planar graphs is minor-closed and, more generally, the class of all graphs embeddable in any fixed surface such as the projective plane or the torus is minor-closed. A graph G which does not belong to $\mathcal{G}$, but all of whose proper minors do, is said to be *minor-minimal* with respect to $\mathcal{G}$. For instance, by virtue of Theorem 10.32, the minor-minimal nonplanar graphs are $K_{3,3}$ and K_5. A straightforward consequence of the theorem of Robertson and Seymour mentioned above is that if $\mathcal{G}$ is any minor-closed family of graphs, then the number of minor-minimal graphs with respect to $\mathcal{G}$ is finite. Thus every minor-closed family of graphs affords an 'excluded-minor' characterization of Kuratowski–Wagner type.

Apart from the proof of Wagner's conjecture, the *Graph Minors* series includes many other remarkable results. Using the theorem of Wagner stated in Exercise 10.5.14 as a prototype, Robertson and Seymour gave a characterization of the graphs which have no K_n-minor. They also described a polynomial-time algorithm for deciding whether a graph has a given graph H as a minor.

LINKAGES

Topological considerations play a significant role in graph theory, even with regard to certain graph-theoretical questions which seemingly have no connection with embeddings. An important and attractive example is the Linkage Problem. Let G be a graph, and let $X := (x_1, x_2, \ldots, x_k)$ and $Y := (y_1, y_2, \ldots, y_k)$ be two ordered subsets of V. An XY-*linkage* in G is a set of k disjoint paths $x_i P_i y_i$, $1 \leq i \leq k$. Such a set is also called a k-*linkage*. The *Linkage Problem* is the problem of deciding whether there exists an XY-linkage for given sets X and Y. Note that this is not quite the same as the Disjoint Paths Problem discussed in Chapter 9; the existence of k disjoint XY-paths does not guarantee the existence of an XY-linkage. For example, in the graph of Figure 9.6, there are two disjoint paths connecting $\{x_1, x_2\}$ and $\{y_1, y_2\}$, but no pair of disjoint $x_1 y_1$- and $x_2 y_2$-paths. Seymour (1980), Shiloach (1980), and Thomassen (1980) proved a theorem which essentially characterizes the graphs which have a 2-linkage, for given pairs (x_1, x_2) and (y_1, y_2). They showed that a 4-connected graph has such a 2-linkage unless the graph is planar and the vertices x_1, x_2, y_1, and y_2 appear on the boundary of some face in that cyclic order (as in Figure 9.6). One of the important byproducts of the theory of graph minors is a polynomial-time algorithm for the k-linkage problem for any fixed value of k; see Robertson and Seymour (1995). In stark contrast, the k-linkage problem for directed graphs is $\mathcal{NP}$-hard, even for $k = 2$; see Fortune et al. (1980).

BRAMBLES

By definition, a graph has a complete minor of order k if and only if it contains k mutually disjoint connected subgraphs, any two of which are linked by at least one edge. Weakening these requirements a little, we obtain a structure called a *bramble*. This is a set of connected subgraphs (the *elements* of the bramble) any two of which either intersect or are linked by at least one edge. Figure 10.29 shows two brambles in $K_{2,3}$,

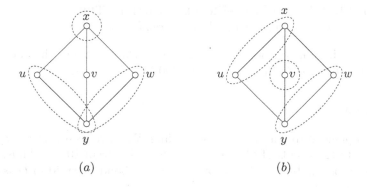

$$(a) \qquad\qquad (b)$$

Fig. 10.29. Two brambles in $K_{2,3}$: (a) one of order two, and (b) one of order three

A *transversal* of a bramble is a set of vertices which meets every element of the bramble. For instance, $\{x, y\}$ is a transversal of the bramble in Figure 10.29a, and $\{u, v, w\}$ is a transversal of the one in Figure 10.29b. The *order* of a bramble is the minimum cardinality of a transversal. It can be seen that the brambles in Figure 10.29 are of orders two and three, respectively. The maximum order of a bramble in a graph is its *bramble number*. The graph $K_{2,3}$ has bramble number three. The bramble of six elements in the (3×3)-grid shown in Figure 10.30 is of order four.

Brambles and tree-decompositions are dual structures in the following sense. Let $\mathcal{B}$ be a bramble of a graph G, and let $(T, \{T_v : v \in V\})$ be a tree-decomposition of G of minimum width. Consider an element B of $\mathcal{B}$. Because B is connected, $T_B := \cup\{T_v : v \in V(B)\}$ is a subtree of T. Furthermore, because any two bramble elements either meet or are adjacent, $\{T_B : B \in \mathcal{B}\}$ is a pairwise intersecting family of subtrees of T. By the Helly Property (Exercise 4.1.20), these trees therefore

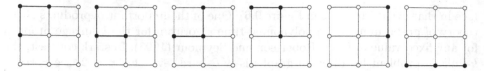

Fig. 10.30. A bramble of order four in the (3×3)-grid

have a vertex x in common. The corresponding subset V_x of V (in the notation of Section 9.8) is thus a transversal of $\mathcal{B}$. This shows that the order of $\mathcal{B}$ is at most $|V_x|$, which in turn is no greater than the tree-width of G. Because this is true for any bramble $\mathcal{B}$, we conclude that the bramble number is bounded above by the tree-width. Observe that, in both $K_{2,3}$ and the (3×3)-grid, these two parameters take exactly the same value. Seymour and Thomas (1993) showed that this is always so.

Theorem 10.41 Tree-Width–Bramble Duality Theorem
The tree-width of any graph is equal to its bramble number.

For a beautiful, unified proof of this and other related duality theorems, we refer the reader to the article by Amini et al. (2007).

Matroids and Duality

Let G be a plane graph, and let G^* be its dual. We have seen that the cycles of G correspond to the bonds of G^*, and conversely (Theorem 10.16). Thus, the cycle matroid of G is the bond matroid of G^* and the bond matroid of G is the cycle matroid of G^*.

Nonisomorphic graphs on the same edge set may have the same cycle matroid (the duals of the graphs in Figure 10.11, for instance). However, it follows from the work of Whitney (1932b) that simple 3-connected graphs with the same cycle matroid are isomorphic (see Welsh (1976) or Oxley (1992)).

Matroid Minors

An element e of a matroid M is a *loop* if it is contained in no basis of M and a *coloop* if it is contained in every basis. When e is neither a loop nor a coloop, let $\mathcal{B} \setminus e$ denote the set of bases of M not containing e and let $\mathcal{B} / e$ denote the set of restrictions to $E \setminus \{e\}$ of the bases of M containing e. When e is either a loop or a coloop, let $\mathcal{B} \setminus e$ and $\mathcal{B} / e$ both denote the set of restrictions of the bases of M to $E \setminus \{e\}$. One may then verify that $(E \setminus \{e\}, \mathcal{B} \setminus e)$ and $(E \setminus \{e\}, \mathcal{B} / e)$ are matroids. The former is said to be obtained from M by the *deletion* of e and is denoted by $M \setminus e$. The latter is said to be obtained from M by the *contraction* of e and is denoted by M / e. These two operations are related via duality in the same manner as they are in the case of graphs (see Propositions 10.12 and 10.13):

$$(M \setminus e)^* = M^* / e, \quad \text{and} \quad (M / e)^* = M^* \setminus e$$

A matroid which may be obtained from another matroid M by a sequence of deletions and contractions is said to be a *minor* of M. There is a vast literature dealing with excluded-minor characterizations of various types of matroids; see, for example, Oxley (1992).

11

The Four-Colour Problem

Contents

11.1 Colourings of Planar Maps

In many areas of mathematics, attempts to find solutions to challenging unsolved problems have led to the advancement of ideas and techniques. In the case of graph theory, it was a seemingly innocuous map-colouring problem that provided the motivation for many of the developments during its first one hundred years.

In a letter written to William Rowan Hamilton in 1852, Augustus De Morgan communicated the following *Four-Colour Problem*, posed by Francis Guthrie.

> *A student of mine [Frederick Guthrie, brother of Francis] asked me today to give him a reason for a fact which I did not know was a fact – and do not yet. He says that if a figure be anyhow divided and the compartments differently coloured so that figures with any portion of common boundary line are differently coloured – four colours may be wanted, but not more – the following is the case in which four colours are wanted. Query cannot a necessity for five or more be invented ...*

This problem, because of its beguilingly simple statement, attracted the attention of many prominent mathematicians of the time. They came to believe that it was indeed possible to colour any map in four colours, and this surmise became known as the *Four-Colour Conjecture*. In the subsequent decades, several attempts were made to settle the conjecture, and some erroneous proofs were published. (For a history of the Four-Colour Problem, see Wilson (2002) or Biggs et al. (1986).)

FACE COLOURINGS

In order to translate the Four-Colour Problem into the language of graph theory, we need the notion of a face colouring of a plane graph. A *k-face colouring* of a plane graph is an assignment of k colours to its faces. The colouring is *proper* if no two adjacent faces are assigned the same colour. A plane graph is *k-face-colourable* if it has a proper k-face colouring. Figure 11.1a shows a proper 4-face-colouring of the triangular prism. Because every map may be regarded as a plane graph without cut edges, the Four-Colour Conjecture is equivalent to the statement:

Conjecture 11.1   THE FOUR-COLOUR CONJECTURE (FACE VERSION)
Every plane graph without cut edges is 4-face-colourable.

Over a century passed before the Four-Colour Conjecture was eventually verified, in 1977, by Appel and Haken (1977b).

Theorem 11.2   THE FOUR-COLOUR THEOREM
Every plane graph without cut edges is 4-face-colourable.

More recently, a somewhat simpler (but still complicated) proof of this theorem, based on the same general approach, was obtained by Robertson et al. (1997a).

One of the remarkable features of the Four-Colour Conjecture is that it has many equivalent formulations, some with no apparent connection to face colourings (see Section 11.3). We describe here two of these reformulations, one which is straightforward, in terms of vertex colourings, and one which is less so, in terms of edge colourings. They have motivated the study of several fundamental questions, to be discussed in subsequent chapters.

VERTEX COLOURINGS

A *k-vertex-colouring* of a graph, or simply a *k-colouring*, is an assignment of k colours to its vertices. The colouring is *proper* if no two adjacent vertices are assigned the same colour. A graph is *k-colourable* if it has a proper k-colouring. Because adjacent pairs of vertices of a plane graph correspond to adjacent pairs of faces of its dual, the Four-Colour Problem is equivalent to the statement that every loopless plane graph is 4-colourable. The 4-face-colouring of the triangular prism shown in Figure 11.1a gives rise in this way to the 4-vertex-colouring of its dual shown in Figure 11.1b.

The advantage of this reformulation in terms of vertex colourings is that one may now restate the problem without reference to any particular embedding:

Conjecture 11.3   THE FOUR-COLOUR CONJECTURE (VERTEX VERSION)
Every loopless planar graph is 4-colourable.

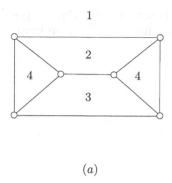

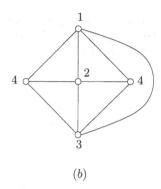

(a)   (b)

Fig. 11.1. (a) A 4-face-colouring of the triangular prism, (b) a 4-vertex-colouring of its dual

In order to show that all loopless planar graphs are 4-colourable, it clearly suffices to show that all simple connected planar graphs are 4-colourable. In fact, it is not hard to reduce the Four-Colour Conjecture to simple 3-connected maximal planar graphs (Exercise 11.1.1). By Corollary 10.21, a planar embedding of such a graph is a 3-connected triangulation. Therefore the Four-Colour Conjecture is equivalent to the assertion that every 3-connected triangulation is 4-colourable, and, by duality, to the assertion that every 3-connected cubic plane graph is 4-face-colourable.

EDGE COLOURINGS: TAIT'S THEOREM

We are now in a position to relate face colourings and edge colourings of plane graphs. A *k-edge-colouring* of a graph is an assignment of k colours to its edges. The colouring is *proper* if no two adjacent edges are assigned the same colour. A graph is *k-edge-colourable* if it has a proper k-edge-colouring. Tait (1880) found a surprising relationship between face colourings and edge colourings of 3-connected cubic plane graphs.

Theorem 11.4   TAIT'S THEOREM
A 3-connected cubic plane graph is 4-face-colourable if and only if it is 3-edge-colourable.

Proof Let G be a 3-connected cubic plane graph. First, suppose that that G has a proper 4-face-colouring. It is of course immaterial which symbols are used as the 'colours'. For mathematical convenience, we denote them by the vectors $\alpha_0 = (0,0)$, $\alpha_1 = (1,0)$, $\alpha_2 = (0,1)$, and $\alpha_3 = (1,1)$ in $\mathbb{Z}_2 \times \mathbb{Z}_2$. We now obtain a 3-edge-colouring of G by assigning to each edge the sum of the colours of the two faces it separates; note that, because G has no cut edges, each edge separates two distinct faces, so no edge is assigned colour α_0 under this scheme. If α_i, α_j, and α_k are the colours assigned to the three faces incident to a vertex v, then

$\alpha_i + \alpha_j$, $\alpha_i + \alpha_k$, and $\alpha_j + \alpha_k$ are the colours assigned to the three edges incident with v (see Figure 11.2). These colours are all different. Thus we have a proper 3-edge-colouring of G (in colours $\alpha_1, \alpha_2, \alpha_3$).

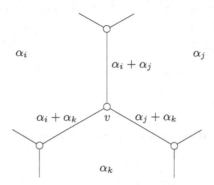

Fig. 11.2. The 3-edge-colouring of a cubic plane graph induced by a 4-face-colouring

Conversely, suppose that G has a proper 3-edge-colouring, in colours $1, 2, 3$. Denote by E_i the set of edges of G of colour i, $1 \le i \le 3$. The subgraph $G[E_i]$ induced by E_i is thus a spanning 1-regular subgraph of G. Set $G_{ij} := G[E_i \cup E_j]$, $1 \le i < j \le 3$. Then each G_{ij} is a spanning 2-regular subgraph of G, and is therefore 2-face-colourable (Exercise 11.1.2). Also, each face of G is the intersection of a face of G_{12} and a face of G_{23} (see Figure 11.3).

Consider 2-face-colourings of G_{12} and G_{23}, each using the colours 0 and 1 (unshaded and shaded, respectively, in Figure 11.3b). We can now obtain a 4-face-colouring of G (in the colours $\alpha_0, \alpha_1, \alpha_2, \alpha_3$ defined above) by assigning to each face f the ordered pair of colours assigned respectively to the faces of G_{12} and G_{23} whose intersection is f. Because $G = G_{12} \cup G_{23}$, this is a proper 4-face-colouring of G. $\square$

By virtue of Tait's Theorem (11.4), the Four-Colour Conjecture can be reformulated in terms of edge colourings as follows.

Conjecture 11.5   THE FOUR-COLOUR CONJECTURE (EDGE VERSION)
Every 3-connected cubic planar graph is 3-edge-colourable.

Recall that a spanning cycle in a graph is referred to as a *Hamilton cycle*. A graph which contains such a cycle is said to be *hamiltonian*. If a cubic graph contains a Hamilton cycle, the edges of that cycle may be coloured alternately in two colours, and the remaining edges, which are mutually nonadjacent, may all be assigned a third colour. Thus every hamiltonian cubic graph is 3-edge-colourable.

Now, if every 3-connected cubic planar graph could be shown to be hamiltonian, the edge-colouring version of the Four-Colour Conjecture would be established. Taking this 'fact' to be self-evident, Tait (1880) convinced himself (and others) that

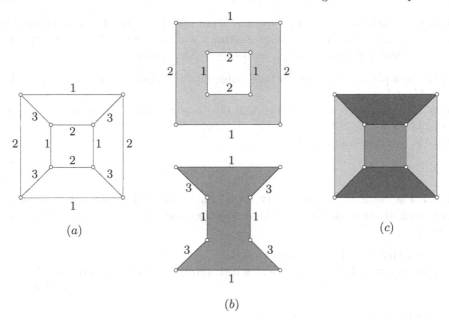

Fig. 11.3. (a) A 3-edge-colouring of the cube, (b) 2-face-colourings of the spanning subgraphs G_{12} and G_{13}, (c) the induced 4-face-colouring of the cube

he had proved the Four-Colour Conjecture. It took over half a century to invalidate Tait's 'proof', when Tutte (1946) constructed a nonhamiltonian 3-connected cubic planar graph. Tutte's construction is described in Chapter 18.

Exercises

⋆**11.1.1** Show that the Four-Colour Conjecture is true provided that it is true for all simple 3-connected maximal planar graphs.

⋆**11.1.2** Show that every even plane graph is 2-face-colourable.

11.1.3 Show that a plane graph is 4-face-colourable if and only if it is the union of two even subgraphs.

11.1.4 Show that a graph is 4-vertex-colourable if and only if it is the union of two bipartite subgraphs.

11.1.5

a) Let G be a graph with a Hamilton bond. Show that G is 4-vertex-colourable.
b) By applying Exercise 10.2.11, deduce that every plane hamiltonian graph is 4-face-colourable.

11.1.6 A *maximal outerplanar graph* is a simple graph which is outerplanar (defined in Exercise 10.2.12) and edge-maximal with respect to this property. Let G be a maximal outerplanar graph with $n \geq 3$. Show that:

  a) G has a planar embedding whose outer face is a Hamilton cycle, all other faces being triangles,
  b) G has a vertex v of degree two and $G - v$ is maximal outerplanar,
  c) $m = 2n - 3$,
  d) G is 3-vertex-colourable.

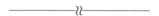

11.1.7 The walls of an art gallery form a polygon with n sides. It is desired to place staff at strategic points so that together they are able to survey the entire gallery.

  a) Show that $\lfloor n/3 \rfloor$ guards always suffice.
  b) For each $n \geq 3$, design a gallery which requires this number of guards.

(V. CHVÁTAL)

$\star$**11.1.8** HEAWOOD'S THEOREM
Show that:

  a) a plane triangulation is 3-vertex-colourable if and only if it is even,
(P.J. HEAWOOD)
  b) a plane graph is 3-vertex-colourable if and only if it it is a subgraph of an even plane triangulation. (M. KRÓL)

11.2 The Five-Colour Theorem

P. G. Tait was not the only mathematician of his time to come up with a false proof of the Four-Colour Conjecture. Kempe (1879), too, published a paper which was believed to contain a proof of the conjecture. This time it was Heawood (1890) who discovered a serious flaw in Kempe's proof. Fortunately, all was not lost, for Heawood showed that Kempe's approach could be used to prove that all planar graphs are 5-colourable. Here is Heawood's proof, in essence.

Theorem 11.6   THE FIVE COLOUR THEOREM
Every loopless planar graph is 5-colourable.

Proof By induction on the number of vertices. As observed in our earlier discussion, it suffices to prove the theorem for 3-connected triangulations. So let G be such a triangulation. By Corollary 10.22, G has a vertex v of degree at most five. Consider the plane graph $H := G - v$.

   By induction, H has a proper 5-colouring. If, in this colouring of H, one of the five colours is assigned to no neighbour of v, we may assign it to v, thereby extending the proper 5-colouring of H to a proper 5-colouring of G. We may assume, therefore, that the five neighbours of v together receive all five colours.

To fix notation, let $C := v_1v_2v_3v_4v_5v_1$ be the facial cycle of H whose vertices are the neighbours of v in G, where v_i receives colour i, $1 \le i \le 5$. We may suppose that the vertex v of G lies in int(C), so that all the bridges of C in H are outer bridges. If there is no bridge of C in H containing both v_1 and v_3, then by swapping the colours of the vertices coloured 1 and 3 in all the bridges of C containing v_1, we obtain a proper 5-colouring of H in which no vertex of C has colour 1. This colour may now be assigned to v, resulting in a proper 5-colouring of G. Thus we may assume that there is a bridge B_1 of C in H having v_1 and v_3 as vertices of attachment. Likewise, there is a bridge B_2 of C in H having v_2 and v_4 as vertices of attachment. But now the bridges B_1 and B_2 overlap, contradicting Theorem 10.26. □

There exist several proofs of the Five-Colour Theorem (11.6). One of these is outlined in Exercise 11.2.1, and another, based on the notion of list colouring, is given in Chapter 14. Colouring a planar graph with just four colours, rather than five, is quite a different story, even though Kempe's idea of interchanging colours turns out to be an important ingredient. The main ideas involved in the proof of the Four-Colour Theorem are presented in Chapter 15.

Exercises

11.2.1 Prove the Five-Colour Theorem (11.6) by induction on n, proceeding as follows.

i) Define G, v, H, and C as in the proof of Theorem 11.6. Show that the subgraph $H[\{v_1, v_2, v_3, v_4, v_5\}]$ is not complete.
ii) Let v_i and v_j be nonadjacent vertices of C. In H, identify v_i and v_j, so as to obtain a graph H'. Show that H' is planar.
iii) Consider (by induction) a proper 5-colouring of H'. Deduce that G is 5-colourable.

11.3 Related Reading

Equivalent Forms of the Four-Colour Problem

One of the reasons that the Four-Colour Problem has played and continues to play a central role in graph theory is its connection with a wide variety of interesting problems. Surprisingly, some of these questions seem to have nothing at all to do with colouring. For instance, as we show in Chapter 21, the Four-Colour Theorem is equivalent to the statement that every 2-edge-connected planar graph may be expressed as the union of two even subgraphs. Still more surprisingly, a number of other questions from other areas of mathematics have been shown to be equivalent to the Four-Colour Problem. For example, consider an expression of the form $\mathbf{v}_1 \times \mathbf{v}_2 \times \cdots \times \mathbf{v}_k$, where $\mathbf{v}_1, \mathbf{v}_2, \ldots, \mathbf{v}_k$ are arbitrary vectors in $\mathbb{R}^3$ and $\times$ denotes the cross product. Because this operation is not associative, the terms of the given expression need to be bracketed appropriately for the expression to be well-defined. Suppose, now, that two different bracketings of $\mathbf{v}_1 \times \mathbf{v}_2 \times \cdots \times \mathbf{v}_k$ are given. Kauffman (1990) considered the problem of deciding whether it is possible to assign unit vectors $\mathbf{e}_1, \mathbf{e}_2, \mathbf{e}_3$ to $\mathbf{v}_1, \mathbf{v}_2, \ldots, \mathbf{v}_k$ so that the two bracketings result in the same nonzero value for the expression. He showed that this problem can be reduced to one of 3-edge-colouring an associated cubic planar graph, and hence is equivalent to the Four-Colour Problem. An even more astonishing connection was discovered by Y. Matiyasevich, who established the existence of a diophantine equation involving thousands of variables, the solvability of which amounts to the validity of the Four-Colour Theorem (see Thomas (1998)). Further examples can be found in Thomas (1998) and Saaty (1972). Chapters 15 and 21 contain discussions of generalizations of the Four-Colour Problem.

12

Stable Sets and Cliques

Contents

12.1 Stable Sets

STABILITY AND CLIQUE NUMBERS

Throughout this chapter, we restrict our attention to simple graphs.

Recall that a *stable set* in a graph is a set of vertices no two of which are adjacent. (Stable sets are also commonly known as *independent* sets.) A stable set in a graph is *maximum* if the graph contains no larger stable set and *maximal* if the set cannot be extended to a larger stable set; a maximum stable set is necessarily maximal, but not conversely. The cardinality of a maximum stable set in a graph G is called the *stability number* of G and is denoted by $\alpha(G)$. Maximal and maximum stable sets in the Petersen graph are shown in Figure 12.1.

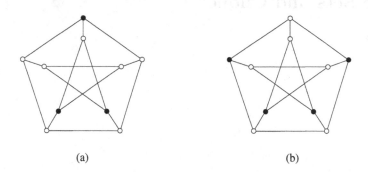

(a)   (b)

Fig. 12.1. (a) A maximal stable set, and (b) a maximum stable set

Recall, also, that an *edge covering* of a graph is a set of edges which together meet all vertices of the graph. Analogously, a *covering* of a graph is a set of vertices which together meet all edges of the graph. The minimum number of vertices in a covering of a graph G is called the *covering number* of G and is denoted by $\beta(G)$. The light vertices in the graphs of Figure 12.1 are examples of coverings. Indeed, stable sets and coverings are related in a very simple way: a set S is a stable set of a graph G if and only if $V \setminus S$ is a covering of G (Exercise 12.1.2). We therefore have the identity, first observed by Gallai (1959):

$$\alpha(G) + \beta(G) = v(G) \tag{12.1}$$

Stable sets and cliques are likewise related in a very simple manner. Recall that a *clique* of a graph is a set of mutually adjacent vertices, and that the maximum size of a clique of a graph G, the *clique number* of G, is denoted $\omega(G)$. Clearly, a set of vertices S is a clique of a simple graph G if and only if it is a stable set of the complement $\overline{G}$. In particular,

$$\omega(G) = \alpha(\overline{G})$$

Thus any assertion about stable sets can be restated in terms of cliques or coverings. We noted in Chapter 8 that the problem of finding a maximum clique in a graph is $\mathcal{NP}$-hard. It follows that the problem of finding a maximum stable set is also $\mathcal{NP}$-hard, as is that of finding a minimum covering. On the other hand, in the case of bipartite graphs, we saw in Chapter 8 that maximum stable sets can be found in polynomial time using linear programming techniques.

SHANNON CAPACITY

A number of real-world problems involve finding maximum stable sets of graphs. The following example, due to Shannon (1956), is one such.

Example 12.1   TRANSMITTING MESSAGES OVER A NOISY CHANNEL
A transmitter over a communication channel is capable of sending signals belonging to a certain finite set (or *alphabet*) A. Some pairs of these signals are so similar to each other that they might be confounded by the receiver because of possible distortion during transmission. Given a positive integer k, what is the greatest number of sequences of signals (or *words*) of length k that can be transmitted with no possibility of confusion at the receiving end?

To translate this problem into graph theory, we need the concept of the *strong product* of two graphs G and H. This is the graph $G \boxtimes H$ whose vertex set is $V(G) \times V(H)$, vertices (u, x) and (v, y) being adjacent if and only if $uv \in E(G)$ and $x = y$, or $u = v$ and $xy \in E(H)$, or $uv \in E(G)$ and $xy \in E(H)$.

Let G denote the graph with vertex set A in which two vertices u and v are adjacent if they represent signals that might be confused with each other, and let G^k be the strong product of k copies of G. Thus G^k is the graph whose vertices are the words of length k over A where two distinct words $(u_1, u_2, \ldots, u_k)$ and $(v_1, v_2, \ldots, v_k)$ are joined by an edge if either $u_i = v_i$ or $u_i v_i \in E(G)$, $1 \le i \le k$. Equivalently, two distinct words are adjacent in G^k if there is a possibility that one of them might be mistaken for the other by the receiver. It follows that the required maximum number of words of length k is simply the stability number of G^k. For example, if $A = \{0, 1, 2, 3, 4\}$ and each signal i may be confused with either $i - 1$ or $i + 1 \pmod 5$, then $G = C_5$. A drawing of $G^2 = C_5^2$ on the torus, together with a stable set of order five indicated by solid dots, is shown in Figure 12.2b; note that, because the graph is drawn on the torus, the four corner vertices represent one and the same vertex, $(0, 0)$. This shows that $\alpha(G^2) \ge 5$. One may in fact verify that $\alpha(G^2) = 5$ (Exercise 12.1.8). Thus, in this case, a maximum of five words of length two, for instance $00, 12, 24, 31, 43$, may be transmitted with no possibility of confusion by the receiver.

Motivated by his seminal work on information theory, Shannon (1956) proposed the parameter

$$\Theta(G) := \lim_{k \to \infty} \sqrt[k]{\alpha(G^k)}$$

now commonly known as the *Shannon capacity* of G, as a measure of the capacity for error-free transmission over a noisy channel whose associated graph is G. (It can be shown that this limit exists and is equal to $\sup_k \sqrt[k]{\alpha(G^k)}$, see Berge (1985).) Because $\alpha^k(G) \le \alpha(G^k)$ for any k (Exercise 12.1.7b), it follows that $\alpha(G) \le \Theta(G)$. Shannon (1956) (see, also, Rosenfeld (1967)) showed that the parameter $\alpha^{**}(G)$ defined in Exercise 8.6.5 is an upper bound for $\Theta(G)$. Therefore, for any graph G,

$$\alpha(G) \le \Theta(G) \le \alpha^{**}(G) \tag{12.2}$$

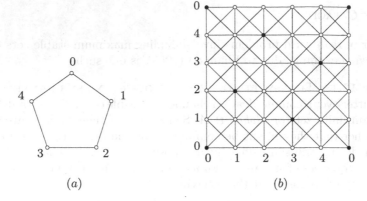

Fig. 12.2. (a) C_5, (b) a stable set of five vertices in C_5^2

If G is a graph in which α is equal to the minimum number of cliques covering all the vertices, then $\alpha = \alpha^{**}$ and hence $\Theta(G) = \alpha(G)$ by (12.2). This is the case when G is bipartite (by the König–Rado Theorem (8.30)), when G is chordal (by Exercise 9.7.5) and, more generally, for a class of graphs known as perfect graphs. A brief discussion of the properties of this important class of graphs is presented in Chapter 14. The smallest graph for which both the inequalities in (12.2) are strict is the 5-cycle. By employing ingenious algebraic techniques, Lovász (1979) showed that $\Theta(C_5) = \sqrt{5}$ (see Exercise 12.1.21).

STABLE SETS IN DIGRAPHS

A *stable set* in a digraph is a stable set in its underlying graph, that is, a set of pairwise nonadjacent vertices. The number of vertices in a largest stable set of a digraph D is denoted by $\alpha(D)$ and called the *stability number* of D.

Rédei's Theorem (2.3) tells us that every tournament has a directed Hamilton path. In other words, the vertex set of a tournament can be covered by a single directed path. In general, one may ask how few disjoint directed paths are needed to cover the vertex set of a digraph. Gallai and Milgram (1960) showed that this number is always bounded above by the stability number.

We refer to a covering of the vertex set of a graph or digraph by disjoint paths or directed paths as a *path partition*, and one with the fewest paths as an *optimal path partition*. The number of paths in an optimal path partition of a digraph D is denoted by $\pi(D)$.

Theorem 12.2   THE GALLAI–MILGRAM THEOREM
For any digraph D, $\pi \leq \alpha$.

Gallai and Milgram (1960) actually proved a somewhat stronger theorem. A directed path P and a stable set S are said to be *orthogonal* if they have exactly one common vertex. By extension, a path partition $\mathcal{P}$ and stable set S are *orthogonal* if each path in $\mathcal{P}$ is orthogonal to S.

Theorem 12.3 *Let $\mathcal{P}$ be an optimal path partition of a digraph D. Then there is a stable set S in D which is orthogonal to $\mathcal{P}$.*

Note that the Gallai–Milgram Theorem is an immediate consequence of Theorem 12.3 because $\pi = |\mathcal{P}| \le |S| \le \alpha$. Theorem 12.3 is established by means of an inductive argument involving the sets of initial and terminal vertices of the constituent paths of a path partition. For a path partition $\mathcal{P}$, we denote these sets by $i(\mathcal{P})$ and $t(\mathcal{P})$, respectively.

Lemma 12.4 *Let $\mathcal{P}$ be a path partition of a digraph D. Suppose that no stable set of D is orthogonal to $\mathcal{P}$. Then there is a path partition $\mathcal{Q}$ of D such that $|\mathcal{Q}| = |\mathcal{P}| - 1$, $i(\mathcal{Q}) \subset i(\mathcal{P})$, and $t(\mathcal{Q}) \subset t(\mathcal{P})$.*

Proof The case $n = 1$ holds vacuously, so we assume that $n \ge 2$ and proceed by induction on n. By hypothesis, $t(\mathcal{P})$ is not a stable set, so there exist vertices $y, z \in t(\mathcal{P})$ such that $(y, z) \in A$. If the vertex z constitutes, by itself, a (trivial) path of $\mathcal{P}$, then we define $\mathcal{Q}$ to be the path partition of D obtained from $\mathcal{P}$ by deleting this path and extending the path of $\mathcal{P}$ that terminates in y by the arc (y, z) (Figure 12.3).

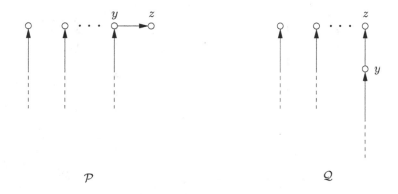

Fig. 12.3. Proof of Lemma 12.4: extending a path

Thus we may assume that z is the terminal vertex of a nontrivial path $P \in \mathcal{P}$. Let x be its predecessor on P, and set $D' := D - z$, $P' := P - z$, and $\mathcal{P}' := (\mathcal{P} \setminus \{P\}) \cup P'$, the restriction of $\mathcal{P}$ to D' (Figure 12.4). Then there is no stable set in D' orthogonal to $\mathcal{P}'$, because such a stable set would also be a stable set in D orthogonal to $\mathcal{P}$, contrary to the hypothesis. Note also that

$$t(\mathcal{P}') = (t(\mathcal{P}) \setminus \{z\}) \cup \{x\} \quad \text{and} \quad i(\mathcal{P}') = i(\mathcal{P})$$

By the induction hypothesis, there is a path partition $\mathcal{Q}'$ of D' such that $|\mathcal{Q}'| = |\mathcal{P}'| - 1$, $i(\mathcal{Q}') \subset i(\mathcal{P}')$ and $t(\mathcal{Q}') \subset t(\mathcal{P}')$. If $x \in t(\mathcal{Q}')$, we define $\mathcal{Q}$ to be the path partition of D obtained from $\mathcal{Q}'$ by extending the path of $\mathcal{Q}'$ that

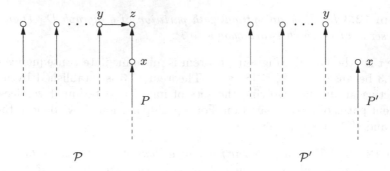

Fig. 12.4. Proof of Lemma 12.4: deleting a vertex

terminates in x by the arc (x, z) (Figure 12.5). If $x \notin t(\mathcal{Q}')$, then $y \in t(\mathcal{Q}')$, and we define $\mathcal{Q}$ to be the path partition of D obtained from $\mathcal{Q}'$ by extending the path of $\mathcal{Q}'$ that terminates in y by the arc (y, z). In both cases, $|\mathcal{Q}| = |\mathcal{P}| - 1$, $i(\mathcal{Q}) \subset i(\mathcal{P})$, and $t(\mathcal{Q}) \subset t(\mathcal{P})$. $\qquad\qquad\qquad\qquad\qquad\qquad\qquad\qquad\qquad\qquad\qquad\qquad\quad\square$

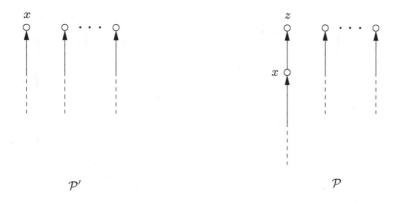

Fig. 12.5. Proof of Lemma 12.4: reinserting the deleted vertex

The inductive proof of Lemma 12.4 gives rise to a polynomial-time recursive algorithm for finding a path partition $\mathcal{P}$ of a digraph D, and a stable set S in D orthogonal to $\mathcal{P}$, of the same cardinality (Exercise 12.1.9).

The Gallai–Milgram Theorem (12.2) can be viewed as a formula for the stability number of an undirected graph in terms of path partitions of its orientations (Exercise 12.1.10). Moreover, it implies the following celebrated theorem of Dilworth (1950) on partially ordered sets.

Theorem 12.5   DILWORTH'S THEOREM
The minimum number of chains into which the elements of a partially ordered set

P can be partitioned is equal to the maximum number of elements in an antichain of P.

Proof Let $P := (X, \prec)$, and denote by $D := D(P)$ the digraph whose vertex set is X and whose arcs are the ordered pairs (u, v) such that $u \prec v$ in P. Chains and antichains in P correspond in D to directed paths and stable sets, respectively. Because no two elements in an antichain of P can belong to a common chain, the minimum number of chains in a chain partition is at least as large as the maximum number of elements in an antichain; that is, $\pi \geq \alpha$. On the other hand, by the Gallai–Milgram Theorem, $\pi \leq \alpha$. Therefore $\pi = \alpha$. $\qquad\qquad$ □

KERNELS

If S is a maximal stable set in a graph G, then every vertex of $G - S$ is adjacent to some vertex of S. In the case of digraphs, it is natural to replace the notion of adjacency by the directed notion of dominance. This results in the concept of a kernel.

A *kernel* in a digraph D is a stable set S of D such that each vertex of $D - S$ dominates some vertex of S. The solid vertices in the digraph of Figure 12.6 constitute a kernel of the digraph.

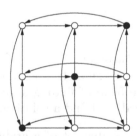

Fig. 12.6. A kernel in a digraph

Kernels arise naturally in the analysis of certain two-person positional games such as Hex (see Berge (1977) or Browne (2000)). Consider the digraph D whose vertices are the positions in the game, one position dominating another if the latter can be reached from the former in one move. Suppose that D has a kernel S and that, according to the rules of play, the last player able to move wins the game. Then a player who starts at a position in $V \setminus S$ can guarantee himself a win or a draw by always moving to a position in S, from which his opponent is obliged to move to a position in $V \setminus S$ (if, indeed, he is able to move). Other applications of kernels can be found in the book by Berge (1985).

Many digraphs fail to have kernels. Directed odd cycles are the simplest examples. Indeed, Richardson (1953) proved that a digraph which has no kernel necessarily contains a directed odd cycle.

Theorem 12.6   RICHARDSON'S THEOREM
Let D be a digraph which contains no directed odd cycle. Then D has a kernel.

Proof   By induction on n. If D is strong, then D is bipartite (Exercise 3.4.11b) and each class of the bipartition is a kernel of D. If D is not strong, let D_1 be a minimal strong component of D (one that dominates no other strong component; see Exercise 3.4.6), and set $V_1 := V(D_1)$. By the induction hypothesis, D_1 has a kernel, S_1. Let V_2 be the set of vertices of D that dominate vertices of S_1, and set $D_2 := D - (V_1 \cup V_2)$. Again by induction, D_2 has a kernel S_2. The set $S_1 \cup S_2$ is then a kernel of D. $\square$

Richardson's Theorem implies that every acyclic digraph has a kernel; in fact, acyclic digraphs have unique kernels (Exercise 12.1.15b). However, as we have seen, not every digraph has a kernel. Moreover, it is an NP-complete problem to decide whether a digraph has a kernel (Exercise 12.1.16). For this reason, the less stringent notion of a semi-kernel was proposed by Chvátal and Lovász (1974). A *semi-kernel* in a digraph D is a stable set S which is reachable from every vertex of $D - S$ by a directed path of length one or two. Chvátal and Lovász (1974) showed that every digraph has a semi-kernel (see Exercise 12.1.17).

We encounter kernels again in Chapter 14, where they play a key role in the solution of certain colouring problems.

Exercises

12.1.1 THE EIGHT QUEENS PROBLEM
Is it possible to place eight queens on a chessboard so that no one queen can take another? Express this as a problem of finding a maximum stable set in a graph associated with the chess board, and find a solution.

$\star$**12.1.2** Show that a set S is a stable set of a graph G if and only if $V \setminus S$ is a covering of G.

12.1.3 Show that a graph G is bipartite if and only if $\alpha(H) \geq \frac{1}{2}v(H)$ for every induced subgraph H of G.

12.1.4 Show that a graph G is bipartite if and only if $\alpha(H) = \beta'(H)$ for every subgraph H of G without isolated vertices, where $\beta'(H)$ is the minimum number of edges in an edge covering of H (see Section 8.6).

12.1.5 Show that $\alpha(KG_{m,n}) \geq \binom{n-1}{m-1}$.
(Erdős et al. (1961) proved that this bound is sharp; see Exercise 13.2.17.)

12.1.6 Show that the strong product is associative.

12.1.7

a) For any two graphs G and H, show that $\alpha(G \boxtimes H) \geq \alpha(G)\alpha(H)$.

b) Deduce that $\alpha(G^k) \geq \alpha^k(G)$.

12.1.8 Show that $\alpha(C_5 \boxtimes C_5) = 5$.

12.1.9 Describe a polynomial-time recursive algorithm for finding in a digraph D a path partition $\mathcal{P}$ and a stable set S, orthogonal to $\mathcal{P}$, such that $|\mathcal{P}| = |S|$.

12.1.10 Let G be an undirected graph. Show that

$$\alpha(G) = \max \{\pi(D) : D \text{ an orientation of } G\}$$

12.1.11 Let P be a longest directed path in a digraph D, and let C be a strong component of $D - P$. Show that $\alpha(C) < \alpha(D)$.

12.1.12 COMPOSITION
The *composition*, or *lexicographic product*, of simple graphs G and H is the simple graph $G[H]$ with vertex set $V(G) \times V(H)$ in which (u, v) is adjacent to (u', v') if and only if either $uu' \in E(G)$ or $u = u'$ and $vv' \in E(H)$. (This amounts to replacing each vertex of G by a copy of H, and linking these copies by complete bipartite graphs according to the edges of G.) The composition of strict digraphs is defined analogously. Show that $\alpha(G[H]) = \alpha(G)\alpha(H)$.

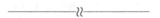

12.1.13 A digraph with stability number α has the *Path-Deletion Property* if the deletion of any set of $\alpha - 1$ directed paths leaves a digraph with stability number α.

a) Find an acyclic digraph of stability number two on six vertices with the Path-Deletion Property.   (P. CHARBIT)
b) An *antichain* of a digraph is a set of vertices no two of which are connected by a directed path. Using Dilworth's Theorem, show that an acyclic digraph D has the Path-Deletion Property if and only if every transversal of its family of maximum stable sets contains an antichain of cardinality α.
c) Let D and H be two digraphs with the Path-Deletion Property, D being acyclic. Deduce from (b) that their composition $D[H]$ has the Path-Deletion Property.
d) Conclude that for all $k \geq 1$ there exists a digraph of stability number 2^k with the Path-Deletion Property.   (J.A. BONDY)

(Hahn and Jackson (1990) conjecture that for all $\alpha \geq 2$ there exists a digraph of stability number α with the Path-Deletion Property.)

12.1.14 Let G be a connected cubic graph on $4k$ vertices. By applying Exercise 2.4.8a, show that G has a stable set S of k vertices such that each component of $G - S$ is unicyclic.   (N. ALON)

⋆12.1.15

a) Show that any digraph which has at least two kernels contains a directed even cycle.

b) Deduce that every acyclic digraph has a unique kernel.

12.1.16 Given a boolean formula f in conjunctive normal form, construct a digraph D from f as follows.

▷ For each clause f_i of f, there is a directed triangle (u_i, v_i, w_i) in D.
▷ For each variable x of f, there is a directed 2-cycle $(x, \bar{x}, x)$ in D.
▷ For each clause f_i of f, there is an arc in D from each of u_i, v_i, w_i to each of the literals appearing in f_i.

a) Show that a set S of vertices of D is a kernel if and only if S consists of literals, includes precisely one of x and $\bar{x}$ for each variable x, and includes at least one literal from each clause of f.

b) Deduce that the problem of deciding whether a digraph has a kernel is NP-complete. (V. CHVÁTAL)

12.1.17 Let D be a digraph. Consider an arbitrary total order $\prec$ of V. Let D' and D'' be the spanning acyclic subgraphs of D induced by $\{(x, y) : x \prec y\}$ and $\{(x, y) : y \prec x\}$, respectively. Let S' be the kernel of D'. Show that the kernel of $D''[S']$ is a semi-kernel of D. (S. THOMASSÉ)

12.1.18 Let D be a digraph, and let $v_1, v_2, \ldots, v_n$ be a fixed linear ordering of its vertex set. An *inductive kernel* of D with respect to this ordering is a stable set S of D such that each vertex of D is reachable from a vertex of S by means of a directed path whose vertices, except possibly the last one, occur in increasing order of index. Show that:

a) every ordered digraph has an inductive kernel, (S. BURCKEL)

b) a set of vertices of a digraph is a kernel if and only if it is an inductive kernel with respect to every linear ordering of the vertex set of the digraph.

12.1.19

a) The *tensor product* of vectors $\mathbf{x} = (x_1, x_2, \ldots, x_m)$ and $\mathbf{y} = (y_1, y_2, \ldots, y_n)$ is the vector

$$\mathbf{x} \circ \mathbf{y} := (x_1 y_1, x_1 y_2, \ldots, x_1 y_n, x_2 y_1, x_2 y_2, \ldots, x_2 y_n, \ldots, x_m y_1, x_m y_2, \ldots, x_m y_n)$$

For vectors $\mathbf{x}, \mathbf{a} \in \mathbb{R}^m$ and $\mathbf{y}, \mathbf{b} \in \mathbb{R}^n$, show that the scalar and tensor products are linked by the following rule:

$$(\mathbf{x} \circ \mathbf{y})(\mathbf{a} \circ \mathbf{b})^t = (\mathbf{x}\mathbf{a}^t)(\mathbf{y}\mathbf{b}^t)$$

b) An *orthonormal representation* of a graph $G = (V, E)$ in the euclidean space $\mathbb{R}^d$ is a mapping $v \mapsto \mathbf{x}_v$ from V into $\mathbb{R}^d$ such that:

▷ $\mathbf{x}_v$ is a unit vector, for all $v \in V$,

▷ $\mathbf{x}_u$ and $\mathbf{x}_v$ are orthogonal whenever $uv \notin E$.

i) Show that every graph has an orthonormal representation in some space $\mathbb{R}^d$.

ii) Let G and H be two graphs, and let $u \mapsto \mathbf{x}_u$, $u \in V(G)$, and $v \mapsto \mathbf{y}_v$, $v \in V(H)$, be orthonormal representations of G and H, respectively, in $\mathbb{R}^d$. Show that $(u, v) \mapsto \mathbf{x}_u \circ \mathbf{y}_v$, $(u, v) \in V(G) \times V(H)$, is an orthonormal representation of $G \boxtimes H$ in $\mathbb{R}^{d^2}$.

12.1.20 Let $G = (V, E)$ be a graph, and let $v \mapsto \mathbf{x}_v$, $v \in V$, be an orthonormal representation of G in $\mathbb{R}^d$. For any stable set S of G and any unit vector $\mathbf{y}$ in $\mathbb{R}^d$, show that

$$\sum_{v \in S} (\mathbf{y}\mathbf{x}_v^t)^2 \leq 1$$

12.1.21

a) Find an orthonormal representation of the 5-cycle $(1, 2, 3, 4, 5, 1)$ in $\mathbb{R}^3$ by five unit vectors $\mathbf{x}_1, \mathbf{x}_2, \mathbf{x}_3, \mathbf{x}_4, \mathbf{x}_5$ such that, for $1 \leq i \leq 5$, the first coordinate of $\mathbf{x}_i$ is $5^{-1/4}$.

b) Using Exercise 12.1.19, deduce that C_5^k has an orthonormal representation in $\mathbb{R}^{3^k}$ in which the first coordinate of each unit vector is $5^{-k/4}$.

c) Taking $\mathbf{y}$ as the unit vector $\mathbf{e}_1$ in $\mathbb{R}^{3^k}$ and applying Exercise 12.1.20, show that $\alpha(C_5^k) \leq 5^{k/2}$.

d) Deduce that the Shannon capacity of C_5 is equal to $\sqrt{5}$.   (L. Lovász)

12.1.22 Show that every oriented graph D contains a family of disjoint directed paths, each of length at least one, which together include all vertices of D of maximum degree Δ.   (V.G. Vizing)

12.2 Turán's Theorem

We have already encountered a number of statements asserting that a simple graph with 'many' edges (in terms of its number of vertices) must contain a subgraph of a certain type. For instance, a simple graph on n vertices contains a cycle if it has at least n edges (Exercise 2.1.3) and contains a triangle if it has more than $n^2/4$ edges (Mantel's Theorem, Exercise 2.1.16). In this section, we generalize Mantel's Theorem by determining the maximum number of edges that a simple graph on n vertices can have without containing a clique of a given size. This theorem, due to Turán (1941), triggered the development of a major branch of graph theory, known as extremal graph theory (see, for example, the monograph with this title by Bollobás (1978)).

If F is a simple graph, we denote by $\mathrm{ex}(n, F)$ the maximum number of edges in a graph G on n vertices which does not contain a copy of F. Such a graph G is called an *extremal graph* (for this particular property), and the set of extremal graphs is denoted by $\mathrm{Ex}(n, F)$. For instance, $\mathrm{ex}(n, K_3) = \lfloor n^2/4 \rfloor$ and $\mathrm{Ex}(n, K_3) = \{K_{\lfloor n/2 \rfloor, \lceil n/2 \rceil}\}$.

The proof of Turán's Theorem given here is due to Zykov (1949). Recall (see Exercise 1.1.11) that the simple complete k-partite graph on n vertices in which all parts are as equal in size as possible is called a *Turán graph* and denoted $T_{k,n}$.

Theorem 12.7   TURÁN'S THEOREM
Let G be a simple graph which contains no K_k, where $k \geq 2$. Then $e(G) \leq e(T_{k-1,n})$, with equality if and only if $G \cong T_{k-1,n}$.

Proof   By induction on k. The theorem holds trivially for $k = 2$. Assume that it holds for all integers less than k, and let G be a simple graph which contains no K_k. Choose a vertex x of degree Δ in G, and set $X := N(x)$ and $Y := V \setminus X$ (see Figure 12.7 for an illustration.) Then

$$e(G) = e(X) + e(X, Y) + e(Y)$$

Because G contains no K_k, $G[X]$ contains no K_{k-1}. Therefore, by the induction hypothesis,
$$e(X) \leq e(T_{k-2,\Delta})$$
with equality if and only if $G[X] \cong T_{k-2,\Delta}$. Also, because each edge of G incident with a vertex of Y belongs to either $E[X, Y]$ or $E(Y)$,

$$e(X, Y) + e(Y) \leq \Delta(n - \Delta)$$

with equality if and only if Y is a stable set all members of which have degree Δ. Therefore $e(G) \leq e(H)$, where H is the graph obtained from a copy of $T_{k-2,\Delta}$ by adding a stable set of $n - \Delta$ vertices and joining each vertex of this set to each vertex of $T_{k-2,\Delta}$. Observe that H is a complete $(k-1)$-partite graph on n vertices. By Exercise 1.1.11, $e(H) \leq e(T_{k-1,n})$, with equality if and only if $H \cong T_{k-1,n}$. It follows that $e(G) \leq e(T_{k-1,n})$, with equality if and only if $G \cong T_{k-1,n}$. $\square$

Many different proofs of Turán's Theorem have been found (see Aigner (1995)). The one given here implies that if G is a graph on n vertices and more than $t_{k-1}(n)$

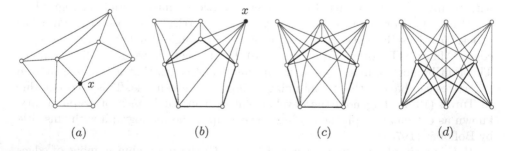

(a)   (b)   (c)   (d)

Fig. 12.7. (a) A graph G with $d(x) = \Delta = 5$, (b) another drawing of G with the subgraph $G[X] \cong C_5$ highlighted, (c) the graph $C_5 \vee \overline{K_3}$, (d) the graph $H \cong T_{2,5} \vee \overline{K_3} \cong T_{3,8}$

edges, where $t_k(n) := e(T_{k,n})$, and if v is a vertex of maximum degree Δ in G, then the subgraph $G[N(v)]$ induced by the neighbours of G has more than $t_{k-2}(\Delta)$ edges. Iterating this procedure, one sees that a clique of size k in G can be found by applying a simple greedy algorithm: select a vertex v_1 of maximum degree in $G_1 := G$, then a vertex v_2 of maximum degree in $G_2 := G_1[N(v_1)]$, then a vertex v_3 of maximum degree in $G_3 := G_2[N(v_2)]$, and so on. The resulting set of vertices $\{v_1, v_2, v_3, \ldots, v_k\}$ is a clique of G (Exercise 12.2.4).

AN APPLICATION TO COMBINATORIAL GEOMETRY

Extremal graph theory has applications to diverse areas of mathematics, including combinatorial number theory and combinatorial geometry. We describe here an application of Turán's Theorem to combinatorial geometry.

The *diameter* of a set of points in the plane is the maximum distance between two points of the set. It should be noted that this is a purely geometric notion and is quite unrelated to the graph-theoretic concepts of diameter and distance.

We discuss sets of diameter one. A set of n points determines $\binom{n}{2}$ distances between pairs of these points. It is intuitively clear that if n is 'large', some of these distances must be 'small'. Therefore, for any d between 0 and 1, it makes sense to ask how many pairs of points in a set $\{x_1, x_2, \ldots, x_n\}$ of diameter one can be at distance greater than d. Here, we present a solution, by Erdős (1955, 1956), of one special case of this problem, namely when $d = 1/\sqrt{2}$.

As an illustration, consider the case $n = 6$. We then have six points x_i, $1 \le i \le 6$. If we place them at the vertices of a regular hexagon with the pairs (x_1, x_4), (x_2, x_5), and (x_3, x_6) at distance one, as shown in Figure 12.8a, these six points clearly constitute a set of diameter one.

It is easily calculated that the pairs

$$(x_1, x_2), \ (x_2, x_3), \ (x_3, x_4), \ (x_4, x_5), \ (x_5, x_6), \text{ and } (x_6, x_1)$$

are at distance $1/2$, and the pairs

$$(x_1, x_3), \ (x_2, x_4), \ (x_3, x_5), \ (x_4, x_6), \ (x_5, x_1), \text{ and } (x_6, x_2)$$

are at distance $\sqrt{3}/2$. Because $\sqrt{3}/2 > \sqrt{2}/2 = 1/\sqrt{2}$, there are nine pairs of points at distance greater than $1/\sqrt{2}$ in this set of diameter one.

However, nine is not the best we can do with six points. By placing the points in the configuration shown in Figure 12.8b, all pairs of points except (x_1, x_2), (x_3, x_4), and (x_5, x_6) are at distance greater than $1/\sqrt{2}$. Thus, we have twelve pairs at distance greater than $1/\sqrt{2}$; in fact, this is the best we can do. The solution to the problem in general is given by the following theorem.

Theorem 12.8 *Let S be a set of diameter one in the plane. Then the number of pairs of points of S whose distance is greater than $1/\sqrt{2}$ is at most $\lfloor n^2/3 \rfloor$, where $n = |S|$. Moreover, for each $n \ge 2$, there is a set of n points of diameter one in which exactly $\lfloor n^2/3 \rfloor$ pairs of points are at distance greater than $1/\sqrt{2}$.*

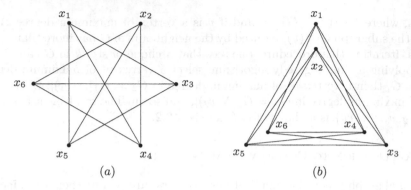

Fig. 12.8. Two sets of diameter one in the plane

Proof Let $S := \{x_1, x_2, \ldots, x_n\}$. Consider the graph G with vertex set S and edge set $\{x_i x_j \mid d(x_i, x_j) > 1/\sqrt{2}\}$, where $d(x_i, x_j)$ denotes the euclidean distance between x_i and x_j. We show that G cannot contain a copy of K_4.

First, note that any four points in the plane determine an angle of at least 90 degrees between three of them: the convex hull of the points is a line, a triangle, or a quadrilateral (see Figure 12.9), and in each case there is an angle $\widehat{x_i x_j x_k}$ of at least 90 degrees.

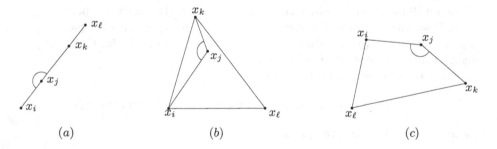

Fig. 12.9. The possible convex hulls of four points in the plane: (a) a line, (b) a triangle, (c) a quadrilateral

Now look at the three points x_i, x_j, x_k which determine this angle. Not all the distances $d(x_i, x_j)$, $d(x_j, x_k)$, and $d(x_i, x_k)$ can be greater than $1/\sqrt{2}$ and less than or equal to 1. For, if $d(x_i, x_j) > 1/\sqrt{2}$ and $d(x_j, x_k) > 1/\sqrt{2}$, then $d(x_i, x_k) > 1$. The set $\{x_1, x_2, \ldots, x_n\}$ is assumed to have diameter one. It follows that, of any four points in G, at least one pair cannot be joined by an edge, and hence that G cannot contain a copy of K_4. By Turán's Theorem (12.7),

$$e(G) \leq e(T_{3,n}) = \lfloor n^2/3 \rfloor$$

This proves the first statement.

One can construct a set $\{x_1, x_2, \ldots, x_n\}$ of diameter one in which exactly $\lfloor n^2/3 \rfloor$ pairs of points are at distance greater than $1/\sqrt{2}$ as follows. Choose r such that $0 < r < (1 - \frac{1}{\sqrt{2}})/4$, and draw three circles of radius r whose centres are at distance $1 - 2r$ from one another (see Figure 12.10). Set $p := \lfloor n/3 \rfloor$. Place points $x_1, x_2, \ldots, x_p$ in one circle, points $x_{p+1}, x_{p+2}, \ldots, x_{2p}$ in another, and points $x_{2p+1}, x_{2p+2}, \ldots, x_n$ in the third, in such a way that $d(x_1, x_n) = 1$. This set clearly has diameter one. Also, $d(x_i, x_j) > 1/\sqrt{2}$ if and only if x_i and x_j are in different circles, and so there are exactly $\lfloor n^2/3 \rfloor$ pairs (x_i, x_j) for which $d(x_i, x_j) > 1/\sqrt{2}$. $\qquad \square$

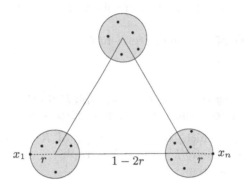

Fig. 12.10. An extremal configuration of diameter one

Many other applications of extremal graph theory to combinatorial geometry and analysis can be found in the articles by Erdős et al. (1971, 1972a,b) and in the survey article by Erdős and Purdy (1995).

Exercises

12.2.1 A certain bridge club has a special rule to the effect that four members may play together only if no two of them have previously partnered each other. At one meeting fourteen members, each of whom has previously partnered five others, turn up. Three games are played and then proceedings come to a halt because of the club rule. Just as the members are preparing to leave, a new member, unknown to any one of them, arrives. Show that at least one more game can now be played.

12.2.2 A flat circular city of radius six miles is patrolled by eighteen police cars, which communicate with one another by radio. If the range of a radio is nine miles, show that, at any time, there are always at least two cars each of which can communicate with at least five others.

12.2.3

a) Show that $\left(\frac{k-1}{2k}\right) n^2 - \frac{1}{8}k \le t_k(n) \le \left(\frac{k-1}{2k}\right) n^2$.

b) Deduce that $t_k(n) = \left\lfloor \left(\frac{k-1}{2k}\right) n^2 \right\rfloor$ for all $k < 8$.

12.2.4 Let G be a graph on n vertices and more than $t_{k-1}(n)$ edges, where $t_{k-1}(n) := e(T_{k-1,n})$. Show that a clique S of k vertices can be found by the following greedy algorithm.

1: set $S := \emptyset$ and $i := 1$
2: **while** $i < k$ **do**
3:     select a vertex v_i of maximum degree in G
4:     replace S by $S \cup \{v_i\}$, G by $G[N(v_i)]$ and i by $i + 1$
5: **end while**
6: select a vertex v_k of maximum degree in G
7: replace S by $S \cup \{v_k\}$
8: return S (J.A. BONDY)

12.2.5 A graph G is *degree-majorized* by a graph H if $v(G) = v(H)$ and the degree sequence of G (in nondecreasing order) is majorized termwise by that of H.

a) Let G be a graph which contains no copy of K_k. Show that G is degree-majorized by a complete $(k-1)$-partite graph.

b) Deduce Turán's Theorem.   (P. ERDŐS)

12.2.6 TURÁN HYPERGRAPH
Let V be an n-set. A k-uniform hypergraph $(V, \mathcal{F})$ is *complete* if $\mathcal{F} = \binom{V}{k}$, the set of all $\binom{n}{k}$ k-subsets of V. This hypergraph is denoted $K_n^{(k)}$.

a) Let $\{X, Y, Z\}$ be a partition of V into three sets which are as nearly equal in size as possible, and let $\mathcal{F}$ be the union of $\{\{x, y, z\} : x \in X, y \in Y, z \in Z\}$, $\{\{x_1, x_2, y\} : x_1 \in X, x_2 \in X, y \in Y\}$, $\{\{y_1, y_2, z\} : y_1 \in Y, y_2 \in Y, z \in Z\}$, and $\{\{z_1, z_2, x\} : z_1 \in Z, z_2 \in Z, x \in X\}$. The 3-uniform hypergraph $(V, \mathcal{F})$ is called the *Turán hypergraph* on n vertices. Verify that this hypergraph does not contain $K_4^{(3)}$.

b) Let $\{X, Y\}$ be a partition of V into two sets which are as nearly equal in size as possible, and let $\mathcal{F}$ be the set of all 3-subsets of V which intersect both X and Y. Verify that the hypergraph $(V, \mathcal{F})$ does not contain $K_5^{(3)}$.

(Turán (1941) conjectured that these hypergraphs are extremal configurations for the two respective extremal problems.)

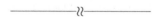

12.2.7

a) Let G be a simple nonbipartite graph with $m > \frac{1}{4}(n-1)^2 + 1$. Show that G contains a triangle.

b) For all odd $n \ge 5$, find a simple triangle-free nonbipartite graph G with $m = \frac{1}{4}(n-1)^2 + 1$.   (P. ERDŐS)

12.2.8 Denote by $t(G)$ the total number of triangles of G, and by $t(e)$ the number of triangles of G containing a given edge e.

a) Let G be a simple graph on n vertices, and let $e = xy \in E$. Show that
$d(x) + d(y) \le n + t(e)$.
b) By summing this inequality over all $e \in E$, deduce that $t(G) \ge m(4m - n^2)/3n$.
(J.W. Moon and L. Moser)
c) Deduce that if $k \ge 3$ and $m \ge \frac{1}{2}(1 - \frac{1}{k})n^2$, then $t(G) \ge \binom{k}{3}(n/k)^3$.
(A.W. Goodman)
d) For $k \ge 3$ and $n \equiv 0 \pmod{k}$, construct a graph G with $m = \frac{1}{2}(1 - \frac{1}{k})n^2$ and
$t(G) = \binom{k}{3}(n/k)^3$.

(Compare with Exercise 2.1.16.)

12.2.9

a) Show that if $m \ge \lfloor n^2/4 \rfloor + 1$, then $t(G) \ge \lfloor n/2 \rfloor$, where $t(G)$ is the number of triangles of G. (H. Rademacher)
b) For each $n \ge 3$, construct a graph G with $m = \lfloor n^2/4 \rfloor + 1$ and $t(G) = \lfloor n/2 \rfloor$.

12.2.10 Let G be a simple graph with average degree d, and let k be a positive integer.

a) Show that if $\sum_{v \in V} \binom{d(v)}{2} > (k - 1)\binom{n}{2}$, then G contains a copy of $K_{2,k}$.
b) Deduce that if $d > (k-1)^{1/2}n^{1/2} + \frac{1}{2}$, then G contains a copy of $K_{2,k}$. (Compare with Exercise 2.1.15, which treats the case $k = 2$.)
c) Let S be a set of n points in the plane. By applying (b), show that the number of pairs of points of S at distance exactly one is at most $\frac{1}{\sqrt{2}}n^{3/2} + \frac{1}{4}n$.

12.2.11 THE KŐVÁRI–SÓS–TURÁN THEOREM
Let G be a simple graph with average degree d, and let k and ℓ be positive integers.

a) Show that if $\sum_{v \in V} \binom{d(v)}{k} > (\ell - 1)\binom{n}{k}$, then G contains a copy of $K_{k,\ell}$
b) Using the identity $\binom{n}{k-1} + \binom{n}{k} = \binom{n+1}{k}$, show that if p and q are integers such that $p > q$, then $\binom{p}{k} + \binom{q}{k} \ge \binom{p-1}{k} + \binom{q+1}{k}$.
c) Deduce from (b) that $\sum_{v \in V} \binom{d(v)}{k} \ge n\binom{r}{k}$, where $r := \lfloor d \rfloor$.
d) Using the bounds $\binom{n}{k} \le n^k/k!$ and $\binom{r}{k} \ge (r - k + 1)^k/k!$, deduce that if $d > (\ell - 1)^{1/k}n^{1-1/k} + k$, then G contains a copy of $K_{k,\ell}$.
(T. Kővári, V.T. Sós, and P. Turán)

12.2.12 POLARITY GRAPH
A *polarity* of a geometric configuration $(P, \mathcal{L})$ is an involution π of $P \cup \mathcal{L}$, mapping points to lines and lines to points, which preserves incidence. Thus a point p and line L are incident if and only if the line $\pi(p)$ and the point $\pi(L)$ are incident. The line $\pi(p)$ is called the *polar* of the point p, and the point $\pi(L)$ the *pole* of the line L. The *polarity graph* of the configuration with respect to a polarity π is the graph G_π whose vertex set is P, two vertices being joined if and only if one lies on the polar of the other.

a) Find a polarity π of the Fano plane and draw its polarity graph G_π.
b) Find a polarity π of the Desargues configuration whose polarity graph G_π is isomorphic to the Petersen graph. (A.B. KEMPE)
c) Show that a polarity graph of a finite projective plane has diameter two.
d) How many vertices does a polarity graph of a finite projective plane of order n have, and what are their degrees?

(Polarity graphs were first described by Artzy (1956), who called them *reduced Levi graphs*. They were rediscovered by Erdős and Rényi (1962).)

12.2.13 Let π be the bijection of points and lines of the projective plane $PG_{2,q}$ (defined in Exercise 1.3.14) which maps each point (a, b, c) to the line $ax + by + cz = 0$. Prove that π is a polarity of $PG_{2,q}$.

12.2.14 An *absolute point* of a polarity is a point which lies on its polar.

a) Show that the polarity π defined in Exercise 12.2.13 has $q + 1$ absolute points. (Baer (1946) proved that every polarity of a finite projective plane of order n has at least $n + 1$ absolute points.)
b) Deduce that the polarity graph of a finite projective plane $PG_{2,q}$ has $q^2 + q + 1$ vertices, $\frac{1}{2}q(q + 1)^2$ edges and no 4-cycles.
(W.G. BROWN; P. ERDŐS, A. RÉNYI, AND V.T. SÓS)
(Füredi (1996) proved that, when q is a prime power, $q > 13$, polarity graphs with $q^2 + q + 1$ vertices are extremal graphs without 4-cycles.)

12.2.15 Let H be a simple graph which contains no 3-cube. By Exercise 2.2.2a, H has a spanning bipartite subgraph G with $e(G) \geq \frac{1}{2}e(H)$. For distinct vertices x and y of G, denote by $p(x, y)$ the number of xy-paths of length three in G, and set $X := N(x)$ and $Y := N(y)$.

a) Show that:
   i) if $X \cap Y \neq \emptyset$, then $p(x, y) = 0$,
   ii) if $X \cap Y = \emptyset$, then the subgraph $B[X, Y]$ whose vertex set is $X \cup Y$ and whose edge set is $\{uv \in E(G) : u \in X, v \in Y\}$ contains no 6-cycle.
b) Deduce that $p(x, y) \leq c_1(d(x) + d(y))^{4/3}$ for all $x, y \in V$, where c_1 is a suitable positive constant.
c) Using the inequality

$$\sum_{i=1}^{n} a_i^\gamma \geq n^{1-\gamma} \left(\sum_{i=1}^{n} a_i \right)^\gamma$$

valid for positive real numbers a_i, $1 \leq i \leq n$, and $\gamma \geq 1$, deduce that G has at most $c_2 n^2 d^{4/3}$ paths of length three, where c_2 is a suitable positive constant and d is the average degree of G.
d) By appealing to Exercise 13.2.6, conclude that $m < cn^{8/5}$ for some positive constant c. (R. PINCHASI AND M. SHARIR)

12.3 Ramsey's Theorem

RAMSEY NUMBERS AND RAMSEY GRAPHS

Because cliques are complements of stable sets, if a graph has no large cliques, one might expect it to have a large stable set. That this is indeed the case was first proved by Ramsey (1930). He showed that, given any positive integers k and ℓ, there exists a smallest integer $r(k, \ell)$ such that every graph on $r(k, \ell)$ vertices contains either a clique of k vertices or a stable set of ℓ vertices. By complementarity, $r(k, \ell) = r(\ell, k)$ (Exercise 12.3.1). Also, it is easily seen that:

$$r(1, \ell) = r(k, 1) = 1 \tag{12.3}$$

and

$$r(2, \ell) = \ell, \qquad r(k, 2) = k \tag{12.4}$$

The numbers $r(k, \ell)$ are known as the *Ramsey numbers*; when $k = \ell$, they are called *diagonal Ramsey numbers*. The following theorem on Ramsey numbers is due to Erdős and Szekeres (1935) and Greenwood and Gleason (1955).

Theorem 12.9 *For any two integers $k \geq 2$ and $\ell \geq 2$,*

$$r(k, \ell) \leq r(k, \ell - 1) + r(k - 1, \ell) \tag{12.5}$$

Furthermore, if $r(k, \ell - 1)$ and $r(k - 1, \ell)$ are both even, strict inequality holds in (12.5).

Proof Let G be a graph on $r(k, \ell - 1) + r(k - 1, \ell)$ vertices, and let $v \in V$. We distinguish two cases:

1. Vertex v is nonadjacent to a set S of at least $r(k, \ell - 1)$ vertices.
2. Vertex v is adjacent to a set T of at least $r(k - 1, \ell)$ vertices.

Note that either case 1 or case 2 must hold because the number of vertices to which v is nonadjacent plus the number of vertices to which v is adjacent is equal to $r(k, \ell - 1) + r(k - 1, \ell) - 1$.

In case 1, $G[S]$ contains either a clique of k vertices or a stable set of $\ell - 1$ vertices, and therefore $G[S \cup \{v\}]$ contains either a clique of k vertices or a stable set of ℓ vertices. Similarly, in case 2, $G[T \cup \{v\}]$ contains either a clique of k vertices or a stable set of ℓ vertices. Because either case 1 or case 2 must hold, it follows that G contains either a clique of k vertices or a stable set of ℓ vertices. This proves (12.5).

Now suppose that $r(k, \ell - 1)$ and $r(k - 1, \ell)$ are both even, and let G be a graph on $r(k, \ell - 1) + r(k - 1, \ell) - 1$ vertices. Because G has an odd number of vertices, it follows from Corollary 1.2 that some vertex v is of even degree; in particular, v cannot be adjacent to precisely $r(k - 1, \ell) - 1$ vertices. Consequently, either case 1 or case 2 above holds, and G therefore contains either a clique of k vertices or a stable set of ℓ vertices. Thus, as asserted,

$$r(k, \ell) \le r(k, \ell-1) + r(k-1, \ell) - 1 \qquad \square$$

The determination of the Ramsey numbers in general is a very difficult unsolved problem. Lower bounds can be obtained by the construction of suitable graphs. Consider, for example, the four graphs in Figure 12.11.

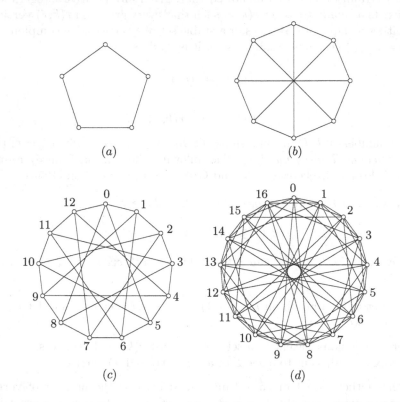

Fig. 12.11. (a) A (3,3)-Ramsey graph, (b) a (3,4)-Ramsey graph, (c) a (3,5)-Ramsey graph, (d) a (4,4)-Ramsey graph

The 5-cycle (Figure 12.11a) contains no clique of three vertices and no stable set of three vertices. It shows, therefore, that

$$r(3, 3) \ge 6 \qquad (12.6)$$

The graph of Figure 12.11b is the Wagner graph (see Section 10.5). It contains no clique of three vertices and no stable set of four vertices. Hence

$$r(3, 4) \ge 9 \qquad (12.7)$$

Similarly, the graph of Figure 12.11c shows that

$$r(3,5) \geq 14 \qquad (12.8)$$

and the graph of Figure 12.11d yields

$$r(4,4) \geq 18 \qquad (12.9)$$

With the aid of Theorem 12.9 and equations (12.4) we can now show that equality in fact holds in (12.6)–(12.9). Firstly, by (12.5) and (12.4),

$$r(3,3) \leq r(3,2) + r(2,3) = 6 \qquad (12.10)$$

and therefore, using (12.6), we have $r(3,3) = 6$. Noting that $r(3,3)$ and $r(2,4)$ are both even, we apply Theorem 12.5 and (12.4) to obtain

$$r(3,4) \leq r(3,3) + r(2,4) - 1 = 9$$

With (12.7) this gives $r(3,4) = 9$. Now we again apply (12.5) and (12.4) to obtain

$$r(3,5) \leq r(3,4) + r(2,5) = 14$$

and

$$r(4,4) \leq r(4,3) + r(3,4) = 18$$

which, together with (12.8) and (12.9), respectively, yield $r(3,5) = 14$ and $r(4,4) = 18$.

The following table shows all the Ramsey numbers $r(k, \ell)$ known to date, where $3 \leq k \leq \ell$.

k	3	3	3	3	3	3	3	4	4
ℓ	3	4	5	6	7	8	9	4	5
$r(k,\ell)$	6	9	14	18	23	28	36	18	25

A (k, ℓ)-*Ramsey graph* is a graph on $r(k, \ell) - 1$ vertices that contains neither a clique of k vertices nor a stable set of ℓ vertices. By definition of $r(k, \ell)$, such graphs exist for all $k \geq 2$ and $\ell \geq 2$. Ramsey graphs often seem to possess interesting structures. All of the graphs in Figure 12.11 are Ramsey graphs; the last two can be obtained from finite fields in the following way. We get the $(3, 5)$-Ramsey graph by regarding the thirteen vertices as elements of the field of integers modulo 13, and joining two vertices by an edge if their difference is a cubic residue of 13 (namely, $1, 5, 8$, or 12); the $(4, 4)$-Ramsey graph is obtained by regarding the vertices as elements of the field of integers modulo 17, and joining two vertices if their difference is a quadratic residue of 17 (namely, $1, 2, 4, 8, 9, 13, 15$, or 16). For $k = 2, 3$, and 4, the (k, k)-Ramsey graphs are self-complementary (that is, isomorphic to their complements) (Exercise 12.3.2). Whether this is true for all values of k is an open question.

BOUNDS ON RAMSEY NUMBERS

Theorem 12.9 yields the following upper bound for the Ramsey numbers.

Theorem 12.10 *For all positive integers k and ℓ,*

$$r(k, \ell) \leq \binom{k + \ell - 2}{k - 1}$$

Proof By induction on $k + \ell$. Using (12.3) and (12.4), we see that the theorem holds when $k + \ell \leq 5$. Let m and n be positive integers, and assume that the theorem is valid for all positive integers k and ℓ such that $5 \leq k + \ell < m + n$. Then, by Theorem 12.9 and the induction hypothesis,

$$r(m, n) \leq r(m, n - 1) + r(m - 1, n)$$
$$\leq \binom{m + n - 3}{m - 1} + \binom{m + n - 3}{m - 2} = \binom{m + n - 2}{m - 1}$$

Thus the theorem holds for all values of k and ℓ. □

On noting that $\binom{k+\ell-2}{k-1}$ is the number of $(k - 1)$-subsets of a $(k + \ell - 2)$-set, whereas $2^{k+\ell-2}$ is the total number of subsets of this set, we have

Corollary 12.11 *For all positive integers k and ℓ, $r(k, \ell) \leq 2^{k+\ell-2}$, with equality if and only if $k = \ell = 1$.* □

Corollary 12.11 shows, in particular, that the diagonal Ramsey numbers grow at most exponentially. We now give an exponential lower bound for these numbers. Due to Erdős (1947), it is obtained by means of a powerful counting technique known as the probabilistic method. This technique, introduced and developed by Erdős in collaboration with other Hungarian mathematicians, including P. Turán, has been applied to remarkable effect in combinatorics, number theory, and computer science, as well as in graph theory. A more formal and detailed account of the method is given in Chapter 13.

Theorem 12.12 *For all positive integers k,*

$$r(k, k) \geq 2^{k/2}$$

Proof Because $r(1, 1) = 1$ and $r(2, 2) = 2$, we may assume that $k \geq 3$. As in Section 1.2, we denote by $\mathcal{G}_n$ the set of simple graphs with vertex set $\{v_1, v_2, \ldots, v_n\}$. Let $\mathcal{G}_n^k$ be the set of these labelled simple graphs which have a clique of k vertices. Observe that

$$|\mathcal{G}_n| = 2^{\binom{n}{2}} \tag{12.11}$$

since each subset of the $\binom{n}{2}$ possible edges $v_i v_j$ determines a graph in $\mathcal{G}_n$. Similarly, the number of graphs in $\mathcal{G}_n$ having a particular set of k vertices as a clique is

$2^{\binom{n}{2}-\binom{k}{2}}$. Because there are $\binom{n}{k}$ distinct k-element subsets of $\{v_1, v_2, \ldots, v_n\}$, we have

$$|\mathcal{G}_n^k| \leq \binom{n}{k} 2^{\binom{n}{2}-\binom{k}{2}} \tag{12.12}$$

(The inequality arises because the graphs in $\mathcal{G}_n^k$ which have more than one k-clique are counted more than once by the expression on the right-hand side.) By (12.11) and (12.12),

$$\frac{|\mathcal{G}_n^k|}{|\mathcal{G}_n|} \leq \binom{n}{k} 2^{-\binom{k}{2}} < \frac{n^k 2^{-\binom{k}{2}}}{k!}$$

Suppose that $n < 2^{k/2}$. Then

$$\frac{|\mathcal{G}_n^k|}{|\mathcal{G}_n|} < \frac{2^{k^2/2} 2^{-\binom{k}{2}}}{k!} = \frac{2^{k/2}}{k!} < \frac{1}{2}$$

In other words, if $n < 2^{k/2}$, then fewer than half of the graphs in $\mathcal{G}_n$ contain a clique of k vertices. Likewise, by complementarity, fewer than half of the graphs in $\mathcal{G}_n$ contain a stable set of k vertices. Hence some graph in $\mathcal{G}_n$ contains neither a clique of k vertices nor a stable set of k vertices. Because this holds for any $n < 2^{k/2}$, we have $r(k, k) \geq 2^{k/2}$.   □

Corollary 12.11 tells us that there exists a (k, k)-Ramsey graph of order less than 2^{2k-2}. However, it does not show us how to find or construct such a graph. All known lower bounds for $r(k, k)$ obtained by constructive arguments are much weaker than the one given by Theorem 12.12. The best is due to Frankl and Wilson (1981); their construction is described in Section 12.5.

The Ramsey numbers $r(k, \ell)$ are sometimes defined in a slightly different manner from the way in which we introduced them. One easily sees that $r(k, \ell)$ can be thought of as the smallest integer n such that every (not necessarily proper) 2-edge-colouring (E_1, E_2) of K_n contains either a complete subgraph on k vertices, all of whose edges are assigned colour 1, or a complete subgraph on ℓ vertices, all of whose edges are assigned colour 2.

Expressed in this form, the Ramsey numbers have a natural generalization. For positive integers t_i, $1 \leq i \leq k$, we define $r(t_1, t_2, \ldots, t_k)$ to be the smallest integer n such that every k-edge-colouring $(E_1, E_2, \ldots, E_k)$ of K_n contains a complete subgraph on t_i vertices all of whose edges belong to E_i, for some i, $1 \leq i \leq k$.

The following theorem and corollary generalize inequality (12.5) and Theorem 12.10, and can be proved in a similar manner (Exercise 12.3.3).

Theorem 12.13 *For all positive integers t_i, $1 \leq i \leq k$,*

$$r(t_1, t_2, \ldots, t_k) \leq r(t_1 - 1, t_2, \ldots, t_k) + r(t_1, t_2 - 1, \ldots, t_k) + \cdots$$
$$+ r(t_1, t_2, \ldots, t_k - 1) - k + 2 \qquad □$$

Corollary 12.14 *For all positive integers t_i, $1 \leq i \leq k$,*

$$r(t_1 + 1, t_2 + 1, \ldots, t_k + 1) \leq \frac{(t_1 + t_2 + \cdots + t_k)!}{t_1! t_2! \ldots t_k!} \qquad □$$

AN APPLICATION TO NUMBER THEORY

We now describe an interesting application of Ramsey's theorem to combinatorial number theory.

Consider the partition $(\{1, 4, 10, 13\}, \{2, 3, 11, 12\}, \{5, 6, 7, 8, 9\})$ of the set of integers $\{1, 2, \ldots, 13\}$. Observe that in no subset of the partition are there three integers x, y, and z (not necessarily distinct) which satisfy the equation

$$x + y = z \tag{12.13}$$

Yet, no matter how we partition $\{1, 2, \ldots, 14\}$ into three subsets, there will always exist a subset of the partition which contains a solution to (12.13) (Exercise 12.3.8a). Schur (1916) proved that, in general, given any positive integer n, there exists an integer r_n such that, for any partition of $\{1, 2, \ldots, r_n\}$ into n subsets, one of the subsets contains a solution to (12.13). We show that the Ramsey number $r_n := r(t_1, t_2, \ldots, t_n)$, where $t_i = 3$, $1 \leq i \leq n$, satisfies this requirement.

Theorem 12.15   SCHUR'S THEOREM
Let $\{A_1, A_2, \ldots, A_n\}$ be a partition of the set of integers $\{1, 2, \ldots, r_n\}$ into n subsets. Then some A_i contains three integers x, y, and z satisfying the equation $x + y = z$.

Proof   Consider the complete graph whose vertex set is $\{1, 2, \ldots, r_n\}$. Colour the edges of this graph in colours $1, 2, \ldots, n$ by the rule that the edge uv is assigned colour i if $|u-v| \in A_i$. By Ramsey's Theorem (12.13), there exists a monochromatic triangle in the graph; that is, there are three vertices a, b, and c such that the edges ab, bc, and ac all have the same colour j. Assume, without loss of generality, that $a > b > c$ and write $x = a - b$, $y = b - c$, and $z = a - c$. Then $x, y, z \in A_j$, and $x + y = z$. $\square$

Let s_n be the least integer such that, in any partition of $\{1, 2, \ldots, s_n\}$ into n subsets, there is a subset which contains a solution to (12.13). It can be checked that $s_1 = 1$, $s_2 = 5$, and $s_3 = 14$ (Exercise 12.3.8a). Also, from Theorem 12.15 and Exercise 12.3.4 we have the upper bound

$$s_n \leq \lfloor n! \, e \rfloor + 1$$

A lower bound for s_n is the topic of Exercise 12.3.8c.

Exercises

$\star$**12.3.1** Show that, for all k and l, $r(k, \ell) = r(\ell, k)$.

12.3.2 Show that the $(4, 4)$-Ramsey graph (Figure 12.11d) is self-complementary.

12.3.3 Prove Theorem 12.13 and Corollary 12.14.

12.3.4

a) Show that $r_n \leq n(r_{n-1} - 1) + 2$.
b) Noting that $r_2 = 6$, use (a) to show that $r_n \leq \lfloor n!e \rfloor + 1$.
c) Deduce that $r_3 \leq 17$.

12.3.5 CLEBSCH GRAPH

a) The *Clebsch graph* has as vertex set the even subsets of $\{1, 2, 3, 4, 5\}$, two being adjacent if their symmetric difference has cardinality four. A drawing of the Clebsch graph is shown in Figure 12.12. Find an appropriate labelling of its vertices.

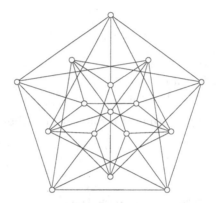

Fig. 12.12. The Clebsch graph

b) Show that the Clebsch graph is vertex-transitive and triangle-free, and that the non-neighbours of any vertex induce a copy of the Petersen graph.
c) Consider two further graphs defined on the same set of vertices as the Clebsch graph. In the first, two vertices are adjacent if their symmetric difference belongs to the set $\{12, 23, 34, 45, 51\}$ (where we write ij for $\{i, j\}$); in the second, two vertices are adjacent if their symmetric difference belongs to the set $\{13, 24, 35, 41, 52\}$. Show that both of these graphs are isomorphic to the Clebsch graph.
d) Using the bound given in Exercise 12.3.4c, deduce that $r_3 = 17$.

(R.E. GREENWOOD AND A.M. GLEASON)

12.3.6 Let $G = K_3 \vee C_5$, the join of K_3 and C_5. Show that:

a) G contains no copy of K_6,
b) every 2-edge colouring of G contains a monochromatic triangle.

(R.L. GRAHAM)

(Folkman (1970) has constructed a (huge) graph which contains no copy of K_4, but every 2-edge colouring of which contains a monochromatic triangle; more generally, Nešetřil and Rödl (1975) have constructed, for every positive integer k, a graph containing no copy of K_4, every k-edge colouring of which contains a monochromatic triangle.)

12.3.7 Let $G_1, G_2, \ldots, G_k$ be simple graphs. The *generalized Ramsey number* $r(G_1, G_2, \ldots, G_k)$ is the smallest integer n such that every k-edge-colouring $(E_1, E_2, \ldots, E_k)$ of K_n contains, for some i, a subgraph isomorphic to G_i in colour i. Let P_3 denote the 3-path, C_4 the 4-cycle, and T_m any tree on m vertices. Show that:

a) $r(P_3, P_3) = 5$, $r(P_3, C_4) = 5$, and $r(C_4, C_4) = 6$,
b) $r(T_m, K_{1,n}) \leq m + n - 1$, with equality if $n - 1 \equiv 0 \,(\mathrm{mod}\,(m-1))$,
c) $r(T_m, K_n) = (m-1)(n-1) + 1$.   (V. CHVÁTAL)

12.3.8

a) Show that:
    i) $s_1 = 2$, $s_2 = 5$, and $s_3 = 14$,
    ii) $s_{n+1} \geq 3s_n - 1$ for all n.
b) Deduce that $s_n \geq \frac{1}{2}(3^n + 1)$ for all n.

(Abbott and Moser (1966) determined a sharper lower bound on s_n.)

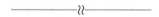

12.3.9

a) By appealing to Exercise 12.1.12, show that

$$r(kl + 1, kl + 1) - 1 \geq (r(k + 1, k + 1) - 1)(r(l + 1, l + 1) - 1)$$

b) Deduce that $r(2^n + 1, 2^n + 1) \geq 5^n + 1$ for all $n \geq 0$.   (H.L. ABBOTT)

12.3.10 Let k, s, and t be positive integers with $s \leq t$. Show that there exists an integer $r := r_k(s, t)$ such that, for every k-edge-colouring $(E_1, E_2, \ldots, E_k)$ of the s-uniform complete hypergraph H on at least r vertices, H contains a monochromatic complete subhypergraph on t vertices.

12.3.11

a) Show that every 2-edge-coloured countably infinite complete graph contains a countably infinite monochromatic complete subgraph.
b) Deduce the finite version of Ramsey's Theorem.

12.3.12 Let V be a set of points in the plane in general position (no three collinear), and let G be the complete geometric graph with vertex set V. Show that, for every 2-edge-colouring of G, there is a monochromatic spanning tree no two edges of which cross.   (G. KÁROLYI, J. PACH, AND G. TÓTH)

12.4 The Regularity Lemma

Somewhat paradoxically, the behaviour of random graphs (studied in Chapter 13) is often highly predictable. For example, when p is a constant, many properties of the random graph $\mathcal{G}_{n,p}$ hold almost surely. One of the difficulties in saying things about concrete graphs is that they are less homogeneous: their edges can be scattered about the graph in unpredictable ways, even if information is known about basic parameters of the graph such as its connectivity or chromatic number

Fortunately, it turns out that any sufficiently large dense graph can be split up in such a way that the resulting subgraphs are joined to one another in an essentially randomlike manner, and this allows one to establish many interesting properties of such graphs. This remarkable and surprising fact is known as the Regularity Lemma. It was developed by Szemerédi (1978) in order to prove a beautiful theorem in number theory: every dense subset of the positive integers contains arbitrarily long arithmetic progressions. In more recent years, it has been used to establish numerous results in extremal graph theory, as well as in number theory, geometry, and other fields, and is now a major tool in combinatorics.

REGULAR PAIRS AND REGULAR PARTITIONS

Let $G = (V, E)$ be a graph and let X and Y be disjoint subsets of V. The *density* $d(X, Y)$ is the proportion of the $|X||Y|$ possible edges of G that are actually present in G:

$$d(X, Y) := \frac{e(X, Y)}{|X||Y|}$$

In the case of a random graph $G = \mathcal{G}_{n,p}$, the expected value of $e(X, Y)$ is $p|X||Y|$, so the expected value of the density is simply p, no matter what the sets X and Y are. Moreover, Chernoff's Inequality shows that $d(X, Y)$ is always close to p with high probability (see Exercise 13.3.7).

Let ϵ be a small fixed positive constant. In what follows, all definitions are with respect to ϵ, and it is convenient not to make explicit mention of this constant each time. We refer to a subset of a set X as a *small* subset of X if it is of cardinality at most $\epsilon|X|$, and a *large* subset of X if it is of cardinality greater than $\epsilon|X|$. With these conventions, a pair (X, Y) of disjoint subsets of V is called *regular* if the densities of (X', Y') and (X, Y) differ by at most ϵ whenever X' is a large subset of X and Y' is a large subset of Y; otherwise, the pair is *irregular*. A *regular partition* of V with *exceptional set* X_0 is a partition $\{X_0, X_1, X_2, \ldots, X_r\}$ such that:

▷ X_0 is a small subset of V,
▷ $|X_i| = |X_j|$, $1 \leq i \leq r$,
▷ (X_i, X_j) is regular for all but at most $\epsilon\binom{r}{2}$ pairs $\{i, j\} \subset \{1, 2, \ldots, r\}$.

A partition $\{X_0, X_1, X_2, \ldots, X_r\}$ satisfying the first two conditions is an *equipartition* of V.

Szemerédi's Regularity Lemma guarantees that every sufficiently large graph admits a regular partition into not too many parts.

Theorem 12.16 THE REGULARITY LEMMA

Let p be an integer and ϵ a positive real number. Then there is an integer q, depending only on p and ϵ, such that any graph G on at least q vertices has a regular partition $\{X_0, X_1, X_2, \ldots, X_r\}$ with $p \leq r \leq q$.

The proof of the Regularity Lemma, although not conceptually difficult, is a bit technical. For this reason, we first of all present one of its important applications, a celebrated theorem due to Erdős and Stone (1946), and another interesting application to Ramsey numbers. These, and many other applications of the Regularity Lemma, rely on the following basic property of regular pairs.

Lemma 12.17 *Let (X, Y) be a regular pair of density d, let Y' be a large subset of Y, and let S be the set of vertices of X which have fewer than $(d-\epsilon)|Y'|$ neighbours in Y'. Then S is a small subset of X.*

Proof Consider a large subset X' of X. Because (X, Y) is a regular pair, $d(X', Y') \geq d - \epsilon$. Hence $e(X', Y') \geq (d - \epsilon)|X'||Y'|$, implying that some vertex of X' has at least $(d - \epsilon)|Y'|$ neighbours in Y'. Thus $X' \neq S$. We conclude that S is not a large subset of X.   □

THE ERDŐS–STONE THEOREM

Turán's Theorem tells us that every graph G with at least $t_{k-1}(n) + 1$ edges contains a copy of the complete graph K_k. The Erdős–Stone Theorem says, roughly speaking, that if n is sufficiently large and G has not too many more edges, then G contains a copy of the Turán graph $T_{k,tk}$, the complete k-partite graph with t vertices in each part.

Theorem 12.18 THE ERDŐS–STONE THEOREM

Let k and t be integers, where $k \geq 3$ and $t \geq 2$, and let d be a real number, where $0 < d < \frac{1}{2}$. Then there is an integer N, depending only on k, t, and d, such that every graph G with at least N vertices and at least $t_{k-1}(n) + dn^2$ edges contains a copy of the Turán graph $T_{k,tk}$.

Proof We apply the Regularity Lemma with

$$\epsilon := \left(\frac{d}{2}\right)^{(k-1)t} \qquad p := \left\lceil \frac{1}{d - 3\epsilon} \right\rceil \quad \text{and} \quad N := \max\left\{q, \frac{k-1}{8\epsilon}\right\} \quad (12.14)$$

where q is the function of p and ϵ defined in the Regularity Lemma. The choices of the parameters ϵ, p, and N in (12.14) are made simply to ensure that the computations below go through smoothly.

Assume that G is a graph on n vertices and at least $t_{k-1}(n) + dn^2$ edges, where $n \geq N$. Then there is a regular partition $\{X_0, X_1, \ldots, X_r\}$ of V such that $p \leq r \leq q$. We set $\ell := |X_i|$, $1 \leq i \leq r$. Thus $r\ell \geq (1 - \epsilon)n$.

We now consider the subgraph H of G obtained by deleting:

▷   the exceptional set X_0,
▷   all edges of $G[X_i]$, $1 \le i \le r$,
▷   all edges between irregular pairs (X_i, X_j),
▷   all edges between regular pairs (X_i, X_j) of density less than d.

Then H is a graph on $r\ell$ vertices, and the number of edges deleted from G to form H is at most

$$\epsilon n(n-1) + r\binom{\ell}{2} + \epsilon\binom{r}{2}\ell^2 + \binom{r}{2}d\ell^2 < \epsilon n^2 - \epsilon n + \frac{1}{2}r\ell^2 + \frac{1}{2}\epsilon r^2\ell^2 + \frac{1}{2}dr^2\ell^2$$

Thus

$$
\begin{aligned}
e(H) - \epsilon n &> e(G) - \left(\epsilon n^2 + \frac{1}{2}r\ell^2 + \frac{1}{2}\epsilon r^2\ell^2 + \frac{1}{2}dr^2\ell^2\right) \\
&\ge t_{k-1}(n) + (d-\epsilon)n^2 - \frac{1}{2}(d+\epsilon)r^2\ell^2 - \frac{1}{2}r\ell^2 \\
&\ge t_{k-1}(n) + (d-\epsilon)r^2\ell^2 - \frac{1}{2}(d+\epsilon)r^2\ell^2 - \frac{1}{2}r\ell^2 \\
&= t_{k-1}(n) + \frac{1}{2}(d-3\epsilon)r^2\ell^2 - \frac{1}{2}r\ell^2
\end{aligned}
$$

The fact that $r \ge p$ and the choice of p in (12.14) yield the inequalities

$$\frac{1}{2}(d-3\epsilon)r^2\ell^2 \ge \frac{1}{2}(d-3\epsilon)pr\ell^2 \ge \frac{1}{2}r\ell^2$$

Combining the bound $\epsilon n \ge \epsilon N \ge (k-1)/8$ implied by (12.14) with the bound (see Exercise 12.2.3a)

$$t_{k-1}(n) \ge \frac{1}{2}\left(\frac{k-2}{k-1}\right)n^2 - \frac{k-1}{8}$$

we now have

$$e(H) > t_{k-1}(n) + \epsilon n \ge t_{k-1}(n) + \frac{k-1}{8} \ge \frac{1}{2}\left(\frac{k-2}{k-1}\right)n^2$$

Let R be the graph with vertex set $\{x_1, x_2, \ldots, x_r\}$, vertices x_i and x_j being joined if (X_i, X_j) is a regular pair of density at least d. Then

$$e(R) \ge \frac{e(H)}{\ell^2} > \frac{1}{2}\left(\frac{k-2}{k-1}\right)\left(\frac{n^2}{\ell^2}\right) \ge \frac{1}{2}\left(\frac{k-2}{k-1}\right)r^2$$

so R contains a copy of K_k, by Turán's Theorem. In other words, H has k sets X_i any two of which form a regular pair of density at least d. Without loss of generality, we may assume that these sets are $X_1, X_2, \ldots, X_k$. We show that $H[\bigcup_{i=1}^{k} X_i]$ contains a copy of the desired Turán graph.

Consider a pair (X_1, X_i). By Lemma 12.17, at most $\epsilon\ell$ vertices of X_1 have fewer than $c\ell$ neighbours in X_i, where $c := d - \epsilon$. Therefore at most $\epsilon(k-1)\ell$ vertices of

X_1 have fewer than $c\ell$ neighbours in one (or more) of the sets X_i, $2 \le i \le k$. In other words, at least $(1-(k-1)\epsilon)\ell$ vertices of X_1 have at least $c\ell$ neighbours in each set X_i, $2 \le i \le k$. Let v_1 be one such vertex, and let X_i^1 be its set of neighbours in X_i, $2 \le i \le k$. Because $c > (d/2) > \epsilon$, this is a large subset of X_i. Therefore, again by Lemma 12.17, at least $(1 - (k - 1)\epsilon)\ell$ vertices of X_1 each has more than $c^2\ell$ neighbours in each set X_i^1, $2 \le i \le k$. Let v_2 be one such vertex, and let X_i^2 be its set of neighbours in X_i^1, $2 \le i \le k$. Continuing in this way, and noting that $c^{t-1} \ge (d/2)^{t-1} \ge \epsilon$, we may find t vertices $v_1, v_2, \ldots, v_t$ in X_1, and a subset X_i^t of more than $c^t\ell$ vertices of X_i, $2 \le i \le r$, such that each vertex v_i is joined to every vertex of X_i^t, $2 \le i \le k$. By induction, because $c^{(k-1)t} \ge (d/2)^{(k-1)t} = \epsilon$, starting with the $k - 1$ sets X_i^t, $2 \le i \le k$, we find a copy of $T_{k-1,t(k-1)}$ in $H[\cup_{i=2}^k X_i^t]$. Together with the vertices $v_1, v_2, \ldots, v_t$ this yields a copy of $T_{k,tk}$ in H, and hence in G. $\qquad\square$

An important consequence of the Erdős–Stone Theorem is that it highlights the intrinsic role played by the chromatic number in extremal graph theory. This striking fact was first noted by Erdős and Simonovits (1966), and the following result is known as the Erdős–Stone–Simonovits Theorem.

Theorem 12.19 *For any simple graph* F

$$lim_{n\to\infty} \frac{ex(n, F)}{n^2} = \frac{1}{2}\left(\frac{\chi(F) - 2}{\chi(F) - 1}\right)$$

Proof   Let $k = \chi(F)$ and let t be the largest size of a colour class in a proper k-colouring of F. Then $F \subseteq T_{k,tk}$. Therefore, by the Erdős–Stone Theorem, for any $d > 0$, there is an integer q, depending on k, t, and d, such that every graph G on at least q vertices and at least $t_{k-1}(n) + dn^2$ edges contains a copy of F. Hence, for all $d > 0$,

$$\frac{ex(n, F)}{n^2} \le \frac{t_{k-1}(n) + dn^2}{n^2} = \frac{t_{k-1}(n)}{n^2} + d$$

On the other hand, because F is k-chromatic and the Turán graph $T_{k-1,n}$ is $(k-1)$-colourable, $F \not\subseteq T_{k-1,n}$. Therefore $ex(n, F) > t_{k-1}(n)$ and

$$\frac{ex(n, F)}{n^2} > \frac{t_{k-1}(n)}{n^2}$$

These two inequalities imply that

$$lim_{n\to\infty} \frac{ex(n, F)}{n^2} = lim_{n\to\infty} \frac{t_{k-1}(n)}{n^2} = \frac{1}{2}\left(\frac{\chi(F) - 2}{\chi(F) - 1}\right) \qquad\square$$

It should be noted that Theorem 12.19 is only of interest for nonbipartite graphs: when F is bipartite, it asserts merely that $ex(n, F) \ll n^2$, a fact which follows directly from the theorem of Kővári, Sós, and Turán (Exercise 12.2.11).

LINEAR RAMSEY NUMBERS

The *Ramsey number* of a simple graph G is the least integer p such that every 2-edge-colouring of K_p yields a monochromatic copy of G (see Exercise 12.3.7). This number is denoted by $r(G, G)$. Because any simple graph G on n vertices is a subgraph of K_n, one has the obvious upper bound $r(G, G) \leq r(n, n)$. However, this exponential bound in n (see Theorem 12.12) is very weak when G is a sparse graph, one with relatively few edges. Indeed, we show here how the Regularity Lemma can be applied to prove that the Ramsey numbers of graphs of bounded maximum degree grow only linearly with their order. This result is due to Chvátal et al. (1983).

Theorem 12.20 *For any graph G of maximum degree Δ,*

$$r(G, G) \leq cn$$

where c depends only on Δ.

Proof We set

$$\epsilon := 4^{-2\Delta} \quad \text{and} \quad p := \epsilon^{-1} = 4^{2\Delta}$$

and take q to be the function of p and ϵ defined in the Regularity Lemma. We now set $c := pq$. Note that c depends only on Δ.

Let H be a graph on cn vertices. Clearly, $cn \geq q$. Therefore, by the Regularity Lemma, H has a regular partition $\{X_0, X_1, X_2, \ldots, X_r\}$, where $p \leq r \leq q$. This partition is also a regular partition of the complement $\overline{H}$ of H, for the same value of ϵ (Exercise 12.4.1). Let R be the graph with vertex set $\{x_1, x_2, \ldots, x_r\}$, vertices x_i and x_j being adjacent in R if (X_i, X_j) is a regular pair. Then $e(R) \geq (1 - \epsilon)\binom{r}{2}$. Because $r > p - 1 = (1 - \epsilon)/\epsilon$,

$$(1 - \epsilon)\binom{r}{2} = (1 - \epsilon)\left(1 - \frac{1}{r}\right)\left(\frac{r^2}{2}\right) > (1 - 2\epsilon)\left(\frac{r^2}{2}\right) \geq t_k(r)$$

where (see Exercise 12.2.3a) $k = (2\epsilon)^{-1} > 2^{2\Delta}$. By Turán's Theorem, R contains a copy of the complete graph K_k. Let F be the spanning subgraph of this complete graph whose edges correspond to the regular pairs (X_i, X_j) of density at least one-half. Because $k > 2^{2\Delta}$, either F or $\overline{F}$ contains a complete subgraph of $\Delta + 1$ vertices, by Corollary 12.11. If this graph is F, we show that H contains a copy of G; if it is $\overline{F}$, the same argument shows that $\overline{H}$ contains a copy of G.

Suppose, then, that F contains a complete subgraph of $\Delta + 1$ vertices. Without loss of generality, we may assume that F has vertex set $\{x_1, x_2, \ldots, x_{\Delta+1}\}$. Let $u_1, u_2, \ldots, u_n$ be an arbitrary ordering of the vertices of G. We show how vertices $v_1, v_2, \ldots, v_n$ can be selected from the sets $X_1, X_2, \ldots, X_{\Delta+1}$ so that $H[\{v_1, v_2, \ldots, v_j\}]$ contains a copy of $G[\{u_1, u_2, \ldots, u_j\}]$, $1 \leq j \leq n$.

Suppose that the vertices $v_1, v_2, \ldots, v_j$ have already been selected, where $0 \leq j \leq n - 1$. We describe how to select the next vertex, v_{j+1}. For $k \geq j + 1$, set

$$N_{j,k} := \{v_i : 1 \leq i \leq j \text{ and } u_i u_k \in E(G)\}$$

We refer to a set X_i which is disjoint from $N_{j,k}$ as being *eligible* for v_k at *stage j*. A *candidate* for v_k, where $k \geq j+1$, is a vertex of H which:

1. belongs to some eligible set,
2. is adjacent in H to every vertex of $N_{j,k}$.

In order to guarantee that a suitable vertex v_{j+1} can be chosen at each stage, we shall require that, for $0 \leq j \leq n-1$, every eligible set for v_k have many candidates at stage j, at least $4^{-d_{j,k}}\ell$ of them, where $d_{j,k} := |N_{j,k}|$ and ℓ is the common cardinality of the sets X_i. This requirement is clearly fulfilled at stage 0. Suppose that it is fulfilled at stage j, where $0 \leq j < n$. We select an eligible set X_s for v_{j+1}, and consider the set Y_s of candidates for v_{j+1} in X_s. We shall choose an element of Y_s as v_{j+1} in such a way that, for all $k \geq j+2$, every eligible set for v_k at stage $j+1$ has at least $4^{-d_{j+1,k}}\ell$ candidates.

If u_k is not adjacent to u_{j+1} in G, then $d_{j+1,k} = d_{j,k}$, and this condition automatically holds if it holds for j. Suppose, then, that u_k is adjacent to u_{j+1} in G, so that $d_{j+1,k} = d_{j,k} + 1$. Consider any set X_t eligible for v_k at stage j; there is at least one such set because $d_{j,k}$ is at most Δ whereas there are $\Delta+1$ sets X_i. Let Y_t be the set of candidates for v_k in X_t at this stage. Observe that Y_t is a large subset of X_t, because $4^{d_{j,k}}\epsilon < 4^{\Delta}\epsilon < 1$. We apply Lemma 12.17 with $X = X_s$, $Y = X_t$, and density at least $\frac{1}{2}$. Thus the set of vertices of X_s which are adjacent to at most $(\frac{1}{2} - \epsilon)|Y_t|$ vertices of Y_t is a small subset of X_s; their number is at most $\epsilon\ell$. Applying this reasoning to all sets X_t eligible for v_k at stage j (there are at most Δ of them), and for all $k \geq j+2$ (at most Δ values of k), we may conclude that there are at least $|Y_s| - \Delta^2 \epsilon\ell$ vertices of Y_s, each adjacent to more than $(\frac{1}{2} - \epsilon)|Y_t|$ vertices of each Y_t. Thus, assuming that $|Y_s| - \Delta^2 \epsilon\ell - j > 0$, some vertex of Y_s different from $v_1, v_2, \ldots, v_j$ is adjacent to more than $(\frac{1}{2} - \epsilon)|Y_t|$ vertices of each Y_t such that X_t is eligible for v_k at stage j. Now $|X_0| \leq \epsilon cn$, so

$$(1 - \epsilon)pqn = (1 - \epsilon)cn \leq \sum_{i=1}^{r} |X_i| = r\ell \leq q\ell$$

whence $(1 - \epsilon)n \leq \epsilon\ell$ and hence $j < n \leq 2\epsilon\ell$. Because $|Y_s| \geq 4^{-d_{j,k}}\ell \geq 4^{-\Delta}\ell$ and $(\Delta^2 + 2)\epsilon < 4^{-\Delta}$,

$$|Y_s| - \Delta^2 \epsilon\ell - j > 4^{-\Delta}\ell - \Delta^2 \epsilon\ell - 2\epsilon\ell = \left(4^{-\Delta} - (\Delta^2 + 2)\epsilon\right)\ell > 0$$

Also, because $\epsilon < \frac{1}{4}$ and $|Y_t| \geq 4^{-d_{j,k}}\ell$,

$$\left(\frac{1}{2} - \epsilon\right)|Y_t| \geq \frac{1}{4}\left(4^{-d_{j,k}}\ell\right) = 4^{-d_{j,k}-1}\ell = 4^{-d_{j+1,k}}\ell$$

We conclude that there is a candidate for v_{j+1} in X_s such that, for all $k \geq j+2$, every eligible set for v_k at stage $j+1$ has at least $4^{-d_{j+1,k}}\ell$ candidates. The theorem follows by induction on j. $\qquad \square$

A PROOF OF THE REGULARITY LEMMA

The proof of the Regularity Lemma given here is based on a measure of how close a given partition is to being regular.

For a pair $\{X, Y\}$ of disjoint sets of vertices of a graph G, we define its *index of regularity* by:

$$\rho(X, Y) := |X||Y|(d(X, Y))^2$$

This index is nonnegative. We extend it to a family $\mathcal{P}$ of disjoint subsets of V by setting:

$$\rho(\mathcal{P}) := \sum_{\substack{X, Y \in \mathcal{P} \\ X \neq Y}} \rho(X, Y)$$

In the case where $\mathcal{P}$ is a partition of V, we have:

$$\rho(\mathcal{P}) = \sum_{\substack{X, Y \in \mathcal{P} \\ X \neq Y}} |X||Y|(d(X, Y))^2 \leq \sum_{\substack{X, Y \in \mathcal{P} \\ X \neq Y}} |X||Y| < \frac{n^2}{2} \qquad (12.15)$$

If $\mathcal{P}$ is an irregular equipartition of V, we show that there is an equipartition $\mathcal{Q}$, a refinement of $\mathcal{P}$, whose index of regularity is significantly greater than that of $\mathcal{P}$. By applying this operation a constant number of times (the constant being a function of ϵ), we end up with a regular partition of V.

The first step is to observe that refining a family of subsets never reduces the index of regularity.

Proposition 12.21 *Let G be a graph, let $\mathcal{P}$ be a family of disjoint subsets of V, and let $\mathcal{Q}$ be a refinement of $\mathcal{P}$. Then $\rho(\mathcal{Q}) \geq \rho(\mathcal{P})$.*

Proof It suffices to show that the conclusion holds when $\mathcal{Q}$ is obtained from $\mathcal{P}$ by partitioning one set $X \in \mathcal{P}$ into two nonempty sets X_1, X_2. We have:

$$\rho(\mathcal{Q}) - \rho(\mathcal{P}) = \rho(X_1, X_2) + \sum_{\substack{Y \in \mathcal{P} \\ Y \neq X}} (\rho(X_1, Y) + \rho(X_2, Y) - \rho(X, Y))$$

Thus it suffices to show that, for $Y \in \mathcal{P}$, $Y \neq X$,

$$\rho(X_1, Y) + \rho(X_2, Y) - \rho(X, Y) \geq 0$$

Set

$$x := |X|, \quad y := |Y|, \quad d := d(X, Y)$$

and, for $i = 1, 2$,

$$x_i := |X_i|, \quad d_i := d(X_i, Y)$$

Then
$$x = x_1 + x_2$$
Also, because $e(X, Y) = e(X_1, Y) + e(X_2, Y)$,
$$xyd = x_1 y d_1 + x_2 y d_2$$
Therefore, after rearranging terms,
$$\rho(X_1, Y) + \rho(X_2, Y) - \rho(X, Y) = x_1 y d_1^2 + x_2 y d_2^2 - xyd^2 = \frac{x_1 x_2 y}{x}(d_1 - d_2)^2 \geq 0$$
The conclusion follows.   □

Next, we show that if (X, Y) is an irregular pair, then there is a refinement $\mathcal{P}$ of $\{X, Y\}$ such that $\rho(\mathcal{P})$ is somewhat bigger than $\rho(X, Y)$.

Lemma 12.22 *Let (X, Y) be an ϵ-irregular pair in a graph G, with $|d(X_1, Y_1) - d(X, Y)| > \epsilon$, where X_1 is a large subset of X, and Y_1 is a large subset of Y. Define $X_2 := X \setminus X_1$ and $Y_2 := Y \setminus Y_1$. Then*
$$\rho(\{X_1, X_2, Y_1, Y_2\}) - \rho(X, Y) \geq \left(\frac{\epsilon^4}{1 - \epsilon^2}\right)|X||Y|$$

Proof Set
$$x := |X|, \quad y := |Y|, \quad d := d(X, Y)$$
and
$$x_i := |X_i|, \quad y_j = |Y_j|, \quad d_{ij} := d(X_i, Y_j), \quad i, j = 1, 2$$
In this notation, the hypotheses of the lemma can (essentially) be written as:
$$x_1 \geq \epsilon x, \quad y_1 \geq \epsilon y, \quad \text{and} \quad (d_{11} - d)^2 \geq \epsilon^2 \tag{12.16}$$
Also $xyd = \sum_{i,j=1,2} x_i y_j d_{ij}$, so
$$xyd - x_1 y_1 d_{11} = x_1 y_2 d_{12} + x_2 y_1 d_{21} + x_2 y_2 d_{22}$$
Applying the Cauchy–Schwarz Inequality $\sum_i a_i^2 \sum_i b_i^2 \geq (\sum_i a_i b_i)^2$ with
$$a_1 := \sqrt{x_1 y_2}, \quad a_2 := \sqrt{x_2 y_1}, \quad a_3 := \sqrt{x_2 y_2}$$
and
$$b_1 := \sqrt{x_1 y_2} d_{12}, \quad b_2 = \sqrt{x_2 y_1} d_{21}, \quad b_3 = \sqrt{x_2 y_2} d_{22}$$
yields
$$x_1 y_2 d_{12}^2 + x_2 y_1 d_{21}^2 + x_2 y_2 d_{22}^2 \geq \frac{(xyd - x_1 y_1 d_{11})^2}{xy - x_1 y_1}$$
Therefore, by virtue of the inequalities (12.16),

$$\rho(\{X_1, X_2, Y_1, Y_2\}) - \rho(X, Y) \geq \sum_{i,j=1,2} x_i y_j d_{ij}^2 - xy d^2$$

$$\geq x_1 y_1 d_{11}^2 + \frac{(xyd - x_1 y_1 d_{11})^2}{xy - x_1 y_1} - xy d^2$$

$$= \frac{xy}{xy - x_1 y_1} x_1 y_1 (d_{11} - d)^2$$

$$\geq \frac{x_1 y_1}{1 - \epsilon^2} (d_{11} - d)^2$$

$$\geq \left(\frac{\epsilon^4}{1 - \epsilon^2}\right) xy \qquad \square$$

Lemma 12.22 is the key to the proof of the Regularity Lemma. Given an irregular equipartition $\mathcal{P}$ of V, we apply it to each irregular pair of $\mathcal{P}$, and in so doing show that $\mathcal{P}$ can be refined to an equipartition $\mathcal{Q}$ whose index of regularity is significantly greater than that of $\mathcal{P}$. For the purposes of the proof, it is convenient to define the index of an equipartition $\{X_0, X_1, \ldots, X_k\}$ with exceptional set X_0 as the index of the partition obtained on splitting X_0 into singletons.

Lemma 12.23 *Let* $\mathcal{P} = \{X_0, X_1, \ldots, X_k\}$ *be an equipartition of* V. *If* $\mathcal{P}$ *is not regular and* $|X_0| \leq (\epsilon - 2^{-k})n$, *then there is an equipartition* $\mathcal{Q} := \{Y_0, Y_1, \ldots, Y_\ell\}$ *of* V *such that*

$$|Y_0| < |X_0| + \frac{n}{2^k}, \qquad k \leq \ell \leq k 4^k \quad and \quad \rho(\mathcal{Q}) - \rho(\mathcal{P}) \geq \left(\frac{1 - \epsilon}{1 + \epsilon}\right) \epsilon^5 n^2$$

Proof By Lemma 12.22, the sets X_i and X_j in any irregular pair (X_i, X_j), can each be partitioned into two sets so that the index of regularity of the resulting partition $\mathcal{P}_{ij}$ satisfies

$$\rho(\mathcal{P}_{ij}) - \rho(X_i, X_j) \geq \left(\frac{\epsilon^4}{1 - \epsilon^2}\right) |X_i||X_j| \geq \left(\frac{\epsilon^4}{1 - \epsilon^2}\right) (1 - \epsilon)^2 \frac{n^2}{k^2} = \left(\frac{1 - \epsilon}{1 + \epsilon}\right) \epsilon^4 \frac{n^2}{k^2}$$

For each i, $1 \leq i \leq k$, each of the $k - 1$ partitions $\mathcal{P}_{ij}$, $j \neq i$, induces a partition of X_i into two parts. Denoting by $\mathcal{P}_i$ the coarsest common refinement of these partitions of X_i, we have $|\mathcal{P}_i| \leq 2^{k-1}$. Let $\mathcal{P}'$ be the partition of V whose members are the members of $\mathcal{P}_i$, $1 \leq i \leq k$, and the singleton elements of X_0. Because $|\mathcal{P}_i| < 2^k$, we have $|\mathcal{P}'| < k 2^k$. Because there are more than ϵk^2 irregular pairs,

$$\rho(\mathcal{P}') - \rho(\mathcal{P}) \geq \epsilon k^2 \left(\frac{1 - \epsilon}{1 + \epsilon}\right) \epsilon^4 \frac{n^2}{k^2} = \left(\frac{1 - \epsilon}{1 + \epsilon}\right) \epsilon^5 n^2$$

From $\mathcal{P}'$ we now construct our desired equipartition $\mathcal{Q}$. Let $\{Y_i : 1 \leq i \leq \ell\}$ be a maximal family of disjoint subsets of V, each of cardinality $\lceil n/(k 4^k) \rceil$, and each contained in some member of $\mathcal{P}'$. Set $Y_0 := V \setminus (\cup_{i=1}^{\ell} Y_i)$ and $\mathcal{Q} := \{Y_0, Y_1, \ldots, Y_\ell\}$. Note that Y_0 contains at most $n/(k 4^k)$ elements of each member of $\mathcal{P}'$, so

$$|Y_0| < |X_0| + (k 2^k) \left(\frac{n}{k 4^k}\right) = |X_0| + \frac{n}{2^k} \leq \epsilon n$$

Also, $k \leq \ell \leq k4^k$. The refinement of $\mathcal{Q}$ obtained by splitting all vertices of Y_0 into singletons is a refinement of $\mathcal{P}'$. Appealing to Proposition 12.21, we have $\rho(\mathcal{Q}) \geq \rho(\mathcal{P}')$. Therefore

$$\rho(\mathcal{Q}) - \rho(\mathcal{P}) = (\rho(\mathcal{Q}) - \rho(\mathcal{P}')) + (\rho(\mathcal{P}') - \rho(\mathcal{P})) \geq \left(\frac{1-\epsilon}{1+\epsilon}\right) \epsilon^5 n^2 \qquad \square$$

The Regularity Lemma now follows on applying Lemma 12.23 iteratively, starting with an equipartition whose exceptional set has fewer than k elements, and noting the upper bound (12.15) on the index of a partition. We leave these technical details as an exercise (12.4.2). A quite different proof of the Regularity Lemma, based on ergodic theory, can be found in the book by Tao and Vu (2006).

A multitude of applications of the Regularity Lemma are described in the excellent survey articles by Komlós and Simonovits (1996) and Komlós et al. (2002). The version of the Regularity Lemma given here applies only to dense graphs, but there exist several variants, including versions for sparse graphs (Kohayakawa and Rödl (2003)) and hypergraphs (Gowers (2006), Rödl et al. (2005), Tao (2006)).

Exercises

12.4.1 Show that a regular partition of a graph G is also a regular partition of its complement $\overline{G}$ (with the same parameter ϵ and the same exceptional set).

12.4.2 Deduce the Regularity Lemma from Lemma 12.23.

12.5 Related Reading

Hypergraph Extremal Problems

The questions treated in this chapter can of course be posed in the context of hypergraphs. Ramsey's Theorem extends quite easily to hypergraphs (Exercise 12.3.10). Indeed, it was in this more general setting that Ramsey (1930) proved his theorem. On the other hand, the evaluation of hypergraph Ramsey numbers remains a totally intractable problem.

The state of affairs with respect to questions of Turán type for hypergraphs is hardly better, even for 3-uniform hypergraphs. Turán's conjectures on the number of triples of an n-set needed to guarantee a $K_4^{(3)}$ or a $K_5^{(3)}$ remain unsolved (see Exercise 12.2.6). The difficulty of the former conjecture can perhaps be attributed to the fact that, should it be true, the number of nonisomorphic extremal configurations would be very large indeed. Kostochka (1982) has constructed 2^{k-2} such 'extremal' examples.

On the positive side, there have been a number of encouraging advances on hypergraph extremal problems in recent years. For instance, when n is sufficiently large, it was shown independently by Füredi and Simonovits (2005) and by Keevash and Sudakov (2005) that, as conjectured by Sós (1976), the only extremal 3-uniform hypergraph not containing a copy of the Fano hypergraph is the unique 2-colourable 3-uniform hypergraph with the maximum number of edges (see Exercise 12.2.6b). There are several informative surveys on this topic; see, for example, Füredi (1991) and Sidorenko (1995).

CONSTRUCTIONS FROM HYPERGRAPHS

We have seen several examples of how hypergraphs can be used to construct graphs with special properties. In particular, the incidence graph of a projective plane of order $k-1$ is a $(k, 6)$-cage (Exercise 3.1.13), and polarity graphs of projective planes provide examples of extremal graphs without 4-cycles (see Exercise 12.2.14). Frankl and Wilson (1981) made use of hypergraphs to obtain constructive lower bounds for Ramsey numbers.

Let S be an n-set, and let q be a prime power with $q^2 \leq n + 1$. Consider the graph whose vertices are the $(q^2 - 1)$-subsets of S, two such subsets X and Y being adjacent if and only if $|X \cap Y| \not\equiv -1 \pmod{q}$. Frankl and Wilson (1981) showed that this graph has neither a stable set nor a clique of cardinality $k := \binom{n}{q-1} + 1$, yielding the Ramsey number bound $r(k, k) \geq \binom{n}{q^2-1} + 1$. Setting $n = q^3$ results in the superpolynomial lower bound $r(k, k) \geq k^{f(k)}$, where $f(k) \sim \log k / 4 \log \log k$.

RAMSEY THEOREMS IN OTHER CONTEXTS

Ramsey's Theorem exemplifies the mathematical dictum, due to T.S. Motzkin, that complete disorder is impossible: if a large enough structure is partitioned arbitrarily into two or more classes, then one of the classes contains a 'regular' substructure of prescribed size. Such theorems appear in many areas of mathematics. One classical example is the theorem due to Van der Waerden (1927), which asserts that if the positive integers are partitioned into a finite number of classes, then one of the classes contains arbitrarily long arithmetic progressions. (This theorem has inspired some deep discoveries in number theory, notably the density theorem of E. Szemerédi mentioned in the section on the Regularity Lemma.)

Ramsey's Theorem was rediscovered by Erdős and Szekeres (1935) while investigating a problem on combinatorial geometry. It led Erdős and his many collaborators to develop over the years an area of geometry known as *euclidean Ramsey theory*. One especially attractive theorem on this topic, due to Kříž (1991), states that if S is the set of points of a regular polygon, and if n is large enough, then in any 2-colouring of $\mathbb{R}^n$, there exists a monochromatic subset which is congruent to S. The book by Graham et al. (1990) contains many other beautiful examples of such Ramsey theorems.

13

The Probabilistic Method

Contents

13.1 Random Graphs

As mentioned in Section 12.3, the lower bound on Ramsey numbers given in Theorem 12.12 was obtained by means of a proof technique known as the *probabilistic method*. Roughly speaking, this technique is based on an understanding of how graphs behave on the average. Theorem 12.12, for instance, was proved by showing that the majority of graphs on $2^{n/2}$ vertices have no clique of cardinality n. (This is evidently not true of all graphs on $2^{n/2}$ vertices.) An understanding of the expected behaviour of graphs turns out to be of immense value in demonstrating the existence of graphs with particular properties. The probabilistic method is also a remarkably effective tool for establishing properties of graphs in general.

A *(finite) probability space* (Ω, P) consists of a finite set Ω, called the *sample space*, and a *probability function* $P : \Omega \rightarrow [0,1]$ satisfying $\sum_{\omega \in \Omega} P(\omega) = 1$. We may regard the set $\mathcal{G}_n$ of all labelled graphs on n vertices (or, equivalently, the set of all spanning subgraphs of K_n) as the sample space of a finite probability space $(\mathcal{G}_n, P)$. The result of selecting an element G of this sample space according to the probability function P is called a *random graph*.

The simplest example of such a probability space arises when all graphs $G \in \mathcal{G}_n$ have the same probability of being chosen. Because $|\mathcal{G}_n| = 2^N$, where $N := \binom{n}{2}$, the probability function in this case is:

$$P(G) = 2^{-N}, \quad \text{for all } G \in \mathcal{G}_n$$

A natural way of viewing this probability space is to imagine the edges of K_n as being considered for inclusion one by one, each edge being chosen with probability one half (for example, by flipping a fair coin), these choices being made independently of one another. The result of such a procedure is a spanning subgraph G of K_n, with all $G \in \mathcal{G}_n$ being equiprobable.

A more refined probability space on the set $\mathcal{G}_n$ may be obtained by fixing a real number p between 0 and 1 and choosing each edge with probability p, these choices again being independent of one another. Because $1 - p$ is the probability that any particular edge is not chosen, the resulting probability function P is given by

$$P(G) = p^m (1 - p)^{N-m}, \quad \text{for each } G \in \mathcal{G}_n$$

where $m := e(G)$. This space is denoted by $\mathcal{G}_{n,p}$. For example, $\mathcal{G}_{3,p}$ has as sample space the $2^{\binom{3}{2}} = 8$ spanning subgraphs of K_3 shown in Figure 13.1, with the probability function indicated.

Note that the smaller the value of p, the higher the probability of obtaining a sparse graph. We are interested in computing or estimating the probability that a random graph has a particular property.

To each graph property, such as connectedness, there corresponds a subset of $\mathcal{G}_n$, namely those members of $\mathcal{G}_n$ which have the given property. The probability that a random graph has this particular property is then just the sum of the probabilities of these graphs. For instance, the probability that a random graph in $\mathcal{G}_{3,p}$ is connected is equal to $3p^2(1 - p) + p^3 = p^2(3 - 2p)$, the probability that it is bipartite is $(1 - p)^3 + 3(1 - p)^2 p + 3(1 - p)p^2 = (1 - p)(1 + p + p^2)$, and the probability that it is both connected and bipartite is $3p^2(1 - p)$.

In a probability space (Ω, P), any subset A of Ω is referred to as an *event*, and the *probability* of the event A is defined by:

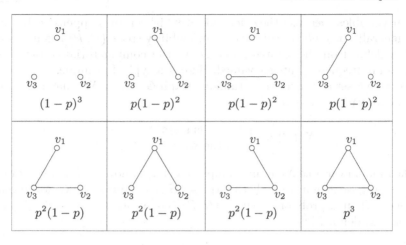

Fig. 13.1. The probability space $\mathcal{G}_{3,p}$

$$P(A) := \sum_{\omega \in A} P(\omega) \tag{13.1}$$

INDEPENDENT EVENTS

Events A and B in a probability space (Ω, P) are *independent* if $P(A \cap B) = P(A)P(B)$; otherwise, they are *dependent*. More generally, events A_i, $i \in I$, are *(mutually) independent* if, for any subset S of I,

$$P(\cap_{i \in S} A_i) = \prod_{i \in S} P(A_i)$$

For example, if A is the event 'G is connected' and B is the event 'G is bipartite' in the space $\mathcal{G}_{3,p}$, then (unless $p = 0$ or $p = 1$)

$$P(A)P(B) = p^2(3 - 2p)(1 - p)(1 + p + p^2) \neq 3p^2(1 - p) = P(A \cap B)$$

Thus these two events are dependent. In other words, the knowledge that G is bipartite has a bearing on the probability of it being connected.

It is important to realize that a set of events may well be dependent even if the events are pairwise independent (see Exercise 13.1.2).

RANDOM VARIABLES

Much of graph theory is concerned with the study of basic parameters, such as the connectivity, the clique number, or the stability number. As we have seen, the values of these parameters provide much information about a graph and its properties. In the context of random graphs, functions such as these are known as random variables, because they depend on which graph happens to be selected. More generally, a *random variable* on a probability space (Ω, P) is any real-valued function defined on the sample space Ω. In the combinatorial context, random variables are frequently integer-valued. Here is a typical example.

Let S be a set of vertices of a random graph $G \in \mathcal{G}_{n,p}$. We may associate with S a random variable X_S defined by:

$$X_S(G) := \begin{cases} 1 & \text{if } S \text{ is a stable set of } G, \\ 0 & \text{otherwise.} \end{cases} \tag{13.2}$$

The random variable X_S is an example of what is known as an indicator random variable, because it indicates whether the set S is a stable set of G. More generally, each event A in a probability space (Ω, P) has an associated *indicator random variable* X_A, defined by:

$$X_A(\omega) := \begin{cases} 1 & \text{if } \omega \in A, \\ 0 & \text{otherwise.} \end{cases}$$

(In the preceding example, A is simply the event that S is a stable set.)

Thus to each event there corresponds a random variable. Conversely, with each random variable X and each real number t, we may associate the event

$$\{\omega \in \Omega : X(\omega) = t\}$$

This event is denoted for short by $X = t$. Analogously, one may define four related events: $X < t$, $X \leq t$, $X \geq t$, and $X > t$. For instance, if X is the number of components of $G \in \mathcal{G}_{3,p}$, the event $X \geq 2$ consists of the first four graphs in Figure 13.1; and if X_S is the random variable defined in (13.2), the event $X_S = 1$ consists of those graphs $G \in \mathcal{G}_{n,p}$ in which S is a stable set. We are interested in the probabilities of such events.

Random variables X_i, $i \in I$, are *(mutually) independent* if the events $X_i = t_i$, $i \in I$, are independent for all real numbers t_i. Random variables are *dependent* if they are not independent.

Exercises

13.1.1 Let $G \in \mathcal{G}_{n,1/2}$. For a subset S of V, let A_S denote the event that S is a stable set of G. Show that if S and T are two distinct k-subsets of V, then $A(S)$ and $A(T)$ are independent if $|S \cap T| = 0$ or $|S \cap T| = 1$, and dependent otherwise.

13.1.2 Let V be an n-set. Consider the probability space (Ω, P), where Ω is the set of k-colourings $(V_1, V_2, \ldots, V_k)$ of V, all colourings being equiprobable (thus each occurring with probability k^{-n}). An element of this space is called a *random k-colouring* of V. Consider a random k-colouring of the vertices of a simple graph G. For an edge e of G, let A_e denote the event that the two ends of e receive the same colour. Show that:

a) for any two edges e and f of G, the events A_e and A_f are independent,

b) if e, f, and g are the three edges of a triangle of G, the events A_e, A_f, and A_g are dependent.

13.1.3

a) Let $\{A_i : i \in I\}$ be a set of events in a probability space (Ω, P). Express the probability of the event $\overline{\cup_{i \in I} A_i}$ in terms of the probabilities of the events $\cap_{i \in S} A_i$, where S ranges over all nonempty subsets of I.

b) Let V be an n-set. Consider the probability space (Ω, P), where Ω is the set of permutations of V, all permutations being equiprobable (thus each occurring with probability $1/n!$). An element of this space is called a *random permutation* or *random linear ordering* of V. A permutation π of V is a *derangement* of V if $\pi(v) \neq v$ for all $v \in V$. Determine the probability of the set of all derangements of V, and the asymptotic value of this probability.

$\star$**13.1.4** Let $\{A_i : i \in I\}$ be a set of independent events in a probability space (Ω, P). For all $S \subseteq I$, show that the events A_i, $i \in S$, and $\overline{A_i}$, $i \in I \setminus S$ are independent.

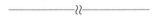

13.2 Expectation

The average value, or mean, of a random variable X is called its *expectation*, and is denoted by $E(X)$. Thus

$$E(X) := \sum_{\omega \in \Omega} X(\omega) P(\omega) \tag{13.3}$$

For instance, if X denotes the number of components of $G \in \mathcal{G}_{3,p}$,

$$E(X) = 3 \times (1-p)^3 + 2 \times 3p(1-p)^2 + 1 \times (3p^2(1-p) + p^3) = 3 - 3p + p^3$$

LINEARITY OF EXPECTATION

We make use of two basic properties of expectation which follow easily from definitions (13.1) and (13.3). Firstly, expectation is a linear function. In other words, for random variables X and Y, and real numbers r and s,

$$E(rX + sY) = rE(X) + sE(Y) \tag{13.4}$$

This property is referred to as *linearity of expectation*. Secondly, if X_A is an indicator random variable,

$$E(X_A) = P(X_A = 1) \tag{13.5}$$

It is important to emphasize that identity (13.4) is valid for any two random variables X and Y, whether or not they are independent. We repeatedly make use of this fact. By contrast, the identity $E(XY) = E(X)E(Y)$ does not hold in general, although it is valid when X and Y are independent random variables (Exercise 13.2.2).

THE CROSSING LEMMA

In order to convince the reader of the power of the probabilistic method, we present a remarkably simple application of this proof technique to crossing numbers of graphs. We obtain a lower bound for the crossing number of a graph in terms of its order and size, and then use this bound to derive two theorems in combinatorial geometry.

Recall that the *crossing number* $cr(G)$ of a graph G is the least number of crossings in a plane embedding of G. This parameter satisfies the trivial lower bound $cr(G) \geq m - 3n$ (in fact $cr(G) \geq m - 3n + 6$ for $n \geq 3$; Exercise 10.3.1). The following much stronger lower bound was given by Ajtai et al. (1982) and, independently, by Leighton (1983). Its very short probabilistic proof is due to N. Alon; see Alon and Spencer (2000).

Lemma 13.1   THE CROSSING LEMMA
Let G be a simple graph with $m \geq 4n$. Then

$$cr(G) \geq \frac{1}{64} \frac{m^3}{n^2}$$

Proof   Consider a planar embedding $\widetilde{G}$ of G with $cr(G)$ crossings. Let S be a random subset of V obtained by choosing each vertex of G independently with probability $p := 4n/m$, and set $H := G[S]$ and $\widetilde{H} := \widetilde{G}[S]$.

Define random variables X, Y, Z on Ω as follows: X is the number of vertices, Y the number of edges, and Z the number of crossings of $\widetilde{H}$. The trivial bound noted above, when applied to H, yields the inequality $Z \geq cr(H) \geq Y - 3X$. By linearity of expectation (13.4), $E(Z) \geq E(Y) - 3E(X)$. Now $E(X) = pn$, $E(Y) = p^2m$ (each edge having two ends) and $E(Z) = p^4 cr(G)$ (each crossing being defined by four vertices). Hence

$$p^4 cr(G) \geq p^2 m - 3pn$$

Dividing both sides by p^4, we have:

$$cr(G) \geq \frac{pm - 3n}{p^3} = \frac{n}{(4n/m)^3} = \frac{1}{64} \frac{m^3}{n^2} \qquad \square$$

Székely (1997) realized that the Crossing Lemma (13.1) could be used to derive very easily a number of theorems in combinatorial geometry, some of which hitherto had been regarded as extremely challenging. We now give proofs of two of them.

Consider a set of n points in the plane. Any two of these points determine a line, but it might happen that some of these lines pass through more than two of the points. Specifically, given a positive integer k, one may ask how many lines there can be which pass through at least k points. For instance, if n is a perfect square and the points are in the form of a square grid, there are $2\sqrt{n} + 2$ lines which pass through $\sqrt{n}$ points. Is there a configuration of points which contains more lines through this number of points? The following theorem of Szemerédi and Trotter (1983) gives a general bound on the number of lines which pass through more than k points.

Theorem 13.2 *Let P be a set of n points in the plane, and let ℓ be the number of lines in the plane passing through at least $k+1$ of these points, where $1 \leq k \leq 2\sqrt{2n}$. Then $\ell < 32n^2/k^3$.*

Proof Form a graph G with vertex set P whose edges are the segments between consecutive points on the lines which pass through at least $k+1$ points of P. This graph has at least $k\ell$ edges and crossing number at most $\binom{\ell}{2}$. Thus either $k\ell < 4n$, in which case $\ell < 4n/k \leq 32n^2/k^3$, or $\ell^2/2 > \binom{\ell}{2} \geq cr(G) \geq (k\ell)^3/64n^2$ by the Crossing Lemma (13.1), and again $\ell < 32n^2/k^3$. □

A second application of the Crossing Lemma (13.1) concerns the number of pairs of points there can be, among a set of n points, whose distance is exactly one. The square grid (see Figure 1.27) shows that this number can grow faster than n, as n tends to infinity. (For this, one has to choose a grid in which the distance between consecutive points is less than one; the calculation relies on a little elementary number theory.) The following theorem, due to Spencer et al. (1984), provides an upper bound on the number of pairs of points at distance one.

Theorem 13.3 *Let P be a set of n points in the plane, and let k be the number of pairs of points of P at unit distance. Then $k < 5n^{4/3}$.*

Proof Draw a unit circle around each point of P. Let n_i be the number of these circles passing through exactly i points of P. Then $\sum_{i=0}^{n-1} n_i = n$ and $k = \frac{1}{2}\sum_{i=0}^{n-1} in_i$. Now form a graph H with vertex set P whose edges are the arcs between consecutive points on the circles that pass through at least three points of P. Then

$$e(H) = \sum_{i=3}^{n-1} in_i = 2k - n_1 - 2n_2 \geq 2k - 2n$$

Some pairs of vertices of H might be joined by two parallel edges. Delete from H one of each pair of parallel edges, so as to obtain a simple graph G with $e(G) \geq k-n$. Now $cr(G) \leq n(n-1)$ because G is formed from at most n circles, and any two circles cross at most twice. Thus either $e(G) < 4n$, in which case $k < 5n < 5n^{4/3}$, or $n^2 > n(n-1) \geq cr(G) \geq (k-n)^3/64n^2$ by the Crossing Lemma (13.1), and $k < 4n^{4/3} + n < 5n^{4/3}$. □

ASYMPTOTIC NOTATION

In the sequel, we are concerned with probability spaces (Ω_n, P_n) which are defined for all positive integers n. Because it is with sparse graphs that we are mostly concerned, we study the behaviour of the probability space $\mathcal{G}_{n,p}$ when p is a function of n and $p(n) \to 0$ as $n \to \infty$. Given a sequence (Ω_n, P_n), $n \geq 1$, of probability spaces, a property A is said to be satisfied *almost surely* if $P_n(A_n) \to 1$ as $n \to \infty$, where $A_n := A \cap \Omega_n$.

The following asymptotic notation is employed. If $f : \mathbb{N} \to \mathbb{R}$ and $g : \mathbb{N} \to \mathbb{R}$ are two functions such that $g(n) > 0$ for n sufficiently large, we write:

$$f \ll g \quad \text{if} \quad f(n)/g(n) \to 0 \quad \text{as } n \to \infty$$
$$f \gg g \quad \text{if} \quad f(n)/g(n) \to \infty \quad \text{as } n \to \infty$$
$$f \sim g \quad \text{if} \quad f(n)/g(n) \to 1 \quad \text{as } n \to \infty$$

MARKOV'S INEQUALITY

The following simple inequality, often used in conjunction with identities (13.4) and (13.5), is one of the basic tools of the probabilistic method.

Proposition 13.4   MARKOV'S INEQUALITY
Let X be a nonnegative random variable and t a positive real number. Then

$$P(X \geq t) \leq \frac{E(X)}{t}$$

Proof

$$E(X) = \sum \{X(\omega)P(\omega) : \omega \in \Omega\} \geq \sum \{X(\omega)P(\omega) : \omega \in \Omega, X(\omega) \geq t\}$$
$$\geq \sum \{tP(\omega) : \omega \in \Omega, X(\omega) \geq t\} = t \sum \{P(\omega) : \omega \in \Omega, X(\omega) \geq t\}$$
$$= tP(X \geq t)$$

Dividing the first and last members by t yields the asserted inequality.   □

Markov's Inequality is frequently applied in the following form in order to show that a random graph in $\mathcal{G}_{n,p}$ almost surely has a particular property for a certain value of p. It is obtained by setting $X = X_n$ and $t = 1$ in Proposition 13.4.

Corollary 13.5 *Let X_n be a nonnegative integer-valued random variable in a probability space (Ω_n, P_n), $n \geq 1$. If $E(X_n) \to 0$ as $n \to \infty$, then $P(X_n = 0) \to 1$ as $n \to \infty$.*   □

As a simple example, let X be the number of triangles in $G \in \mathcal{G}_{n,p}$. We may express X as the sum $X = \sum \{X_S : S \subseteq V, |S| = 3\}$, where X_S is the indicator random variable for the event A_S that $G[S]$ is a triangle. Clearly $P(A_S) = p^3$. By linearity of expectation, we have

$$E(X) = \sum \{E(X_S) : S \subseteq V, |S| = 3\} = \binom{n}{3} p^3 < (pn)^3$$

Thus if $pn \to 0$ as $n \to \infty$, then $E(X) \to 0$ and, by Corollary 13.5, $P(X = 0) \to 1$; in other words, if $pn \to 0$ as $n \to \infty$, then G will almost surely be triangle-free.

Using these same tools, we now establish a fundamental and very useful bound on the stability number of random graphs due to Erdős (1961a). Unless otherwise specified, log stands for the natural logarithm (that is, to the base e).

Theorem 13.6 *A random graph in $\mathcal{G}_{n,p}$ almost surely has stability number at most $\lceil 2p^{-1} \log n \rceil$.*

Proof Let $G \in \mathcal{G}_{n,p}$ and let S be a given set of $k+1$ vertices of G, where $k \in \mathbb{N}$. The probability that S is a stable set of G is $(1-p)^{\binom{k+1}{2}}$, this being the probability that none of the $\binom{k+1}{2}$ pairs of vertices of S is an edge of the random graph G.

Let A_S denote the event that S is a stable set of G, and let X_S denote the indicator random variable for this event. By equation (13.5), we have

$$E(X_S) = P(X_S = 1) = P(A_S) = (1-p)^{\binom{k+1}{2}} \qquad (13.6)$$

Let X be the number of stable sets of cardinality $k+1$ in G. Then

$$X = \sum \{X_S : S \subseteq V, |S| = k+1\}$$

and so, by (13.4) and (13.6),

$$E(X) = \sum \{E(X_S) : S \subseteq V, |S| = k+1\} = \binom{n}{k+1}(1-p)^{\binom{k+1}{2}}$$

We bound the right-hand side by invoking two elementary inequalities (Exercise 13.2.1):

$$\binom{n}{k+1} \leq \frac{n^{k+1}}{(k+1)!} \qquad \text{and} \qquad 1 - p \leq e^{-p}$$

This yields the following upper bound on $E(X)$.

$$E(X) \leq \frac{n^{k+1}e^{-p\binom{k+1}{2}}}{(k+1)!} = \frac{\left(ne^{-pk/2}\right)^{k+1}}{(k+1)!} \qquad (13.7)$$

Suppose now that $k = \lceil 2p^{-1} \log n \rceil$. Then $k \geq 2p^{-1} \log n$, so $ne^{-pk/2} \leq 1$. Because k grows at least as fast as the logarithm of n, (13.7) implies that $E(X) \to 0$ as $n \to \infty$. Because X is integer-valued and nonnegative, we deduce from Corollary 13.5 that $P(X = 0) \to 1$ as $n \to \infty$. Consequently, a random graph in $\mathcal{G}_{n,p}$ almost surely has stability number at most k. $\qquad \square$

In the case where $p = \frac{1}{2}$, a slightly sharper bound on α than the one provided by Theorem 13.6 can be obtained, yielding the lower bound on the Ramsey numbers given in Theorem 12.12 (see Exercise 13.2.11). We encounter further interesting and surprising applications of Theorem 13.6 in Chapter 14.

This first excursion into the probabilistic method should have given the reader an idea of its remarkable power. A number of the exercises in this and subsequent sections provide further applications of the method. Many more can be found in the monographs by Spencer (1987), Alon and Spencer (2000), and Molloy and Reed (2002).

Exercises

$\star$**13.2.1** Prove the inequalities:

a) $n^k/k^k \leq \binom{n}{k} \leq n^k/k!$, for $n \geq k \geq 0$,
b) $1 + x \leq e^x$, for all $x \in \mathbb{R}$.

$\star$**13.2.2**

a) Let X_i, $i \in I$, be independent random variables. Show that $E(\prod_{i \in I} X_i) = \prod_{i \in I} E(X_i)$.
b) Give an example of dependent random variables X and Y such that $E(XY) = E(X)E(Y)$.

$\star$**13.2.3** Let X be a nonnegative integer-valued random variable. Using the Cauchy–Schwarz Inequality, show that $E(X^2)P(X \geq 1) \geq E^2(X)$.

13.2.4 Find an infinite family of graphs G with $cr(G) = cm^3/n^2$, where c is a suitable positive constant.

13.2.5 Let $G := (V, E)$ be a simple graph, let S be a random subset of V obtained by choosing each vertex of G independently with probability p, and let $F := G[S]$. Consider the random variables $X := v(F)$ and $Y := e(F)$.

a) Show that $\alpha(F) \geq X - Y$.
b) By calculating $E(X)$ and $E(Y)$, and selecting the value of p appropriately, deduce that $\alpha(G) \geq n^2/4m$, provided that $m \geq n/2$.

13.2.6 Let $G = (V, E)$ be a simple graph.

a) If $m = n$, show that G contains a 3-path unless it has a specific structure.
b) Deduce that G contains at least $m - n$ 3-paths.
c) If $m \geq 3n/2$, show that G contains at least $4m^3/27n^2$ 3-paths.

(R. PINCHASI AND M. SHARIR)

d) By adopting the same approach, and appealing to Exercise 4.1.9b, show that if $m \geq kn$, where k is a positive integer, then G contains at least $k^{-k}m^k/n^{k-1}$ copies of every tree on $k + 1$ vertices.

13.2.7

a) Let $G := (V, E)$ be a graph. Consider a random 2-colouring of V. Show that the expected number of edges of G whose ends receive distinct colours is $m/2$.
b) Deduce that every (loopless) graph G contains a spanning bipartite subgraph F with $e(F) \geq \frac{1}{2}e(G)$ (compare Exercise 2.2.2a).   (P. ERDŐS)

13.2.8 Let $G = (V, E)$ be a complete graph on n vertices. Consider the probability space (Ω, P), where Ω is the set of orientations of G, all orientations being equiprobable (thus each occurring with probability 2^{-N}, where $N := \binom{n}{2}$). An element of this space is called a *random tournament*.

a) Show that the expected number of directed Hamilton paths in a random tournament is $2^{-(n-1)}n!$
b) Deduce that, for all $n \geq 1$, there is a tournament on n vertices which has at least $2^{-(n-1)}n!$ directed Hamilton paths.   (T. SZELE)

13.2.9 A hypergraph is 2-*colourable* if there is a 2-colouring of its vertex set with respect to which no edge is monochromatic.

a) Show that the Fano hypergraph is not 2-colourable and is minimal with respect to this property.

b) Let $H := (V, \mathcal{F})$ be a k-uniform hypergraph.

   i) Consider a random 2-colouring of V. For each edge F of H, denote by A_F the event that F is monochromatic. Show that $P(A_F) = 2^{1-k}$.

   ii) Deduce that if $|\mathcal{F}| < 2^{k-1}$, then H is 2-colourable.   (P. ERDŐS)

c) By considering an appropriate hypergraph defined on the edge set of K_n, deduce from (b)(ii) that if $\binom{n}{k}2^{1-\binom{k}{2}} < 1$, then $r(k,k) > n$.

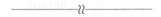

13.2.10 Let $G = (V, E)$ be a graph.

a) Let σ be a linear ordering of V. For $x, y \in V$, write $x \prec_\sigma y$ if x precedes y in the ordering σ. Show that $S_\sigma := \{x \in V : x \prec_\sigma y \text{ for all } y \in N(x)\}$ is a stable set of G.

b) Consider a random linear ordering σ of V. For each vertex v, let X_v be the indicator random variable for the event $v \in S_\sigma$. Show that $E(X_v) = 1/(d(v) + 1)$, where $d(v)$ is the degree of v.

c) Determine $E(X)$, where $X := \sum_{v \in V} X_v$.

d) Deduce that $\alpha(G) \geq \sum_{v \in V} 1/(d(v) + 1)$.

e) Prove that equality holds in (d) if and only if G is a disjoint union of complete graphs.

f) Deduce Turán's Theorem (12.7).   (N. ALON AND J. SPENCER)

$\star$**13.2.11** Let n be a positive integer. For $0 \leq k \leq n$, set $f(k) := \binom{n}{k}2^{-\binom{k}{2}}$.

a) Denote by k^* the least value of k for which $f(k)$ is less than one. Show that:

   i) $k^* \leq \lceil 2\log_2 n \rceil \leq k^* + \log_2 k^* - 1$,

   ii) $f(k^* + 1) \to 0$ as $n \to \infty$,

   iii) $f(k^*) \ll f(k^* - 1)$,

   iv) $f(k^* - 2) \geq n/4$ for $k^* \geq 2$.

b) Deduce from (a)(i) and (a)(ii) that:

   i) if $G \in \mathcal{G}_{n,1/2}$, then almost surely $\alpha(G) \leq \lceil 2\log_2 n \rceil$,

   ii) the Ramsey number $r(k,k)$ is at least $2^{k/2}$.   (P. ERDŐS)

13.2.12 A *dominating set* in a graph $G := (V, E)$ is a subset S of V such that each vertex of G either belongs to S or is adjacent to some element of S; that is, $S \cup N(S) = V$. Let $G = (V, E)$ be a graph with minimum degree δ, and let S be a random subset of V obtained by selecting each vertex of G independently with probability p. Set $T := V \setminus (S \cup N(S))$.

a) Show that:

   i) $E(|S|) = pn$,

   ii) $E(|T|) \leq (1 - p)^{\delta+1}n$,

iii) $S \cup T$ is a dominating set of G.

b) Deduce that G contains a dominating set of at most $(\log(\delta + 1) + 1)n/(\delta + 1)$ vertices.

13.2.13

a) i) Let $\mathcal{F} := \{(X_i, Y_i) : 1 \le i \le m\}$ be a family of pairs of sets such that:
   ▷ $|X_i| = k$ and $|Y_i| = \ell$, $1 \le i \le m$,
   ▷ $X_i \cap Y_j = \emptyset$ if and only if $i = j$.
   By considering a random linear ordering of the set $\cup_{i=1}^m (X_i \cup Y_i)$, show that $m \le \binom{k+\ell}{k}$.   (B. BOLLOBÁS)
   ii) Give an example of such a family $\mathcal{F}$ with $m = \binom{k+\ell}{k}$.

b) i) Suppose that each edge-deleted subgraph of a graph G has more stable sets on k vertices than G itself. Using (a), show that $m \le \binom{n-k+2}{2}$.
   ii) Give an example of such a graph G with $m = \binom{n-k+2}{2}$.

13.2.14 Let $G := G[V_1, V_2, \ldots, V_k]$ be a k-partite graph on n vertices. Denote by d_{ij} the density $d(V_i, V_j)$ (as defined in Section 12.4), and by d_k the smallest value of d such that every k-partite graph in which all densities d_{ij} are greater than d contains a triangle.

a) By considering the graph derived from $K_{k,k}$ by deleting a perfect matching, prove that $d_k \ge \frac{1}{2}$ for all $k \ge 3$.

b) A *transversal* of G is a set $S \subseteq V$ such that $|S \cap V_i| = 1$ for all i, $1 \le i \le k$. Let S be a transversal obtained by selecting one vertex from each set V_i uniformly and at random. Denote by X the number of edges of $G[S]$, and by X_{ij} the number of edges of $G[S]$ linking V_i and V_j. Prove that $E(X_{ij}) = d_{ij}$ and that $E(X) = \sum_{1 \le i < j \le k} d_{ij}$.

c) Deduce that G has a transversal with at least $\sum_{1 \le i < j \le k} d_{ij}$ edges.

d) By applying Turán's Theorem (12.7), conclude that $d_k \to \frac{1}{2}$ as $k \to \infty$.
                        (J.A. BONDY, J. SHEN, S. THOMASSÉ, AND C. THOMASSEN)

13.2.15 Let t be a positive integer. Set $k := 2t$, $n := 2t^2$, and $p := 2\binom{t^2}{2t}/\binom{2t^2}{2t}$. Consider a set V of n elements and a 2-colouring c of V.

a) Let S be a random k-subset of V. Denote by A_S the event that the colouring c assigns the same colour to all vertices of S. Show that $P(A_S) \ge p$.

b) Let $\mathcal{F}$ be a family of m random k-subsets of V. Denote by $A_{\mathcal{F}}$ the event that c is a proper 2-colouring of the hypergraph $(V, \mathcal{F})$. Show that $P(A_{\mathcal{F}}) \le (1-p)^m$.

c) Deduce that there exists a non-2-colourable hypergraph $(V, \mathcal{F})$ with $|V| = n$ and $|\mathcal{F}| = \lceil n \log 2/p \rceil$.

d) Estimate p by using the asymptotic formula, valid for $k \sim \gamma n^{1/2}$ (γ being a positive constant):
$$\binom{n}{k} \sim \frac{n^k}{k!} e^{-k^2/2n}$$

e) Deduce that there exists a non-2-colourable k-uniform hypergraph $H := (V, \mathcal{F})$ with $|V| = k^2/2$ and $|\mathcal{F}| \le \alpha k^2 2^k$, for some constant α.

13.2.16 THE LYM INEQUALITY AND SPERNER'S THEOREM
A *clutter* is a hypergraph no edge of which is properly contained in another.

a) Let $(V, \mathcal{F})$ be a clutter on n vertices. Consider a random permutation σ of V. For $F \in \mathcal{F}$, let A_F denote the event that the first $|F|$ symbols of σ are precisely the elements of F. Show that:
   i) the events A_F, $F \in \mathcal{F}$, are pairwise disjoint,
   ii) if $|F| = k$, then $P(A_F) = 1/\binom{n}{k}$.
b) Deduce:
   i) *The LYM Inequality*: if $(V, \mathcal{F})$ is a clutter on n vertices, where $n \geq 1$, and $\mathcal{F}_k := \{F \in \mathcal{F} : |F| = k\}$, $1 \leq k \leq n$, then $\sum_{k=1}^{n} |\mathcal{F}_k| / \binom{n}{k} \leq 1$.
   (D. LUBELL; K. YAMAMOTO; I.D. MESHALKIN)
   ii) *Sperner's Theorem*: for all $n \geq 1$, a clutter on n vertices has at most $\binom{n}{\lfloor n/2 \rfloor}$ edges. (E. SPERNER)
c) For all $n \geq 1$, give an example of a clutter on n vertices with exactly $\binom{n}{\lfloor n/2 \rfloor}$ edges.

13.2.17 THE ERDŐS–KO–RADO THEOREM
Let $(V, \mathcal{F})$ be an intersecting k-uniform hypergraph, where $V := \{1, 2, \ldots, n\}$ and $n \geq 2k$.

a) Set $F_i := \{i, i+1, \ldots, i+k-1\}$, $1 \leq i \leq n$, addition being modulo n. Show that $\mathcal{F}$ contains at most k of the sets F_i, $1 \leq i \leq n$.
b) Consider a random permutation σ of V and a random element i of V, and set $F := \{\sigma(i), \sigma(i+1), \ldots, \sigma(i+k-1)\}$.
   i) Deduce from (a) that $P(F \in \mathcal{F}) \leq k/n$.
   ii) Show, on the other hand, that $P(F \in \mathcal{F}) = |\mathcal{F}|/\binom{n}{k}$.
c) Deduce the *Erdős–Ko–Rado Theorem*: an intersecting k-uniform hypergraph on n vertices, where $n \geq 2k$, has at most $\binom{n-1}{k-1}$ edges. (G.O.H. KATONA)
d) For $n \geq 2k$ and $k \geq 1$, give an example of an intersecting k-uniform hypergraph on n vertices with exactly $\binom{n-1}{k-1}$ edges.

13.2.18 THE COUNTABLE RANDOM GRAPH
Consider a random graph G on a countably infinite set of vertices in which each potential edge is selected independently with probability $\frac{1}{2}$. Show that:

a) with probability one G satisfies the following adjacency property: given any two disjoint finite sets X and Y of vertices of G, there exists a vertex z which is adjacent to every vertex of X and to no vertex of Y,
b) any two countable graphs satisfying the above property are isomorphic.
(P. ERDŐS AND A. RÉNYI; R. RADO)

(The unique countable graph which satisfies this adjacency property is called the *Rado graph* or *countable random graph*; many of its remarkable properties are discussed by Cameron (1997, 2001).)

13.3 Variance

For a random variable X, it is often useful to know not only its expectation $E(X)$ but also how concentrated is its distribution about this value. A basic measure of this degree of concentration is the *variance* $V(X)$ of X, defined by

$$V(X) := E((X - E(X))^2)$$

Thus the smaller the variance, the more concentrated is the random variable about its expectation.

The variance is evidently nonnegative. Using linearity of expectation, it can be expressed in the form

$$V(X) = E(X^2) - E^2(X)$$

In particular, if X is an indicator random variable, then $E(X^2) = E(X)$, so $V(X) = E(X) - E^2(X) \leq E(X)$.

CHEBYSHEV'S INEQUALITY

The following inequality bounds the divergence of a random variable from its mean. It plays, in some sense, a complementary role to that of Markov's Inequality.

Theorem 13.7   CHEBYSHEV'S INEQUALITY
Let X be a random variable and let t be a positive real number. Then

$$P(|X - E(X)| \geq t) \leq \frac{V(X)}{t^2}$$

Proof By Markov's Inequality,

$$P(|X - E(X)| \geq t) = P((X - E(X))^2 \geq t^2) \leq \frac{E((X - E(X))^2)}{t^2} = \frac{V(X)}{t^2} \qquad \square$$

Chebyshev's Inequality is frequently applied in the following form.

Corollary 13.8 *Let X_n be a random variable in a probability space (Ω_n, P_n), $n \geq 1$. If $E(X_n) \neq 0$ and $V(X_n) \ll E^2(X_n)$, then*

$$P(X_n = 0) \to 0 \quad as\ n \to \infty$$

Proof Set $X := X_n$ and $t := |E(X_n)|$ in Chebyshev's Inequality, and observe that $P(X_n = 0) \leq P(|X_n - E(X_n)| \geq |E(X_n)|)$ because $|X_n - E(X_n)| = |E(X_n)|$ when $X_n = 0$. $\qquad \square$

Let us illustrate the use of Chebyshev's Inequality by considering triangles in random graphs. Let $G \in \mathcal{G}_{n,p}$ be a random graph. We showed earlier, with the aid of Markov's Inequality, that G almost surely is triangle-free when $pn \to 0$. We prove here, on the other hand, that if $pn \to \infty$, then G almost surely has at least one triangle.

As before, we denote by A_S the event that $G[S]$ is a triangle, where S is a 3-subset of V, and by X_S the indicator random variable for A_S. As before, we set

$$X := \sum \{X_S : S \subseteq V, |S| = 3\}$$

so that X is the number of triangles in G. Recall that $E(X) = \binom{n}{3}p^3$. We now apply Corollary 13.8. The following concept is useful in this regard.

The *covariance* $C(X, Y)$ of two random variables X and Y is defined by

$$C(X, Y) := E(XY) - E(X)E(Y)$$

Because X is a sum of indicator random variables X_S, its variance can be bounded in terms of covariances as follows (Exercise 13.3.1).

$$V(X) \leq E(X) + \sum_{S \neq T} C(X_S, X_T) \tag{13.8}$$

The value of $C(X_S, X_T)$ depends only on $|S \cap T|$. If either $|S \cap T| = 0$ or $|S \cap T| = 1$, then $G[S]$ and $G[T]$ can have no common edges, so $E(X_S X_T) = p^6 = E(X_S)E(X_T)$ and $C(X_S, X_T) = 0$. However, if $|S \cap T| = 2$, then $G[S]$ and $G[T]$ have one potential edge in common, so $C(X_S, X_T) = E(X_S X_T) - E(X_S)E(X_T) = p^5 - p^6$. There are $\binom{n}{2}(n-2)(n-3)$ such pairs (S, T). Thus

$$V(X) \leq E(X) + \sum_{S \neq T} C(X_S, X_T) < \binom{n}{3}p^3 + \binom{n}{2}(n-2)(n-3)p^5$$

It follows that if $pn \to \infty$, then $V(X)/E^2(X) \to 0$ as $n \to \infty$. Corollary 13.8 now tells us that $P(X = 0) \to 0$ as $n \to \infty$; in other words, G almost surely has at least one triangle.

The bound in Chebyshev's Inequality can be sharpened significantly when the random variable X has a particular structure, for instance when X is a sum of independent random variables X_i, $1 \leq i \leq n$, such that $P(X_i = +1) = P(X_i = -1) = 1/2$ (see Exercise 13.3.4).

STABILITY NUMBERS OF RANDOM GRAPHS

We showed in Theorem 13.6 that a random graph in $\mathcal{G}_{n,p}$ almost surely has stability number at most $\lceil 2p^{-1} \log n \rceil$. We also noted that this bound can be refined to $\lceil 2 \log_2 n \rceil$ when $p = \frac{1}{2}$ (see Exercise 13.2.11). Here we use Corollary 13.8 to derive a very much sharper result, due independently to Bollobás and Erdős (1976) and Matula (1976).

Theorem 13.9 *Let $G \in \mathcal{G}_{n,1/2}$. For $0 \leq k \leq n$, set $f(k) := \binom{n}{k}2^{-\binom{k}{2}}$ and let k^* be the least value of k for which $f(k)$ is less than one. Then almost surely $\alpha(G)$ takes one of the three values $k^* - 2, k^* - 1, k^*$.*

Proof As in the proof of Theorem 13.6, let X_S denote the indicator random variable for the event A_S that a given subset S of V is a stable set of G, and set $X := \sum\{X_S : S \subseteq V, |S| = k\}$ so that $E(X) = f(k)$. Almost surely $\alpha(G) \leq k^*$ (Exercise 13.2.11b). Consequently, by virtue of Corollary 13.8, it will suffice to prove that $V(X) \ll E^2(X)$ when $k = k^* - 2$. We assume from now on that k takes this value. By Exercise 13.2.11a, we have:

$$k < 2\log_2 n \quad \text{and} \quad f(k) \geq n/4 \tag{13.9}$$

As above, we bound the variance $V(X)$ by applying inequality (13.8).

Let S and T be two sets of k vertices. The value of the covariance $C(X_S, X_T)$ depends only on $|S \cap T|$. If either $|S \cap T| = 0$ or $|S \cap T| = 1$, then $C(X_S, X_T) = 0$ because no edge has both ends in $S \cap T$. If $|S \cap T| = i$, where $2 \leq i \leq k - 1$, then

$$C(X_S, X_T) \leq E(X_S X_T) = P(A_S \cap A_T) = 2^{\binom{i}{2} - 2\binom{k}{2}}$$

There are $\binom{n}{k}$ choices for S, $\binom{k}{i}$ choices for $S \cap T$, and $\binom{n-k}{k-i}$ choices for T. Thus, by inequality (13.8),

$$V(X) \leq E(X) + \sum_{S \neq T} C(X_S, X_T) \leq E(X) + \sum_{i=2}^{k-1} \binom{n}{k}\binom{k}{i}\binom{n-k}{k-i} 2^{\binom{i}{2} - 2\binom{k}{2}}$$

Because $E(X) \ll E^2(X)$, it remains to show that

$$\sum_{i=2}^{k-1} \binom{n}{k}\binom{k}{i}\binom{n-k}{k-i} 2^{\binom{i}{2} - 2\binom{k}{2}} \ll E^2(X) = \binom{n}{k}^2 2^{-2\binom{k}{2}}$$

or equivalently that

$$\binom{n}{k}^{-1} \sum_{i=2}^{k-1} g(i) \to 0 \quad \text{as } n \to \infty \tag{13.10}$$

where

$$g(i) := \binom{k}{i}\binom{n-k}{k-i} 2^{\binom{i}{2}}$$

We have:

$$g(2) = 2\binom{k}{2}\binom{n-k}{k-2} < k^2\binom{n}{k-2}$$

and, for $2 \leq i \leq k - 2$,

$$\frac{g(i+1)}{g(i)} = \frac{(k-i)^2 2^i}{(i+1)(n-2k+i+1)} < \frac{k^2 2^i}{i(n-2k)} = \left(\frac{2^i}{i}\right)\left(\frac{k^2}{n-2k}\right)$$

Set $t := \lfloor c\log_2 n \rfloor$, where $0 < c < 1$. Then, for $2 \leq i \leq t - 1$ and n sufficiently large,

$$\frac{g(i+1)}{g(i)} < \left(\frac{2^t}{t}\right)\left(\frac{k^2}{n-2k}\right) < \left(\frac{n^c}{c\log_2 n}\right)\left(\frac{(2\log_2 n)^2}{n-4\log_2 n}\right) = \frac{4n^c \log_2 n}{c(n-4\log_2 n)} \le 1$$

so

$$\binom{n}{k}^{-1}\sum_{i=2}^{t} g(i) < t\binom{n}{k}^{-1} g(2) \sim \frac{tk^4}{n^2} \to 0 \text{ as } n \to \infty \qquad (13.11)$$

We now consider the remaining terms of the sum in (13.10). We have:

$$\sum_{i=t+1}^{k-1} g(i) = \sum_{i=t+1}^{k-1}\binom{k}{i}\binom{n-k}{k-i}2^{\binom{i}{2}} = 2^{\binom{k}{2}}\sum_{i=t+1}^{k-1}\binom{k}{k-i}\binom{n-k}{k-i}2^{-(k-i)(k+i-1)/2}$$

$$= 2^{\binom{k}{2}}\sum_{j=1}^{k-t-1}\binom{k}{j}\binom{n-k}{j}2^{-j(2k-j-1)/2}$$

$$< 2^{\binom{k}{2}}\sum_{j=1}^{k-t-1}\left(k(n-k)2^{-(k+t)/2}\right)^j$$

In order to bound the right-hand side, we use the fact that $k^* + \log_2 k^* - 1 \ge 2\log_2 n$ (Exercise 13.2.11a) and hence that $2^{-k/2} < (2k+4)^{1/2}n^{-1}$. We deduce that for n sufficiently large,

$$k(n-k)2^{-(k+t)/2} < k(n-k)(2k+4)^{1/2}n^{-1}n^{-c/2} \le 1$$

and so

$$\sum_{i=t+1}^{k-1} g(i) < 2^{\binom{k}{2}}(k-t-1)$$

Using the bound (13.9) on $f(k)$, we now have:

$$\binom{n}{k}^{-1}\sum_{i=t+1}^{k-1} g(i) < \binom{n}{k}^{-1}2^{\binom{k}{2}}(k-t-1) = \frac{(k-t-1)}{f(k)} \to 0 \text{ as } n \to \infty \qquad (13.12)$$

The limits (13.11) and (13.12) imply (13.10). □

Theorem 13.9, striking though it is, can be sharpened still further, to a 'two-point' concentration theorem (Exercise 13.3.2).

Corollary 13.10 *Let $G \in \mathcal{G}_{n,1/2}$, and let f and k^* be as defined in Theorem 13.9. Then either:*

1. *$f(k^*) \ll 1$, in which case almost surely $\alpha(G)$ is equal to either $k^* - 2$ or $k^* - 1$, or*
2. *$f(k^* - 1) \gg 1$, in which case almost surely $\alpha(G)$ is equal to either $k^* - 1$ or k^*.* □

Observe that if $f(k^*) \ll 1$ and $f(k^* - 1) \gg 1$, then Corollary 13.10(i) and (ii) together imply that almost surely $\alpha(G) = k^* - 1$. This is indeed the case for most values of n.

Exercises

$\star$**13.3.1** Let $X = \sum_{i=1}^{n} X_i$ be a sum of random variables.

a) Show that $V(X) = \sum_{i=1}^{n} V(X_i) + \sum_{i \neq j} C(X_i, X_j)$.
b) Deduce that if the X_i are indicator random variables, then $V(X) \leq E(X) + \sum_{i \neq j} C(X_i, X_j)$.

$\star$**13.3.2** Using Exercise 13.2.11a(iii), deduce Corollary 13.10 from Theorem 13.9.

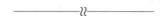

13.3.3

a) Show that, for every positive integer n and every real number $p \in (0, 1]$, there exists a graph on n vertices with at most $(np)^3$ triangles and stability number at most $\lceil 2p^{-1} \log n \rceil$.
b) Deduce that the Ramsey number $r(3, k)$ satisfies the inequality $r(3, k) > n - (np)^3$ for every positive integer n and every real number $p \in (0, 1]$, where $k = \lceil 2p^{-1} \log n \rceil + 1$.
c) Deduce, further, that $r(3, k) > 2n/3$, where $k = \lceil 2p^{-1} \log n \rceil + 1$ and $p^{-1} = 3^{1/3} n^{2/3}$.
d) Conclude that $r(3, k) > c (k / \log k)^{3/2}$ for a suitable positive constant c.
(It is known that $c_1 k^2 / \log k < r(3, k) < c_2 k^2 / \log k$ for suitable positive constants c_1 and c_2. The lower bound is due to Kim (1995), the upper bound to Ajtai et al. (1980).)

13.3.4 CHERNOFF'S INEQUALITY
Let X_i, $1 \leq i \leq n$, be independent random variables such that $P(X_i = +1) = P(X_i = -1) = 1/2$, $1 \leq i \leq n$, and let $X := \sum_{i=1}^{n} X_i$.

a) Show that:
   i) for any real number α, the random variables $e^{\alpha X_i}$, $1 \leq i \leq n$, are independent,
   ii) $E(e^{\alpha X_i}) \leq e^{\alpha^2/2}$, $1 \leq i \leq n$.
b) Deduce that $E(e^{\alpha X}) \leq e^{\alpha^2 n/2}$.
c) By applying Markov's Inequality and choosing an appropriate value of α, derive the following concentration bound, valid for all $t > 0$, known as *Chernoff's Inequality*.

$$P(X \geq t) \leq e^{-t^2/2n} \qquad \text{(H. CHERNOFF)}$$

13.3.5 Let $H := (V, \mathcal{F})$ be a hypergraph. Given a 2-vertex-colouring $c : V \to \{+1, -1\}$ of H, set $c(F) := \sum \{c(v) : v \in F\}$, $F \in \mathcal{F}$. The *discrepancy* of the colouring c is the maximum value of $|c(F)|$ taken over all edges F of H. (The discrepancy is thus a measure of how far the colouring is from being 'balanced'.) By applying Chernoff's Inequality (Exercise 13.3.4), show that every hypergraph with n vertices and n edges has a 2-vertex-colouring whose discrepancy is at most $(2n \log 2n)^{1/2}$.

13.3.6 Let T be a tournament with vertex set $V := \{1, 2, \ldots, n\}$. Given an ordering σ of V and an arc $a = (i, j)$ of T, define $f(a, \sigma) := +1$ if $\sigma(i) < \sigma(j)$ and $f(a, \sigma) := -1$ if $\sigma(i) > \sigma(j)$. Now set $f(T, \sigma) := \sum\{f(a, \sigma) : a \in A(T)\}$. The *fit* of T is the maximum value of $f(T, \sigma)$, taken over all orderings σ of V. (Thus the fit of a tournament is a measure of how close it is to being a transitive tournament, whose fit is equal to $\binom{n}{2}$.) By applying Chernoff's Inequality (Exercise 13.3.4), show that there is a tournament on n vertices whose fit is at most $(n^3 \log n)^{1/2}$.

(J. SPENCER)

13.3.7 . Let $G = G[X, Y]$ be a random bipartite graph obtained by selecting each edge xy with $x \in X$ and $y \in Y$ independently with probability p. Let ϵ be a positive real number. Show that, almost surely:

a) $|d(X, Y) - p| \leq \epsilon$
b) (X, Y) is a regular pair (as defined in Section 12.4).

13.4 Evolution of Random Graphs

THRESHOLD FUNCTIONS

We have seen that the behaviour of a random graph $G \in \mathcal{G}_{n,p}$ changes abruptly at the threshold $p = n^{-1}$: if $p \ll n^{-1}$, then G almost surely has no triangles, whereas if $p \gg n^{-1}$, then G almost surely has at least one triangle. The function n^{-1} is called a *threshold function* for the property of containing a triangle. More generally, if **P** is any *monotone* property of graphs (one which is preserved when edges are added), a *threshold function* for **P** is a function $f(n)$ such that:

▷ if $p \ll f(n)$, then $G \in \mathcal{G}_{n,p}$ almost surely does not have **P**,
▷ if $p \gg f(n)$, then $G \in \mathcal{G}_{n,p}$ almost surely has **P**.

Note that we say 'a threshold function', and not 'the threshold function'. This is because the function is not unique. For example, $10^{10}n^{-1}$ and $n^{-1} + n^{-2}$ are also threshold functions for the property of containing a triangle.

BALANCED GRAPHS

It turns out that every monotone property of graphs has a threshold function (Exercise 13.4.1). For instance, if F is a fixed graph, there is a threshold function for the property of containing a copy of F as a subgraph. We determine such a function in the special case where F is *balanced*, that is, where the average degrees of the proper subgraphs of F do not exceed the average degree $d(F) = 2e(F)/v(F)$ of the graph F itself. Balanced graphs can be recognized in polynomial time (see Exercise 21.4.5 and its hint). They include trees and regular graphs (for instance, cycles and complete graphs).

The following theorem is due to Erdős and Rényi (1960). As in the case of triangles, the proof relies on both Markov's Inequality and Chebyshev's Inequality.

Theorem 13.11 *Let F be a nonempty balanced graph with k vertices and l edges. Then $n^{-k/l}$ is a threshold function for the property of containing F as a subgraph.*

Proof Let $G \in \mathcal{G}_{n,p}$. For each k-subset S of V, let A_S be the event that $G[S]$ contains a copy of F, and let X_S be the indicator random variable for A_S. Set

$$X := \sum \{X_S : S \subseteq V, \ |S| = k\}$$

so that X is the number of k-subsets which span copies of F, and thus is no greater than the total number of copies of F in G.

We first bound the expectation of X. Consider a k-subset S of V. If $G[S]$ contains a copy of F, there is a bijection $f : V(F) \rightarrow S$ such that $f(u)f(v)$ is an edge of $G[S]$ whenever uv is an edge of F. The probability that all these l edges $f(u)f(v)$ are present in $G[S]$ is p^l. Thus $E(X_S) = P(A_S) \geq p^l$. On the other hand, because there are $k!$ bijections $f : V(F) \rightarrow S$, hence $k!$ possible copies of F in $G[S]$ in all, $E(X_S) \leq k!p^l$. (The inequality arises here because copies of F in $G[S]$ may have edges in common, so are not independent.) By linearity of expectation and Exercise 13.2.1a, it follows that

$$\frac{n^k p^l}{k^k} \leq \binom{n}{k} p^l \leq E(X) \leq \binom{n}{k} k! p^l < n^k p^l \tag{13.13}$$

If $p \ll n^{-k/l}$, then $E(X) < n^k p^l \rightarrow 0$ as $n \rightarrow \infty$ and, by Markov's Inequality, G almost surely contains no copy of F.

We now bound the variance of X with the aid of (13.8). As before, the value of $C(X_S, X_T)$ depends only on $|S \cap T|$. If $|S \cap T| = 0$ or $|S \cap T| = 1$, then again $C(X_S, X_T) = 0$. If $|S \cap T| = i$, where $2 \leq i \leq k - 1$, then each copy F_S of F in $G[S]$ meets each copy F_T of F in $G[T]$ in i vertices. Because F is balanced, the intersection $F_S \cap F_T$ of these two copies of F has at most il/k edges, so their union $F_S \cup F_T$ has at least $2l - (il/k)$ edges. The probability that both copies are present in G is therefore at most $p^{2l-(il/k)}$. Because there are $k!$ possible copies F_S of F in $G[S]$ and $k!$ possible copies F_T of F in $G[T]$,

$$C(X_S, X_T) \leq E(X_S X_T) = P(A_S \cap A_T) \leq (k!)^2 p^{2l-(il/k)}$$

Altogether, there are $\binom{n}{k}\binom{n-k}{k-i}$ pairs (S, T) of k-subsets with $|S \cap T| = i$. Since $\binom{n}{k} \leq n^k$ and $\binom{n-k}{k-i} \leq n^{k-i}$,

$$\sum_{S \neq T} C(X_S, X_T) \leq \sum_{i=2}^{k-1} n^{2k-i} (k!)^2 p^{2l-(il/k)} = (k!)^2 \sum_{i=2}^{k-1} (n^k p^l)^2 (np^{l/k})^{-i} \tag{13.14}$$

If $p \gg n^{-k/l}$, then $(np^{l/k})^{-i} \rightarrow 0$ for $i \geq 1$. Moreover, by (13.13), $E(X) \geq n^k p^l / k^k \rightarrow \infty$ as $n \rightarrow \infty$. Thus $E(X) \ll E^2(X)$ and $(n^k p^l)^2 \leq k^{2k} E^2(X)$. Inequalities (13.8) and (13.14) now yield:

$$V(X) \leq E(X) + \sum_{S \neq T} C(X_S, X_T) \ll E^2(X)$$

Applying Corollary 13.8, we conclude that the random graph G almost surely contains a copy of F.   □

THE GIANT COMPONENT

Erdős and Rényi (1960) showed that as the probability $p := p(n)$ increases (while n stays fixed), a typical random graph $G \in \mathcal{G}_{n,p}$ passes through a number of critical phases during which its structure changes abruptly. Besides being very interesting in itself, an understanding of this behaviour can be of great help when applying the probabilistic method. We are content here to give a broad description of the phenomenon without entering into the intricate technical details.

Choosing F to be a tree on k vertices in Theorem 13.11, we see that $n^{-k/(k-1)}$ is a threshold function for G to contain such a tree. Because the number of non-isomorphic trees on k vertices is certainly less than k^{k-2} (the number of labelled trees, Theorem 4.8), this implies that when $p \ll n^{-k/(k-1)}$, G has no component on k or more vertices (as such a component would contain a tree on k vertices), but when $p \gg n^{-k/(k-1)}$, G has such components. Moreover, these components are trees because, again by Theorem 13.11, cycles only appear at the threshold $p = 1/n$. Therefore, at $p = n^{-k/(k-1)}$, G is a forest all of whose components have at most k vertices. These components become larger and larger as k increases. By a more sophisticated probabilistic analysis using branching processes, one can show that when $p = c/n$ with $c < 1$, the largest component of G has size roughly $\log n$, whereas at $p = 1/n$ it already has size about $n^{2/3}$, and there are many components of this size. When $p = c/n$ with $c > 1$, another major transformation takes place, with the emergence of a 'giant component' containing a positive fraction of all n vertices. This dramatic change at the threshold $p = 1/n$ is referred to variously as the *Double Jump* or the *Big Bang*.

Another remarkable evolution occurs at the threshold $p = \log n/n$. At this stage, G may still have isolated vertices (Exercise 13.4.2). When these disappear, G becomes connected and then, almost immediately, hamiltonian (see Section 18.5).

For a thorough account of the evolution of random graphs, we refer the reader to Bollobás (2001) or Janson et al. (2000).

Exercises

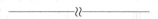

13.4.1 Let **P** be a monotone property of graphs. Assume that **P** is nontrivial, so that for large n, $\overline{K_n}$ does not have **P** whereas K_n does have **P**.

a) Set $P(p) := P(G \in G_{n,p}$ has **P**).
  i) Show that, for every fixed large n, $P(p)$ is a monotone increasing polynomial in p satisfying $P(0) = 0$ and $P(1) = 1$.

ii) Deduce that, for every r, $0 \leq r \leq 1$, there is a p, $0 \leq p \leq 1$, such that $P(p) = r$.

b) Suppose that $P(G \in G_{n,p}$ has $\mathbf{P}) = r$. Let $G_i \in G_{n,p}$ be independent members of $G_{n,p}$, $1 \leq i \leq k$. Show that:

$$P(G \in G_{n,kp} \text{ has } \mathbf{P}) \geq P(\cup_{i=1}^{k} G_i \text{ has } \mathbf{P})$$
$$\geq P(G_i \text{ has } \mathbf{P} \text{ for some } i, 1 \leq i \leq k) = 1 - (1 - r)^k$$

c) For large n, let $f(n)$ satisfy $P(G \in G_{n,f(n)}$ has $\mathbf{P}) = 1/2$, and suppose that $k(n)$ satisfies $f(n)k(n) \leq 1$. Show that:

$$P(G \in G_{n,f(n)/k(n)} \text{ has } \mathbf{P}) \leq 1/k(n)$$

and that

$$P(G \in G_{n,f(n)k(n)} \text{ has } \mathbf{P}) \geq 1 - \frac{1}{2^k}$$

d) Conclude that $f(n)$ is a threshold function for $\mathbf{P}$.   (N. Alon)

13.4.2 Let $G \in \mathcal{G}(n,p)$.

a) Calculate $E(X)$ and $V(X)$, where X is the number of isolated vertices of G.

b) Suppose that $p = (\log n + f(n))/n$, where $f(n) \to \infty$ as $n \to \infty$. Show that almost surely G has no isolated vertices.

c) Suppose that $p = (\log n - f(n))/n$, where $f(n) \to \infty$ as $n \to \infty$. Show that almost surely G has at least one isolated vertex.

d) Obtain similar estimates on p for the nonexistence or existence (almost surely) of vertices of degree one in G.

13.4.3 Let $G \in \mathcal{G}(n,p)$.

a) Calculate $E(X)$, where X is the number of spanning trees of G.

b) Determine a function $p := p(n)$ such that $E(X) \to \infty$ and $P(X = 0) \to 1$ as $n \to \infty$.

c) Conclude that, for this function $p(n)$, $E^2(X) \not\sim E(X^2)$.

13.5 The Local Lemma

In this section, we discuss an important and much used probabilistic tool known as the *Local Lemma*. In order to motivate it, let us consider the problem of colouring a graph G in k colours. We would like to know if G has a proper k-colouring. A naive approach would be to randomly colour G in k colours and then examine whether this random k-colouring is a proper colouring. This will be the case if the ends of each edge of G receive distinct colours. Therefore, if we denote by A_e the event that the ends of e are assigned the same colour, we are interested in the probability $P(\cap_{e \in E} \overline{A_e})$ that none of these 'bad' events occurs. If we can show that this probability is positive, we have a proof that G is k-colourable.

More generally, let $\{A_i : i \in N\}$, be a set of events in a probability space (Ω, P), where $N := \{1, 2, \ldots, n\}$. We regard these events as undesirable or 'bad' events, and are interested in the probability that none of them occurs, namely $P(\cap_{i \in N} \overline{A_i})$. This probability will be positive if the events A_i are independent and each occurs with probability less than one, because then (see Exercise 13.2.2)

$$P(\cap_{i \in N} \overline{A_i}) = \prod_{i \in N} P(\overline{A_i}) = \prod_{i \in N} (1 - P(A_i)) > 0$$

Usually, however, the events under consideration are not independent. In the above example, for instance, if e, f, and g are the edges of a triangle and $k \geq 2$,

$$P(A_e \cap A_f \cap A_g) = k^{-2} > k^{-3} = P(A_e)P(A_f)P(A_g)$$

But all is not lost. Erdős and Lovász (1975) showed that the probability $P(\cap_{i \in N} \overline{A_i})$ will still be positive provided that the events A_i occur with low probability and are, to a sufficient extent, independent of one another.

It is convenient to adopt the following notation. If $\{A_i : i \in S\}$ is a set of events, we denote their intersection $\cap_{i \in S} A_i$ by A_S. Using this notation, an event A_i is *independent* of a set of events $\{A_j : j \in J\}$ if, for all subsets S of J, $P(A_i \cap A_S) = P(A_i)P(A_S)$.

Theorem 13.12   THE LOCAL LEMMA
Let A_i, $i \in N$, be events in a probability space (Ω, P), and let N_i, $i \in N$, be subsets of N. Suppose that, for all $i \in N$,

i) A_i is independent of the set of events $\{A_j : j \notin N_i\}$,
ii) there exists a real number p_i such that $0 < p_i < 1$ and $P(A_i) \leq p_i \prod_{j \in N_i}(1 - p_j)$.

Set $B_i := \overline{A_i}$, $i \in N$. Then, for any two disjoint subsets R and S of N,

$$P(B_R \cap B_S) \geq P(B_R) \prod_{i \in S} (1 - p_i) \tag{13.15}$$

In particular (when $R = \emptyset$ and $S = N$)

$$P(\cap_{i \in N} \overline{A_i}) \geq \prod_{i \in N} (1 - p_i) > 0 \tag{13.16}$$

Remark 13.13 *When the events A_i are independent, the quantities p_i in condition (ii) of the Local Lemma can be regarded as probabilities. When the A_i are not independent, each such 'probability' is reduced by a 'compensation factor' $\prod_{j \in N_i}(1 - p_j)$, according to the 'probabilities' p_j of the events upon which A_i is dependent.*

Proof   We prove (13.15) by induction with respect to the lexicographic order of the pair $(|R \cup S|, |S|)$.

If $S = \emptyset$, then $B_S = \Omega$ and $\prod_{i \in S}(1 - p_i) = 1$, so

$$P(B_R \cap B_S) = P(B_R) \geq P(B_R) \prod_{i \in S}(1 - p_i)$$

If $S = \{i\}$, then $B_S = B_i$ and $\prod_{j \in S}(1 - p_j) = 1 - p_i$. Setting $R_1 := R \setminus N_i$ and $S_1 := R \cap N_i$, we have:

$$P(A_i \cap B_R) \leq P(A_i \cap B_{R_1}) = P(A_i)P(B_{R_1})$$

By assumption, and the fact that $S_1 \subseteq N_i$,

$$P(A_i) \leq p_i \prod_{j \in N_i}(1 - p_j) \leq p_i \prod_{j \in S_1}(1 - p_j)$$

Because $|R_1| + |S_1| = |R| < |R| + |S|$, we have by induction,

$$P(B_{R_1}) \prod_{j \in S_1}(1 - p_j) \leq P(B_{R_1} \cap B_{S_1})$$

Therefore,

$$P(A_i \cap B_R) \leq p_i\, P(B_{R_1} \cap B_{S_1}) = p_i\, P(B_R)$$

and so

$$\begin{aligned}
P(B_R \cap B_S) = P(B_R \cap B_i) &= P(B_R) - P(A_i \cap B_R) \\
&\geq P(B_R) - p_i\, P(B_R) = P(B_R)(1 - p_i)
\end{aligned}$$

If $|S| \geq 2$, we set $R_1 \cup S_1 := S$, where $R_1 \cap S_1 = \emptyset$ and $R_1, S_1 \neq \emptyset$. Then

$$P(B_R \cap B_S) = P(B_R \cap B_{R_1 \cup S_1}) = P(B_R \cap B_{R_1} \cap B_{S_1}) = P(B_{R \cup R_1} \cap B_{S_1})$$

We now apply induction twice. Because $|S_1| < |S|$,

$$P(B_{R \cup R_1} \cap B_{S_1}) \geq P(B_{R \cup R_1}) \prod_{i \in S_1}(1 - p_i) = P(B_R \cap B_{R_1}) \prod_{i \in S_1}(1 - p_i)$$

and since $|R \cup R_1| < |R \cup S|$,

$$P(B_R \cap B_{R_1}) \geq P(B_R) \prod_{i \in R_1}(1 - p_i)$$

Therefore

$$P(B_R \cap B_S) \geq P(B_R) \prod_{i \in R_1}(1 - p_i) \prod_{i \in S_1}(1 - p_i) = P(B_R) \prod_{i \in S}(1 - p_i) \qquad \square$$

Given events A_i, $i \in N$, in a probability space, and subsets N_i of N such that A_i is independent of $\{A_j : j \notin N_i\}$, one may form the digraph D with vertex set N and arc set $\{(i,j) : i \in N, \; j \in N_i\}$. Such a digraph is called a *dependency digraph* (or, if symmetric, a *dependency graph*) for $\{A_i : i \in N\}$. For instance, in our colouring example the event A_e is clearly independent of $\{A_f : f \text{ nonadjacent to } e\}$. Thus the line graph of G is a dependency graph for $\{A_e : e \in E\}$. In general, there are many possible choices of dependency digraph (or graph) for a given set of events; usually, however, one natural choice presents itself, as in this illustration.

For many applications, the following simpler version of the Local Lemma, in which the probabilities of the events A_i have a common upper bound, is sufficient. (Here e denotes the base of natural logarithms.)

Theorem 13.14 THE LOCAL LEMMA – SYMMETRIC VERSION
Let A_i, $i \in N$, be events in a probability space (Ω, P) having a dependency graph with maximum degree d. Suppose that $P(A_i) \leq 1/(e(d+1))$, $i \in N$. Then $P(\cap_{i \in N} \overline{A_i}) > 0$.

Proof Set $p_i := p$, $i \in N$, in the Local Lemma. Now set $p := 1/(d+1)$ in order to maximize $p(1-p)^d$ and apply the inequality $(d/(d+1))^d = (1 - 1/(d+1))^d > e^{-1}$. $\qquad\square$

TWO-COLOURABLE HYPERGRAPHS

Although the proof of the Local Lemma is rather subtle, applying it is frequently a routine matter. We give three examples. Whereas the one below is straightforward (it was, in fact, one of the original applications of the lemma in Erdős and Lovász (1975)), the others require additional ideas.

Theorem 13.15 *Let $H := (V, \mathcal{F})$ be a hypergraph in which each edge has at least k elements and meets at most d other edges. If $e(d + 1) \leq 2^{k-1}$, then H is 2-colourable.*

Proof Consider a random 2-colouring of V. For each edge F, denote by A_F the event that F is monochromatic. Then $P(A_F) = 2 \cdot 2^{-k} = 2^{1-k}$. The result now follows from Theorem 13.14. $\qquad\square$

Corollary 13.16 *Let $H := (V, \mathcal{F})$ be a k-uniform k-regular hypergraph, where $k \geq 9$. Then H is 2-colourable.*

Proof Set $d := k(k - 1)$ in Theorem 13.15. $\qquad\square$

EVEN CYCLES IN DIRECTED GRAPHS

The disarmingly simple question of which digraphs contain directed cycles of even length turns out to be surprisingly hard to answer. Indeed, it remained open for many years before being settled both by McCuaig (2000) and by Robertson et al. (1999). On the other hand, by a clever application of the Local Lemma, Alon and Linial (1989) showed very easily that all diregular digraphs of sufficiently high degree contain such cycles. (Note that the 3-diregular Koh–Tindell digraph of Figure 1.26a has no directed even cycles.)

Theorem 13.17 *Let D be a strict k-diregular digraph, where $k \geq 8$. Then D contains a directed even cycle.*

Proof Consider a random 2-colouring c of V. For each vertex v of D, denote by A_v the event that $c(u) = c(v)$ for all $u \in N^+(v)$. For each colour i, we have $P(A_v) = P(A_v \mid c(v) = i) = 2^{-d}$. Thus A_v is independent of all A_u such that $(\{u\} \cup N^+(u)) \cap N^+(v) = \emptyset$. Setting $d := k^2$ in Theorem 13.14, it follows that there is a 2-colouring of V in which each vertex has an outneighbour of the opposite colour. With respect to this colouring, let uPv be a maximal properly 2-coloured directed path, and let w be an outneighbour of v of the opposite colour. The directed cycle $wPvw$ is then a cycle of even length in D.   □

LINEAR ARBORICITY

A *linear forest* in a graph $G = (V, E)$ is a subgraph each component of which is a path. Particular instances are Hamilton paths and 1-factors. Our aim here is to decompose a graph G into as few linear forests as possible. This number is called the *linear arboricity* of G, denoted $\mathrm{la}(G)$. In the case of a complete graph K_{2n}, the linear arboricity is equal to n, because K_{2n} admits a decomposition into Hamilton paths (Exercise 2.4.6). For an arbitrary graph, the linear arboricity is bounded above by the edge chromatic number, the minimum number of 1-factors into which the graph can be decomposed. We show in Chapter 17 that this number is at most $\Delta + 1$ for any simple graph G. On the other hand, a lower bound for the linear arboricity can be obtained very simply, by counting edges. For example, if G is $2r$-regular, then $m = rn$ and, because no linear forest has more than $n - 1$ edges,

$$\mathrm{la}\,(G) \geq \left\lceil \frac{rn}{n-1} \right\rceil = r + 1 \tag{13.17}$$

We shall apply the Local Lemma to show that this lower bound is tight for $2r$-regular graphs of large enough girth. This result, due to Alon (1988), is based on the following lemma.

Lemma 13.18 *Let $G = (V, E)$ be a simple graph, and let $\{V_1, V_2, \ldots, V_k\}$ be a partition of V into k sets, each of cardinality at least $2e\Delta$. Then there is a stable set S in G such that $|S \cap V_i| = 1$, $1 \leq i \leq k$.*

Proof By deleting vertices from G if necessary, we may assume that $|V_i| = t := \lceil 2e\Delta \rceil$, $1 \leq i \leq k$. We select one vertex v_i at random from each set V_i, $1 \leq i \leq k$, and set $S := \{v_1, v_2, \ldots, v_n\}$.

For an edge e of G, let A_e denote the event that both ends of e belong to S. Then $P(A_e) = 1/t^2$ for all $e \in E$, and A_e is dependent only on those events A_f such that an end of f lies in the same set V_i as an end of e. There are fewer than $2t\Delta$ such events. Setting $d := 2t\Delta - 1$ in Theorem 13.14, we see that with positive probability the set S is stable, provided that $1/t^2 \leq 1/(e2t\Delta)$; that is, $t \geq 2e\Delta$. Because $t = \lceil 2e\Delta \rceil$, we conclude that there does indeed exist such a stable set S.   □

Theorem 13.19 *Let* $G = (V, E)$ *be a simple $2r$-regular graph with girth at least* $2e(4r - 2)$. *Then* $la(G) = r + 1$.

Proof By (13.17), we must show that $la(G) \leq r + 1$. We make use of the fact that every regular graph of even degree admits a decomposition into 2-factors (see Exercise 16.4.16).

Consider such a decomposition $\{F_1, F_2, \ldots, F_r\}$ of G, and let C_i, $1 \leq i \leq k$, be the constituent cycles of these 2-factors. Set $V_i := E(C_i)$, $1 \leq i \leq k$. The line graph H of G is $(4r - 2)$-regular. Because G has girth at least $2e(4r - 2)$, $\{V_1, V_2, \ldots, V_k\}$ is a partition of $V(H)$ into k sets, each of cardinality at least $2e(4r - 2)$. Applying Lemma 13.18 to H, we deduce that H has a stable set S meeting each set V_i in one vertex. The subgraphs $L_i := F_i \setminus M$, $1 \leq i \leq r$, are therefore linear forests in G, as is $L_0 := G[M]$, so $\{L_0, L_1, \ldots, L_r\}$ is a decomposition of G into $r + 1$ linear forests. $\qquad \square$

We present an application of the asymmetric version of the Local Lemma in Chapter 18.

Exercises

13.5.1

a) Suppose that $e(\binom{k}{2}\binom{n}{k-2} + 1) \leq 2^{\binom{k}{2}} - 1$. Show that $r(k, k) > n$.
b) Deduce that $r(k, k) > c\,k\,2^{(k+1)/2}$, where $c \to e^{-1}$ as $k \to \infty$.

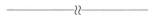

13.5.2 Let D be a strict digraph with maximum indegree Δ^- and minimum out-degree δ^+. Suppose that $e(\Delta^- \delta^+ + 1) \leq (k/(k-1))^{\delta^+}$. Show that D contains a directed cycle of length congruent to 0 (mod k).   (N. ALON AND N. LINIAL)

13.6 Related Reading

Probabilistic Models

In this chapter, we have focussed on properties of the probability space $\mathcal{G}_{n,p}$, in which each edge is chosen independently with probability p. We have also encountered the model in which each vertex of a particular graph is chosen independently with a given probability p. There are several other natural models of random graphs. For instance, one may consider the space $\mathcal{G}_{n,m}$ consisting of all labelled graphs on n vertices and m edges, each such graph being equiprobable. This is the model studied by Erdős and Rényi (1959, 1960) in their pioneering work on the evolution of random graphs. As might be expected, there is a close connection between the properties of $\mathcal{G}_{n,m}$ and $\mathcal{G}_{n,p}$ when $p = m/\binom{n}{2}$. Another much-studied model is that of random k-regular graphs, introduced by Bollobás (1980).

In discussing the evolution of random graphs, the concept of a *graph process* is illuminating. Here, one starts with an empty graph on n vertices, and edges are added one at a time, each potential new edge having the same probability of being chosen. One is then interested in the time (as measured by the number of edges) at which the evolving graph acquires a particular monotone property, such as being connected, or hamiltonian. This is a very fine measure of the evolution of a random graph. It can be shown, for example, that almost surely a graph becomes connected as soon as it loses its last isolated vertex, and becomes hamiltonian as soon as all of its vertices have degree two or more; see Bollobás (2001).

More recently, the Internet, molecular biology, and various other applied areas, have fuelled interest and research into diverse models of random graphs, designed specifically to reflect the particular structures and evolution of such Web graphs, biological networks and random geometric graphs; see, for example, Dousse et al. (2006), Kumar et al. (2000), or Leonardi (2004).

SHARP THRESHOLD FUNCTIONS

In order to gain a clearer picture of the behaviour of a random graph $G \in \mathcal{G}_{n,p}$ at a critical threshold such as $p = (\log n)/n$, it is common to introduce and study sharper threshold functions. Erdős and Rényi (1960) showed, for example, that if $p = (\log n + c)/n$, where $c \in \mathbb{R}$, then the probability that G is connected tends to $e^{-e^{-c}}$ as n tends to infinity. A thorough discussion of sharp threshold functions can be found in Bollobás (2001), Janson et al. (2000), or Palmer (1985).

CONCENTRATION INEQUALITIES

We have discussed in this chapter the most basic concentration bound, namely Chebyshev's Inequality. Another such bound, Chernoff's Inequality, is described in Exercise 13.3.4. There exist several variants of these bounds, and indeed many further concentration inequalities, each designed to handle a certain class of problems. *Azuma's Inequality* and *Talagrand's Inequality*, for instance, are especially useful in connection with colouring problems. We refer the reader to Alon and Spencer (2000) or Molloy and Reed (2002) for the statements of these inequalities and examples of their application.

14

Vertex Colourings

Contents

14.1 Chromatic Number

Recall that a *k-vertex-colouring*, or simply a *k-colouring*, of a graph $G = (V, E)$ is a mapping $c : V \to S$, where S is a set of k *colours*; thus, a k-colouring is an assignment of k colours to the vertices of G. Usually, the set S of colours is taken to be $\{1, 2, \ldots, k\}$. A colouring c is *proper* if no two adjacent vertices are assigned the same colour. Only loopless graphs admit proper colourings.

Alternatively, a k-colouring may be viewed as a partition $\{V_1, V_2, \ldots, V_k\}$ of V, where V_i denotes the (possibly empty) set of vertices assigned colour i. The sets

V_i are called the *colour classes* of the colouring. A proper k-colouring is then a k-colouring in which each colour class is a stable set. In this chapter, we are only concerned with proper colourings. It is convenient, therefore, to refer to a proper colouring as a 'colouring' and to a proper k-colouring as a 'k-colouring'.

A graph is *k-colourable* if it has a k-colouring. Thus a graph is 1-colourable if and only if it is empty, and 2-colourable if and only if it is bipartite. Clearly, a loopless graph is k-colourable if and only if its underlying simple graph is k-colourable. Therefore, in discussing vertex colourings, we restrict our attention to simple graphs.

The minimum k for which a graph G is k-colourable is called its *chromatic number*, and denoted $\chi(G)$. If $\chi(G) = k$, the graph G is said to be *k-chromatic*. The triangle, and indeed all odd cycles, are easily seen to be 3-colourable. On the other hand, they are not 2-colourable because they are not bipartite. They therefore have chromatic number three: they are 3-chromatic. A 4-chromatic graph known as the *Hajós graph* is shown in Figure 14.1. The complete graph K_n has chromatic number n because no two vertices can receive the same colour. More generally, every graph G satisfies the inequality

$$\chi \geq \frac{n}{\alpha} \tag{14.1}$$

because each colour class is a stable set, and therefore has at most α vertices.

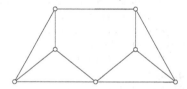

Fig. 14.1. The Hajós graph: a 4-chromatic graph

Colouring problems arise naturally in many practical situations where it is required to partition a set of objects into groups in such a way that the members of each group are mutually compatible according to some criterion. We give two examples of such problems. Others will no doubt occur to the reader.

Example 14.1   EXAMINATION SCHEDULING
The students at a certain university have annual examinations in all the courses they take. Naturally, examinations in different courses cannot be held concurrently if the courses have students in common. How can all the examinations be organized in as few parallel sessions as possible? To find such a schedule, consider the graph G whose vertex set is the set of all courses, two courses being joined by an edge if they give rise to a conflict. Clearly, stable sets of G correspond to conflict-free groups of courses. Thus the required minimum number of parallel sessions is the chromatic number of G.

Example 14.2 CHEMICAL STORAGE

A company manufactures n chemicals $C_1, C_2, \ldots, C_n$. Certain pairs of these chemicals are incompatible and would cause explosions if brought into contact with each other. As a precautionary measure, the company wishes to divide its warehouse into compartments, and store incompatible chemicals in different compartments. What is the least number of compartments into which the warehouse should be partitioned? We obtain a graph G on the vertex set $\{v_1, v_2, \ldots, v_n\}$ by joining two vertices v_i and v_j if and only if the chemicals C_i and C_j are incompatible. It is easy to see that the least number of compartments into which the warehouse should be partitioned is equal to the chromatic number of G.

If H is a subgraph of G and G is k-colourable, then so is H. Thus $\chi(G) \geq \chi(H)$. In particular, if G contains a copy of the complete graph K_r, then $\chi(G) \geq r$. Therefore, for any graph G,

$$\chi \geq \omega \tag{14.2}$$

The odd cycles of length five or more, for which $\omega = 2$ and $\chi = 3$, show that this bound for the chromatic number is not sharp. More surprisingly, as we show in Section 14.3, there exist graphs with arbitrarily high girth and chromatic number.

A GREEDY COLOURING HEURISTIC

Because a graph is 2-colourable if and only if it is bipartite, there is a polynomial-time algorithm (for instance, using breadth-first search) for deciding whether a given graph is 2-colourable. In sharp contrast, the problem of 3-colourability is already $\mathcal{NP}$-complete. It follows that the problem of finding the chromatic number of a graph is $\mathcal{NP}$-hard. In practical situations, one must therefore be content with efficient heuristic procedures which perform reasonably well. The most natural approach is to colour the vertices in a greedy fashion, as follows.

Heuristic 14.3 THE GREEDY COLOURING HEURISTIC

  INPUT: a graph G
  OUTPUT: a colouring of G
  1. *Arrange the vertices of G in a linear order: $v_1, v_2, \ldots, v_n$.*
  2. *Colour the vertices one by one in this order, assigning to v_i the smallest positive integer not yet assigned to one of its already-coloured neighbours.*

It should be stressed that the number of colours used by this greedy colouring heuristic depends very much on the particular ordering chosen for the vertices. For example, if $K_{n,n}$ is a complete bipartite graph with parts $X := \{x_1, x_2, \ldots, x_n\}$ and $Y := \{y_1, y_2, \ldots, y_n\}$, then the bipartite graph $G[X, Y]$ obtained from this graph by deleting the perfect matching $\{x_i y_i : 1 \leq i \leq n\}$ would require n colours if the vertices were listed in the order $x_1, y_1, x_2, y_2, \ldots, x_n, y_n$. On the other hand, only two colours would be needed if the vertices were presented in the order $\{x_1, x_2, \ldots, x_n, y_1, y_2, \ldots, y_n\}$; indeed there is always an ordering which yields an optimal colouring (Exercise 14.1.9). The problem is that it is hard to know in advance which orderings will produce optimal colourings.

Nevertheless, the number of colours used by the greedy heuristic is never greater than $\Delta + 1$, regardless of the order in which the vertices are presented. When a vertex v is about to be coloured, the number of its neighbours already coloured is clearly no greater than its degree $d(v)$, and this is no greater than the maximum degree, Δ. Thus one of the colours $1, 2, \ldots, \Delta + 1$ will certainly be available for v. We conclude that, for any graph G,

$$\chi \le \Delta + 1 \tag{14.3}$$

In other words, every k-chromatic graph has a vertex of degree at least $k - 1$. In fact, every k-chromatic graph has at least k vertices of degree at least $k - 1$ (Exercise 14.1.3b).

The bound (14.3) on the chromatic number gives essentially no information on how many vertices of each colour there are in a $(\Delta + 1)$-colouring. A far-reaching strengthening of inequality (14.3) was obtained by Hajnal and Szemerédi (1970), who showed that every graph G admits a balanced $(\Delta + 1)$-colouring, that is, one in which the numbers of vertices of each colour differ by at most one. A shorter proof of this theorem was found by Kierstead and Kostochka (2008).

BROOKS' THEOREM

Although the bound (14.3) on the chromatic number is best possible, being attained by odd cycles and complete graphs, Brooks (1941) showed that these are the only connected graphs for which equality holds.

Our proof of Brooks' Theorem is similar in spirit to one given by Lovász (1975b), but makes essential use of DFS-trees. In particular, we appeal to a result of Chartrand and Kronk (1968), who showed that cycles, complete graphs, and complete bipartite graphs whose parts are of equal size are the only graphs with the property that every DFS-tree is a Hamilton path rooted at one of its ends (see Exercise 6.1.11).

Theorem 14.4   BROOKS' THEOREM
If G is a connected graph, and is neither an odd cycle nor a complete graph, then $\chi \le \Delta$.

Proof   Suppose first that G is not regular. Let x be a vertex of degree δ and let T be a search tree of G rooted at x. We colour the vertices with the colours $1, 2, \ldots, \Delta$ according to the greedy heuristic, selecting at each step a leaf of the subtree of T induced by the vertices not yet coloured, assigning to it the smallest available colour, and ending with the root x of T. When a vertex v different from x is about to be coloured, it is adjacent in T to at least one uncoloured vertex, and so is adjacent in G to at most $d(v) - 1 \le \Delta - 1$ coloured vertices. It is therefore assigned one of the colours $1, 2, \ldots, \Delta$. Finally, when x is coloured, it, too, is assigned one of the colours $1, 2, \ldots, \Delta$, because $d(x) = \delta \leq \Delta - 1$. The greedy heuristic therefore produces a Δ-colouring of G.

Suppose now that G is regular. If G has a cut vertex x, then $G = G_1 \cup G_2$, where G_1 and G_2 are connected and $G_1 \cap G_2 = \{x\}$. Because the degree of x in G_i is less than $\Delta(G)$, neither subgraph G_i is regular, so $\chi(G_i) \leq \Delta(G_i) = \Delta(G)$, $i = 1, 2$, and $\chi(G) = \max\{\chi(G_1), \chi(G_2)\} \leq \Delta(G)$ (Exercise 14.1.2). We may assume, therefore, that G is 2-connected.

If every depth-first search tree of G is a Hamilton path rooted at one of its ends, then G is a cycle, a complete graph, or a complete bipartite graph $K_{n,n}$ (Exercise 6.1.11). Since, by hypothesis, G is neither an odd cycle nor a complete graph, $\chi(G) = 2 \leq \Delta(G)$.

Suppose, then, that T is a depth-first search tree of G, but not a path. Let x be a vertex of T with at least two children, y and z. Because G is 2-connected, both $G - y$ and $G - z$ are connected. Thus y and z are either leaves of T or have proper descendants which are joined to ancestors of x. It follows that $G' := G - \{y, z\}$ is connected. Consider a search tree T' with root x in G'. By colouring y and z with colour 1, and then the vertices of T' by the greedy heuristic as above, ending with the root x, we obtain a Δ-colouring of G.   □

COLOURINGS OF DIGRAPHS

A (proper) *vertex colouring* of a digraph D is simply a vertex colouring of its underlying graph G, and its *chromatic number* $\chi(D)$ is defined to be the chromatic number $\chi(G)$ of G. Why, then, consider colourings of digraphs? It turns out that the chromatic number of a digraph provides interesting information about its subdigraphs. The following theorem of Gallai (1968a) and Roy (1967) tells us that digraphs with high chromatic number always have long directed paths. It can be viewed as a common generalization of a theorem about chains in posets (see Exercise 2.1.22) and Rédei's Theorem on directed Hamilton paths in tournaments (Theorem 2.3).

Theorem 14.5 THE GALLAI–ROY THEOREM
Every digraph D contains a directed path with χ vertices.

Proof Let k be the number of vertices in a longest directed path of D. Consider a maximal acyclic subdigraph D' of D. Because D' is a subdigraph of D, each directed path in D' has at most k vertices. We k-colour D by assigning to vertex v the colour $c(v)$, where $c(v)$ is the number of vertices of a longest directed path in D' starting at v. Let us show that this colouring is proper.

Consider any arc (u, v) of D. If (u, v) is an arc of D', let vPw be a longest directed v-path in D'. Then $u \notin V(P)$, otherwise $vPuv$ would be a directed cycle in D'. Thus $uvPw$ is a directed u-path in D', implying that $c(u) > c(v)$.

If (u, v) is not an arc of D', then $D' + (u, v)$ contains a directed cycle, because the subdigraph D' is maximally acyclic, so D' contains a directed (v, u)-path P. Let Q be a longest directed u-path in D'. Because D' is acyclic, $V(P) \cap V(Q) = \{u\}$.

Thus PQ is a directed v-path in D', implying that $c(v) > c(u)$. In both cases, $c(u) \neq c(v)$. □

The Gallai–Milgram and Gallai-Roy Theorems (12.2 and 14.5) bear a striking formal resemblance. By interchanging the roles of directed paths and stable sets, one theorem is transformed into the other: stable sets become directed paths, and path partitions become vertex colourings (which are partitions into stable sets). This correspondence does not, however, extend to the stronger versions of these theorems in terms of orthogonality. Although the proof of the Gallai-Roy Theorem shows that there exists in any digraph some colouring $\mathcal{C}$ and some directed path P which are orthogonal (meaning that P is orthogonal to every colour class of $\mathcal{C}$), it is not true that every minimum colouring is orthogonal to some directed path (Exercise 14.1.21). Nonetheless, a possible common generalization of the two theorems was proposed by Linial (1981) (see Exercise 14.1.22) and conjectured in a stronger form by Berge (1982).

Let k be a positive integer. A path partition $\mathcal{P}$ is k-optimal if it minimizes the function $\sum\{\min\{v(P), k\} : P \in \mathcal{P}\}$, and a partial k-colouring of a graph or digraph is a family of k disjoint stable sets. In particular, a 1-optimal path partition is one that is optimal, and a partial 1-colouring is simply a stable set.

The concept of orthogonality of paths and stable sets is extended as follows. A path partition $\mathcal{P}$ and partial k-colouring $\mathcal{C}$ are orthogonal if every directed path $P \in \mathcal{P}$ meets $\min\{v(P), k\}$ different colour classes of $\mathcal{C}$. We can now state the conjecture proposed by Berge.

THE PATH PARTITION CONJECTURE

Conjecture 14.6 *Let D be a digraph, k a positive integer, and $\mathcal{P}$ a k-optimal path partition of D. Then there is a partial k-colouring of D which is orthogonal to $\mathcal{P}$.*

The Path Partition Conjecture has been proved for $k = 1$ by Linial (1978) and for $k = 2$ by Berger and Hartman (2008). It has also been established for acyclic digraphs, by Aharoni et al. (1985) and Cameron (1986), and for digraphs containing no directed path with more than k vertices, by Berge (1982) We refer the reader to the survey by Hartman (2006) for a full discussion of this conjecture and of related questions.

Exercises

14.1.1 CHVÁTAL GRAPH
The *Chvátal graph*, shown in Figure 14.2, is a 4-regular graph of girth four on twelve vertices. Show that this graph is 4-chromatic.   (V. CHVÁTAL)

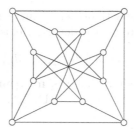

Fig. 14.2. The Chvátal graph: a 4-chromatic 4-regular graph of girth four

⋆**14.1.2** Show that $\chi(G) = \max\{\chi(B) : B$ a block of $G\}$.

⋆**14.1.3**

a) In a k-colouring of a k-chromatic graph, show that there is a vertex of each colour which is adjacent to vertices of every other colour.

b) Deduce that every k-chromatic graph has at least k vertices of degree at least $k - 1$.

14.1.4 Show that a graph is $k_1 k_2$-colourable if and only if it may be expressed as the union of a k_1-colourable graph and a k_2-colourable graph.   (S.A. BURR)

14.1.5 k-DEGENERATE GRAPH

A graph is k-*degenerate* if it can be reduced to K_1 by repeatedly deleting vertices of degree at most k.

a) Show that a graph is k-degenerate if and only if every subgraph has a vertex of degree at most k.

b) Characterize the 1-degenerate graphs.

c) Show that every k-degenerate graph is $(k + 1)$-colourable.

d) Using Exercise 14.1.4, deduce that the union of a k-degenerate graph and an ℓ-degenerate graph is $(k + 1)(\ell + 1)$-colourable.

14.1.6 Establish the following bounds on the chromatic number of the Kneser graph $KG_{m,n}$.
$$\frac{n}{m} \leq \chi(KG_{m,n}) \leq n - 2m + 2$$

(Lovász (1978) proved the conjecture of Kneser (1955) that the upper bound is sharp; see, also, Bárány (1978) and Greene (2002).)

14.1.7 Show that, for any graph G, $\chi \geq n^2/(n^2 - 2m)$.

14.1.8 Let G be a graph in which any two odd cycles intersect. Show that:

a) $\chi \leq 5$,
b) if $\chi = 5$, then G contains a copy of K_5.

14.1.9 Given any graph G, show that there is an ordering of its vertices such that the greedy heuristic, applied to that ordering, yields a colouring with χ colours.

14.1.10 Let G have degree sequence $(d_1, d_2, \ldots, d_n)$, where $d_1 \geq d_2 \geq \cdots \geq d_n$.

a) Using a greedy heuristic, show that $\chi \leq \max\{\min\{d_i + 1, i\} : 1 \leq i \leq n\}$.
b) Deduce that $\chi \leq \lceil (2m)^{1/2} \rceil$.   (D.J.A. WELSH AND M.B. POWELL)

14.1.11

a) Show that $\chi(G)\chi(\overline{G}) \geq n$.
b) Using Exercise 14.1.10, deduce that $2\sqrt{n} \leq \chi(G) + \chi(\overline{G}) \leq n + 1$.
            (E.A. NORDHAUS AND J.W. GADDUM)

14.1.12 Let k be a positive integer, and let G be a graph which contains no cycle of length $1 \pmod{k}$. Show that G is k-colourable.   (ZS. TUZA)

14.1.13 CATLIN GRAPH
The *composition* $G[H]$ was defined in Exercise 12.1.12.

a) Show that $\chi(G[H]) \leq \chi(G)\chi(H)$, for any two graphs G and H.
b) The graph $C_5[K_3]$ shown in Figure 14.3 is known as the *Catlin graph*. Show that $\chi(C_5[K_3]) < \chi(C_5)\chi(K_3)$.   (P. CATLIN)

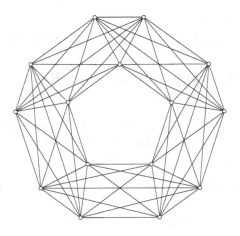

Fig. 14.3. The Catlin graph $C_5[K_3]$

14.1.14 Let G be the graph $C_5[K_n]$.

a) Show that $\chi = \lceil \frac{5n}{2} \rceil$.
b) Deduce that $\chi = \lceil (\omega + \Delta + 1)/2 \rceil$.   (A. KOSTOCHKA)

⋆**14.1.15**

a) Show that every graph G has an orientation each of whose induced subdigraphs has a kernel.
b) Consider any such orientation D. Show that G is $(\Delta^+(D)+1)$-colourable.
c) Deduce inequality (14.3).

14.1.16 THE ERDŐS–SZEKERES THEOREM

a) Let D be a digraph with $\chi \geq kl+1$, and let f be a real-valued function defined on V. Show that D contains either a directed path $(u_0, u_1, \ldots, u_k)$ with $f(u_0) \leq f(u_1) \leq \cdots \leq f(u_k)$ or a directed path $(v_0, v_1, \ldots, v_l)$ with $f(v_0) > f(v_1) > \cdots > f(v_l)$.   (V. CHVÁTAL AND J. KOMLÓS)
b) Deduce that any sequence of $kl+1$ distinct integers contains either an increasing subsequence of $k+1$ terms or a decreasing sequence of $l+1$ terms.
                              (P. ERDŐS AND G. SZEKERES)

14.1.17 Let G be an undirected graph. Show that

$$\chi(G) = \min\,\{\lambda(D) : D \text{ an orientation of } G\}$$

where $\lambda(D)$ denotes the number of vertices in a longest directed path of D. (Compare with Exercise 12.1.10.)

14.1.18 WEAK PRODUCT
The *weak product* of graphs G and H is the graph $G \times H$ with vertex set $V(G) \times V(H)$ and edge set $\{((u, u'), (v, v')) : (u, v) \in E(G), (u', v') \in E(H)\}$. Show that, for any two graphs G and H, $\chi(G \times H) \leq \min\{\chi(G), \chi(H)\}$.   (S. HEDETNIEMI)

14.1.19 CHROMATIC NUMBER OF A HYPERGRAPH
The *chromatic number* $\chi(H)$ of a hypergraph $H := (V, \mathcal{F})$ is the least number of colours needed to colour its vertices so that no edge of cardinality more than one is monochromatic. (This is one of several ways of defining the chromatic number of a hypergraph; it is often referred to as the *weak chromatic number*.) Determine the chromatic number of:

a) the Fano hypergraph (Figure 1.15a),
b) the Desargues hypergraph (Figure 1.15b).

14.1.20

a) Show that the Hajós graph (Figure 14.1) is a unit-distance graph.
   (P. O'Donnell has shown that there exists a 4-chromatic unit-distance graph of arbitrary girth.)
b) Let G be a unit-distance graph. Show that $\chi \leq 7$ by considering a plane hexagonal lattice and finding a suitable 7-face colouring of it.

14.1.21

a) A directed path P and colouring C of a digraph are *orthogonal* if P is orthogonal to every colour class of C. Show that, in any digraph, there exists a directed path P and a colouring C which are orthogonal.
b) By considering an appropriate orientation of a 5-cycle, show that there does not necessarily exist a directed path P which is orthogonal to every minimum colouring.
(K. CAMERON)

14.1.22 LINIAL'S CONJECTURE
A partial k-colouring C of a digraph D is *optimal* if the number of coloured vertices, $\sum_{C \in \mathcal{C}} |C|$, is as large as possible. Let $\mathcal{P}$ be a k-optimal path partition and $\mathcal{C}$ an optimal k-colouring of a digraph D. Define

$$\pi_k(D) := \sum_{P \in \mathcal{P}} \min\{v(P), k\} \quad \text{and} \quad \alpha_k(D) := \sum_{C \in \mathcal{C}} |C|$$

(In particular, $\pi_1 = \pi$, the number of paths in an optimal path partition of D, and $\alpha_1 = \alpha$.) Linial's Conjecture asserts that $\pi_k \leq \alpha_k$ for all digraphs D and all positive integers k. Deduce Linial's Conjecture from Berge's Path Partition Conjecture (14.6).

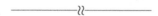

14.1.23 Show that:

a) if $\chi(G) = 2k$, then G has a bipartite subgraph with at least $mk/(2k-1)$ edges,
b) if $\chi(G) = 2k+1$, then G has a bipartite subgraph with at least $m(k+1)/(2k+1)$ edges.
(L.D. ANDERSEN, D. GRANT AND N. LINIAL)

14.1.24 Let $G := (V, E)$ be a graph, and let $f(G)$ be the number of proper k-colourings of G. By applying the inequality of Exercise 13.2.3, show that

$$k^n \left(1 - \frac{m}{k}\right) \leq f(G) \leq k^n \left(1 - \frac{m}{k+m-1}\right)$$

14.1.25 Let G be a 5-regular graph on $4k$ vertices, the union of a Hamilton cycle C and k disjoint copies $G_1, G_2, \ldots, G_k$ of K_4. Let F and F' be the two 1-factors of G contained in C, and let F_i be a 1-factor of G_i, $1 \leq i \leq k$. By combining a 2-vertex colouring of $F \cup_i F_i$ with a 2-vertex colouring of $F' \cup_i F_i'$, where F_i' is an appropriately chosen 1-factor of G_i, $1 \leq i \leq k$, deduce that $\chi(G) = 4$.
(N. ALON)

14.1.26 Let G be a 3-chromatic graph on n vertices. Show how to find, in polynomial time, a proper colouring of G using no more than $3\sqrt{n}$ colours.
(A. WIGDERSON)
(Blum and Karger (1997) have described a polynomial-time algorithm for colouring a 3-chromatic graph on n vertices using $O(n^{3/14})$ colours.)

14.1.27 Let G be a simple connected claw-free graph with $\alpha \geq 3$.

a) Show that $\Delta \leq 4(\omega - 1)$ by induction on n, proceeding as follows.
   - ▷ If G is separable, apply induction.
   - ▷ If G is 2-connected, let x be a vertex of degree Δ and set $X := N(x) \cup \{x\}$. Show that $\alpha(G[X]) = 2$. Deduce that $Y := V \setminus X \neq \emptyset$.
   - ▷ If $\alpha(G - v) \geq 3$ for some $v \in Y$, apply induction.
   - ▷ If $\alpha(G - v) = 2$ for all $v \in Y$, show that Y consists either of a single vertex or of two nonadjacent vertices.
   - ▷ Show that, in the former case, $N(x)$ is the union of four cliques, and in the latter case, the union of two cliques.
   - ▷ Conclude.

b) Deduce that $\chi \leq 4(\omega - 1)$.   (M. CHUDNOVSKY AND P.D. SEYMOUR)

(Chudnovsky and Seymour have in fact shown that $\chi \leq 2\omega$.)

14.1.28

a) Show that every digraph D contains a spanning branching forest F in which the sets of vertices at each level are stable sets of D (the vertices at level zero being the roots of the components of F).

b) Deduce the Gallai–Roy Theorem (14.5).

c) A (k, l)-*path* is an oriented path of length $k + l$ obtained by identifying the terminal vertices of a directed path of length k and a directed path of length l. Let D be a digraph and let k and l be positive integers such that $k + l = \chi$. Deduce from (a) that D contains either a $(k, l - 1)$-path or a $(k - 1, l)$-path.

(A. EL-SAHILI AND M. KOUIDER)

14.1.29 Let k be a positive integer. Show that every infinite k-chromatic graph contains a finite k-chromatic subgraph.   (N.G. DE BRUIJN AND P. ERDŐS)

14.2 Critical Graphs

When dealing with colourings, it is helpful to study the properties of a special class of graphs called colour-critical graphs. We say that a graph G is *colour-critical* if $\chi(H) < \chi(G)$ for every proper subgraph H of G. Such graphs were first investigated by Dirac (1951). Here, for simplicity, we abbreviate the term 'colour-critical' to 'critical'. A k-*critical* graph is one that is k-chromatic and critical. Note that a minimal k-chromatic subgraph of a k-chromatic graph is k-critical, so every k-chromatic graph has a k-critical subgraph. The *Grötzsch graph*, a 4-critical graph discovered independently by Grötzsch (1958/1959) and Mycielski (1955) is shown in Figure 14.4 (see Exercise 14.3.1).

Theorem 14.7 *If G is k-critical, then $\delta \geq k - 1$.*

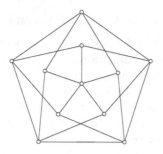

Fig. 14.4. The Grötzsch graph: a 4-critical graph

Proof By contradiction. Let G be a k-critical graph with $\delta < k - 1$, and let v be a vertex of degree δ in G. Because G is k-critical, $G - v$ is $(k-1)$-colourable. Let $\{V_1, V_2, \ldots, V_{k-1}\}$ be a $(k-1)$-colouring of $G - v$. The vertex v is adjacent to $\delta < k - 1$ vertices. It therefore must be nonadjacent in G to every vertex in some V_j. But then $\{V_1, V_2, \ldots, V_j \cup \{v\}, \ldots, V_{k-1}\}$ is a $(k-1)$-colouring of G, a contradiction. Thus $\delta \geq k - 1$. $\square$

Theorem 14.7 implies that every k-chromatic graph has at least k vertices of degree at least $k - 1$, as noted already in Section 14.1.

Let S be a vertex cut of a connected graph G, and let the components of $G - S$ have vertex sets $V_1, V_2, \ldots, V_t$. Recall that the subgraphs $G_i := G[V_i \cup S]$ are the *S-components* of G. We say that colourings of $G_1, G_2, \ldots, G_t$ *agree* on S if, for every $v \in S$, vertex v is assigned the same colour in each of the colourings.

Theorem 14.8 *No critical graph has a clique cut.*

Proof By contradiction. Let G be a k-critical graph. Suppose that G has a clique cut S. Denote the S-components of G by $G_1, G_2, \ldots, G_t$. Because G is k-critical, each G_i is $(k-1)$-colourable. Furthermore, because S is a clique, the vertices of S receive distinct colours in any $(k-1)$-colouring of G_i. It follows that there are $(k-1)$-colourings of $G_1, G_2, \ldots, G_t$ which agree on S. These colourings may be combined to yield a $(k-1)$-colouring of G, a contradiction. $\square$

Corollary 14.9 *Every critical graph is nonseparable.* $\square$

By Theorem 14.8, if a k-critical graph has a 2-vertex cut $\{u, v\}$, then u and v cannot be adjacent. We say that a $\{u, v\}$-component G_i of G is of *type 1* if every $(k-1)$-colouring of G_i assigns the same colour to u and v, and of *type 2* if every $(k-1)$-colouring of G_i assigns distinct colours to u and v. Figure 14.5 depicts the $\{u, v\}$-components of the Hajós graph with respect to a 2-vertex cut $\{u, v\}$. Observe that there are just two $\{u, v\}$-components, one of each type. Dirac (1953) showed that this is always so in critical graphs.

Theorem 14.10 *Let G be a k-critical graph with a 2-vertex cut $\{u, v\}$, and let e be a new edge joining u and v. Then:*

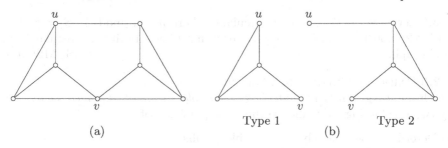

Fig. 14.5. (a) A 2-vertex cut $\{u, v\}$ of the Hajós graph, (b) its two $\{u, v\}$-components

1. $G = G_1 \cup G_2$, where G_i is a $\{u, v\}$-component of G of type i, $i = 1, 2$,
2. both $H_1 := G_1 + e$ and $H_2 := G_2 / \{u, v\}$ are k-critical.

Proof

1. Because G is critical, each $\{u, v\}$-component of G is $(k-1)$-colourable. Now there cannot exist $(k-1)$-colourings of these $\{u, v\}$-components all of which agree on $\{u, v\}$, as such colourings would together yield a $(k-1)$-colouring of G. Therefore there are two $\{u, v\}$-components G_1 and G_2 such that no $(k-1)$-colouring of G_1 agrees with any $(k-1)$-colouring of G_2. Clearly one, say G_1, must be of type 1, and the other, G_2, of type 2. Because G_1 and G_2 are of different types, the subgraph $G_1 \cup G_2$ of G is not $(k-1)$-colourable. The graph G being critical, we deduce that $G = G_1 \cup G_2$.
2. Because G_1 is of type 1, H_1 is k-chromatic. We prove that H_1 is critical by showing that, for every edge f of H_1, the subgraph $H_1 \setminus f$ is $(k-1)$-colourable. This is clearly so if $f = e$, since in this case $H_1 \setminus e = G_1$. Let f be some other edge of H_1. In any $(k-1)$-colouring of $G \setminus f$, the vertices u and v receive different colours, because G_2 is a subgraph of $G \setminus f$. The restriction of such a colouring to the vertices of G_1 is a $(k-1)$-colouring of $H_1 \setminus f$. Thus H_1 is k-critical. An analogous argument shows that H_2 is k-critical. $\qquad\square$

Exercises

14.2.1 Show that $\chi(G) \leq 1 + \max \{\delta(F) : F \subseteq G\}$.

14.2.2 Show that the only 1-critical graph is K_1, the only 2-critical graph is K_2, and the only 3-critical graphs are the odd cycles of length three or more.

14.2.3 Show that the Chvátal graph (Figure 14.2) is 4-critical.

14.2.4 Let G be the 4-regular graph derived from the cartesian product of a triangle $x_1 x_2 x_3 x_1$ and a path $y_1 y_2 y_3 y_4 y_5$ by identifying the vertices (x_1, y_1) and (x_1, y_5), (x_2, y_1) and (x_3, y_5), and (x_3, y_1) and (x_2, y_5). Show that G is 4-critical.

(T. Gallai)

14.2.5 Let $G = \mathrm{CG}\,(\mathbb{Z}_n, S)$ be a circulant, where $n \equiv 1\,(\mathrm{mod}\,3)$, $|S| = k$, $1 \in S$, and $i \equiv 2\,(\mathrm{mod}\,3)$ for all $i \in S$, $i \neq 1$. Show that G is a 4-critical k-regular k-connected graph. (L.S. MELNIKOV)

14.2.6 UNIQUELY COLOURABLE GRAPH
A k-chromatic graph G is *uniquely k-colourable*, or simply *uniquely colourable*, if any two k-colourings of G induce the same partition of V.

a) Determine the uniquely 2-colourable graphs.
b) Generalize Theorem 14.8 by showing that no vertex cut of a critical graph induces a uniquely colourable subgraph.

14.2.7

a) Show that if u and v are two vertices of a critical graph G, then $N(u) \not\subseteq N(v)$.
b) Deduce that no k-critical graph has exactly $k + 1$ vertices.

14.2.8 Show that:

a) $\chi(G_1 \vee G_2) = \chi(G_1) + \chi(G_2)$,
b) $G_1 \vee G_2$ is critical if and only if both G_1 and G_2 are critical.

14.2.9 HAJÓS JOIN
Let G_1 and G_2 be disjoint graphs, and let $e_1 := u_1v_1$ and $e_2 := u_2v_2$ be edges of G_1 and G_2, respectively. The graph obtained from G_1 and G_2 by identifying u_1 and u_2, deleting e_1 and e_2, and adding a new edge v_1v_2 is called a *Hajós join* of G_1 and G_2. Show that the Hajós join of two graphs is k-critical if and only if both graphs are k-critical. (G. HAJÓS)

14.2.10 For $n = 4$ and all $n \geq 6$, construct a 4-critical graph on n vertices.

14.2.11 SCHRIJVER GRAPH
Let $S := \{1, 2, \ldots, n\}$. The *Schrijver graph* $SG_{m,n}$ is the subgraph of the Kneser graph $KG_{m,n}$ induced by the m-subsets of S which contain no two consecutive elements in the cyclic order $(1, 2, \ldots, n, 1)$.

a) Draw the Schrijver graph $SG_{3,8}$.
b) Show that this graph is 4-chromatic, whereas every vertex-deleted subgraph of it is 3-chromatic.

(Schrijver (1978) has shown that $SG_{m,n}$ is $(n - 2m + 2)$-chromatic, and that every vertex-deleted subgraph of it is $(n - 2m + 1)$-chromatic.)

14.2.12

a) Let G be a k-critical graph with a 2-vertex cut $\{u, v\}$. Show that $d(u) + d(v) \geq 3k - 5$.

b) Deduce Brooks' Theorem (14.4) for graphs with 2-vertex cuts.

14.2.13 Show that Brooks' Theorem (14.4) is equivalent to the following statement: if G is k-critical ($k \geq 4$) and not complete, then $2m \geq (k - 1)n + 1$. (Dirac (1957) sharpened this bound to $2m \geq (k - 1)n + (k - 3)$.)

14.2.14 A hypergraph H is k-*critical* if $\chi(H) = k$, but $\chi(H') < k$ for every proper subhypergraph H' of H. Show that:

a) the only 2-critical hypergraph is K_2,

b) the Fano hypergraph (depicted in Figure 1.15a) is 3-critical.

14.2.15 Let $H := (V, \mathcal{F})$ be a 3-critical hypergraph, where $V := \{v_1, v_2, \ldots, v_n\}$ and $\mathcal{F} := \{F_1, F_2, \ldots, F_m\}$, and let $\mathbf{M}$ be the incidence matrix of H.

a) Suppose that the rows of $\mathbf{M}$ are linearly dependent, so that there are real numbers λ_i, $1 \leq i \leq n$, not all zero, such that $\sum\{\lambda_i : v_i \in F_j\} = 0$, $1 \leq j \leq m$. Set $Z := \{i : \lambda_i = 0\}$, $P := \{i : \lambda_i > 0\}$, and $N := \{i : \lambda_i < 0\}$. Show that:

   i) $H' := H[Z]$ has a 2-colouring $\{R, B\}$,
   ii) H has the 2-colouring $\{R \cup P, B \cup N\}$.

b) Deduce that the rows of $\mathbf{M}$ are linearly independent.

c) Conclude that $|\mathcal{F}| \geq |V|$. (P.D. SEYMOUR)

14.2.16 Let G be a k-chromatic graph which has a colouring in which each colour is assigned to at least two vertices. Show that G has a k-colouring with this property.

(T. GALLAI)

14.2.17

a) By appealing to Theorem 2.5, show that a bipartite graph with average degree $2k$ or more contains a path of length $2k + 1$. (A. GYARFÁS AND J. LEHEL)

b) An *antidirected path* in a digraph is a path whose edges alternate in direction. Using Exercise 14.1.23, deduce that every digraph D contains an antidirected path of length at least $\chi/4$.

14.2.18 An *antidirected cycle* in a digraph is a cycle of even length whose edges alternate in direction.

a) Find a tournament on five vertices which contains no antidirected cycle.

b) Show that every 8-chromatic digraph contains an antidirected cycle.

(D. GRANT, F. JAEGER, AND C. PAYAN)

14.3 Girth and Chromatic Number

As we noted in the previous section, a graph which contains a large clique necessarily has a high chromatic number. On the other hand, and somewhat surprisingly, there exist triangle-free graphs with arbitrarily high chromatic number. Recursive constructions of such graphs were first described by (Blanche) Descartes (see Ungar and Descartes (1954) and Exercise 14.3.3). Later, Erdős (1961a) applied the probabilistic method to demonstrate the existence of graphs with arbitrarily high girth and chromatic number.

Theorem 14.11 *For each positive integer k, there exists a graph with girth at least k and chromatic number at least k.*

Proof Consider $G \in \mathcal{G}_{n,p}$, and set $t := \lceil 2p^{-1} \log n \rceil$. By Theorem 13.6, almost surely $\alpha(G) \leq t$. Let X be the number of cycles of G of length less than k. By linearity of expectation (13.4),

$$E(X) = \sum_{i=3}^{k-1} \frac{(n)_i}{2i} p^i < \sum_{i=0}^{k-1} (np)^i = \frac{(np)^k - 1}{np - 1}$$

where $(n)_i$ denotes the falling factorial $n(n-1)\cdots(n-i+1)$. Markov's Inequality (13.4) now yields:

$$P(X > n/2) < \frac{E(X)}{n/2} < \frac{2((np)^k - 1)}{n(np - 1)}$$

Therefore, if $p := n^{-(k-1)/k}$,

$$P(X > n/2) < \frac{2(n-1)}{n(n^{1/k} - 1)} \to 0 \quad \text{as } n \to \infty$$

in other words, G almost surely has no more than $n/2$ cycles of length less than k.

It follows that, for n sufficiently large, there exists a graph G on n vertices with stability number at most t and no more than $n/2$ cycles of length less than k. By deleting one vertex of G from each cycle of length less than k, we obtain a graph G' on at least $n/2$ vertices with girth at least k and stability number at most t. By inequality (14.1),

$$\chi(G') \geq \frac{v(G')}{\alpha(G')} \geq \frac{n}{2t} \sim \frac{n^{1/k}}{8 \log n}$$

It suffices, now, to choose n large enough to guarantee that $\chi(G') \geq k$.   □

MYCIELSKI'S CONSTRUCTION

Note that the above proof is nonconstructive: it merely asserts the *existence* of graphs with arbitrarily high girth and chromatic number. Recursive *constructions* of such graphs were given by Lovász (1968a) and also by Nešetřil and Rödl (1979). We describe here a simpler construction of triangle-free k-chromatic graphs, due to Mycielski (1955).

Theorem 14.12 *For any positive integer k, there exists a triangle-free k-chromatic graph.*

Proof For $k = 1$ and $k = 2$, the graphs K_1 and K_2 have the required property. We proceed by induction on k. Suppose that we have already constructed a triangle-free graph G_k with chromatic number $k \geq 2$. Let the vertices of G_k be $v_1, v_2, \ldots, v_n$. Form the graph G_{k+1} from G_k as follows: add $n + 1$ new vertices $u_1, u_2, \ldots, u_n, v$, and then, for $1 \leq i \leq n$, join u_i to the neighbours of v_i in G_k, and also to v. For example, if $G_2 := K_2$, then G_3 is the 5-cycle and G_4 the Grötzsch graph (see Figure 14.6).

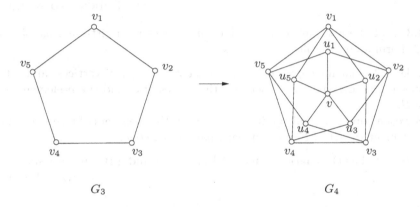

G_3 $\qquad\qquad\qquad\qquad\qquad$ G_4

Fig. 14.6. Mycielski's construction

The graph G_{k+1} certainly has no triangles. For, because $u_1, u_2, \ldots, u_n$ is a stable set in G_{k+1}, no triangle can contain more than one u_i; and if $u_i v_j v_k u_i$ were a triangle in G_{k+1}, then $v_i v_j v_k v_i$ would be a triangle in G_k, contrary to our assumption.

We now show that G_{k+1} is $(k + 1)$-chromatic. Note, first, that G_{k+1} is $(k + 1)$-colourable, because any k-colouring of G_k can be extended to a $(k + 1)$-colouring of G_{k+1} by assigning the colour of v_i to u_i, $1 \leq i \leq n$, and then assigning a new colour to v. Therefore, it remains to show that G_{k+1} is not k-colourable.

Suppose that G_{k+1} has a k-colouring. This colouring, when restricted to $\{v_1, v_2, \ldots, v_n\}$, is a k-colouring of the k-chromatic graph G_k. By Exercise 14.1.3, for each colour j, there exists a vertex v_i of colour j which is adjacent in G_k to vertices of every other colour. Because u_i has precisely the same neighbours in G_k as v_i, the vertex u_i must also have colour j. Therefore, each of the k colours appears on at least one of the vertices u_i. But no colour is now available for the vertex v, a contradiction. We infer that G_{k+1} is indeed $(k + 1)$-chromatic, and the theorem follows by induction. $\qquad\square$

Other examples of triangle-free graphs with arbitrarily high chromatic number are the shift graphs (see Exercise 14.3.2).

Exercises

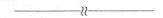

14.3.1 Let $G_2 := K_2$, and let G_k be the graph obtained from G_{k-1} by Mycielski's construction, $k \geq 3$. Show that G_k is a k-critical graph on $3 \cdot 2^{k-2} - 1$ vertices.

14.3.2 SHIFT GRAPH

The *shift graph* SG_n is the graph whose vertex set is the set of 2-subsets of $\{1, 2, \ldots, n\}$, there being an edge joining two pairs $\{i, j\}$ and $\{k, l\}$, where $i < j$ and $k < l$, if and only if $j = k$. Show that SG_n is a triangle-free graph of chromatic number $\lceil \log_2 n \rceil$.   (P. ERDŐS AND A. HAJNAL)

14.3.3 Let G be a k-chromatic graph on n vertices with girth at least six, where $k \geq 2$. Form a new graph H as follows.

▷ Take $\binom{kn}{n}$ disjoint copies of G and a set S of kn new vertices, and set up a one-to-one correspondence between the copies of G and the n-element subsets of S.

▷ For each copy of G, pair up its vertices with the members of the corresponding n-element subset of S and join each pair by an edge.

Show that H has chromatic number at least $k + 1$ and girth at least six.
                    (B. DESCARTES)

14.4 Perfect Graphs

Inequality (14.2), which states that $\chi \geq \omega$, leads one to ask which graphs G satisfy it with equality. One soon realizes, however, that this question as it stands is not particularly interesting, because if H is any k-colourable graph and G is the disjoint union of H and K_k, then $\chi(G) = \omega(G) = k$. Berge (1963) noted that such artificial examples may be avoided by insisting that inequality (14.2) hold not only for G but also for all of its induced subgraphs. He called such graphs G 'perfect', and observed that the graphs satisfying this property include many basic families of graphs, such as bipartite graphs, line graphs of bipartite graphs, chordal graphs, and comparability graphs. He also noted that well-known min–max theorems concerning these seemingly disparate families of graphs simply amount to saying that they are perfect.

A graph G is *perfect* if $\chi(H) = \omega(H)$ for every induced subgraph H of G; otherwise, it is *imperfect*. An imperfect graph is *minimally imperfect* if each of its proper induced subgraphs is perfect. The triangular prism and the octahedron are examples of perfect graphs (Exercise 14.4.1), whereas the odd cycles of length five or more, as well as their complements, are minimally imperfect (Exercise 14.4.2). The cycle C_7 and its complement $\overline{C_7}$ are shown in Figure 14.7.

Being 2-colourable, bipartite graphs are clearly perfect. The fact that their line graphs are perfect is a consequence of a theorem concerning edge colourings of

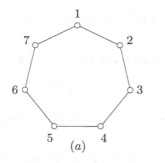
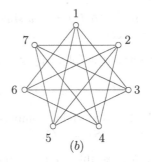

Fig. 14.7. The minimally imperfect graphs (a) C_7, and (b) $\overline{C_7}$

bipartite graphs (see Exercise 17.1.15). By Theorem 9.20, every chordal graph has a simplicial decomposition, and this property can be used to show that chordal graphs are perfect (Exercise 14.4.3). Comparability graphs are perfect too. That this is so may be deduced from a basic property of partially ordered sets (see Exercise 14.4.4).

THE PERFECT GRAPH THEOREM

Berge (1963) observed that all the perfect graphs in the above classes also have perfect complements. For example, the König–Rado Theorem (8.30) implies that the complement of a bipartite graph is perfect, and Dilworth's Theorem (12.5) implies that the complement of a comparability graph is perfect. Based on this empirical evidence, Berge (1963) conjectured that a graph is perfect if and only if its complement is perfect. This conjecture was verified by Lovász (1972b), resulting in what is now known as the *Perfect Graph Theorem*.

Theorem 14.13   THE PERFECT GRAPH THEOREM
A graph is perfect if and only if its complement is perfect.   □

Shortly thereafter, A. Hajnal (see Lovász (1972a)) proposed the following beautiful characterization of perfect graphs. This, too, was confirmed by Lovász (1972a).

Theorem 14.14 *A graph G is perfect if and only if every induced subgraph H of G satisfies the inequality*

$$v(H) \leq \alpha(H)\omega(H)$$

Observe that the above inequality is invariant under complementation, because $v(\overline{H}) = v(H)$, $\alpha(\overline{H}) = \omega(H)$, and $\omega(\overline{H}) = \alpha(H)$. Theorem 14.14 thus implies the Perfect Graph Theorem (14.13).

The proof that we present of Theorem 14.14 is due to Gasparian (1996). It relies on an elementary rank argument (the proof technique of Linear Independence discussed in Section 2.4). We need the following property of minimally imperfect graphs.

Proposition 14.15 *Let S be a stable set in a minimally imperfect graph G. Then $\omega(G - S) = \omega(G)$.*

Proof We have the following string of inequalities (Exercise 14.4.5).

$$\omega(G - S) \le \omega(G) \le \chi(G) - 1 \le \chi(G - S) = \omega(G - S)$$

Because the left and right members are the same, equality holds throughout. In particular, $\omega(G - S) = \omega(G)$. □

We can now establish a result on the structure of minimally imperfect graphs. This plays a key role in the proof of Theorem 14.14.

Lemma 14.16 *Let G be a minimally imperfect graph with stability number α and clique number ω. Then G contains $\alpha\omega + 1$ stable sets $S_0, S_1, \ldots, S_{\alpha\omega}$ and $\alpha\omega + 1$ cliques $C_0, C_1, \ldots, C_{\alpha\omega}$ such that:*

▷ *each vertex of G belongs to precisely α of the stable sets S_i,*
▷ *each clique C_i has ω vertices,*
▷ *$C_i \cap S_i = \emptyset$, for $0 \le i \le \alpha\omega$,*
▷ *$|C_i \cap S_j| = 1$, for $0 \le i < j \le \alpha\omega$.*

Proof Let S_0 be a stable set of α vertices of G, and let $v \in S_0$. The graph $G - v$ is perfect because G is minimally imperfect. Thus $\chi(G - v) = \omega(G - v) \le \omega(G)$. This means that for any $v \in S_0$, the set $V \setminus \{v\}$ can be partitioned into a family $\mathcal{S}_v$ of ω stable sets. Denoting $\{\cup \mathcal{S}_v : v \in S_0\}$ by $\{S_1, S_2, \ldots, S_{\alpha\omega}\}$, it can be seen that $\{S_0, S_1, \ldots, S_{\alpha\omega}\}$ is a family of $\alpha\omega + 1$ stable sets of G satisfying the first property above.

By Proposition 14.15, $\omega(G - S_i) = \omega(G)$, $0 \le i \le \alpha\omega$. Therefore there exists a maximum clique C_i of G that is disjoint from S_i. Because each of the ω vertices in C_i lies in α of the stable sets S_j, $0 \le i \le \alpha\omega$, and because no two vertices of C_i can belong to a common stable set, $|C_i \cap S_j| = 1$, for $0 \le i < j \le \alpha\omega$. □

Let us illustrate Lemma 14.16 by taking G to be the minimally imperfect graph $\overline{C_7}$, labelled as shown in Figure 14.7b. Here $\alpha = 2$ and $\omega = 3$. Applying the procedure described in the proof of the lemma, we obtain the following seven stable sets and seven cliques.

$$S_0 = 12, \quad S_1 = 23, \quad S_2 = 45, \quad S_3 = 67, \quad S_4 = 34, \quad S_5 = 56, \quad S_6 = 17$$
$$C_0 = 357, \quad C_1 = 146, \quad C_2 = 136, \quad C_3 = 135, \quad C_4 = 257, \quad C_5 = 247, \quad C_6 = 246$$

(where we write 12 for the set $\{1, 2\}$, and so on.) The incidence matrices $\mathbf{S}$ and $\mathbf{C}$ of these families are shown in Figure 14.8.

We are now ready to prove Theorem 14.14.

Proof Suppose that G is perfect, and let H be an induced subgraph of G. Because G is perfect, H is $\omega(H)$-colourable, implying that $v(H) \le \alpha(H)\omega(H)$. We prove the converse by showing that if G is minimally imperfect, then $v(G) \ge \alpha(G)\omega(G) + 1$.

$$\mathbf{S}: \quad \begin{array}{c|ccccccc} & S_0 & S_1 & S_2 & S_3 & S_4 & S_5 & S_6 \\ \hline 1 & 1 & 0 & 0 & 0 & 0 & 0 & 1 \\ 2 & 1 & 1 & 0 & 0 & 0 & 0 & 0 \\ 3 & 0 & 1 & 0 & 0 & 1 & 0 & 0 \\ 4 & 0 & 0 & 1 & 0 & 1 & 0 & 0 \\ 5 & 0 & 0 & 1 & 0 & 0 & 1 & 0 \\ 6 & 0 & 0 & 0 & 1 & 0 & 1 & 0 \\ 7 & 0 & 0 & 0 & 1 & 0 & 0 & 1 \end{array} \qquad \mathbf{C}: \quad \begin{array}{c|ccccccc} & C_0 & C_1 & C_2 & C_3 & C_4 & C_5 & C_6 \\ \hline 1 & 0 & 1 & 1 & 1 & 0 & 0 & 0 \\ 2 & 0 & 0 & 0 & 0 & 1 & 1 & 1 \\ 3 & 1 & 0 & 1 & 1 & 0 & 0 & 0 \\ 4 & 0 & 1 & 0 & 0 & 0 & 1 & 1 \\ 5 & 1 & 0 & 0 & 1 & 1 & 0 & 0 \\ 6 & 0 & 1 & 1 & 0 & 0 & 0 & 1 \\ 7 & 1 & 0 & 0 & 0 & 1 & 1 & 0 \end{array}$$

Fig. 14.8. Incidence matrices of families of stable sets and cliques of $\overline{C_7}$

Consider the families $\{S_i : 0 \le i \le \alpha\omega\}$ and $\{C_i : 0 \le i \le \alpha\omega\}$ of stable sets and cliques described in Lemma 14.16. Let $\mathbf{S}$ and $\mathbf{C}$ be the $n \times (\alpha\omega + 1)$ incidence matrices of these families. It follows from Lemma 14.16 that $\mathbf{S}^t\mathbf{C} = \mathbf{J} - \mathbf{I}$, where $\mathbf{J}$ is the square matrix of order $\alpha\omega + 1$ all of whose entries are 1 and $\mathbf{I}$ is the identity matrix of order $\alpha\omega + 1$. Now $\mathbf{J} - \mathbf{I}$ is a nonsingular matrix (with inverse $(1/\alpha\omega)\mathbf{J} - \mathbf{I}$). Its rank is thus equal to its order, $\alpha\omega + 1$. Hence both $\mathbf{S}$ and $\mathbf{C}$ are also of rank $\alpha\omega + 1$. But these matrices have n rows, so $n \ge \alpha\omega + 1$. $\qquad\square$

Two consequences of the Perfect Graph Theorem are (Exercise 14.4.6):

Corollary 14.17 *A graph G is perfect if and only if, for any induced subgraph H of G, the maximum number of vertices in a stable set of H is equal to the minimum number of cliques required to cover all the vertices of H.* $\qquad\square$

Corollary 14.18 *The Shannon capacity of a perfect graph G is equal to its stability number: $\Theta(G) = \alpha(G)$.* $\qquad\square$

Corollary 14.18 prompts the problem of determining the Shannon capacities of the minimally imperfect graphs. Of these, only $\Theta(C_5)$ is known (see Exercise 12.1.21). It would be interesting to determine $\Theta(C_7)$.

THE STRONG PERFECT GRAPH THEOREM

If a graph is perfect, then so are all of its induced subgraphs. This means that one can characterize perfect graphs by describing all minimally imperfect graphs. We have remarked that the odd cycles of length five or more are minimally imperfect, as are their complements. Berge (1963) proposed the conjecture that these are the only minimally imperfect graphs; equivalently, that a graph is perfect if and only if it contains no odd cycle of length at least five, or its complement, as an induced subgraph. He named this conjecture, whose truth would imply the Perfect Graph Theorem, the *Strong Perfect Graph Conjecture*. Some forty years later, it was proved by Chudnovsky et al. (2006).

Theorem 14.19 THE STRONG PERFECT GRAPH THEOREM
A graph is perfect if and only if it contains no odd cycle of length at least five, or its complement, as an induced subgraph.   □

This theorem was a major achievement, as much effort had been expended over the years on attempts to settle the Strong Perfect Graph Conjecture. Furthermore, a polynomial-time recognition algorithm for perfect graphs was developed shortly thereafter by Chudnovsky et al. (2005).

Perfect graphs play an important role in combinatorial optimization and poly-hedral combinatorics. Schrijver (2003) dedicates three chapters of his scholarly treatise to this widely studied area. The survey article by Chudnovsky et al. (2003) includes an excellent account of some of the recent developments in the subject. The original motivations for the study of perfect graphs, and its early history, are described by Berge (1996, 1997).

Exercises

14.4.1 Show that the triangular prism and the octahedron are perfect graphs.

14.4.2 For each $k \geq 2$, show that both C_{2k+1} and $\overline{C_{2k+1}}$ are minimally imperfect graphs.

14.4.3

a) Let G be a chordal graph and $(X_1, X_2, \ldots, X_k)$ a simplicial decomposition of G. Show that $\chi = \max \{|X_i| : 1 \leq i \leq k\}$.
b) Deduce that every chordal graph is perfect.

14.4.4 Using the result stated in Exercise 2.1.22, show that every comparability graph is perfect.

$\star$**14.4.5** Verify the three inequalities in the proof of Proposition 14.15.

14.4.6 Prove Corollaries 14.17 and 14.18.

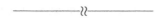

14.4.7 Without appealing to the Strong Perfect Graph Theorem, show that every minimally imperfect graph G satisfies the relation $n = \alpha\omega + 1$.

14.4.8 Deduce from Theorem 14.14 that the problem of recognizing perfect graphs belongs to co-$\mathcal{NP}$.   (K. CAMERON; V. CHVÁTAL)

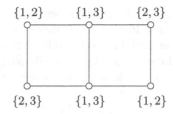

Fig. 14.9. A bipartite graph whose list chromatic number is three

14.5 List Colourings

In most practical colouring problems, there are restrictions on the colours that may be assigned to certain vertices. For example, in the chemical storage problem of Example 14.2, radioactive substances might require special storage facilities. Thus in the corresponding graph there is a list of colours (appropriate storage compartments) associated with each vertex (chemical). In an admissible colouring (assignment of compartments to chemicals), the colour of a vertex must be chosen from its list. This leads to the notion of list colouring.

Let G be a graph and let L be a function which assigns to each vertex v of G a set $L(v)$ of positive integers, called the *list* of v. A colouring $c : V \to \mathbb{N}$ such that $c(v) \in L(v)$ for all $v \in V$ is called a *list colouring* of G with respect to L, or an *L-colouring*, and we say that G is *L-colourable*. Observe that if $L(v) = \{1, 2, \ldots, k\}$ for all $v \in V$, an *L*-colouring is simply a k-colouring. For instance, if G is a bipartite graph and $L(v) = \{1, 2\}$ for all vertices v, then G has the *L*-colouring which assigns colour 1 to all vertices in one part and colour 2 to all vertices in the other part. Observe, also, that assigning a list of length one to a vertex amounts to precolouring the vertex with that colour. As for colouring, the notion of list colouring extends in a straightforward manner to digraphs.

List Chromatic Number

At first glance, one might believe that a k-chromatic graph in which each list $L(v)$ is of length at least k necessarily has an *L*-colouring. However, this is not so. It can be checked that the bipartite graph shown in Figure 14.9 has no list colouring with respect to the indicated lists. On the other hand, if arbitrary lists of length three are assigned to the vertices of this graph, it will have a compatible list colouring (Exercise 14.5.1).

A graph G or digraph D is said to be k-*list-colourable* if it has a list colouring whenever all the lists have length k. Every graph G is clearly n-list-colourable. The smallest value of k for which G is k-list-colourable is called the *list chromatic number* of G, denoted $\chi_L(G)$. For example, the list chromatic number of the graph shown in Figure 14.9 is equal to three, whereas its chromatic number is two. (More generally, there exist 2-chromatic graphs whose list chromatic number is arbitrarily large, see Exercise 14.5.5.)

Bounds on the list chromatic numbers of certain graphs can be found by means of kernels. This might seem odd at first, because the kernel (introduced in Section 12.1) is a notion concerning directed graphs, whereas the list chromatic number is one concerning undirected graphs. The following theorem (a strengthening of Exercise 14.1.15) provides a link between kernels and list colourings.

Theorem 14.20 *Let $D = (V, A)$ be a digraph each of whose induced subdigraphs has a kernel. For $v \in V$, let $L(v)$ be an arbitrary list of at least $d^+(v) + 1$ colours. Then D admits an L-colouring.*

Proof By induction on n, the statement being trivial for $n = 1$. Let V_1 be the set of vertices of D whose lists include colour 1. We may assume that $V_1 \neq \emptyset$ by renaming colours if necessary. By assumption, $D[V_1]$ has a kernel S_1. Colour the vertices of S_1 with colour 1, and set $D' := D - S_1$ and $L'(v) := L(v) \setminus \{1\}$, $v \in V(D')$. For any vertex v of D' whose list did not contain colour 1,

$$|L'(v)| = |L(v)| \geq d_D^+(v) + 1 \geq d_{D'}^+(v) + 1$$

and for any vertex v of D' whose list did contain colour 1,

$$|L'(v)| = |L(v)| - 1 \geq d_D^+(v) \geq d_{D'}^+(v) + 1$$

The last inequality holds because, in D, the vertex v dominates some vertex of the kernel S_1, so its outdegree in D' is smaller than in D. By induction, D' has an L'-colouring. When combined with the colouring of S_1, this yields an L-colouring of D. □

As a simple illustration of Theorem 14.20, consider a digraph which contains no directed odd cycle. By Theorem 12.6, such a digraph has a kernel, as do all its induced subdigraphs, so it satisfies the hypothesis of the theorem. We therefore have the following corollary.

Corollary 14.21 *Let $D = (V, A)$ be a digraph which contains no directed odd cycle. For $v \in V$, let $L(v)$ be an arbitrary list of at least $d^+(v) + 1$ colours. Then D admits an L-colouring. In particular, every graph G is $(\Delta + 1)$-list-colourable.*

A similar approach can be applied to list colourings of interval graphs. Woodall (2001) showed that every interval graph G has an acyclic orientation D with $\Delta^+ \leq \omega - 1$ (Exercise 14.5.10). Appealing to Theorem 14.20 yields the following result.

Corollary 14.22 *Every interval graph G has list chromatic number ω.* □

Exercises

★**14.5.1** Show that the list chromatic number of the graph shown in Figure 14.9 is equal to three.

14.5.2

a) Show that $\chi_L(K_{3,3}) = 3$.
b) Using the Fano plane, obtain an assignment of lists to the vertices of $K_{7,7}$ which shows that $\chi_L(K_{7,7}) > 3$.

14.5.3 Generalize Brooks' Theorem (14.4) by proving that if G is a connected graph, and is neither an odd cycle nor a complete graph, then G is Δ-list-colourable. (P. ERDŐS, A.L. RUBIN, AND H. TAYLOR; V.G. VIZING)

14.5.4 Show that $K_{m,n}$ is k-list-colourable for all $k \geq \min\{m, n\} + 1$.

$\star$**14.5.5** Show that $\chi_L(K_{n,n^n}) = n + 1$.   (N. ALON AND M. TARSI)

14.5.6 By choosing as lists the edges of the non-2-colourable hypergraph whose existence was established in Exercise 13.2.15, show that $\chi_L(K_{n,n}) \geq c_n \log_2 n$, where $c_n \sim 1$.

14.5.7 Let S be a set of cardinality $2k - 1$, where $k \geq 1$. Consider the complete bipartite graph $K_{n,n}$, where $n = \binom{2k-1}{k}$, in which the lists attached to the vertices in each part are the k-subsets of S. Show that $K_{n,n}$ has no list colouring with this assignment of lists. (P. ERDŐS, A.L. RUBIN, AND H. TAYLOR)

14.5.8 A *theta graph* $TG_{k,l,m}$ is a graph obtained by joining two vertices by three internally disjoint paths of lengths k, l, and m. Show that:

a) $TG_{2,2,2k}$ is 2-list-colourable for all $k \geq 1$,
b) a connected simple graph is 2-list-colourable if and only if the subgraph obtained by recursively deleting vertices of degree one is an isolated vertex, an even cycle, or a theta graph $TG_{2,2,2k}$, where $k \geq 1$. (P. ERDŐS, A.L. RUBIN, AND H. TAYLOR)

14.5.9 Let $G = (V, E)$ be a simple graph. For $v \in V$, let $L(v)$ be a list of k or more colours. Suppose that, for each vertex v and each colour in $L(v)$, no more than $k/2e$ neighbours of v have that same colour in their lists (where e is the base of natural logarithms). By applying the Local Lemma (Theorem 13.12), show that G has a list colouring with respect to L. (B.A. REED)

$\star$**14.5.10** Let G be an interval graph.

a) Show that G has an acyclic orientation D with $\Delta^+ = \omega - 1$.
b) Deduce that $\chi_L = \chi = \omega$. (D.R. WOODALL)

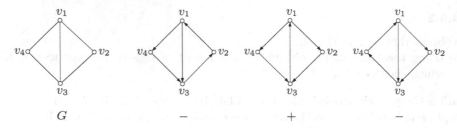

Fig. 14.10. A labelled graph G and the three orientations corresponding to the term $x_1^2 x_2 x_3 x_4$ of its adjacency polynomial

14.6 The Adjacency Polynomial

We have already seen how linear algebraic techniques can be used to prove results in graph theory, for instance by means of rank arguments (see the inset in Chapter 2) or by studying the eigenvalues of the adjacency matrix of the graph (see the inset in Chapter 3). In this section, we develop yet another algebraic tool, this time related to polynomials, and apply it to obtain results on list colouring. To this end, we define a natural polynomial associated with a graph, indeed so natural that it is often referred to as the *graph polynomial*.

Let G be a graph with vertex set $V := \{v_1, v_2, \ldots, v_n\}$. Set $\mathbf{x} := (x_1, x_2, \ldots, x_n)$. The *adjacency polynomial* of G is the multivariate polynomial

$$A(G, \mathbf{x}) := \prod_{i<j} \{(x_i - x_j) : v_i v_j \in E\}$$

The relevance of the adjacency polynomial to vertex colouring should be clear. If the value of $A(G, \mathbf{x})$ at $\mathbf{x} = \mathbf{c}$ is nonzero, where the entries of $\mathbf{c}$ are elements of some field F, then $\mathbf{c}$, regarded as a function $c : V \to F$, is a proper colouring of G. Conversely, if $c : V \to F$, is a proper colouring of G, then $A(G, \mathbf{c}) \neq 0$.

We now show how individual terms in the expansion of the adjacency polynomial of a graph are related to orientations of the graph. Upon expanding $A(G, \mathbf{x})$ we obtain 2^m monomials (some of which might cancel out). Each of these monomials is obtained by selecting exactly one variable from every factor $x_i - x_j$, and thus corresponds to an orientation of G: we orient the edge $v_i v_j$ of G in such a way that the vertex corresponding to the chosen variable is designated to be the tail of the resulting arc.

For example, if G is the graph shown in Figure 14.10, its adjacency polynomial is given by

$$A(G, \mathbf{x}) = (x_1 - x_2)(x_1 - x_3)(x_1 - x_4)(x_2 - x_3)(x_3 - x_4) \tag{14.4}$$

There are $2^5 = 32$ terms in the expansion of this expression before cancellation, whereas after cancellation only 24 terms remain:

$$
\begin{aligned}
A(G, \mathbf{x}) = \quad & x_1^3 x_2 x_3 - x_1^3 x_2 x_4 - x_1^3 x_3^2 + x_1^3 x_3 x_4 - x_1^2 x_2^2 x_3 + x_1^2 x_2^2 x_4 \\
& - x_1^2 x_2 x_3 x_4 + x_1^2 x_2 x_4^2 + x_1^2 x_3^3 - x_1^2 x_3 x_4^2 + x_1 x_2^2 x_3^2 - x_1 x_2^2 x_4^2 \\
& - x_1 x_2 x_3^3 + x_1 x_2 x_3^2 x_4 - x_1 x_3^3 x_4 + x_1 x_3^2 x_4^2 - x_2^2 x_3^2 x_4 + x_2^2 x_3 x_4^2 \\
& + x_2 x_3^3 x_4 - x_2 x_3^2 x_4^2 + x_2 x_3 x_4^3 - x_2 x_4^4 - x_3^2 x_4^3 + x_3 x_4^4
\end{aligned}
$$

The graph G has the three orientations with outdegree sequence $(2, 1, 1, 1)$ shown in Figure 14.10. These orientations are precisely the ones which correspond to the monomial $x_1^2 x_2 x_3 x_4$. Observe that the coefficient of this term in A(G,x) is -1. This is because two of the three terms in the expansion of the product (14.4) have a negative sign, whereas the remaining one has a positive sign.

As a second example, consider the complete graph K_n. We have

$$
A(K_n, \mathbf{x}) = \prod_{1 \le i < j \le n} (x_i - x_j) =
\begin{vmatrix}
x_1^{n-1} & x_2^{n-1} & \dots & x_n^{n-1} \\
x_1^{n-2} & x_2^{n-2} & \dots & x_n^{n-2} \\
\cdot & \cdot & \dots & \cdot \\
\cdot & \cdot & \dots & \cdot \\
x_1 & x_2 & \dots & x_n \\
1 & 1 & \dots & 1
\end{vmatrix}
$$

The number of monomials in the expansion of this *Vandermonde determinant* is $n!$ (Exercise 14.6.1) which is much smaller (due to cancellation of terms) than $2^{\binom{n}{2}}$, the number of monomials in the expansion of the adjacency polynomial.

In order to express the adjacency polynomial of a graph in terms of its orientations, we need a little notation. In the expansion of $A(G, \mathbf{x})$, each monomial occurs with a given sign. We associate this same sign with the corresponding orientation D of G by defining

$$
\sigma(D) := \prod \{ \sigma(e) : a \in A(D) \}
$$

where

$$
\sigma(a) :=
\begin{cases}
+1 & \text{if } a = (v_i, v_j) \text{ with } i < j \\
-1 & \text{if } a = (v_i, v_j) \text{ with } i > j
\end{cases}
$$

For example, the three orientations of the graph G in Figure 14.10 have the signs indicated.

Now let $\mathbf{d} := (d_1, d_2, \dots, d_n)$ be a sequence of nonnegative integers whose sum is m. We define the *weight* of $\mathbf{d}$ by

$$
w(\mathbf{d}) := \sum \sigma(D)
$$

where the sum is taken over all orientations D of G whose outdegree sequence is $\mathbf{d}$. Setting

$$
\mathbf{x}^{\mathbf{d}} := \prod_{i=1}^{n} x_i^{d_i}
$$

we can now express the adjacency polynomial as:

$$A(G, \mathbf{x}) = \sum_{\mathbf{d}} w(\mathbf{d}) \mathbf{x}^{\mathbf{d}}$$

In order to understand the relevance of the latter expression to list colourings, we need an algebraic tool developed by Alon (1999) and known as the *Combinatorial Nullstellensatz*, by analogy with a celebrated theorem of D. Hilbert.

PROOF TECHNIQUE: THE COMBINATORIAL NULLSTELLENSATZ

The Combinatorial Nullstellensatz is based on the following proposition, a generalization to n variables of the fact that a polynomial of degree d in one variable has at most d distinct roots.

Proposition 14.23 *Let f be a nonzero polynomial over a field F in the variables $\mathbf{x} = (x_1, x_2, \ldots, x_n)$, of degree d_i in x_i, $1 \leq i \leq n$. Let L_i be a set of $d_i + 1$ elements of F, $1 \leq i \leq n$. Then there exists $\mathbf{t} \in L_1 \times L_2 \times \cdots \times L_n$ such that $f(\mathbf{t}) \neq 0$.*

Proof As noted above, the case $n = 1$ simply expresses the fact that a polynomial of degree d in one variable has at most d distinct roots. We proceed by induction on n, where $n \geq 2$.

We first express f as a polynomial in x_n whose coefficients f_j are polynomials in the variables $x_1, x_2, \ldots, x_{n-1}$:

$$f = \sum_{j=0}^{d_n} f_j x_n^j$$

Because f is nonzero by hypothesis, f_j is nonzero for some j, $0 \leq j \leq d_n$. By induction, there exist $t_i \in L_i$, $1 \leq i \leq n-1$, such that $f_j(t_1, t_2, \ldots, t_{n-1}) \neq 0$. Therefore the polynomial $\sum_{j=0}^{d_n} f_j(t_1, t_2, \ldots, t_{n-1}) x_n^j$ is nonzero. Applying the case $n = 1$ to this polynomial, we deduce that $f(t_1, t_2, \ldots, t_n) \neq 0$ for some $t_n \in L_n$. $\square$

Theorem 14.24 THE COMBINATORIAL NULLSTELLENSATZ
Let f be a polynomial over a field F in the variables $\mathbf{x} = (x_1, x_2, \ldots, x_n)$. Suppose that the total degree of f is $\sum_{i=1}^{n} d_i$ and that the coefficient in f of $\prod_{i=1}^{n} x_i^{d_i}$ is nonzero. Let L_i be a set of $d_i + 1$ elements of F, $1 \leq i \leq n$. Then there exists $\mathbf{t} \in L_1 \times L_2 \times \cdots \times L_n$ such that $f(\mathbf{t}) \neq 0$.

THE COMBINATORIAL NULLSTELLENSATZ (CONTINUED)

Proof For $1 \leq i \leq n$, set

$$f_i := \prod_{t \in L_i} (x_i - t)$$

Then f_i is a polynomial of degree $|L_i| = d_i + 1$, with leading term $x_i^{d_i+1}$, so we may write $f_i = g_i + x_i^{d_i+1}$, where g_i is a polynomial in x_i of degree at most d_i. By repeatedly substituting $-g_i$ for $x_i^{d_i+1}$ in the polynomial f, we obtain a new polynomial in which the degree of x_i does not exceed d_i. Performing this substitution operation for all i, $1 \leq i \leq n$, results in a polynomial g of degree at most d_i in x_i, $1 \leq i \leq n$.

Moreover, because $f_i(t) = 0$ for all $t \in L_i$, we have $t^{d_i+1} = -g_i(t)$ for all $t \in L_i$, $1 \leq i \leq n$. It follows that

$$g(\mathbf{t}) = f(\mathbf{t}) \quad \text{for all } \mathbf{t} \in L_1 \times L_2 \times \cdots \times L_n$$

Observe that every monomial of g is of total degree strictly less than $\sum_{i=1}^{n} d_i$, apart from the monomial $\prod_{i=1}^{n} x_i^{d_i}$, which is unchanged. Thus g is nonzero. By Proposition 14.23, applied to g, there exists $\mathbf{t} \in L_1 \times L_2 \times \cdots \times L_n$ such that $g(\mathbf{t}) \neq 0$. This implies that $f(\mathbf{t}) \neq 0$. $\qquad\square$

Now let G be a graph, and let D be an orientation of G without directed odd cycles. If the outdegree sequence of D is $\mathbf{d}$, then every orientation of G with outdegree sequence $\mathbf{d}$ has the same sign as D (Exercise 14.6.2a), so $w(\mathbf{d}) \neq 0$. Applying Theorem 14.24 with $f(\mathbf{x}) = A(G, \mathbf{x})$ yields that G is $(\mathbf{d}+1)$-list-colourable; this is an alternative proof of Corollary 14.21. Another immediate consequence of the Combinatorial Nullstellensatz is the following one.

Corollary 14.25 *If G has an odd number of orientations D with outdegree sequence $\mathbf{d}$, then G is $(\mathbf{d}+1)$-list-colourable.*

Proof In this case $w(\mathbf{d})$ is also odd, thus nonzero. $\qquad\square$

Further applications of the Combinatorial Nullstellensatz are given in the exercises which follow.

Exercises

14.6.1 Show that the number of monomials in the expansion of the Vandermonde determinant of order n is $n!$

14.6.2

a) Let G be a graph, and let D be an orientation of G with outdegree sequence **d**.

    i) If D' is an orientation of G with outdegree sequence **d**, show that $\sigma(D') = \sigma(D)$ if and only if $|A(D) \setminus A(D')|$ is even.

    ii) Deduce that if D has no directed odd cycles, then all orientations of G with outdegree sequence **d** have the same sign.

b) For a graph G, denote by $G(\mathbf{d})$ the graph whose vertices are the orientations of G with outdegree sequence **d**, two such orientations D and D' being adjacent in $G(\mathbf{d})$ if and only if $A(D) \setminus A(D')$ is the arc set of a directed cycle. Denote by $B(\mathbf{d})$ the spanning subgraph of $G(\mathbf{d})$ whose edges correspond to directed odd cycles. Show that:

    i) $G(\mathbf{d})$ is connected,

    ii) $B(\mathbf{d})$ is bipartite.

14.6.3 Let T be a transitive tournament on n vertices with outdegree sequence $\mathbf{d} := (d_1, d_2, \ldots, d_n)$, where $d_1 \leq d_2 \leq \cdots \leq d_n$.

a) Express the number of directed triangles of T in terms of n and d.

b) Deduce that if $G = K_n$ and $\mathbf{d} \neq (0, 1, 2, \ldots, n-1)$, then the bipartite graph $B(\mathbf{d})$ (defined in Exercise 14.6.2) has parts of equal size.

c) Deduce that $(0, 1, 2, \ldots, n-1)$ is the only sequence **d** such that $w(\mathbf{d}) \neq 0$.

14.6.4 Let $G(x, y)$ be a graph, where $N(x) \setminus \{y\} = N(y) \setminus \{x\}$, and let D be an orientation of G with $d^+(x) = d^+(y)$. Show that $w(\mathbf{d}) = 0$, where **d** is the outdegree sequence of D. (S. CEROI)

14.6.5 THE FLEISCHNER–STIEBITZ THEOREM
Let G be a 4-regular graph on $3k$ vertices, the union of a cycle of length $3k$ and k pairwise disjoint triangles.

a) Show that the number of eulerian orientations of G with a given sign is even.

b) Fleischner and Stiebitz (1992) have shown (by induction on n) that the total number of eulerian orientations of G is congruent to $2 \,(\mathrm{mod}\ 4)$. Deduce that G is 3-list-colourable and thus 3-colourable.

(H. FLEISCHNER AND M. STIEBITZ)

(Sachs (1993) has shown that the number of 3-colourings of G is odd.)

14.6.6

a) For a graph G, as in Exercise 21.4.5, define $d^*(G) := \max\{d(F) : F \subseteq G\}$, the maximum of the average degrees of the subgraphs of G. Show that every bipartite graph G is $(\lceil d^*/2 \rceil + 1)$-list-colourable.

b) Deduce that every planar bipartite graph is 3-list-colourable.

c) Find a planar bipartite graph whose list chromatic number is three.

(N. ALON AND M. TARSI)

———————— ⟩⟩ ————————

14.6.7 THE CAUCHY–DAVENPORT THEOREM
Let A and B be nonempty subsets of $\mathbb{Z}_p$, where p is a prime. Define the *sum* $A + B$
of A and B by $A + B := \{a + b : a \in A, b \in B\}$.

a) If $|A| + |B| > p$, show that $A + B = \mathbb{Z}_p$.
b) Suppose that $|A| + |B| \le p$ and also that $|A+B| \le |A| + |B| - 2$. Let C be a set
   of $|A| + |B| - 2$ elements of $\mathbb{Z}_p$ that contains $A + B$. Consider the polynomial
   $f(x, y) := \prod_{c \in C}(x + y - c)$. Show that:
   i) $f(a, b) = 0$ for all $a \in A$ and all $b \in B$,
   ii) the coefficient of $x^{|A|-1}y^{|B|-1}$ in $f(x, y)$ is nonzero.
c) By applying the Combinatorial Nullstellensatz, deduce the *Cauchy–Davenport*
   Theorem: if A and B are nonempty subsets of $\mathbb{Z}_p$, where p is a prime, then
   either $A + B = \mathbb{Z}_p$ or $|A + B| \ge |A| + |B| - 1$.
                    (N. ALON, M.B. NATHANSON, AND I.Z. RUSZA)

14.6.8 Let $G = (V, E)$ be a loopless graph with average degree greater than $2p - 2$
and maximum degree at most $2p - 1$, where p is a prime. Show that G has a
p-regular subgraph by proceeding as follows.
      Consider the polynomial f over $\mathbb{Z}_p$ in the variables $\mathbf{x} = (x_e : e \in E)$ defined by

$$f(\mathbf{x}) := \prod_{v \in V}\left(1 - \left(\sum_{e \in E} m_{ve}x_e\right)^{p-1}\right) - \prod_{e \in E}(1 - x_e)$$

a) Show that:
   i) the degree of f is $e(G)$,
   ii) the coefficient of $\prod_{e \in E} x_e$ in f is nonzero.
b) Deduce from the Combinatorial Nullstellensatz that $f(\mathbf{c}) \ne 0$ for some vector
   $\mathbf{c} = (c_e : e \in E) \in \{0, 1\}^E$.
c) Show that $\mathbf{c} \ne \mathbf{0}$ and $\mathbf{Mc} = \mathbf{0}$.
d) By considering the spanning subgraph of G with edge set $\{e \in E : c_e = 1\}$,
   deduce that G has a p-regular subgraph.
e) Deduce, in particular, that every 4-regular loopless graph with one additional
   link contains a 3-regular subgraph.
                          (N. ALON, S. FRIEDLAND, AND G. KALAI)
   (Tashkinov (1984) proved that every 4-regular simple graph contains a 3-
   regular subgraph.)

14.7 The Chromatic Polynomial

We have seen how the adjacency polynomial provides insight into the complex topic
of graph colouring. Here, we discuss another polynomial related to graph colouring,
the chromatic polynomial. In this final section, we permit loops and parallel edges.

In the study of colourings, some insight can be gained by considering not only the existence of k-colourings but the number of such colourings; this approach was developed by Birkhoff (1912/13) as a possible means of attacking the Four-Colour Conjecture.

We denote the number of distinct k-colourings $c : V \to \{1, 2, \ldots, k\}$ of a graph G by $C(G, k)$. Thus $C(G, k) > 0$ if and only if G is k-colourable. In particular, if G has a loop then $C(G, k) = 0$. Two colourings are to be regarded as distinct if some vertex is assigned different colours in the two colourings; in other words, if $\{V_1, V_2, \ldots, V_k\}$ and $\{V_1', V_2', \ldots, V_k'\}$ are two k-colourings, then $\{V_1, V_2, \ldots, V_k\} = \{V_1', V_2', \ldots, V_k'\}$ if and only if $V_i = V_i'$ for $1 \le i \le k$. A triangle, for example, has six distinct 3-colourings.

If G is empty, then each vertex can be independently assigned any one of the k available colours, so $C(G, k) = k^n$. On the other hand, if G is complete, then there are k choices of colour for the first vertex, $k - 1$ choices for the second, $k - 2$ for the third, and so on. Thus, in this case, $C(G, k) = k(k - 1) \cdots (k - n + 1)$.

There is a simple recursion formula for $C(G, k)$, namely:

$$C(G, k) = C(G \setminus e, k) - C(G / e, k) \tag{14.5}$$

where e is any link of G. Formula (14.5) bears a close resemblance to the recursion formula for $t(G)$, the number of spanning trees of G (Proposition 4.9). We leave its proof as an exercise (14.7.1). The formula gives rise to the following theorem.

Theorem 14.26 *For any loopless graph G, there exists a polynomial $P(G, x)$ such that $P(G, k) = C(G, k)$ for all nonnegative integers k. Moreover, if G is simple and e is any edge of G, then $P(G, x)$ satisfies the recursion formula:*

$$P(G, x) = P(G \setminus e, x) - P(G / e, x) \tag{14.6}$$

The polynomial $P(G, x)$ is of degree n, with integer coefficients which alternate in sign, leading term x^n, and constant term zero.

Proof By induction on m. If $m = 0$, then $C(G, k) = k^n$, and the polynomial $P(G, x) = x^n$ satisfies the conditions of the theorem trivially.

Suppose that the theorem holds for all graphs with fewer than m edges, where $m \ge 1$, and let G be a loopless graph with m edges. If G is not simple, define $P(G, x) := P(H, x)$, where H is the underlying simple graph of G. By induction, H satisfies the conditions of the theorem, so G does also. If G is simple, let e be an edge of G. Both $G \setminus e$ and G / e have $m - 1$ edges and are loopless. By induction, there exist polynomials $P(G \setminus e, x)$ and $P(G / e, x)$ such that, for all nonnegative integers k,

$$P(G \setminus e, k) = C(G \setminus e, k) \quad \text{and} \quad P(G / e, k) = C(G / e, k) \tag{14.7}$$

Furthermore, there are nonnegative integers $a_1, a_2, \ldots, a_{n-1}$ and $b_1, b_2, \ldots, b_{n-1}$ such that:

$$P(G\backslash e, x) = \sum_{i=1}^{n-1}(-1)^{n-i}a_i x^i + x^n \quad \text{and} \quad P(G/e, x) = \sum_{i=1}^{n-1}(-1)^{n-i-1}b_i x^i \quad (14.8)$$

Define $P(G, x) := P(G \backslash e, x) - P(G/e, x)$, so that the desired recursion (14.6) holds. Applying (14.6), (14.7), and (14.5), we have:

$$P(G, k) = P(G \backslash e, k) - P(G/e, k) = C(G \backslash e, k) - C(G/e, k) = C(G, k)$$

and applying (14.6) and (14.8) yields

$$P(G, x) = P(G \backslash e, x) - P(G/e, x) = \sum_{i=1}^{n-1}(-1)^{n-i}(a_i + b_i)x^i + x^n$$

Thus $P(G, x)$ satisfies the stated conditions.   □

The polynomial $P(G, x)$ is called the *chromatic polynomial* of G. Formula (14.6) provides a means of calculating chromatic polynomials recursively. It can be used in either of two ways:

i) by repeatedly applying the recursion $P(G, x) = P(G \backslash e, x) - P(G/e, x)$, thereby expressing $P(G, x)$ as an integer linear combination of chromatic polynomials of empty graphs,

ii) by repeatedly applying the recursion $P(G \backslash e, x) = P(G, x) + P(G/e, x)$, thereby expressing $P(G, x)$ as an integer linear combination of chromatic polynomials of complete graphs.

Method (i) is more suited to graphs with few edges, whereas (ii) can be applied more efficiently to graphs with many edges (see Exercise 14.7.2).

The calculation of chromatic polynomials can sometimes be facilitated by the use of a number of formulae relating the chromatic polynomial of a graph to the chromatic polynomials of certain subgraphs (see Exercises 14.7.6a, 14.7.7, and 14.7.8). However, no polynomial-time algorithm is known for finding the chromatic polynomial of a graph. (Such an algorithm would clearly provide a polynomial-time algorithm for computing the chromatic number.)

Although many properties of chromatic polynomials have been found, no one has yet discovered which polynomials are chromatic. It has been conjectured by Read (1968) that the sequence of coefficients in any chromatic polynomial must first rise in absolute value and then fall; in other words, that no coefficient may be flanked by two coefficients having greater absolute value. But even if true, this property together with the properties listed in Theorem 14.26 would not be enough to characterize chromatic polynomials. For example, the polynomial $x^4 - 3x^3 + 3x^2$ satisfies all of these properties but is not the chromatic polynomial of any graph (Exercise 14.7.3b).

By definition, the value of the chromatic polynomial $P(G, x)$ at a positive integer k is the number of k-colourings of G. Surprisingly, evaluations of the polynomial at certain other special values of x also have interesting interpretations.

For example, it was shown by Stanley (1973) that $(-1)^n P(G, -1)$ is the number of acyclic orientations of G (Exercise 14.7.11).

Roots of chromatic polynomials, or *chromatic roots*, exhibit a rather curious behaviour. Using the recursion (14.6), one can show that 0 is the only real chromatic root less than 1 (Exercise 14.7.9); note that 0 is a chromatic root of every graph and 1 is a chromatic root of every nonempty loopless graph. Jackson (1993b) extended these observations by proving that no chromatic polynomial can have a root in the interval $(1, 32/27]$. Furthermore, Thomassen (1997c) showed that the only real intervals that are free of chromatic roots are $(-\infty, 0)$, $(0, 1)$, and $(1, 32/27]$. Thomassen (2000) also established an unexpected link between chromatic roots and Hamilton paths.

In the context of plane triangulations, the values of $P(G, x)$ at the *Beraha numbers* $B_k := 2 + 2\cos(2\pi/k)$, $k \geq 1$, are remarkably small, suggesting that the polynomial might have roots close to these numbers (see Tutte (1970)).

For a survey of this intriguing topic, we refer the reader to Read and Tutte (1988).

Exercises

$\star$**14.7.1** Prove the recursion formula (14.5).

14.7.2

a) Calculate the chromatic polynomial of the 3-star $K_{1,3}$ by using the recursion $P(G, x) = P(G \backslash e, x) - P(G / e, x)$ to express it as an integer linear combination of chromatic polynomials of empty graphs.

b) Calculate the chromatic polynomial of the 4-cycle C_4 by using the recursion $P(G \backslash e, x) = P(G, x) + P(G / e, x)$ to express it as an integer linear combination of chromatic polynomials of complete graphs.

14.7.3

a) Show that if G is simple, then the coefficient of x^{n-1} in $P(G, x)$ is $-m$.

b) Deduce that no graph has chromatic polynomial $x^4 - 3x^3 + 3x^2$.

14.7.4 Show that:

a) if G is a tree, then $P(G, x) = x(x - 1)^{n-1}$,

b) if G is connected and $P(G, x) = x(x - 1)^{n-1}$, then G is a tree.

14.7.5 Show that if G is a cycle of length n, then $P(G, x) = (x-1)^n + (-1)^n(x-1)$.

14.7.6

a) Show that $P(G \vee K_1, x) = xP(G, x - 1)$.

b) Using (a) and Exercise 14.7.5, show that if G is a wheel with n spokes, then $P(G, x) = x(x - 2)^n + (-1)^n x(x - 2)$.

14.7.7

a) Show that if G and H are disjoint, then $P(G \cup H, x) = P(G, x)P(H, x)$.

b) Deduce that the chromatic polynomial of a graph is equal to the product of the chromatic polynomials of its components.

14.7.8 If $G \cap H$ is complete, show that $P(G \cup H, x)P(G \cap H, x) = P(G, x)P(H, x)$.

14.7.9 Show that zero is the only real root of $P(G, x)$ smaller than one.

———————————⟫———————————

14.7.10 Show that no real root of $P(G, x)$ can exceed n. (L. LOVÁSZ)

⋆**14.7.11** Show that the number of acyclic orientations of a graph G is equal to $(-1)^n P(G, -1)$. (R.P. STANLEY)

⋆**14.7.12** Let G be a graph. For a subset S of E, denote by $c(S)$ the number of components of the spanning subgraph of G with edge set S. Show that $P(G, x) = \sum_{S \subseteq E} (-1)^{|S|} x^{c(S)}$. (H. WHITNEY)

14.8 Related Reading

FRACTIONAL COLOURINGS

A vertex colouring $\{V_1, V_2, \ldots, V_k\}$ of a graph $G = (V, E)$ can be viewed as expressing the incidence vector $\mathbf{1} := (1, 1, \ldots, 1)$ of V as the sum of the incidence vectors of the stable sets $V_1, V_2, \ldots, V_k$. This suggests the following relaxation of the notion of vertex colouring.

A *fractional colouring* of a graph $G = (V, E)$ is an expression of $\mathbf{1}$ as a non-negative rational linear combination of incidence vectors of stable sets of G. The least sum of the coefficients in such an expression is called the *fractional chromatic number* of G, denoted $\chi^*(G)$. Thus

$$\chi^* := \min \left\{ \sum \lambda_S : \sum \lambda_S f_S = \mathbf{1} \right\}$$

where the sums are taken over all stable sets S of G. The fractional chromatic number is clearly a lower bound on the chromatic number. However, it is still $\mathcal{NP}$-hard to compute this parameter.

By applying linear programming duality and using the fact that the stable sets of a graph G are the cliques of its complement $\overline{G}$, it can be shown that $\chi^*(G) = \alpha^{**}(\overline{G})$. Thus $\chi^*(G)$ is an upper bound for the Shannon capacity of $\overline{G}$ (see (12.2)).

The fractional chromatic number is linked to list colourings in a simple way. A graph G is (k, l)-*list-colourable* if, from arbitrary lists $L(v)$ of k colours, sets $C(v)$ of l colours can be chosen so that $C(u) \cap C(v) = \emptyset$ whenever $uv \in E$. It was shown by Alon et al. (1997) that $\chi^* = \inf\{k/l : G \text{ is } (k, l)\text{-list-colourable}\}$.

Further properties of the fractional chromatic number can be found in Scheinerman and Ullman (1997) and Schrijver (2003).

HOMOMORPHISMS AND CIRCULAR COLOURINGS

A *homomorphism* of a graph G into another graph H is a mapping $f : V(G) \to V(H)$ such that $f(u)f(v) \in E(H)$ for all $uv \in E(G)$. When H is the complete graph K_k, a homomorphism from G into H is simply a k-colouring of G. Thus the concept of a homomorphism may be regarded as a generalization of the notion of vertex colouring studied in this chapter. Many intriguing unsolved problems arise when one considers homomorphisms of graphs into graphs which are not necessarily complete (see Hell and Nešetřil (2004)). One particularly interesting instance is described below.

Let k and d be two positive integers such that $k \geq 2d$. A (k, d)-*colouring* of a graph G is a function $f : V \to \{1, 2, \ldots, k\}$ such that $d \leq |f(u) - f(v)| \leq k - d$ for all $uv \in E$. Thus a $(k, 1)$-colouring of a graph is simply a proper k-colouring, and a (k, d)-colouring is a homomorphism from the graph into $\overline{C_k^{d-1}}$, the complement of the $(d - 1)$st power of a k-cycle. Vince (1988) (see also Bondy and Hell (1990)) showed that, for any graph G, $\min\{k/d\colon G \text{ has a } (k, d)\text{-colouring}\}$ exists. This minimum, denoted by $\chi_c(G)$, is known as the *circular chromatic number* of G. (The name of this parameter derives from an alternative definition, due to X. Zhu, in which the vertices are associated with arcs of a circle, adjacent vertices corresponding to disjoint arcs.) One can easily show that $\chi(G) - 1 < \chi_c(G) \leq \chi(G)$, so $\chi(G) = \lceil \chi_c(G) \rceil$. However, there are graphs whose chromatic numbers are the same but whose circular chromatic numbers are different. For example, $\chi_c(K_3) = 3$ whereas $\chi_c(C_5) = 5/2$. One challenging unsolved problem in this area is to characterize the graphs for which these two parameters are equal. This question remains unsolved even for planar graphs. The comprehensive survey by Zhu (2001) contains many other intriguing problems.

15

Colourings of Maps

Contents

15.1 Chromatic Numbers of Surfaces

The Four-Colour Theorem tells us that every graph embeddable on the sphere has chromatic number at most four. More generally, for every closed surface Σ, there is a least integer k such that every graph embeddable on Σ has chromatic number at most k.

To see this, consider an arbitrary graph G embedded on Σ. In seeking an upper bound on the chromatic number of G, we may clearly assume G to be colour-critical. Let $d := d(G)$ denote the average degree of G. Noting that $\delta \leq d$, Theorem 14.7 yields:

$$\chi \leq d + 1$$

On the other hand, by Corollary 10.39, the number m of edges of G is at most $3n - 3c$, where $c := c(\Sigma)$ is the Euler characteristic of Σ. Because $d = 2m/n$, we have:

$$d \leq 6 - \frac{6c}{n}$$

These two inequalities now imply the following upper bound for the chromatic number of G.

$$\chi \leq 7 - \frac{6c}{n}$$

For a surface Σ, the least integer k such that every graph embeddable on Σ is k-colourable is called the *chromatic number* of Σ, denoted $\chi(\Sigma)$. We thus have:

$$\chi(\Sigma) \leq 7 - \frac{6c}{n} \tag{15.1}$$

The table in Figure 15.1 shows the resulting upper bounds on $\chi(\Sigma)$ for the four closed surfaces of smallest genus.

Σ	$c(\Sigma)$	$\chi(\Sigma)$
sphere	2	≤ 6
projective plane	1	≤ 6
torus	0	≤ 7
Klein bottle	0	≤ 7

Fig. 15.1. Bounds on the chromatic numbers of various surfaces

HEAWOOD'S INEQUALITY

When $c \leq 0$, a general upper bound for the chromatic number $\chi := \chi(\Sigma)$ of a surface Σ in terms of its Euler characteristic can be derived from (15.1). Noting that $\chi \leq n$, (15.1) yields the inequality $\chi \leq 7 - 6c/\chi$, and so $\chi^2 - 7\chi + 6c \leq 0$; that is, $(\chi - \alpha)(\chi - \beta) \leq 0$, where $\alpha, \beta = \frac{1}{2}(7 \pm \sqrt{49 - 24c})$. This results in the following bound, due to Heawood (1890).

Theorem 15.1 HEAWOOD'S INEQUALITY
For any surface Σ with Euler characteristic $c \leq 0$:

$$\chi(\Sigma) \leq \frac{1}{2}\left(7 + \sqrt{49 - 24c}\right) \qquad \square$$

As Heawood observed, in order to show that the bound in Theorem 15.1 is attainable for a given surface Σ, it suffices to find just one graph which is embeddable on the surface and requires the appropriate number of colours. Because the inequality $\chi \leq n$ was employed in deriving Heawood's Inequality, the graphs G for which $\chi = n$ (that is, the complete graphs) are natural candidates. Heawood himself found an embedding of K_7 on the torus (see Figure 3.9a), and deduced that the chromatic number of the torus is seven. On the other hand, although the Klein bottle has characteristic zero, Franklin (1934) showed that this surface is not 7-chromatic (the bound given by the table in Figure 15.1) but only 6-chromatic. Figure 15.2a depicts a 6-chromatic triangulation of the Klein bottle (an embedding of the graph obtained from K_6 by duplicating three pairwise nonadjacent edges), and Figure 15.2b shows its dual, known as the *Franklin graph*. Another drawing of the Franklin graph is given in Figure 15.3.

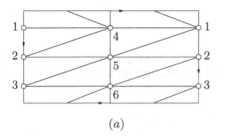
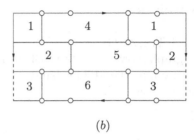

(a)   (b)

Fig. 15.2. (a) A 6-chromatic triangulation of the Klein bottle, and (b) its dual, the Franklin graph

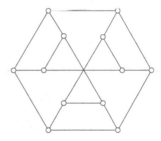

Fig. 15.3. Another drawing of the Franklin graph

THE MAP COLOUR THEOREM

With the exception of the Klein bottle, it has been proved that equality holds in Heawood's bound (Theorem 15.1) for every surface Σ of characteristic at most zero. This result, due to Ringel and Youngs (1968), is known as the *Map Colour Theorem* (see Ringel (1974)). There remain the two surfaces of positive characteristic, namely the projective plane and the sphere. Figure 10.25a shows an embedding of K_6 on the projective plane. Thus, the chromatic number of the projective plane is six, the bound given by the table in Figure 15.1. On the other hand, as the Four-Colour Theorem shows, Heawood's bound for the chromatic number of the sphere (see Figure 15.1) is off by two. We sketch a proof of the Four-Colour Theorem in the next section.

Although in this chapter we are interested in closed surfaces, one may define in an analogous manner chromatic numbers of surfaces with boundaries, such as the Möbius band (see Exercise 15.1.1).

Exercises

15.1.1 TIETZE GRAPH
The graph shown in Figure 15.4 is known as the *Tietze graph*. Find an embedding of this graph on the Möbius band showing that the chromatic number of the Möbius band is at least six.

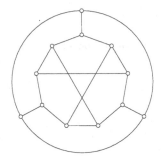

Fig. 15.4. The Tietze graph

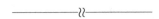

15.1.2 Let G be a triangulation of a closed surface.

a) Suppose that G has a proper 4-vertex-colouring $c : V \rightarrow \{1, 2, 3, 4\}$.
   i) Show that the parity of the number of faces of G whose vertices are coloured i, j, k is the same for each of the four triples $\{i, j, k\} \subset \{1, 2, 3, 4\}$.
   ii) Deduce from (a) that the parity of the number of vertices of odd degree coloured i is the same for each $i \in \{1, 2, 3, 4\}$.
b) Suppose that G has exactly two vertices of odd degree, and that these two vertices are adjacent. Deduce that G is not 4-colourable.

(J.P. BALLANTINE; S. FISK)

15.1.3 Let G be a quadrangulation of the projective plane.

a) Show that if G is not bipartite, then $\chi(G) \geq 4$, by proceeding as follows.

▷ Assume that $c : V \to \mathbb{Z}_3$ is a 3-colouring of G. Orient each edge uv of G, where $c(v) = c(u) + 1$, from u to v. Show that, in the resulting digraph, each 4-cycle of G has two arcs oriented in each sense.

▷ Because G is nonbipartite, it must have an odd cycle $C := v_1 v_2 \ldots v_{2k+1}$ and, because each face of G is a quadrangle, C must be a noncontractible cycle. Cutting the projective plane along C produces a planar near quadrangulation G' in which C corresponds to the $(4k + 2)$-cycle $C' := v_1' v_2' \ldots v_{2k+1}' v_1'' v_2'' \ldots v_{2k+1}''$, where v_i' and v_i'' are the two copies of v_i resulting from cutting along C. Obtain a contradiction by arguing that, in the derived orientation of G', the numbers of arcs of C' oriented in the two senses cannot be equal.

b) Using Euler's Formula for the projective plane (see Theorem 10.37), show that every subgraph of G has a vertex of degree at most three.

c) Deduce that $\chi = 2$ or $\chi = 4$.   (D.A. YOUNGS)

15.1.4 Let G_k denote the graph with vertex set $\{(i, S) : 1 \le i \le k, \emptyset \subset S \subseteq \{1, 2, \ldots, k\}$, and $i \notin S\}$, by joining two vertices (i, S) and (j, T) if and only if $i \in T, j \in S$ and $S \cap T = \emptyset$.

a) Draw G_2 and G_3.

b) Find an embedding of G_4 as a quadrangulation of the projective plane.

c) For $k = 2, 3, 4$, find a k-colouring of G_k in which the neighbour set of each of the k colour classes is a stable set.

d) Using Exercise 15.1.3, deduce that $\chi(G_4) = 4$.

(Gyárfás et al. (2004) have shown that, for $k \ge 2$, G_k is the unique smallest k-chromatic graph with a k-colouring in which the neighbour set of each of the k colour classes is a stable set.)

15.1.5

a) Show that the circulant $CG(\mathbb{Z}_{13}, \{1, -1, 5, -5\})$ is a triangle-free 4-chromatic graph.

b) Find a quadrangular embedding of this graph on the torus.

    (D. ARCHDEACON, J. HUTCHINSON, A. NAKAMOTO, S. NEGAMI AND K. OTA)

15.2 The Four-Colour Theorem

The Four-Colour Theorem, introduced in Section 11.1, states that every loopless planar graph is 4-colourable. Here, we give a brief outline of the ideas involved in its proof, due to Appel and Haken (1977a) and Appel et al. (1977).

The proof is by contradiction. Suppose that the Four-Colour Theorem is false. Then there is a smallest loopless plane graph which is not 4-colourable. The idea is to study the properties of such a hypothetical smallest counterexample, and eventually arrive at a contradiction. Throughout this section, therefore, G denotes a smallest counterexample to the Four-Colour Theorem, in the following sense.

i) G is not 4-colourable.
ii) Subject to (i), $v(G) + e(G)$ is as small as possible.

The starting point of our analysis consists of several basic observations concerning any such graph G.

Proposition 15.2 *Let G be a smallest counterexample to the Four-Colour Theorem. Then:*

i) G is 5-critical,
ii) G is a triangulation,
iii) G has no vertex of degree less than four.

Proof

i) Clearly, G must be 5-critical, otherwise there would exist a proper subgraph of G that is not 4-colourable, contradicting the minimality of $v(G) + e(G)$.
ii) To see that G is a triangulation, suppose that it has a face whose boundary is a cycle C of length greater than three. Because G is planar, there are two vertices x and y of C which are nonadjacent in G. The graph $G \, / \, \{x, y\}$ obtained by identifying x and y into a single vertex z is a planar graph with fewer vertices than G, and the same number of edges, hence has a 4-colouring c. The colouring of G derived from c by assigning the colour $c(z)$ to both x and y is then a 4-colouring of G, a contradiction.
iii) Because G is 5-critical, Theorem 14.7 implies that $\delta \geq 4$. $\square$

KEMPE CHAINS

By Corollary 10.22, every simple plane graph has a vertex of degree at most five. To obtain the needed contradiction, it would therefore suffice to show that a smallest counterexample G has no such vertex. Proposition 15.2(iii) states that G has no vertex of degree less than four. Kempe (1879) extended this result to vertices of degree four.

Theorem 15.3 *G has no vertex of degree four.*

Proof By contradiction. Let v be a vertex of degree four in G, and let $\{V_1, V_2, V_3, V_4\}$ be a 4-colouring of $G - v$; such a colouring exists because G is 5-critical. Because G itself is not 4-colourable, v must be adjacent to a vertex of each colour. Therefore, we can assume that the neighbours of v in clockwise order around v are v_1, v_2, v_3, and v_4, where $v_i \in V_i$ for $1 \leq i \leq 4$.

Denote by G_{ij} the subgraph of G induced by $V_i \cup V_j$. The vertices v_i and v_j belong to the same component of G_{ij}. If not, consider the component of G_{ij} that contains v_i. By interchanging the colours i and j in this component, we obtain a new 4-colouring of $G - v$ in which only three colours (all but i) are assigned to the neighbours of v. We have already shown that this situation cannot arise.

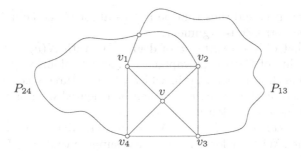

Fig. 15.5. Kempe's proof of the case $d(v) = 4$

Therefore, v_i and v_j indeed belong to the same component of G_{ij}. Let P_{ij} be a v_iv_j-path in G_{ij}, and let C denote the cycle $vv_1P_{13}v_3v$ (see Figure 15.5).

Because C separates v_2 and v_4 (in Figure 15.5, $v_2 \in \text{int}(C)$ and $v_4 \in \text{ext}(C)$), it follows from the Jordan Curve Theorem (10.1) that the path P_{24} meets C in some point. Because G is a plane graph, this point must be a vertex. But this is impossible, because the vertices of P_{24} have colours 2 and 4, whereas no vertex of C has either of these colours. $\qquad\square$

The bicoloured paths P_{ij} considered in the proof of Theorem 15.3 are known as *Kempe chains*, and the procedure of switching the two colours on a Kempe chain is referred to as a *Kempe interchange*. These ideas can be employed to establish the following more general theorem. We leave its proof, and the proof of its corollary, as an exercise (15.2.1).

Theorem 15.4 G *contains no separating 4-cycle.* $\qquad\square$

Corollary 15.5 G *is 5-connected.* $\qquad\square$

By virtue of Theorem 15.3, any smallest counterexample to the Four-Colour Theorem has a vertex of degree five, so its neighbours induce a 5-cycle, and this cycle is a separating 5-cycle. Birkhoff (1913) showed, moreover, that every separating 5-cycle in a smallest counterexample is one of this type, namely one induced by the set of neighbours of a vertex of degree five. A 5-connected graph with this property is said to be *essentially 6-connected*. (Essential connectivity is the vertex analogue of essential edge connectivity, treated in Section 9.3.) Combining Birkhoff's result with Proposition 15.2 and Corollary 15.5, we now have:

Theorem 15.6 G *is an essentially 6-connected triangulation.* $\qquad\square$

KEMPE'S ERRONEOUS PROOF

Proceeding in much the same way as in the proof of Theorem 15.3, Kempe (1879) believed that he had also proved that a smallest counterexample can contain no vertex of degree five, and thus that he had established the Four-Colour Theorem. Here is Kempe's erroneous argument.

Suppose that G has a vertex v of degree five with $N(v) = \{v_1, v_2, v_3, v_4, v_5\}$. By the minimality of G, the subgraph $G - v$ has a 4-colouring (V_1, V_2, V_3, V_4). Our aim is to find such a 4-colouring in which at most three colours are assigned to the neighbours of v; the vertex v can then be coloured with one of the remaining colours, resulting in a 4-colouring of G.

Consider a 4-colouring of $G - v$. As before, v is adjacent to a vertex of each of the four colours. Without loss of generality, suppose that $v_i \in V_i$, $1 \leq i \leq 4$, and $v_5 \in V_2$ (see Figure 15.6).

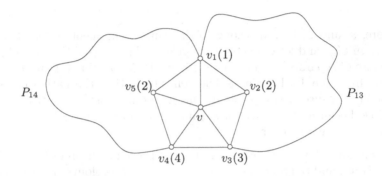

Fig. 15.6. Kempe's erroneous proof of the case $d(v) = 5$

We may assume that v_1 and v_3 belong to the same component of G_{13} and that v_1 and v_4 belong to the same component of G_{14}, otherwise the colours could be switched in the component of G_{13} or G_{14} containing v_1, resulting in a 4-colouring in which only three colours are assigned to the neighbours of v.

Let P_{13} be a v_1v_3-path in G_{13} and P_{14} a v_1v_4-path in G_{14}. The cycle $vv_1P_{13}v_3v$ separates vertices v_2 and v_4; thus v_2 and v_4 belong to different components of G_{24}. Similarly, the cycle $vv_1P_{14}v_4v$ separates v_3 and v_5, so v_3 and v_5 belong to different components of G_{23}. In light of these observations, Kempe argued that the colours 2 and 4 in the component of G_{24} containing v_2, and the colours 2 and 3 in the component of G_{23} containing v_5, could be interchanged to produce a 4-colouring of $G - v$ in which just the three colours 1, 3, and 4, are assigned to the neighbours of v. On assigning colour 2 to v, a 4-colouring of G would then be obtained.

At first sight, Kempe's line of argument seems perfectly reasonable, and his 'proof' remained unchallenged for over a decade. But eventually, after carefully analysing Kempe's argument, Heawood (1890) discovered that the double Kempe interchange does not necessarily result in a 4-colouring of $G - v$ in which just three colours are used on the neighbours of v (see Exercise 15.2.2). Heawood noted, however, that Kempe interchanges could be used to prove the Five-Colour Theorem. (In essence, the proof presented in Section 11.1.)

Reducibility

Although Kempe's proof was incorrect, it did contain the two main ingredients — reducibility and unavoidability — that led eventually to a proof of the Four-Colour Theorem. These notions both involve the concept of a configuration.

Let C be a cycle in a simple plane triangulation G. If C has no inner chords and exactly one inner bridge B, we call $B \cup C$ a *configuration of* G. The cycle C is the *bounding cycle* of the configuration, and B is its *bridge*. By a *configuration*, we mean a configuration of some simple plane triangulation. For example, the wheel W_k with k spokes ($k \geq 2$) is a configuration.

A configuration is *reducible* if it cannot be a configuration of a smallest counterexample to the Four-Colour Conjecture. It follows from Proposition 15.2(iii) and Theorem 15.3 that W_2, W_3, and W_4 are reducible. (By Theorem 15.6, every configuration bounded by a 5-cycle, except possibly W_5, is reducible. Kempe's failed proof was an attempt to show that W_5, also, is reducible.) Another example of a reducible configuration is the *Birkhoff diamond*, shown in Figure 15.7.

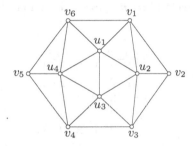

Fig. 15.7. The Birkhoff diamond

Theorem 15.7 *The Birkhoff diamond is reducible.*

Proof If possible, let G be a smallest counterexample that contains this configuration. Because G is essentially 6-connected by Theorem 15.6, no edge of G can join nonconsecutive vertices on the bounding cycle (Exercise 15.2.3). Consider the plane graph G' derived from G by deleting the four internal bridge vertices, u_1, u_2, u_3, and u_4, identifying v_1 and v_3 to form a new vertex v_0, joining v_0 and v_5 by a new edge, and deleting one of the two multiple edges between v_0 and v_2 (see Figure 15.8). (Observe that the operation of deriving G' from G does not create loops, since G is essentially 6-connected.)

Because $v(G') + e(G') < v(G) + e(G)$ and G is a smallest counterexample, there exists a 4-colouring c' of G'. The colouring c' gives rise to a partial 4-colouring of G (a 4-colouring of $G - \{u_1, u_2, u_3, u_4\}$) in which:

▷ v_1 and v_3 receive the same colour, say 1,
▷ v_5 receives a colour different from 1, say 2,

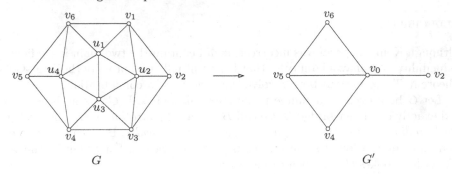

Fig. 15.8. Reduction of a graph containing the Birkhoff diamond

▷ v_2 receives a colour different from 1, without loss of generality either 2 or 3,
▷ v_4 and v_6 each receives a colour different from 1 and 2, namely either 3 or 4.

Thus, up to permutations of colours and symmetry, there are five possible ways in which the bounding cycle $C := v_1v_2v_3v_4v_5v_6v_1$ may be coloured, as indicated in the following table.

	v_1	v_2	v_3	v_4	v_5	v_6
c_1	1	2	1	3	2	3
c_2	1	2	1	4	2	3
c_3	1	3	1	4	2	3
c_4	1	3	1	4	2	4
c_5	1	3	1	3	2	3

In each of the first four cases, the given colouring can be extended to a 4-colouring of G by assigning appropriate colours to u_1, u_2, u_3, and u_4. For instance, the first colouring c_1 can be extended by assigning colour 4 to u_1, colour 3 to u_2, colour 2 to u_3, and colour 1 to u_4. We leave it to the reader to find appropriate extensions of the colourings c_2, c_3, and c_4 (Exercise 15.2.4a).

Consider, now, the colouring c_5, in which v_1 and v_3 have colour 1, v_5 has colour 2, and v_2, v_4, and v_6 have colour 3. In this case, we shall see that a Kempe interchange may be applied so that the resulting partial 4-colouring of G can be extended to a 4-colouring.

Firstly, consider the bipartite subgraph G_{34} induced by the vertices coloured 3 or 4. We may assume that v_2, v_4, and v_6 belong to the same component H of G_{34}. Suppose, for example, that some component of G_{34} contains v_2 but neither v_4 nor v_6. By swapping the colours 3 and 4 in this component, we obtain a colouring of type c_4 (with the colours 3 and 4 interchanged). The other cases can be disposed of similarly. We leave their verification to the reader (Exercise 15.2.4b).

It follows that H is an outer bridge of C in G, with v_2, v_4, and v_6 as its vertices of attachment. Now consider the bipartite subgraph G_{12} induced by the vertices coloured 1 and 2. If there were a component of G_{12} which contained both v_3 and v_5, this component would be an outer bridge of C overlapping H, which is

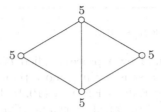

Fig. 15.9. The Heesch representation of the Birkhoff diamond

impossible. Thus the component H' of G_{12} which contains v_3 does not contain v_5. By interchanging the colours 1 and 2 in H', we obtain a new partial 4-colouring of G. In this colouring, v_1 has colour 1, v_3 and v_5 colour 2, and v_2, v_4, and v_6 colour 3. This colouring may now be extended to a 4-colouring of G by assigning colour 2 to u_1, colour 4 to u_2 and u_4, and colour 1 to u_3. This contradicts the minimality of G. We conclude that the Birkhoff diamond is reducible.   □

UNAVOIDABILITY

A set $\mathcal{U}$ of configurations is *unavoidable* if every essentially 6-connected triangulation necessarily contains at least one member of $\mathcal{U}$. As a simple example, the singleton $\{W_5\}$ is an unavoidable set.

It follows from the above definitions that a smallest counterexample can contain no reducible configuration but must contain at least one configuration from each unavoidable set. To obtain a contradiction to the existence of such a counterexample, it thus suffices to find an unavoidable set of configurations, each of which is reducible. The original set constructed by Appel and Haken (1977a) had 1936 members. Robertson et al. (1997a), using more refined techniques, constructed a smaller set, consisting of 633 configurations.

We have seen that the relation $m = 3n - 6$ between the numbers of edges and vertices of a triangulation implies that every essentially 6-connected triangulation contains a vertex of degree five, and hence that $\{W_5\}$ is an unavoidable set. By exploiting the same identity in subtler ways, one may derive further constraints on the degrees of vertices in triangulations, and thereby obtain other unavoidable sets of configurations. For this purpose, it is convenient to represent each configuration $F := B \cup C$ by the subgraph H induced by the internal vertices of its bridge B, together with the function d on $V(H)$, where $d(v)$ is the degree of v in G. Heesch (1969) was the first to propose this representation. For this reason, we refer to the pair (H, d) as the *Heesch representation* of F. Figure 15.9 shows the Heesch representation of the Birkhoff diamond.

PROOF TECHNIQUE: DISCHARGING

Unavoidable sets are found by a process called *discharging*. In the context of plane graphs, this process is nothing more than an ingenious and highly effective way of applying Euler's Formula (10.2). Initially, each vertex v is assigned a weight of $6 - d(v)$, called its *charge*. Thus the charge on v is positive if $d(v) < 6$, zero if $d(v) = 6$, and negative if $d(v) > 6$. We then attempt to *discharge* G (that is, make the charge at each vertex negative or zero) by redistributing the charge in some methodical fashion. Each such *discharging algorithm* defines a set $\mathcal{U}$ of configurations such that any triangulation which contains no member of $\mathcal{U}$ will be discharged by the algorithm. Observe, however, that no discharging algorithm can completely discharge a triangulation, because the sum of the charges, which remains constant throughout the procedure, is positive:

$$\sum_{v \in V} (6 - d(v)) = 6v(G) - \sum_{v \in V} d(v) = 6v(G) - 2e(G) = 12$$

We conclude that every triangulation must contain at least one member of $\mathcal{U}$; in other words, $\mathcal{U}$ is an unavoidable set of configurations.

The following is a simple example of a discharging algorithm: for each vertex of degree five, distribute its charge of 1 equally amongst its five neighbours. Every vertex of degree eight or more is discharged by this algorithm, because the maximum charge that a vertex v can receive from its neighbours is $\frac{1}{5}d(v)$, and if $d(v) \geq 8$,

$$6 - d(v) + \frac{1}{5}d(v) < 0$$

Also, each vertex of degree seven with no more than five neighbours of degree five is discharged, as is each vertex of degree five or six with no neighbour of degree five. Thus every essentially 6-connected triangulation must contain either a vertex of degree five that is adjacent to a vertex of degree five or six, or else a vertex of degree seven with at least five neighbours of degree five. However, a vertex of degree seven with five neighbours of degree five clearly has two consecutive neighbours of degree five, and these are adjacent in G. Thus the set $\mathcal{U}$ consisting of the two configurations shown below is the unavoidable set corresponding to this discharging algorithm.

In recent years, this discharging technique has been used with success to attack a variety of other colouring problems for graphs embeddable on the plane and other surfaces. Here is one example.

DISCHARGING (CONTINUED)

In 1975, R. Steinberg conjectured that every planar graph without cycles of length four or five is 3-colourable (see Steinberg (1993)). Abbott and Zhou (1991) employed discharging to prove the weaker statement that a planar graph is 3-colourable if it contains no cycles of length k for $4 \leq k \leq 11$. To see this, observe first that a smallest counterexample to the assertion must be a 2-connected planar graph with minimum degree at least three. Let G be a planar embedding of such a graph. Assign a charge of $d(v) - 6$ to each vertex $v \in V$, and a charge of $2d(f) - 6$ to each face $f \in F$. Using Euler's Formula, it can be verified that the total charge assigned to vertices and faces is -12. Now, for each face of degree twelve or more, transfer a charge of $3/2$ to each of the vertices incident with it. Because G is 2-connected, all faces of G are bounded by cycles, by Theorem 10.7. Also, because G has no 4-cycles, no edge of G is incident with two triangles. Thus each vertex v is incident with at least $\lceil d(v)/2 \rceil$ distinct faces of degree twelve or more. A simple computation shows that, after the transfers of charges, all vertices and faces have nonnegative charges. This contradiction establishes the result. By using more complicated discharging rules, Borodin et al. (2005) showed that every planar graph without cycles of lengths between four and seven is 3-colourable. An analogous result for surfaces was established by Zhao (2000). He showed that, given any surface Σ, there exists a constant $f(\Sigma)$ such that any graph embeddable on Σ and containing no k-cycles, $4 \leq k \leq f(\Sigma)$, is 3-colourable. For a survey of applications of the discharging technique, see Salavatipour (2003).

The original proof of the Four-Colour Theorem by Appel and Haken (1977a) and Appel et al. (1977) relies heavily on the computer for checking details involved in finding an unavoidable set and verifying that all configurations in that set are reducible. It employs no fewer than 487 discharging rules, resulting in a set of over 1400 unavoidable configurations. The smaller configurations in this list could be shown to be reducible by using the fact that a smallest counterexample to the Four Colour Conjecture is essentially 6-connected. However, in order to handle some of the larger configurations, results of Bernhart (1947) on the reducibility of certain special configurations were crucial.

The more recent proof by Robertson et al. (1997a), although also dependent on the computer, is simpler in many ways. Only thirty-two discharging rules are needed, generating a list of 633 unavoidable configurations. (The definition of a configuration used there differs slightly from the one given here.)

Further information about the Four-Colour Theorem and its proof can be found in the expository article by Woodall and Wilson (1978) and the book by Wilson (2002).

Exercises

$\star$**15.2.1** Give proofs of Theorem 15.4 and Corollary 15.5.

15.2.2 By considering the partially coloured plane triangulation depicted in Figure 15.10, show that the 'double switching' of colours proposed by Kempe leads to an improper colouring.   (W.T. TUTTE AND H. WHITNEY)

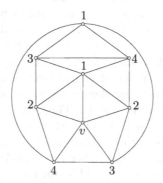

Fig. 15.10. An example illustrating the flaw in Kempe's argument

$\star$**15.2.3** Show that if G is a smallest counterexample that contains the Birkhoff diamond, no two nonconsecutive vertices on the bounding cycle of that configuration can be adjacent in G.

$\star$**15.2.4** Complete the proof that the Birkhoff diamond is reducible by:

   i) finding appropriate extensions of the colourings c_2, c_3, and c_4,
  ii) verifying the cases in which some component of G_{34} contains v_4 but neither v_2 nor v_6, or v_6 but neither v_2 nor v_4.

15.2.5

   a) Determine the configuration whose Heesch representation is shown in Figure 15.11.
   b) Show that this configuration is reducible.

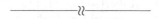

15.2.6 Consider the discharging algorithm in which, for each vertex v of degree five, the unit charge of v is distributed equally among its neighbours of degree nine or more. Find the unavoidable set (consisting of seven configurations) determined by this algorithm.

15.2.7 Invent a discharging algorithm and determine the set $\mathcal{U}$ of unavoidable configurations to which it gives rise.

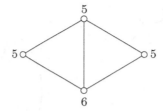

Fig. 15.11. The Heesch representation for Exercise 15.2.5

15.3 List Colourings of Planar Graphs

THOMASSEN'S PROOF OF THE FIVE-COLOUR THEOREM

List colourings are of interest not only because they extend the usual notion of colouring, but also because, when judiciously used, they can provide remarkably simple proofs of theorems concerning standard colourings. The following theorem on list colourings of planar graphs, due to Thomassen (1994), implies the Five-Colour Theorem.

A *near-triangulation* is a plane graph all of whose inner faces have degree three.

Theorem 15.8 *Let G be a near-triangulation whose outer face is bounded by a cycle C, and let x and y be consecutive vertices of C. Suppose that $L : V \to 2^{\mathbb{N}}$ is an assignment of lists of colours to the vertices of G such that:*

i) $|L(x)| = |L(y)| = 1$*, where* $L(x) \neq L(y)$*,*
ii) $|L(v)| \geq 3$ *for all* $v \in V(C) \setminus \{x, y\}$*,*
iii) $|L(v)| \geq 5$ *for all* $v \in V(G) \setminus V(C)$*.*

Then G is L-colourable.

Proof By induction on $v(G)$. If $v(G) = 3$, then $G = C$ and the statement is trivial. So we may assume that $v(G) > 3$.

Let z and x' be the immediate predecessors of x on C. Consider first the case where x' has a neighbour y' on C other than x and z (see Figure 15.12a.) In this case, $C_1 := x'Cy'x'$ and $C_2 := x'y'Cx'$ are two cycles of G, and G is the union of the near-triangulation G_1 consisting of C_1 together with its interior and the near-triangulation G_2 consisting of C_2 together with its interior. Let L_1 denote the restriction of L to $V(G_1)$. By induction, G_1 has an L_1-colouring c_1. Now let L_2 be the function on $V(G_2)$ defined by $L_2(x') := \{c_1(x')\}$, $L_2(y') := \{c_1(y')\}$, and $L_2(v) := L(v)$ for $v \in V(G_2) \setminus \{x', y'\}$. Again by induction (with x' and y' playing the roles of x and y, respectively), there is an L_2-colouring c_2 of G_2. By the definition of L_2, the colourings c_1 and c_2 assign the same colours to x' and y', the two vertices common to G_1 and G_2. Thus the function c defined by $c(v) := c_1(v)$ for $v \in V(G_1)$ and $c(v) := c_2(v)$ for $v \in V(G_2) \setminus V(G_1)$ is an L-colouring of G.

Suppose now that the neighbours of x' lie on a path xPz internally disjoint from C, as shown in Figure 15.12b. In this case, $G' := G - x'$ is a near-triangulation

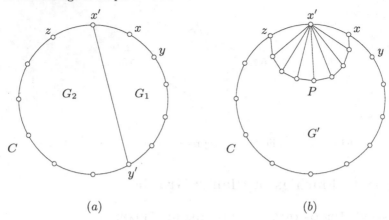

Fig. 15.12. Proof of Theorem 15.8

whose outer face is bounded by the cycle $C' := xCz\overleftarrow{P}x$. Let α and β be two distinct colours in $L(x') \setminus L(x)$. Consider the function L' on $V(G')$ defined by $L'(v) := L(v) \setminus \{\alpha, \beta\}$ for $v \in V(P) \setminus \{x, z\}$, and $L'(v) := L(v)$ for all other vertices v of G'. By induction, there exists an L'-colouring c' of G'. One of the colours α and β is different from $c'(z)$. By assigning that colour to x', the colouring c' is extended to an L-colouring c of G. $\qquad\square$

An immediate consequence of Theorem 15.8 is the following strengthening of the Five-Colour Theorem.

Corollary 15.9 *Every planar graph is 5-list-colourable.*

This is one of the more illuminating proofs of the Five-Colour Theorem. Unfortunately, no list colouring analogue of the Four-Colour Theorem is known. Indeed, Voigt (1993) found examples of planar graphs which are not 4-list-colourable. Even so, it is conceivable that an appropriate list colouring version of the Four-Colour Theorem will provide a more transparent (and shorter) proof of that theorem as well. (For instance, Kündgen and Ramamurthi (2002) have suggested that every planar graph admits a list colouring when the available colours come in pairs and each list consists of two of these pairs.)

Voigt (1995) also gave examples of triangle-free planar graphs that are not 3-list-colourable. These show that there is no natural list-colouring extension of the following 'three-colour theorem', due to Grötzsch (1958/1959).

Theorem 15.10   GRÖTZSCH'S THEOREM
Every triangle-free planar graph is 3-colourable. $\qquad\square$

Nonetheless, it turns out that every planar graph of girth five is 3-list-colourable. This result was established by Thomassen (1994) using similar but more involved arguments than those he employed to establish Theorem 15.8. It may legitimately be regarded as a list-colouring extension of Grötzsch's Theorem, because the latter can be reduced quite easily to planar graphs of girth five.

Grötzsch's Theorem can also be proved in much the same way as the Four-Colour Theorem, but the arguments are considerably simpler. The 4-chromatic Grötzsch graph (Figure 14.4) shows that Grötzsch's Theorem does not extend to nonplanar graphs. Indeed, it was constructed by Grötzsch (1958/1959) for this very purpose.

Exercises

15.3.1

a) Show that every k-degenerate graph is $(k + 1)$-list-colourable.
b) Deduce that:
    i) every simple outerplanar graph is 3-list-colourable,
    ii) every simple triangle-free planar graph is 4-list-colourable,
                                          (J. KRATOCHVÍL AND ZS. TUZA)
    iii) every simple planar graph is 6-list-colourable.

————————⟩⟩————————

15.3.2 The *list chromatic number* of a surface Σ is the smallest positive integer k such that every loopless graph embeddable on Σ is k-list-colourable. Show that, except for the sphere, the list chromatic number of any closed surface (whether orientable or not) is equal to its chromatic number.

15.4 Hadwiger's Conjecture

As we saw in Section 14.3, there exist graphs of arbitrarily high girth and chromatic number. What, then, can one say about the structure of graphs with high chromatic number? A longstanding conjecture due to Hadwiger (1943) asserts that any such graph necessarily contains a large clique, not as a subgraph but as a minor. (Recall that a *minor* of a graph G is a graph which can be obtained from G by deleting vertices and deleting or contracting edges.)

HADWIGER'S CONJECTURE

Conjecture 15.11 *Every k-chromatic graph has a K_k-minor.*

For $k = 1$ and $k = 2$, the validity of Hadwiger's conjecture is obvious. It is also easily verified for $k = 3$, because a 3-chromatic graph necessarily contains an odd cycle, and every odd cycle contains K_3 as a minor. Hadwiger (1943) settled the case $k = 4$, and Dirac (1952a) proved the following somewhat stronger theorem.

Theorem 15.12 *Every 4-chromatic graph contains a K_4-subdivision.*

Proof Let G be a 4-chromatic graph, and let F be a 4-critical subgraph of G. By Theorem 14.7, $\delta(F) \geq 3$. By Exercise 10.1.5, F contains a subdivision of K_4, so G does too. □

Wagner (1964) showed that the case $k = 5$ of Hadwiger's Conjecture is equivalent to the Four-Colour Theorem; thus it also is true. The case $k = 6$ was verified by Robertson et al. (1993); this proof, too, relies on the Four-Colour Theorem. However, the conjecture has not yet been settled in general, and some mathematicians now believe that it might even be false. Although it has been verified for small values of k, the conjecture appears to be hard to prove when k is large relative to n, in particular when $k = n/2$ (and n is even). More precisely, if $\alpha = 2$, then $\chi \geq n/2$ by (14.1), so G should have a $K_{\lceil n/2 \rceil}$-minor according to Hadwiger's Conjecture. It is not hard to show that every graph G with $\alpha = 2$ has a $K_{\lceil n/3 \rceil}$-minor (Exercise 15.4.4), but no one has yet succeeded in bridging the gap between $n/3$ and $n/2$.

On the other hand, a weaker form of Hadwiger's Conjecture can be proved. Due to Mader (1967), it states that every graph of sufficiently high chromatic number has a K_k-minor. As is frequently the case with results about the chromatic number, this theorem is really one about graphs with high average degree; the link with the chromatic number is made via Theorem 14.7 (as in the proof of Theorem 15.12).

Theorem 15.13 *Every simple graph G with $m \geq 2^{k-3}n$ has a K_k-minor.*

Proof By induction on m. The validity of the theorem is readily verified when $k \leq 3$. Thus we may assume that $k \geq 4$ and $m \geq 10$. Let G be a graph with n vertices and m edges, where $m \geq 2^{k-3}n$. If G has an edge e which lies in at most $2^{k-3} - 1$ triangles, the underlying simple graph of $G \,/\, e$ has $n - 1$ vertices and at least $2^{k-3}n - 2^{k-3} = 2^{k-3}(n-1)$ edges, and so has a K_k-minor by induction. Thus G, too, has a K_k-minor.

We may assume, therefore, that each edge of G lies in at least 2^{k-3} triangles. For $e \in E$, let us denote by $t(e)$ the number of triangles containing e. Because an edge e lies in the subgraph $G[N(v)]$ induced by the neighbours of a vertex v if and only if v is the 'apex' of a triangle whose 'base' is e, we have:

$$\sum_{v \in V} |E(G[N(v)])| = \sum_{e \in E} t(e) \geq 2^{k-3}m = \sum_{v \in V} 2^{k-4}d(v)$$

We deduce that G has a vertex v whose neighbourhood subgraph $H := G[N(v)]$ satisfies the inequality:

$$e(H) \geq 2^{k-4}d(v) = 2^{k-4}v(H)$$

By induction, H has a K_{k-1}-minor. Therefore G has a K_k-minor. □

Corollary 15.14 *For $k \geq 2$, every $(2^{k-2} + 1)$-chromatic graph has a K_k-minor.*

Proof Let G be a $(2^{k-2}+1)$-chromatic graph, and let F be a $(2^{k-2}+1)$-critical subgraph of G. By Theorem 14.7, $\delta(F) \geq 2^{k-2}$, and so $e(F) \geq 2^{k-3}v(F)$. Theorem 15.13 implies that F has a K_k-minor. Therefore G, too, has a K_k-minor. $\qquad\square$

If, instead of the chromatic number, one considers the fractional chromatic number, Corollary 15.14 can be sharpened to a linear bound, as was shown by Reed and Seymour (1998): every graph with fractional chromatic number greater than $2k-2$ has a K_k-minor.

Hajós' Conjecture

An even stronger conjecture than Hadwiger's was proposed by G. Hajós, probably in the early 1950s (see Dirac (1952a)). It asserted that every k-chromatic graph contains a subdivision of K_k. (For $k = 4$, this is Theorem 15.12.) Hajós' Conjecture was disproved by Catlin (1979), who found the 8-chromatic graph shown in Figure 14.3, which contains no subdivision of K_8 (see Exercise 15.4.3). Shortly thereafter, Erdős and Fajtlowicz (1981) totally demolished the conjecture by proving that almost every graph (in the probabilistic sense) is a counterexample. This is but one more illustration of the power of the probabilistic method.

Theorem 15.15 *Almost every graph is a counterexample to Hajós' Conjecture.*

Proof Let $G \in \mathcal{G}_{n,1/2}$. Then almost surely $\alpha \leq \lceil 2\log_2 n \rceil$ (Exercise 13.2.11). Thus almost surely

$$\chi \geq \frac{n}{\alpha} \geq \frac{n}{\lceil 2\log_2 n \rceil} \tag{15.2}$$

Now the expected number of subgraphs of G with s vertices and $t := \binom{s}{2} - n$ edges is

$$\binom{n}{s}\binom{\binom{s}{2}}{t}2^{-t} = \binom{n}{s}\binom{\binom{s}{2}}{n}2^{n-\binom{s}{2}} \leq n!\frac{(s^2/2)^n}{n!}2^{n-\binom{s}{2}} = s^{2n}2^{-\binom{s}{2}}$$

If $s := \lceil n^c \rceil$, where $\frac{1}{2} < c < 1$, this latter quantity tends to zero as n tends to infinity. Therefore, by Markov's Inequality (Proposition 13.4), almost surely every subgraph of G on s vertices has fewer than $\binom{s}{2} - n$ edges. Now this implies that G almost surely contains no subdivision of K_s. For, in order to form such a subdivision, at least $n+1$ edges would need to be be subdivided, and there are simply not enough vertices available for this. Because s is much smaller than the lower bound on χ given by (15.2) for n sufficiently large, we conclude that almost every graph G is a counterexample to Hajós' Conjecture. $\qquad\square$

More recently, Thomassen (2005) made the surprising observation that several classical results in extremal graph theory (discussed in Chapter 12) furnish counterexamples to Hajós' Conjecture (see Exercise 15.4.5). Ironically, these extremal results were known long before Catlin announced his counterexamples to Hajós' Conjecture, indeed even before Hajós proposed the conjecture in the first place!

Hadwiger's Conjecture, whether true or false, will not suffer the sorry fate of Hajós' Conjecture. In contrast to Theorem 15.15, it has been shown by Bollobás et al. (1980) that almost no graph is a counterexample to Hadwiger's Conjecture.

Exercises

15.4.1

a) Show that $K_{4,4}$ contains no K_5-subdivision.
b) Find a K_5-subdivision in $K_{4,5}$.

15.4.2

a) Show that if G is simple with $n \geq 4$ and $m \geq 2n - 2$, then G contains a K_4-subdivision.
b) For $n \geq 4$, find a simple graph G with $m = 2n - 3$ that contains no K_4-subdivision.

15.4.3 By verifying that it is 8-chromatic but contains no K_8-subdivision, show that the Catlin graph (Figure 14.3) is a counterexample to Hajós' Conjecture.

15.4.4 Let G be a graph with $\alpha = 2$.

a) If G is not connected, show that G contains $K_{\lceil n/2 \rceil}$ (as a subgraph).
b) If G is connected, show that G contains a path uvw with $uw \notin E$ and that every vertex of $G - \{u, v, w\}$ is adjacent to either u or w (or both).
c) Deduce, by induction, that G has a $K_{\lceil n/3 \rceil}$-minor.

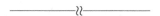

15.4.5 Let G be a graph on $n := 2k^3$ vertices with $\alpha \leq k$ and $\omega \leq k$, where $k \geq 40$.

a) Using Theorem 12.12, show that there exists such a graph G.
b) Show that $\chi \geq 2k^2$.
c) Suppose that G contains a subdivision of K_{2k^2}.
   i) By applying Turán's Theorem (12.7), show that at least $2k^3 - k^2$ 'edges' of this subdivision are subdivided by vertices of G (that is, correspond to paths of length at least two in G).
   ii) Deduce that $n \geq 2k^3 + k^2$.
d) Conclude that G is a counterexample to Hajós' Conjecture. (C. THOMASSEN)

15.4.6 Show that any graph G has $K_{\lceil n/(2\alpha-1) \rceil}$ as a minor.
                                                    (P. DUCHET AND H. MEYNIEL)

15.4.7 Show that Hadwiger's Conjecture is true for graphs with stability number two if and only if every graph G with $\alpha = 2$ has $K_{\lceil n/2 \rceil}$ as a minor.

15.5 Related Reading

NEAR 4-COLOURINGS OF GRAPHS ON SURFACES

The Map Colour Theorem, discussed in Section 11.1, implies that the chromatic number of S_k, the sphere with k handles, increases with k. On the other hand, Albertson and Hutchinson (1978) showed that if G is a graph embeddable on S_k, then one can always obtain a planar graph by deleting no more than $k\sqrt{2n}$ of its vertices. This implies that all but $k\sqrt{2n}$ vertices of any graph G embeddable on S_k can be 4-coloured. A far stronger assertion was conjectured by Albertson (1981): given any surface Σ, there exists an integer $f(\Sigma)$ such that that all but $f(\Sigma)$ vertices of any graph embeddable on Σ can be 4-coloured. He conjectured, in particular, that all but three vertices of any toroidal graph can be 4-coloured. Thomassen (1994) proved that there are precisely four 6-critical graphs on the torus, and deduced that all but two vertices of any toroidal graph can be 5-coloured. Thomassen (1997b) showed, furthermore, that for any fixed surface Σ, the number of 6-critical graphs embeddable on Σ is finite. By contrast, Fisk (1978) constructed infinitely many 5-critical graphs on every surface but the sphere. For a discussion of Albertson's conjecture and related topics, see Jensen and Toft (1995).

16

Matchings

Contents

16.1 Maximum Matchings

A *matching* in a graph is a set of pairwise nonadjacent links. If M is a matching, the two ends of each edge of M are said to be *matched* under M, and each vertex incident with an edge of M is said to be *covered* by M. A *perfect matching* is one which covers every vertex of the graph, a *maximum matching* one which covers as many vertices as possible. A graph is *matchable* if it has a perfect matching. Not all graphs are matchable. Indeed, no graph of odd order can have a perfect matching, because every matching clearly covers an even number of vertices. Recall that the number of edges in a maximum matching of a graph G is called the *matching number* of G and denoted $\alpha'(G)$. A *maximal matching* is one which cannot be extended to a larger matching. Equivalently, it is one which may be obtained by choosing edges in a greedy fashion until no further edge can be incorporated. Such a matching is not necessarily a maximum matching. Examples of maximal and perfect matchings in the pentagonal prism are indicated in Figures 16.1 and 16.1b, respectively.

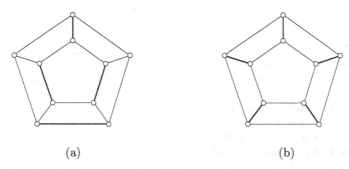

(a)   (b)

Fig. 16.1. (a) A maximal matching, (b) a perfect matching

The main question we address in this chapter is:

Problem 16.1   THE MAXIMUM MATCHING PROBLEM
GIVEN: *a graph G,*
FIND: *a maximum matching M^* in G.*

There are many questions of practical interest which, when translated into the language of graph theory, amount to finding a maximum matching in a graph. One such is:

Problem 16.2   THE ASSIGNMENT PROBLEM
A certain number of jobs are available to be filled. Given a group of applicants for these jobs, fill as many of them as possible, assigning applicants only to jobs for which they are qualified.

This situation can be represented by means of a bipartite graph $G[X, Y]$ in which X represents the set of applicants, Y the set of jobs, and an edge xy with $x \in X$ and $y \in Y$ signifies that applicant x is qualified for job y. An assignment of applicants to jobs, one person per job, corresponds to a matching in G, and the problem of filling as many vacancies as possible amounts to finding a maximum matching in G.

As we show in Section 16.5, the Assignment Problem can be solved in polynomial time. Indeed, we present there a polynomial-time algorithm for finding a maximum matching in an arbitrary graph. The notions of alternating and augmenting paths with respect to a given matching, defined below, play an essential role in these algorithms.

AUGMENTING PATHS

Let M be a matching in a graph G. An M-*alternating path* or *cycle* in G is a path or cycle whose edges are alternately in M and $E \setminus M$. An M-alternating path might or might not start or end with edges of M (see Figure 16.2).

Fig. 16.2. Types of M-alternating paths

If neither its origin nor its terminus is covered by M (as in the top-left path in Figure 16.2) the path is called an M-*augmenting path*. Figure 16.3a shows an M-augmenting path in the pentagonal prism, where M is the matching indicated in Figure 16.1a.

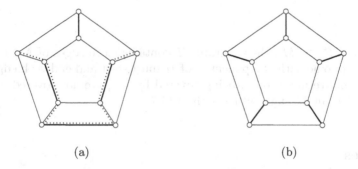

(a)   (b)

Fig. 16.3. (a) An M-augmenting path P, (b) the matching $M \triangle E(P)$

BERGE'S THEOREM

The following theorem, due to Berge (1957), points out the relevance of augmenting paths to the study of maximum matchings.

Theorem 16.3   BERGE'S THEOREM
A matching M in a graph G is a maximum matching if and only if G contains no M-augmenting path.

Proof   Let M be a matching in G. Suppose that G contains an M-augmenting path P. Then $M' := M \triangle E(P)$ is a matching in G, and $|M'| = |M| + 1$ (see Figure 16.3). Thus M is not a maximum matching.

Conversely, suppose that M is not a maximum matching, and let M^* be a maximum matching in G, so that $|M^*| > |M|$. Set $H := G[M \triangle M^*]$, as illustrated in Figure 16.4.

Each vertex of H has degree one or two in H, for it can be incident with at most one edge of M and one edge of M^*. Consequently, each component of H is either an even cycle with edges alternately in M and M^*, or else a path with edges alternately in M and M^*.

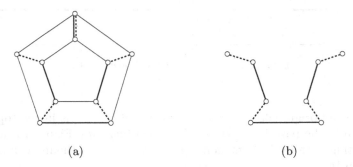

(a)   (b)

Fig. 16.4. (a) Matchings M (heavy) and M^* (broken), and (b) the subgraph $H := G[M \triangle M^*]$

Because $|M^*| > |M|$, the subgraph H contains more edges of M^* than of M, and therefore some path-component P of H must start and end with edges of M^*. The origin and terminus of P, being covered by M^*, are not covered by M. The path P is thus an M-augmenting path in G.   □

Exercises

16.1.1

a) Show that the Petersen graph has exactly six perfect matchings.
b) Determine $pm(K_{2n})$ and $pm(K_{n,n})$, where $pm(G)$ denotes the number of perfect matchings in graph G.

16.1.2 Show that it is impossible, using 1×2 rectangles (dominoes), to tile an 8×8 square (chessboard) from which two opposite 1×1 corner squares have been removed.

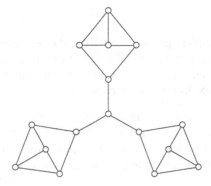

Fig. 16.5. The Sylvester graph: a 3-regular graph with no perfect matching

16.1.3 Show that if G is triangle-free, then $\alpha'(G) = n - \chi(\overline{G})$.

16.1.4 Find a maximal matching M and a perfect matching M^* in the pentagonal prism such that the subgraph induced by $M \triangle M^*$ has two components, one a cycle and the other an M-augmenting path.

⋆16.1.5

a) Let M and M' be maximum matchings of a graph G. Describe the structure of the subgraph $H := G[M \triangle M']$.

b) Let M and M' be perfect matchings of a graph G. Describe the structure of the subgraph $H := G[M \triangle M']$.

c) Deduce from (b) that a tree has at most one perfect matching.

16.1.6 Let M and N be matchings of a graph G, where $|M| > |N|$. Show that there are disjoint matchings M' and N' of G such that $|M'| = |M| - 1, |N'| = |N| + 1$ and $M' \cup N' = M \cup N$.

⋆16.1.7

a) Let M be a perfect matching in a graph G and S a subset of V. Show that $|M \cap \partial(S)| \equiv |S| \pmod 2$.

b) Deduce that if M is a perfect matching of the Petersen graph, and C is the edge set of one of its 5-cycles, then $|M \cap C|$ is even.

16.1.8

a) Let M be a perfect matching in a graph G, all of whose vertices are of odd degree. Show that M includes every cut edge of G.

b) Deduce that the 3-regular graph of Figure 16.5 has no perfect matching.

c) For each $k \geq 2$, find a $(2k+1)$-regular simple graph with no perfect matching.

16.1.9 Let M be a maximal matching in a graph G, and let M^* be a maximum matching in G. Show that $|M| \geq \frac{1}{2}|M^*|$.

———————♆———————

16.1.10 Consider a complete graph K on $2n$ vertices embedded in the plane, with n vertices coloured red, n vertices coloured blue, and each edge a straight-line segment. Show that K has a perfect matching whose edges do not cross, with each edge joining a red vertex and a blue vertex.

16.1.11 The game of *Slither* is played as follows. Two players alternately select distinct vertices $v_0, v_1, v_2, \ldots$ of a graph G, where, for $i \geq 0$, v_{i+1} is required to be adjacent to v_i. The last player able to select a vertex wins the game. Show that the first player has a winning strategy if and only if G has no perfect matching.

(W.N. ANDERSON, JR.)

16.1.12 Let G be a simple graph with $n \geq 2\delta$. Show that $\alpha' \geq \delta$.

16.1.13 Let G be a nonempty graph which has a unique perfect matching M.

a) Show that G has no M-alternating cycle, and that the first and last edges of every M-alternating path belong to M.
b) Deduce that if $G := G[X, Y]$ is bipartite, then X and Y each contain a vertex of degree one.
c) Give an example of a graph with a unique perfect matching and no vertex of degree one.

16.1.14

a) Let M be a matching in a graph G. Show that there is a maximum matching in G which covers every vertex covered by M.
b) Deduce that every vertex of a connected nontrivial graph is covered by some maximum matching.
c) Let $G[X, Y]$ be a bipartite graph and let $A \subseteq X$ and $B \subseteq Y$. Suppose that G has a matching which covers every vertex in A and also one which covers every vertex in B. Show that G has a matching which covers every vertex in $A \cup B$. (L. DULMAGE AND N.S. MENDELSOHN)

★16.1.15 ESSENTIAL VERTEX
A vertex v of a graph G is *essential* if v is covered by every maximum matching in G, that is, if $\alpha'(G - v) = \alpha'(G) - 1$.

a) Describe an infinite family of connected graphs which contain no essential vertices.
b) Show that every nonempty bipartite graph has an essential vertex.

(D. DE CAEN)

16.1.16 A factory has n jobs $1, 2, \ldots, n$, to be processed, each requiring one day of processing time. There are two machines available. One can handle one job at a time and process it in one day, whereas the other can process two jobs simultaneously and complete them both in one day. The jobs are subject to precedence constraints represented by a binary relation $\prec$, where $i \prec j$ signifies that job i

must be completed before job j is started. The objective is to complete all the jobs while minimizing $d_1 + d_2$, where d_i is the number of days during which machine i is in use. Formulate this problem as one of finding a maximum matching in a suitably defined graph.   (M. FUJII, T. KASAMI, AND N. NINOMIYA)

16.2 Matchings in Bipartite Graphs

HALL'S THEOREM

In many applications, one wishes to find a matching in a bipartite graph $G[X,Y]$ which covers every vertex in X. Necessary and sufficient conditions for the existence of such a matching were first given by Hall (1935). Recall that if S is a set of vertices in a graph G, the set of all neighbours of the vertices in S is denoted by $N(S)$.

Theorem 16.4 HALL'S THEOREM
A bipartite graph $G := G[X,Y]$ has a matching which covers every vertex in X if and only if

$$|N(S)| \geq |S| \quad \text{for all } S \subseteq X \tag{16.1}$$

Proof Let $G := G[X,Y]$ be a bipartite graph which has a matching M covering every vertex in X. Consider a subset S of X. The vertices in S are matched under M with distinct vertices in $N(S)$. Therefore $|N(S)| \geq |S|$, and (16.1) holds.

Conversely, let $G := G[X,Y]$ be a bipartite graph which has no matching covering every vertex in X. Let M^* be a maximum matching in G and u a vertex in X not covered by M^*. Denote by Z the set of all vertices reachable from u by M^*-alternating paths. Because M^* is a maximum matching, it follows from Theorem 16.3 that u is the only vertex in Z not covered by M^*. Set $R := X \cap Z$ and $B := Y \cap Z$ (see Figure 16.6).

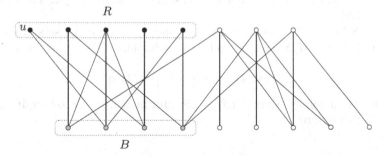

Fig. 16.6. Proof of Hall's Theorem (16.4)

Clearly the vertices of $R \setminus \{u\}$ are matched under M^* with the vertices of B. Therefore $|B| = |R| - 1$ and $N(R) \supseteq B$. In fact $N(R) = B$, because every vertex in $N(R)$ is connected to u by an M^*-alternating path. These two equations imply that

$$|N(R)| = |B| = |R| - 1$$

Thus Hall's condition (16.1) fails for the set $S := R$.   □

Theorem 16.4 is also known as the *Marriage Theorem*, because it can be restated more picturesquely as follows: if every group of girls in a village collectively like at least as many boys as there are girls in the group, then each girl can marry a boy she likes.

Hall's Theorem has proved to be a valuable tool both in graph theory and in other areas of mathematics. It has several equivalent formulations, including the following one in terms of set systems.

Let $\mathcal{A} := (A_i : i \in I)$ be a finite family of (not necessarily distinct) subsets of a finite set A. A *system of distinct representatives* (SDR) for the family $\mathcal{A}$ is a set $\{a_i : i \in I\}$ of distinct elements of A such that $a_i \in A_i$ for all $i \in I$. In this language, Hall's Theorem says that $\mathcal{A}$ has a system of distinct representatives if and only if $|\cup_{i \in J} A_i| \geq |J|$ for all subsets J of I. (To see that this is indeed a reformulation of Hall's Theorem, let $G := G[X, Y]$, where $X := I$, $Y := A$, and $N(i) := A_i$ for all $i \in I$.) This was, in fact, the form in which Hall presented his theorem. He used it to answer a question in group theory (see Exercise 16.2.21).

Hall's Theorem provides a criterion for a bipartite graph to have a perfect matching.

Corollary 16.5 *A bipartite graph $G[X, Y]$ has a perfect matching if and only if $|X| = |Y|$ and $|N(S)| \geq |S|$ for all $S \subseteq X$.*   □

This criterion is satisfied by all nonempty regular bipartite graphs.

Corollary 16.6 *Every nonempty regular bipartite graph has a perfect matching.*

Proof Let $G[X, Y]$ be a k-regular bipartite graph, where $k \geq 1$. Then $|X| = |Y|$ (Exercise 1.1.9).

Now let S be a subset of X and let E_1 and E_2 denote the sets of edges of G incident with S and $N(S)$, respectively. By definition of $N(S)$, we have $E_1 \subseteq E_2$. Therefore

$$k|N(S)| = |E_2| \geq |E_1| = k|S|$$

Because $k \geq 1$, it follows that $|N(S)| \geq |S|$ and hence, by Corollary 16.5, that G has a perfect matching.   □

MATCHINGS AND COVERINGS

Recall that a *covering* of a graph G is a subset K of V such that every edge of G has at least one end in K. A covering K^* is a *minimum covering* if G has no covering K with $|K| < |K^*|$. The number of vertices in a minimum covering of G

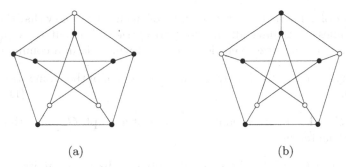

(a)   (b)

Fig. 16.7. (a) A minimal covering, (b) a minimum covering

is called the *covering number* of G, and is denoted by $\beta(G)$. A covering is *minimal* if none of its proper subsets is itself a covering. Minimal and minimum coverings of the Petersen graph are indicated (by solid vertices) in Figure 16.7.

If M is a matching of a graph G, and K is a covering of G, then at least one end of each edge of M belongs to K. Because all these ends are distinct, $|M| \leq |K|$. Moreover, if equality holds, then M is a maximum matching and K is a minimum covering (Exercise 16.2.2):

Proposition 16.7 *Let M be a matching and K a covering such that $|M| = |K|$. Then M is a maximum matching and K is a minimum covering.* □

The König–Egerváry Theorem (8.32) tells us that equality always holds when G is bipartite: for all bipartite graphs G,

$$\alpha'(G) = \beta(G)$$

This identity can be derived with ease from the theory of alternating paths. Let $G := G[X, Y]$ be a bipartite graph, let M^* be a maximum matching in G, and let U denote the set of vertices in X not covered by M^*. Denote by Z the set of all vertices in G reachable from some vertex in U by M^*-alternating paths, and set $R := X \cap Z$ and $B := Y \cap Z$. Then $K^* := (X \setminus R) \cup B$ is a covering with $|K^*| = |M^*|$ (Exercise 16.2.8). By Proposition 16.7, K^* is a minimum covering.

Exercises

16.2.1

a) Show that a bipartite graph G has a perfect matching if and only if $|N(S)| \geq |S|$ for all $S \subseteq V$.

b) Give an example to show that this condition does not guarantee the existence of a perfect matching in an arbitrary graph.

16.2.2 Prove Proposition 16.7.

16.2.3 A *line* of a matrix is a row or column of the matrix. Show that the minimum number of lines containing all the nonzero entries of a matrix is equal to the maximum number of nonzero entries, no two of which lie in a common line.

16.2.4 Using Exercise 16.1.15, give an inductive proof of the König–Egerváry Theorem (8.32).   (D. DE CAEN)

16.2.5 Let S and T be maximum stable sets of a graph G. Show that $G[S \triangle T]$ has a perfect matching.

$\star$**16.2.6** Let $\mathcal{A} := (A_i : i \in I)$ be a finite family of subsets of a finite set A, and let $f : I \to \mathbb{N}$ be a nonnegative integer-valued function. An f-*SDR* of $\mathcal{A}$ is a family $(S_i : i \in I)$ of disjoint subsets of A such that $S_i \subseteq A_i$ and $|S_i| = f(i)$, $i \in I$. (Thus, when $f(i) = 1$ for all $i \in I$, an f-SDR of $\mathcal{A}$ is simply an SDR of $\mathcal{A}$.)

a) Consider the family $\mathcal{B}$ of subsets of A consisting of $f(i)$ copies of A_i, $i \in I$. Show that $\mathcal{A}$ has an f-SDR if and only if $\mathcal{B}$ has an SDR.

b) Deduce, with the aid of Hall's Theorem (16.4), that $\mathcal{A}$ has an f-SDR if and only if

$$\left| \bigcup_{i \in J} A_i \right| \geq \sum_{i \in J} f(i) \quad \text{for all } J \subseteq I$$

16.2.7

a) Show that every minimal covering of a bipartite graph $G[X, Y]$ is of the form $N(S) \cup (X \setminus S)$ for some subset S of X.

b) Deduce Hall's Theorem (16.4) from the König–Egerváry Theorem (8.32).

$\star$**16.2.8** Let $G := G[X, Y]$ be a bipartite graph, let M^* be a maximum matching in G, and let U be the set of vertices in X not covered by M^*. Denote by Z the set of all vertices in G reachable from some vertex in U by M^*-alternating paths, and set $R := X \cap Z$ and $B := Y \cap Z$. Show that:

a) $K^* := (X \setminus R) \cup B$ is a covering of G,

b) $|K^*| = |M^*|$.

$\star$**16.2.9** THE KÖNIG–ORE FORMULA

a) Let $G := G[X, Y]$ be a bipartite graph, M a matching in G, and U the set of vertices in X not covered by M. Show that:

i) for any subset S of X, $|U| \geq |N(S)| - |S|$,

ii) $|U| = |N(S)| - |S|$ if and only if M is a maximum matching of G.

b) Prove the following generalization of Hall's Theorem (16.4):

The matching number of a bipartite graph $G := G[X, Y]$ is given by:

$$\alpha' = |X| - \max\{|S| - |N(S)| : S \subseteq X\}$$

This expression for α' is known as the *König-Ore Formula*.

16.2.10 Deduce from the König–Egerváry Theorem (8.32) that if $G := G[X, Y]$ is a simple bipartite graph, with $|X| = |Y| = n$ and $m > (k-1)n$, then $\alpha' \geq k$.

16.2.11

a) Let G be a graph and let (X, Y) be a partition of V such that $G[X]$ and $G[Y]$ are both k-colourable. If the edge cut $[X, Y]$ has at most $k - 1$ edges, show that G also is k-colourable.   (P. KAINEN)
b) Deduce that every k-critical graph is $(k-1)$-edge-connected.   (G.A. DIRAC)

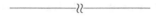

$\star$**16.2.12** Recall that an *edge covering* of a graph without isolated vertices is a set of edges incident with all the vertices, and that the number of edges in a minimum edge covering of a graph G is denoted by $\beta'(G)$. Show that $\alpha' + \beta' = n$ for any graph G without isolated vertices.   (T. GALLAI)

16.2.13 Let $G := G[X, Y]$ be a bipartite graph in which each vertex of X is of odd degree. Suppose that any two vertices of X have an even number of common neighbours. Show that G has a matching covering every vertex of X.   (N. ALON)

16.2.14 Let $G := G[X, Y]$ be a bipartite graph such that $d(x) \geq 1$ for all $x \in X$ and $d(x) \geq d(y)$ for all $xy \in E$, where $x \in X$ and $y \in Y$. Show that G has a matching covering every vertex of X.   (N. ALON)

16.2.15 Show that a bipartite graph $G[X, Y]$ has an f-factor with $f(x) = 1$ for all $x \in X$ and $f(y) \leq k$ for all $y \in Y$ if and only if $|N(S)| \geq |S|/k$ for all $S \subseteq X$.

16.2.16 A *2-branching* is a branching in which each vertex other than the root has outdegree at most two. Let T be a tournament, and let v be a vertex of maximum outdegree in T. Set $Y := N^+(v)$ and $X := V \setminus (Y \cup \{v\})$, and denote by $G[X, Y]$ the bipartite graph in which $x \in X$ is adjacent to $y \in Y$ if and only if y dominates x in T. For $S \subseteq X$, denote by $N(S)$ the set of neighbours of S in G.

a) Show that $|N(S)| \geq \frac{1}{2}|S|$, for all $S \subseteq X$.
b) By applying Exercise 16.2.15, deduce that T has a spanning 2-branching of depth at most two with root x.   (X. LU)

16.2.17 Let $\mathcal{C} = \{C_i : 1 \leq i \leq n\}$ be a family of n directed cycles in a digraph D. Show that there exist arcs $a_i \in A(C_i)$, $1 \leq i \leq n$, such that $D[\{a_i : 1 \leq i \leq n\}]$ contains a directed cycle.   (A. FRANK AND L. LOVÁSZ)

16.2.18

a) Let G be a graph in which each vertex is of degree either k or $k+1$, where $k \geq 1$. Prove that G has a spanning subgraph H in which:
   i) each vertex is of degree either k or $k+1$,
   ii) the vertices of degree $k+1$ form a stable set.

b) Let H be a graph satisfying conditions (i) and (ii) of (a), where $k \geq 1$. Denote by X the set of vertices in H of degree $k + 1$ and by Y the set of vertices in H of degree k. Prove that H has a spanning bipartite subgraph $B(X, Y)$ in which:

   i) each vertex of X has degree $k + 1$,
   ii) each vertex of Y has degree at most k.

c) Let $B(X, Y)$ be a bipartite graph satisfying conditions (i) and (ii) of (b). Prove that there is a matching M in B which covers every vertex of X.

d) Deduce from (a), (b), and (c) that if G is a graph in which each vertex is of degree either k or $k + 1$, where $k \geq 1$, then G contains a spanning subgraph in which each vertex is of degree either $k - 1$ or k.

(W.T. TUTTE; C. THOMASSEN)

16.2.19 THE BIRKHOFF–VON NEUMANN THEOREM
A nonnegative real matrix is *doubly stochastic* if each of its line sums is 1. A *permutation matrix* is a $(0, 1)$-matrix which has exactly one 1 in each line. (Thus every permutation matrix is doubly stochastic.) Let $\mathbf{Q}$ be a doubly stochastic matrix. Show that:

a) $\mathbf{Q}$ is a square matrix,
b) $\mathbf{Q}$ can be expressed as a convex linear combination of permutation matrices, that is,

$$\mathbf{Q} = c_1 \mathbf{P}_1 + c_2 \mathbf{P}_2 + \cdots + c_k \mathbf{P}_k$$

where each $\mathbf{P}_i$ is a permutation matrix, each c_i is a nonnegative real number, and $\sum_{i=1}^{k} c_i = 1$. (G. BIRKHOFF; J. VON NEUMANN)

16.2.20 Let $\mathcal{A} := (A_i : i \in I)$ and $\mathcal{B} := (B_i : i \in I)$ be two finite families of subsets of a finite set A. Construct a digraph $D(x, y)$ with the property that some SDR of $\mathcal{A}$ is also an SDR of $\mathcal{B}$ if and only if there are $|I|$ internally disjoint directed (x, y)-paths in D.

16.2.21 Let H be a finite group and let K be a subgroup of H. Show that there exist elements $h_1, h_2, \ldots, h_n \in H$ such that $h_1 K, h_2 K, \ldots, h_n K$ are the left cosets of K and $K h_1, K h_2, \ldots, K h_n$ are the right cosets of K. (P. HALL)

16.2.22 Let $G[X, Y]$ be a bipartite graph, and let S_1 and S_2 be subsets of X. Show that
$$|N(S_1)| + |N(S_2)| \geq |N(S_1 \cup S_2)| + |N(S_1 \cap S_2)|$$

16.2.23 Let $G[X, Y]$ be a bipartite graph in which $|N(S)| \geq |S|$ for all $S \subseteq X$.

a) A subset S of X is said to be *tight* if $|N(S)| = |S|$. Deduce from Exercise 16.2.22 that the union and intersection of tight subsets are tight also.
b) Deduce Hall's Theorem (16.4), that G has a matching covering X, by induction on n, proceeding as follows.

i) Suppose, first, that there are no nonempty proper tight subsets of X. Let xy be an edge of $G[X, Y]$ with $x \in X$ and $y \in Y$. Show that, for every subset S of $X \setminus \{x\}$, $|N_{G'}(S)| \geq |S|$, where $G' = G - \{x, y\}$. (In this case, by induction, G' has a matching M' that covers $X \setminus \{x\}$, and $M' \cup \{xy\}$ is a matching of G that covers X.)

ii) Suppose, now, that T is a nonempty proper tight subset of X. Let G_1 denote the subgraph of G induced by $T \cup N(T)$ and let $G_2 := G - (T \cup N(T))$. Show that $|N_{G_1}(S)| \geq |S|$, for all $S \subseteq T$ and $|N_{G_2}(S)| \geq |S|$, for all $S \subseteq X \setminus T$. (In this case, by induction, G_1 has a matching M_1 that covers T, and G_2 has a matching M_2 that covers $X \setminus T$, so $M_1 \cup M_2$ is a matching of G that covers X.)   (P.R. HALMOS AND H.E. VAUGHN)

16.2.24 A nonempty connected graph is *matching-covered* if every edge belongs to some perfect matching. Let $G := G[X, Y]$ be a connected bipartite graph with a perfect matching. Show that:

a) G is matching-covered if and only if X has no nonempty proper tight subsets,

b) if G is matching covered, then $G - \{x, y\}$ has a perfect matching for all $x \in X$ and all $y \in Y$.

16.2.25 DULMAGE–MENDELSOHN DECOMPOSITION

Let $G[X, Y]$ be a bipartite graph with a perfect matching. Show that there exist a positive integer k and partitions $(X_1, X_2, \ldots, X_k)$ of X and $(Y_1, Y_2, \ldots, Y_k)$ of Y such that, for $1 \leq i \leq k$,

i) the subgraph $G[X_i \cup Y_i]$ of $G[X, Y]$ induced by $X_i \cup Y_i$ is matching-covered,

ii) $N(X_i) \subseteq Y_1 \cup Y_2 \cup \cdots \cup Y_i$.   (L. DULMAGE AND N.S. MENDELSOHN)

16.2.26 Let G be a matching-covered bipartite graph.

a) Show that G has an *odd-ear decomposition*, that is, a nested sequence of subgraphs $(G_0, G_1, \ldots, G_k)$ such that $G_0 \cong K_2$, $G_k = G$, and $G_{i+1} = G_i \cup P_i$, $0 \leq i < k$, where P_i is an ear of G_i of odd length.

b) Show that, in any such decomposition, G_i is matching-covered for all i, $0 \leq i \leq k$.

c) Deduce that G has $m - n + 2$ perfect matchings whose incidence vectors are linearly independent.

d) The *matching space* of G is the vector space generated by the set of incidence vectors of perfect matchings of G. Show that the dimension of this space is $m - n + 2$.   (J. EDMONDS, L. LOVÁSZ, AND W.R. PULLEYBLANK)

(The corresponding results for nonbipartite matching-covered graphs are considerably more difficult; see Carvalho et al. (2002).)

16.2.27

a) Let $G[X, Y]$ be a bipartite graph on $2n$ vertices in which all vertices have degree three except for one vertex in X and one vertex in Y, which have degree two. Show that $pm(G) \geq 2(4/3)^{n-1}$.

b) Deduce that if G is a cubic bipartite graph on $2n$ vertices, then $pm(G) \geq (4/3)^n$.   (M. Voorhoeve)

16.2.28

a) Let $G[X, Y]$ be an infinite bipartite graph. Show that the condition $|N(S)| \geq |S|$, for every finite subset S of X, is a necessary condition for G to have a matching covering every vertex of X.

b) Give an example of a countable bipartite graph $G[X, Y]$ for which this condition is not sufficient for the existence of such a matching.

16.3 Matchings in Arbitrary Graphs

In this section, we derive a min–max formula for the number of edges in a maximum matching of an arbitrary graph, analogous to the König–Ore Formula for bipartite graphs (see Exercise 16.2.9). We begin by establishing an upper bound for this number.

BARRIERS

If M is a matching in a graph G, each odd component of G must clearly include at least one vertex not covered by M. Therefore $|U| \geq o(G)$, where U denotes the set of such vertices and $o(G)$ the number of odd components of G. This inequality can be extended to all induced subgraphs of G as follows.

Let S be a proper subset of V and let M be a matching in G. Consider an odd component H of $G - S$. If every vertex of H is covered by M, at least one vertex of H must be matched with a vertex of S. Because no more than $|S|$ vertices of $G - S$ can be matched with vertices of S, at least $o(G - S) - |S|$ odd components of G must contain vertices not covered by M. This observation yields the following inequality, valid for all proper subsets S of V.

$$|U| \geq o(G - S) - |S| \tag{16.2}$$

From this inequality we may deduce, for example, that the Sylvester graph (of Figure 16.5) has no perfect matching, because three odd components are obtained upon deleting its central cut vertex. Likewise, the indicated set S of three vertices in the graph G of Figure 16.8a shows that any matching M must leave at least $5 - 3 = 2$ uncovered vertices because $G - S$ has five odd components (and one even component), as shown in Figure 16.8b.

Note that if equality should hold in (16.2) for some matching M and some subset $S := B$ of V, that is, if

$$|U| = o(G - B) - |B| \tag{16.3}$$

where $|U| = v(G) - 2|M|$, then the set B would show that the matching M leaves as few uncovered vertices as possible, and hence is a maximum matching

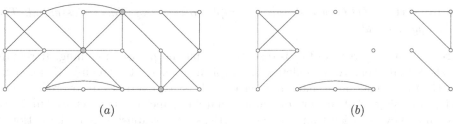

Fig. 16.8. A set S with $o(G - S) > |S|$

(Exercise 16.3.1). Thus, B would serve as a succinct certificate of the optimality of M. Such a set B is called a *barrier* of G. The set of three vertices indicated in the graph of Figure 16.8 is a barrier, because this graph has a matching covering all but two of its vertices (see Figure 16.15a).

A matchable graph (one with a perfect matching) has both the empty set and all singletons as barriers. The empty set is also a barrier of a graph when some vertex-deleted subgraph is matchable. Graphs which are very nearly matchable, in the sense that every vertex-deleted subgraph is matchable, are said to be *hypomatchable* or *factor-critical*. In particular, trivial graphs are hypomatchable. For future reference, we state as a lemma the observation that all hypomatchable graphs have the empty set as a barrier. (Indeed, the empty set is their only barrier, see Exercise 16.3.8.)

Lemma 16.8 *The empty set is a barrier of every hypomatchable graph.*   □

THE TUTTE–BERGE THEOREM

In a bipartite graph, a minimum covering constitutes a barrier of the graph (Exercise 16.3.4). More generally, every graph has a barrier. This fact is known as the *Tutte–Berge Theorem*. We present a proof by Gallai (1964a) of this theorem. It proceeds by induction on the number of vertices. By Lemma 16.8, a trivial graph has the empty set as barrier.

Recall that a vertex v of a graph G is *essential* if every maximum matching covers v, and *inessential* otherwise. Thus v is essential if $\alpha'(G - v) = \alpha'(G) - 1$ and inessential if $\alpha'(G - v) = \alpha'(G)$. We leave the proof of the following lemma as an exercise (16.3.5).

Lemma 16.9 *Let v be an essential vertex of a graph G and let B be a barrier of $G - v$. Then $B \cup \{v\}$ is a barrier of G.*   □

By Lemma 16.9, in order to show that every graph has a barrier, it suffices to consider graphs with no essential vertices. It turns out that such graphs always have the empty set as a barrier. We establish this fact for connected graphs. Its validity for all graphs can be deduced without difficulty from this special case (Exercise 16.3.6).

Lemma 16.10 *Let G be a connected graph no vertex of which is essential. Then G is hypomatchable.*

Proof Since no vertex of G is essential, G has no perfect matching. It remains to show that every vertex-deleted subgraph has a perfect matching. If this is not so, then each maximum matching leaves at least two vertices uncovered. Thus it suffices to show that for any maximum matching and any two vertices in G, the matching covers at least one of these vertices. We establish this by induction on the distance between these two vertices.

Consider a maximum matching M and two vertices x and y in G. Let xPy be a shortest xy-path in G. Suppose that neither x nor y is covered by M. Because M is maximal, P has length at least two. Let v be an internal vertex of P. Since xPv is shorter than P, the vertex v is covered by M, by induction. On the other hand, because v is inessential, G has a maximum matching M' which does not cover v. Furthermore, because xPv and vPy are both shorter than P, the matching M' covers both x and y, again by induction.

The components of $G[M \triangle M']$ are even paths and cycles whose edges belong alternately to M and M' (Exercise 16.1.5). Each of the vertices x, v, y is covered by exactly one of the two matchings and thus is an end of one of the paths. Because the paths are even, x and y are not ends of the same path. Moreover, the paths starting at x and y cannot both end at v. We may therefore suppose that the path Q that starts at x ends neither at v nor at y. But then the matching $M' \triangle E(Q)$ is a maximum matching which covers neither x nor v, contradicting the induction hypothesis and establishing the lemma. $\square$

One may now deduce (Exercise 16.3.7) the following fundamental theorem and corollary. These results, obtained by Berge (1958), can also be derived from a theorem of Tutte (1947a) on perfect matchings (Theorem 16.13).

Theorem 16.11   THE TUTTE–BERGE THEOREM
Every graph has a barrier. $\square$

Corollary 16.12   THE TUTTE–BERGE FORMULA
For any graph G:

$$\alpha'(G) = \frac{1}{2}\min\{v(G) - (o(G-S) - |S|) : S \subset V\}$$ $\square$

A refinement of Theorem 16.11 states that every graph G has a barrier B such that each odd component of $G - B$ is hypomatchable and each even component of $G - B$ has a perfect matching. Such a barrier is known as a *Gallai barrier*. In Section 16.5, we present a polynomial-time algorithm which finds not only a maximum matching in a graph, but also a succinct certificate for the optimality of the matching, namely a Gallai barrier.

Exercises

$\star$**16.3.1** Let M be a matching in a graph G, and let B be a set of vertices of G such that $|U| = o(G - B) - |B|$, where U is the set of vertices of G not covered by M. Show that M is a maximum matching of G.

$\star$**16.3.2** Let G be a graph and S a proper subset of V. Show that $o(G - S) - |S| \equiv v(G) \pmod 2$.

$\star$**16.3.3** Show that the union of barriers of the components of a graph is a barrier of the entire graph.

$\star$**16.3.4** Show that, in a bipartite graph, any minimum covering is a barrier of the graph.

$\star$**16.3.5** Give a proof of Lemma 16.9.

$\star$**16.3.6** Deduce from Lemma 16.10 that the empty set is a barrier of every graph without essential vertices.

$\star$**16.3.7**

a) Prove the Tutte–Berge Theorem (Theorem 16.11) by induction on the number of vertices.
b) Deduce the Tutte–Berge Formula (Corollary 16.12) from the Tutte–Berge Theorem.

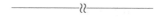

16.3.8

a) Show that:
   i) a graph is hypomatchable if and only if each of its blocks is hypomatchable,
   ii) a graph G is hypomatchable if and only if $o(G - S) \le |S| - 1$ for every nonempty proper subset S of V.
b) Deduce that a graph is hypomatchable if and only if the empty set is its only barrier.

16.3.9 Let B be a maximal barrier of a graph G. Show that each component of $G - B$ is hypomatchable.

16.3.10 Let G be a graph and let (X, Y) be a partition of V with $X, Y \ne \emptyset$. Show that if both G / X and G / Y are hypomatchable, then G is hypomatchable.

16.3.11 Let G be a nonseparable graph which has an odd-ear decomposition starting with an odd cycle (instead of K_2). Show that G is hypomatchable.

16.3.12

a) Let x and y be adjacent inessential vertices of a graph G and let M and N be maximum matchings of $G - x$ and $G - y$, respectively. Show that G has an xy-path of even length whose edges belong alternately to N and M.

b) Deduce that every nontrivial hypomatchable graph G contains an odd cycle C such that $G \, / \, C$ is hypomatchable.

c) Prove the converse of the statement in Exercise 16.3.11: show that every nontrivial nonseparable hypomatchable graph has an odd-ear decomposition starting with an odd cycle (instead of K_2).   (L. LOVÁSZ)

16.3.13 Let G be a k-chromatic graph containing no stable set of three vertices and no clique of k vertices, where $k \geq 3$. Let $k_1 + k_2$ be a partition of $k + 1$ such that k_1, $k_2 \geq 2$. By appealing to Exercises 16.1.3 and 16.3.9, show that G has disjoint subgraphs G_1 and G_2 such that $\chi(G_1) = k_1$ and $\chi(G_2) = k_2$.

(L. LOVÁSZ AND P.D. SEYMOUR)

16.4 Perfect Matchings and Factors

TUTTE'S THEOREM

If a graph G has a perfect matching M, then it follows from (16.2) that $o(G - S) \leq |S|$ for all $S \subseteq V$, because the set U of uncovered vertices is empty. The following fundamental theorem due to Tutte (1947a) shows that the converse is true. It is a special case of the Tutte–Berge Formula (Corollary 16.12).

Theorem 16.13   TUTTE'S THEOREM
A graph G has a perfect matching if and only if

$$o(G - S) \leq |S| \quad \text{for all} \ \ S \subseteq V \tag{16.4}$$

Proof   As already noted, (16.4) holds if G has a perfect matching. Conversely, let G be a graph which has no perfect matching. Consider a maximum matching M^* of G, and denote by U the set of vertices in G not covered by M^*. By Theorem 16.11, G has a barrier, that is, a subset B of V such that $o(G - B) - |B| = |U|$. Because M^* is not perfect, $|U|$ is positive. Thus

$$o(G - B) = |B| + |U| \geq |B| + 1$$

and Tutte's condition (16.4) fails for the set $S := B$. $\square$

The first significant result on perfect matchings in graphs was obtained by Petersen (1891) in connection with a problem about factoring homogeneous polynomials into irreducible factors (see Biggs et al. (1986) and Sabidussi (1992)). In this context, perfect matchings correspond to factors of degree one; it is for this reason that they are also referred to as '1-factors'; it is also the origin of the term 'degree'. Petersen was particularly interested in the case of polynomials of degree three; these correspond to 3-regular graphs.

Theorem 16.14   PETERSEN'S THEOREM
Every 3-regular graph without cut edges has a perfect matching.

Proof   We derive Petersen's Theorem from Tutte's Theorem (16.13).

Let G be a 3-regular graph without cut edges, and let S be a subset of V. Consider the vertex sets $S_1, S_2, \ldots, S_k$, of the odd components of $G - S$. Because G has no cut edges, $d(S_i) \geq 2$, $1 \leq i \leq k$. But because $|S_i|$ is odd, $d(S_i)$ is odd also (Exercise 2.5.5). Thus, in fact,

$$d(S_i) \geq 3, \quad 1 \leq i \leq k$$

Now the edge cuts $\partial(S_i)$ are pairwise disjoint, and are contained in the edge cut $\partial(S)$, so we have:

$$3k \leq \sum_{i=1}^{k} d(S_i) = d(\cup_{i=1}^{k} S_i) \leq d(S) \leq 3|S|$$

Therefore $o(G - S) = k \leq |S|$, and it follows from Theorem 16.13 that G has a perfect matching.   □

The condition in Petersen's Theorem that the graph be free of cut edges cannot be omitted: the Sylvester graph of Figure 16.5, for instance, has no perfect matching. However, a stronger form of the theorem may be deduced from Tutte's Theorem (16.13), namely that each edge of a 3-regular graph without cut edges belongs to some perfect matching (Exercise 16.4.8).

FACTORS

Let G be a graph and let f be a nonnegative integer-valued function on V. An *f-factor* of G is a spanning subgraph F of G such that $d_F(v) = f(v)$ for all $v \in V$. A *k-factor* of G is an f-factor with $f(v) := k$ for all $v \in V$; in particular, a 1-factor is a spanning subgraph whose edge set is a perfect matching and a 2-factor is a spanning subgraph whose components are cycles.

Many interesting graph-theoretical problems can be solved in polynomial time by reducing them to problems about 1-factors. One example is the question of deciding whether a given graph G has an f-factor. Tutte (1954b) showed how this problem can be reduced to the problem of deciding whether a related graph G' has a 1-factor. We now describe this reduction procedure. We may assume that $d(v) \geq f(v)$ for all $v \in V$; otherwise G obviously has no f-factor. For simplicity, we assume that our graph G is loopless.

For each vertex v of G, we first replace v by a set Y_v of $d(v)$ vertices, each of degree one. We then add a set X_v of $d(v) - f(v)$ vertices and form a complete bipartite graph H_v by joining each vertex of X_v to each vertex of Y_v. In effect, the resulting graph H is obtained from G by replacing each vertex v by a complete bipartite graph $H_v[X_v, Y_v]$ and joining each edge incident to v to a separate element of Y_v. Figure 16.9 illustrates this construction in the case of 2-factors. Note that

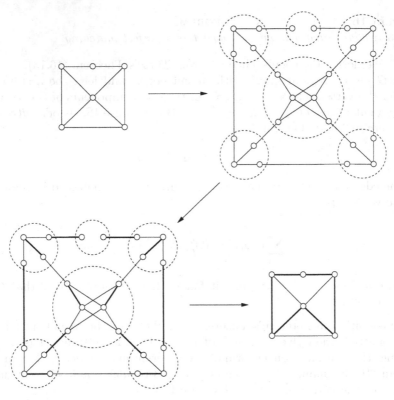

Fig. 16.9. Polynomial reduction of the 2-factor problem to the 1-factor problem

G can be recovered from H simply by shrinking each bipartite subgraph H_v to a single vertex v.

In H, the vertices of X_v are joined only to the vertices of Y_v. Thus if F is a 1-factor of H, the $d(v) - f(v)$ vertices of X_v are matched by F with $d(v) - f(v)$ of the $d(v)$ vertices in Y_v. The remaining $f(v)$ vertices of Y_v are therefore matched by F with $f(v)$ vertices of $V(H) \setminus V(H_v)$. Upon shrinking H to G, the 1-factor F of H is therefore transformed into an f-factor of G. Conversely, any f-factor of G can easily be converted into a 1-factor of H.

This reduction of the f-factor problem to the 1-factor problem is a polynomial reduction (Exercise 16.4.2).

T-JOINS

A number of problems in graph theory and combinatorial optimization amount to finding a spanning subgraph H of a graph G (or a spanning subgraph of minimum weight, in the case of weighted graphs) whose degrees have prescribed parities (rather than prescribed values, as in the f-factor problem). Precise statements of such problems require the notion of a T-join.

Let G be a graph and let T be an even subset of V. A spanning subgraph H of G is called a *T-join* if $d_H(v)$ is odd for all $v \in T$ and even for all $v \in V \setminus T$. For example, a 1-factor of G is a V-join; and if P is an xy-path in G, the spanning subgraph of G with edge set $E(P)$ is an $\{x, y\}$-join.

Problem 16.15   THE WEIGHTED T-JOIN PROBLEM
  GIVEN: *a weighted graph $G := (G, w)$ and a subset T of V,*
  FIND: *a minimum-weight T-join in G (if one exists).*

As remarked above, the Shortest Path Problem may be viewed as a particular case of the Weighted T-Join Problem. Another special case is the *Postman Problem*, described in Exercise 16.4.22, whose solution involves finding a minimum-weight T-join when T is the set of vertices of odd degree in the graph (see Exercises 16.4.21 and 16.4.22).

By means of a construction similar to Tutte's reduction of the 2-factor problem to the 1-factor problem, one may obtain a polynomial reduction of the Weighted T-Join Problem to the following problem (see Exercise 16.4.21).

Problem 16.16   THE MINIMUM-WEIGHT MATCHING PROBLEM
  GIVEN: *a weighted complete graph $G := (G, w)$ of even order,*
  FIND: *a minimum-weight perfect matching in G.*

This latter problem can be seen to include the maximum matching problem: it suffices to embed the input graph G in a complete graph of even order, and assign weight zero to each edge of G and weight one to each of the remaining edges. Edmonds (1965b) found a polynomial-time algorithm for solving the Minimum-Weight Matching Problem. His algorithm relies on techniques from the theory of linear programming, and on his characterization of the perfect matching polytope (see Exercise 17.4.5).

Exercises

16.4.1 Show that a tree G has a perfect matching if and only if $o(G - v) = 1$ for all $v \in V$.   (V. CHUNGPHAISAN)

$\star$**16.4.2**

  a) Show that Tutte's reduction of the f-factor problem to the 1-factor problem is a polynomial reduction.
  b) Describe how this polynomial reduction of the f-factor problem to the 1-factor problem can be generalized to handle graphs with loops.

16.4.3 Let G be a graph, and let $F := G[X]$ be an induced subgraph of V. Form a graph H from G as follows.

▷   Add edges between nonadjacent pairs of vertices in $V \setminus X$.
▷   If n is odd, add a new vertex and join it to all vertices in $V \setminus X$.

a) Show that there is a matching in G covering all vertices in X if and only if H has a perfect matching.
b) By applying Tutte's Theorem (16.13), deduce that G has a matching covering every vertex in X if and only, for all $S \subseteq V$, the number of odd components of $G - S$ which are subgraphs of $F - S$ is at most $|S|$.
c) Now suppose that F is bipartite. Using Exercise 16.3.9, strengthen the statement in (b) to show that there is a matching covering all vertices in X if and only if, for all $S \subseteq V$, the number of isolated vertices of $G - S$ which belong to $F - S$ is at most $|S|$.

16.4.4 Using Exercise 16.4.3b, derive Hall's Theorem (16.4) from Tutte's Theorem (16.13).

16.4.5 Let G be a graph whose vertices of degree Δ induce a bipartite subgraph. Using Exercise 16.4.3c, show that there is a matching in G covering all vertices of degree Δ.   (H. KIERSTEAD)

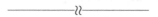

16.4.6 Derive the Tutte–Berge Formula (Corollary 16.12) from Tutte's Theorem (16.13).

16.4.7 Let G be a graph with a perfect matching and let $x, y \in V$.

a) Show that there is a barrier of G containing both x and y if and only if $G - \{x, y\}$ has no perfect matching.
b) Suppose that x and y are adjacent. Deduce from (a) that there is a perfect matching containing the edge xy if and only if no barrier of G contains both x and y.

16.4.8 Deduce from Tutte's Theorem (16.13) that every edge of a 3-regular graph without cut edges belongs to some perfect matching.

16.4.9

a) For $k \geq 1$, show that every $(k-1)$-edge-connected k-regular graph on an even number of vertices has a perfect matching.
b) For each $k \geq 2$, give an example of a $(k-2)$-edge-connected k-regular graph on an even number of vertices with no perfect matching.

16.4.10 A graph G on at least three vertices is *bicritical* if, for any two vertices u and v of G, the subgraph $G - \{u, v\}$ has a perfect matching.

a) Show that a graph is bicritical if and only if it has no barriers of cardinality greater than one.
b) Deduce that every essentially 4-edge-connected cubic nonbipartite graph is bicritical.

16.4.11

a) Show that every connected claw-free graph on an even number of vertices has a perfect matching.   (M. LAS VERGNAS; D. SUMNER)

b) Deduce that every 2-connected claw-free graph on an odd number of vertices is hypomatchable.

16.4.12 Let G be a simple graph with $\delta \geq 2(n-1)$ and containing no induced $K_{1,n}$, where $n \geq 3$. Show that G has a 2-factor.   (K. OTA AND T. TOKUDA)

16.4.13 Show that every 2-connected graph that has one perfect matching has at least two perfect matchings.   (A. KOTZIG)

16.4.14

a) Show that a simple graph on $2n$ vertices with exactly one perfect matching has at most n^2 edges.

b) For all $n \geq 1$, construct a simple graph on $2n$ vertices with exactly one perfect matching and exactly n^2 edges.

16.4.15 Let $H = (V, \mathcal{F})$ be a hypergraph. A *cycle* in H is a sequence $v_1 F_1 v_2 F_2 \ldots v_k F_k v_{k+1}$, where v_i, $1 \leq i \leq k$, are distinct vertices, F_i, $1 \leq i \leq k$, are distinct edges, $v_{k+1} = v_1$, and $\{v_i, v_{i+1}\} \subseteq F_i$, $1 \leq i \leq k$. The hypergraph H is *balanced* if every odd cycle $v_1 F_1 v_2 F_2 \ldots v_{2k+1} F_{2k+1} v_1$ includes an edge containing at least three vertices of the cycle. A *perfect matching* in H is a set of disjoint edges whose union is V. Show that a balanced hypergraph $(V, \mathcal{F})$ has a perfect matching if and only if there exist disjoint subsets X and Y of V such that $|X| > |Y|$ and $|X \cap F| \leq |Y \cap F|$ for all $F \in \mathcal{F}$

$\star$**16.4.16** A graph G is *k-factorable* if it admits a decomposition into k-factors. Show that:

a) every k-regular bipartite graph is 1-factorable,   (D. KÖNIG)

b) every $2k$-regular graph is 2-factorable.   (J. PETERSEN)

16.4.17

a) Show that the thickness of a $2k$-regular graph is at most k.

b) Find a 4-regular graph of thickness two and a 6-regular graph of thickness three.

16.4.18 Show that every triangulation with m edges contains a spanning bipartite subgraph with $2m/3$ edges.   (F. HARARY AND D. MATULA)

16.4.19 Show that every 2-connected 3-regular graph on four or more vertices admits a decomposition into paths of length three.

16.4.20 Let G be a graph. For $v \in V$, let $L(v) \subseteq \{0, 1, \ldots, d(v)\}$ be a list of integers associated with v.

a) Show that G has an f-factor with $f(v) \in L(v)$ for all $v \in V$ if $|L(v)| \geq d^+(v)+1$ for all $v \in V$, where D is an orientation of G.

(A. FRANK, L. LAO, AND J. SZABÓ; J.A. BONDY)

b) Deduce that G has an f-factor with $f(v) \in L(v)$ for all $v \in V$ if $|L(v)| > \lceil d(v)/2 \rceil$, for all $v \in V$.

(H. SHIRAZI AND J. VERSTRAËTE)

16.4.21

a) Find a reduction of the Minimum-Weight T-Join Problem (16.15) to the Minimum-Weight Matching Problem (16.16).

b) Let G be a graph, let T be the set of vertices of odd degree in G, and let H be a T-join of G. Show that the graph obtained from G by duplicating the edges of H is an even graph.

16.4.22 THE POSTMAN PROBLEM

In his job, a postman picks up mail at the post office, delivers it, and then returns to the post office. He must, of course, cover each street at least once. Subject to this condition, he wishes to choose a route entailing as little walking as as possible.

a) Show that the Postman Problem is equivalent to the following graph-theoretic problem.

Problem 16.17 MINIMUM-WEIGHT EULERIAN SPANNING SUBGRAPH

GIVEN: *a weighted connected graph $G := (G, w)$ with nonnegative weights,*

FIND: *by duplicating edges (along with their weights), an eulerian weighted spanning supergraph H of G whose weight $w(H)$ is as small as possible.*

(An Euler tour in H can then be found by applying Fleury's Algorithm (3.3).)

(M. GUAN)

b) In the special case where G has just two vertices of odd degree, explain how the above problem can be solved in polynomial time.

16.5 Matching Algorithms

This final section of the chapter is devoted to the description of a polynomial-time algorithm for finding a maximum matching in an arbitrary graph. We first consider the easier case of bipartite graphs, and then show how that algorithm can be refined to yield one applicable to all graphs.

AUGMENTING PATH SEARCH

Berge's Theorem (16.3) suggests a natural approach to finding a maximum matching in a graph. We start with some matching M (for example, the empty matching) and search for an M-augmenting path. If such a path P is found, we replace M by $M \triangle E(P)$. We repeat the procedure until no augmenting path with respect to the current matching can be found. This final matching is then a maximum matching.

The challenge here is to carry out an exhaustive search for an M-augmenting path in an efficient manner. This can indeed be achieved. In this section, we describe a polynomial-time algorithm which either finds an M-augmenting path in a bipartite graph, or else supplies a succinct certificate that there is no such path. This algorithm, as well as its extension to arbitrary graphs, is based on the notion of an M-alternating tree.

Let G be a graph, M a matching in G, and u a vertex not covered by M. A tree T of G is an M-*alternating* u-*tree* if $u \in V(T)$ and, for any $v \in V(T)$, the path uTv is an M-alternating path. An M-alternating u-tree T is M-*covered* if the matching $M \cap E(T)$ covers all vertices of T except u (see Figures 16.10a and 16.10c).

There is a simple tree-search algorithm, which we refer to as *Augmenting Path Search*, that finds either an M-augmenting u-path or else a maximal M-covered u-tree (that is, an M-covered u-tree which can be grown no further). We call such a tree an *APS-tree* (rooted at u).

Augmenting Path Search begins with the trivial M-covered tree consisting of just the vertex u. At each stage, it attempts to extend the current M-covered u-tree T to a larger one. We refer to those vertices at even distance from u in T as *red* vertices and those at odd distance as *blue* vertices; these sets will be denoted by $R(T)$ and $B(T)$, respectively (so that $(R(T), B(T))$ is a bipartition of T, with $u \in R(T)$). In the M-covered u-tree T displayed in Figure 16.10a, the red vertices are shown as solid dots and the blue vertices as shaded dots.

Consider an M-covered u-tree T. Being M-covered, T contains no M-augmenting u-path. If there is such a path in G, the edge cut $\partial(T)$ necessarily includes an edge of this path. Accordingly, we attempt to extend T to a larger M-alternating u-tree by adding to it an edge from its associated edge cut $\partial(T)$. This is possible only if there is an edge xy with $x \in R(T)$ and $y \in V(G) \setminus V(T)$. If there is no such edge, then T is an APS-tree rooted at u, and the procedure terminates. If, on the other hand, there is such an edge, two possibilities arise. Either y is not covered by M, in which case we have found our M-augmenting path (Figure 16.10b), or y is incident to an edge yz of M, and we grow T into a larger M-covered u-tree by adding the two vertices y and z and the two edges xy and yz (Figure 16.10c).

This tree-growing operation is repeated until either an M-augmenting u-path P is found, in which case the matching M is replaced by $M \triangle E(P)$, or the procedure terminates with an APS-tree T. It can be summarized as follows.

Algorithm 16.18 AUGMENTING PATH SEARCH: APS(G, M, u)

INPUT: a graph G, a matching M in G, and an uncovered vertex u of G

OUTPUT: a matching M with one more edge than the input matching, or an APS-tree T with root $u(T)$, a bipartition $(R(T), B(T))$ of T, and the set $M(T)$ of matching edges in T

 1: set $V(T) := \{u\}$, $E(T) := \emptyset$, $R(T) := \{u\}$

 2: **while** there is an edge xy with $x \in R(T)$ and $y \in V(G) \setminus V(T)$ **do**

 3:   replace $V(T)$ by $V(T) \cup \{y\}$ and $E(T)$ by $E(T) \cup \{xy\}$.

 4:   **if** y is not covered by M **then**

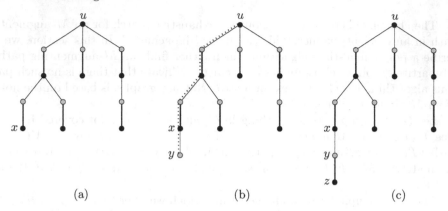

(a)   (b)   (c)

Fig. 16.10. Augmenting Path Search: growing an M-covered tree

5:      *replace M by $M \triangle E(P)$, where $P := uTy$*

6:      *return M*

7:   **else**

8:      *replace $V(T)$ by $V(T) \cup \{z\}$, $E(T)$ by $E(T) \cup \{yz\}$, and $R(T)$ by $R(T) \cup \{z\}$, where $yz \in M$*

9:   **end if**

10: **end while**

11: *set: $T := (V(T), E(T))$, $u(T) := u$, $B(T) := V(T) \setminus R(T)$, and $M(T) := M \cap E(T)$*

12: *return $(T, u(T), R(T), B(T), M(T))$*

In the event that APS outputs an APS-tree T, we note for future reference that:

▷ because T is M-covered, the vertices of $R(T) \setminus u(T)$ are matched with the vertices of $B(T)$, so

$$|B(T)| = |R(T)| - 1 \tag{16.5}$$

and

$$B(T) \subseteq N(R(T)) \tag{16.6}$$

(where $N(R(T))$ denotes the set of neighbours of $R(T)$ in G).

▷ because T is maximal, no vertex of $R(T)$ is adjacent in G to any vertex of $V(G) \setminus V(T)$; that is,

$$N(R(T)) \subseteq R(T) \cup B(T) \tag{16.7}$$

If APS finds an M-augmenting path uPy, well and good. We simply apply APS once more, replacing M by the augmented matching $M \triangle E(P)$ returned by APS. But what if APS returns an APS-tree T? Can we then be sure that G contains no M-augmenting u-path? Unfortunately we cannot, as the example in Figure 16.11b illustrates. However, if it so happens that no two red vertices of T are adjacent in G, that is, if $N(R(T)) \cap R(T) = \emptyset$, then (16.6) and (16.7) imply that

$$N(R(T)) = B(T) \tag{16.8}$$

In this case, we may restrict our search for an M-augmenting path to the subgraph $G - T$. Indeed, the following stronger statement is true (Exercise 16.5.3).

Proposition 16.19 *Let T be an APS-tree returned by $APS(G, M, u)$. Suppose that no two red vertices of T are adjacent in G. Then no M-augmenting path in G can include any vertex of T.* ☐

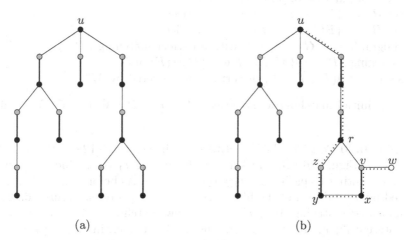

(a)   (b)

Fig. 16.11. (a) An APS-tree, (b) an M-augmenting u-path

One important instance where condition (16.8) is satisfied is when G is bipartite. In this case, no two red vertices of T can be adjacent in G because they all belong to the same part of the bipartition of G. Thus $APS(G, M, u)$ finds all the vertices of G that can be reached by M-alternating u-paths. This observation is the basis of the following algorithm for finding a maximum matching in a bipartite graph. It was conceived by the Hungarian mathematician Egerváry (1931), and for this reason is sometimes referred to as the *Hungarian Algorithm*.

EGERVÁRY'S ALGORITHM

Let $G[X, Y]$ be a bipartite graph. In searching for a maximum matching of G, we start with an arbitrary matching M of G (for instance, the empty matching) and apply APS to search for an M-augmenting u-path, where u is an uncovered vertex. (If there are no such vertices, M is a perfect matching.) The output of APS is either an M-augmenting u-path P, or else an APS-tree T rooted at u. In the former case, we replace the matching M by $M \triangle E(P)$ and apply APS once more, starting with an uncovered vertex of G with respect to the new matching M, if there is one. In the latter eventuality, by Proposition 16.19 we may restrict our attention to the subgraph $G - V(T)$ in continuing our search for an M-augmenting path. We simply record the set $M(T) := M \cap E(T)$ (this set will be part of our maximum matching), replace the matching M by the residual matching $M \setminus E(T)$ and the graph G by the subgraph $G - V(T)$, and then apply APS once more, starting with an uncovered vertex of this subgraph, if there is one. We proceed in this way until the subgraph we are left with has no uncovered vertex (so has a perfect matching). The output of this algorithm is as follows.

▷ A set T of pairwise disjoint APS-trees.
▷ A set $R := \cup\{R(T) : T \in \mathcal{T}\}$ of red vertices.
▷ A set $B := \cup\{B(T) : T \in \mathcal{T}\}$ of blue vertices.
▷ A subgraph $F := G - (R \cup B)$ with a perfect matching $M(F)$.
▷ A matching $M^* := \cup\{M(T) : T \in \mathcal{T}\} \cup M(F)$ of G.
▷ A set $U := \{u(T) : T \in \mathcal{T}\}$ of vertices not covered by M^*.

(When the initial matching M is perfect, $\mathcal{T} = R = B = \emptyset$, $F = G$, $M^* = M$, and $U = \emptyset$.)

Example 16.20 Consider the bipartite graph in Figure 16.12a, with the indicated matching M. Figure 16.12b shows an M-alternating x_1-tree, which is grown until the M-augmenting x_1-path $P := x_1 y_2 x_2 y_1$ is found. As before, the red vertices are indicated by solid dots and the blue vertices by shaded dots. Figure 16.12c shows the augmented matching $M \triangle E(P)$ (the new matching M), and Figure 16.12d an M-alternating x_4-tree which contains no M-augmenting x_4-path and can be grown no further, and thus is an APS-tree T_1 with $R(T_1) = \{x_1, x_3, x_4\}$ and $B(T_1) = \{y_2, y_3\}$. The set of all vertices reachable in G from x_4 by M-alternating paths is therefore $V(T_1) = \{x_1, x_3, x_4, y_2, y_3\}$. This set does not include y_4, the only other vertex not covered by M. Thus we may conclude that $M^* := M$ is a maximum matching. However, for the purpose of illustrating the entire algorithm, we continue, deleting $V(T_1)$ from G and growing an M-alternating y_4-tree in the resulting subgraph (see Figure 16.12e), thereby obtaining the APS-tree T_2 with $R(T_2) = \{y_1, y_4, y_5\}$ and $B(T_2) = \{x_2, x_5\}$, as shown in Figure 16.12f. The procedure ends there, because every vertex of the graph $F := G - V(T_1 \cup T_2)$, which consists of the vertices x_6 and y_6 and the edge $x_6 y_6$, is covered by M. The output of the algorithm is therefore:

$$\mathcal{T} = \{T_1, T_2\}, \quad R = \{x_1, x_3, x_4, y_1, y_4, y_5\}, \quad B = \{y_2, y_3, x_2, x_5\}$$

$$V(F) = \{x_6, y_6\}, \quad E(F) = \{x_6 y_6\}$$

$$M^* = \{x_1 y_2, x_2 y_1, x_3 y_3, x_5 y_5, x_6 y_6\}, \quad U = \{x_4, y_4\}$$

We now verify the correctness of Egerváry's Algorithm.

Theorem 16.21 *The matching M^* returned by Egerváry's Algorithm is a maximum matching.*

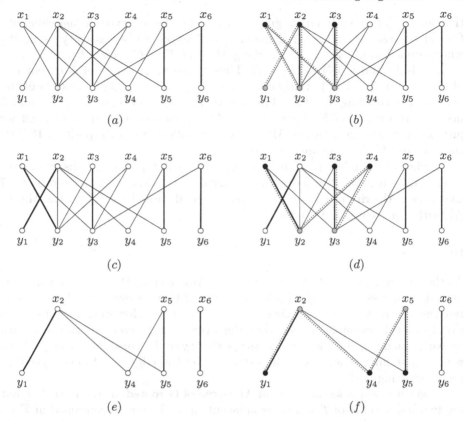

Fig. 16.12. Egerváry's Algorithm: finding a maximum matching in a bipartite graph

Proof Let $\mathcal{T}$, R, B, and U be the sets of trees, red vertices, blue vertices, and uncovered vertices returned by Egerváry's Algorithm. Because each tree $T \in \mathcal{T}$ contains exactly one uncovered vertex, namely its root $u(T)$, we have $|U| = |\mathcal{T}|$. Also, by (16.5), $|B(T)| = |R(T)| - 1$ for each tree $T \in \mathcal{T}$. Summing this identity over all $T \in \mathcal{T}$ gives

$$|B| = |R| - |\mathcal{T}|$$

Therefore

$$|U| = |R| - |B| \tag{16.9}$$

Because red vertices are adjacent in G only to blue vertices, in any matching M of G, red vertices can only be matched with blue vertices. There are therefore at least $|R| - |B|$ red vertices not covered by M. Thus, by (16.9), there are at least $|U|$ such vertices, no matter what the matching M. Because there are exactly this number of vertices not covered by M^*, we conclude that M^* is a maximum matching. $\square$

Egerváry's Algorithm returns not only a maximum matching M^* but also a covering K^* of the same size, which is consequently a minimum covering. To see this, let $G[X, Y]$ be a bipartite graph, and let M^* be a maximum matching of G. Consider the sets R and B of red and blue vertices output by Egerváry's Algorithm when applied to G with input matching M^*. Set $F := G - (R \cup B)$.

By (16.6) and (16.7), $N(R) = B$. Thus B covers all edges of G except those of F. Because $X \cap V(F)$ clearly covers $E(F)$, the union $B \cup (X \cap V(F))$ of these two sets is a covering K^* of G. Moreover, there is a bijection between M^* and K^* because each vertex of K^* is covered by M^* and each edge of M^* is incident with just one vertex of K^*. Hence $|M^*| = |K^*|$. It follows from Proposition 16.7 that the covering K^* is a minimum covering.

In view of the relationship between matchings in bipartite graphs and families of internally disjoint directed paths in digraphs (as described in Exercise 8.6.7), Egerváry's Algorithm may be viewed as a special case of the Max-Flow Min-Cut Algorithm presented in Chapter 7.

BLOSSOMS

As the example in Figure 16.11b illustrates, Augmenting Path Search is not guaranteed to find an M-augmenting u-path, even if there is one, should there be two red vertices in the APS-tree that are adjacent in G. However, if we look more closely at this example, we see that the cycle $rvxyzr$ contains two alternating rv-paths, namely, the edge rv and the path $rzyxv$. Because the latter path ends with a matching edge, it may be extended by the edge vw, thereby yielding a uw-alternating path.

In general, suppose that T is an APS-tree of G rooted at u, and that x and y are two red vertices of T which are adjacent in G. The cycle contained in $T + xy$ is then called a *blossom*. A blossom C is necessarily of odd length, because each blue vertex is matched with a red vertex and there is one additional red vertex, which we call the *root* of C and denote by $r := r(C)$ (see Figure 16.13a). Note that $M \cap E(C)$ is a perfect matching of $C - r$. Note, also, that the path uTr is M-alternating, and terminates with a matching edge (unless $r = u$). Moreover, this path is internally disjoint from C.

The key to finding a maximum matching in an arbitrary graph is to *shrink* blossoms (that is, contract them to single vertices) whenever they are encountered during APS. By shrinking a blossom and continuing to apply APS to the resulting graph, one might be able to reach vertices by M-alternating u-paths which could not have been reached before. For example, if T is an APS-tree with a blossom C, and if there happens to be an edge vw with $v \in V(C)$ and $w \in V(G) \setminus V(T)$, as in Figure 16.13a), then w is reachable from u by an M-alternating path P' in $G' := G / C$ (see Figure 16.13b, where the shrunken blossom $C = rvxyzr$ is indicated by a large solid dot), and this path P' can be modified to an M-alternating path P in G by inserting the rv-segment of C that ends with a matching edge (Figure 16.13c). In particular, if P' is an M-augmenting path in G / C, then the modified path P is an M-augmenting path of G. We refer to this process of obtaining an M-alternating path of G from an M-alternating path of G / C as *unshrinking* C.

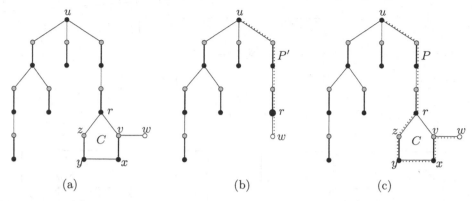

Fig. 16.13. (a) A blossom C, (b) an M-alternating u-path P' in $G' := G/C$, (c) an M-alternating u-path P in G

If C is a blossom with root r, we denote the vertex resulting from shrinking C by r also (and keep a record of the blossom C). The effect of shrinking a blossom C is to replace the graph G by G/C, the tree T by $(T + xy)/C$, where x and y are the adjacent red vertices of C, and the matching M by $M \setminus E(C)$. When we incorporate this blossom-shrinking operation into APS, we obtain a modified search procedure, APS$^+$.

By way of illustration, consider the graph G and the matching M shown in Figure 16.14a.

We grow an M-alternating tree T rooted at the uncovered vertex u. A blossom $C = uvwu$ is found (Figure 16.14b) and shrunk to its root u. In Figure 16.14c, the contracted tree (now a single vertex) is grown further, and an M-augmenting path $uxyz$ is found in the contracted graph G/C, giving rise (after unshrinking the blossom C) to the M-augmenting path $uvwxyz$ in G, as shown in Figure 16.14d. The augmented matching is indicated in Figure 16.14e.

Starting with this new matching M and the vertex a not covered by it, the above procedure is now repeated, and evolves as illustrated in Figure 16.15, terminating with the APS-tree depicted in Figure 16.15g. Note that, because the vertex a of Figure 16.15c was obtained by shrinking the blossom $abca$, the blossom $adea$ is, in fact, a 'compound' blossom. We now examine the structure of such compound blossoms.

FLOWERS

As the above example illustrates, during the execution of APS$^+$, the graph G is repeatedly modified by the operation of shrinking blossoms. Suppose that $(C_0, C_1, \ldots, C_{k-1})$ is the sequence of blossoms shrunk, in that order, during the execution of APS$^+$. The original graph G is thus progressively modified, yielding a sequence of graphs $(G_0, G_1, \ldots, G_k)$, where $G_0 := G$ and, for $0 \leq i \leq k-1$, $G_{i+1} := G_i / C_i$. If APS$^+$ fails to find an M-augmenting u-path, it terminates

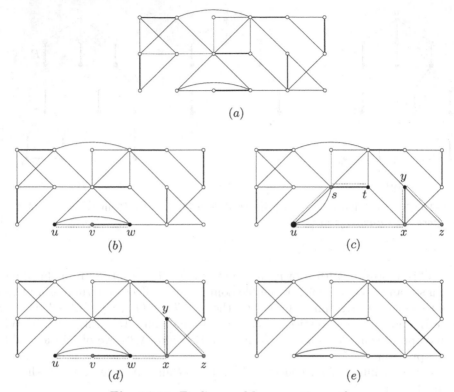

Fig. 16.14. Finding an M-augmenting path

with an APS-tree T_k in G_k, no two red vertices of which are adjacent in G_k. Let us suppose that APS$^+$ returns such a tree.

Because a blossom C_i is always shrunk to its root, a red vertex, the blue vertices in each of the graphs G_i are simply vertices of the original input graph. However, the red vertices might well correspond to nontrivial induced subgraphs of the input graph. The subgraphs of G corresponding to the red vertices of G_i are called the *flowers* of G associated with T_i. For instance, the three flowers of the graph G of Figure 16.14a associated with the APS-tree shown in Figure 16.15g are the subgraphs of G induced by $\{a, b, c, d, e\}$, $\{g, h, i\}$, and $\{u, v, w\}$.

Flowers satisfy two basic properties, described in the following proposition.

Proposition 16.22 *Let F be a flower of G. Then:*

i) F *is connected and of odd order,*
ii) *for any vertex v of F, there is an M-alternating uv-path in G of even length (that is, one terminating in an edge of M).*

Proof The proof is by induction on i, where F is a flower associated with T_i. We leave the details to the reader (Exercise 16.5.8).   □

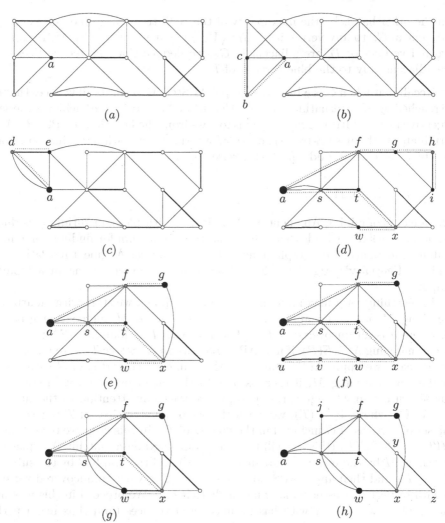

Fig. 16.15. Growing an APS-tree

We are now ready to prove the validity of Algorithm APS$^+$.

Corollary 16.23 *Let T_k be an APS-tree of G_k no two red vertices of which are adjacent in G_k. Then the red vertices of T_k are adjacent in G_k only to blue vertices of T_k. Equivalently, the flowers of G associated with T_k are adjacent only to the blue vertices of G in T_k.*

Proof It follows from Proposition 16.22(ii) that if G_k has an M-augmenting u-path, then so has G. Thus if G has no M-augmenting u-path, no red vertex of T_k can be adjacent in G_k to any vertex in $V(G_k) \setminus V(T_k)$ that is not covered by M.

On the other hand, by the maximality of the APS-tree T_k, no red vertex of T_k is adjacent in G_k to any vertex in $V(G_k) \setminus V(T_k)$ that is covered by M. Because no two red vertices of T_k are adjacent in G_k, we deduce that the red vertices of T_k are adjacent only to the blue vertices of T_k. $\qquad\square$

Recall that when G is a bipartite graph, APS(G, M, u) finds all vertices that can be reached by M-alternating u-paths. Algorithm APS$^+(G, M, u)$ achieves the same objective in an arbitrary graph G. This follows from the fact that if APS$^+(G, M, u)$ terminates with an APS-tree T_k, every M-alternating u-path in G that terminates in a blue vertex is of odd length (Exercise 16.5.10).

EDMONDS' ALGORITHM

The idea of combining Augmenting Path Search with blossom-shrinking is due to Edmonds (1965d). It leads to a polynomial-time algorithm for finding a maximum matching in an arbitrary graph, in much the same way as Augmenting Path Search leads to Egerváry's Algorithm for finding a maximum matching in a bipartite graph.

In searching for a maximum matching of a graph G, we start with an arbitrary matching M of G, and apply APS$^+$ to search for an M-augmenting u-path in G, where u is an uncovered vertex. If such a path P is found, APS$^+$ returns the larger matching $M \triangle E(P)$; if not, APS$^+$ returns an APS-tree T rooted at u. In the former case, we apply APS$^+$ starting with an uncovered vertex of G with respect to the new matching M, if there is one. In the latter eventuality, in continuing our search for an M-augmenting path, we restrict our attention to the subgraph $G - V(T)$ (where, by $V(T)$, we mean the set of blue vertices of T together with the set of vertices of G included in the flowers of T). In this case, we record the set $M(T) := M \cap E(T)$ (this set will be part of our maximum matching), replace the matching M by the residual matching $M \setminus E(T)$ and the graph G by the subgraph $G - V(T)$, and then apply APS$^+$ once more, starting with an uncovered vertex of G with respect to this new matching, if there is one. We proceed in this way until the graph F we are left with has no uncovered vertices (and thus has a perfect matching).

For example, after having found the APS-tree in Figure 16.15g, there remains one uncovered vertex, namely j. The APS-tree grown from this vertex is just the trivial APS-tree. The subgraph F consists of the vertices y and z, together with the edge linking them. The red and blue vertices in the two APS-trees are indicated in Figure 16.15h.

The output of Edmonds' Algorithm is as follows.

▷   A set $\mathcal{T}$ of pairwise disjoint APS-trees.
▷   A set $R := \cup\{R(T) : T \in \mathcal{T}\}$ of red vertices.
▷   A set $B := \cup\{B(T) : T \in \mathcal{T}\}$ of blue vertices.
▷   A subgraph $F := G - (R \cup B)$ of G with a perfect matching $M(F)$.
▷   A matching $M^* := \cup\{M(T) : T \in \mathcal{T}\} \cup M(F)$ of G.
▷   A set $U := \{u(T) : T \in \mathcal{T}\}$ of vertices not covered by M^*.

(As in Egerváry's Algorithm, when the initial matching M is perfect, $T = R = B = \emptyset$, $F = G$, $M^* = M$, and $U = \emptyset$.)

The proof that Edmonds' Algorithm does indeed return a maximum matching closely resembles the proof of Theorem 16.21. We leave it as an exercise (16.5.9).

Theorem 16.24 *The set B returned by Edmonds' Algorithm is a barrier and the matching M^* returned by the algorithm is a maximum matching.*   □

To conclude, we note that Edmonds' Algorithm, combined with the polynomial reduction of the f-factor problem to the 1-factor problem described in Section 16.4, yields a polynomial-time algorithm for solving the f-factor problem.

Exercises

16.5.1 Apply Egerváry's Algorithm to find a maximum matching in the bipartite graph of Figure 16.16a.

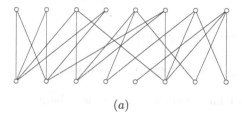
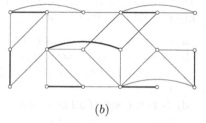

(a) (b)

Fig. 16.16. Find maximum matchings in these graphs (Exercises 16.5.1 and 16.5.7)

16.5.2 Show that Egerváry's Algorithm is a polynomial-time algorithm.

★**16.5.3** Prove Proposition 16.19.

16.5.4 Describe how the output of Egerváry's Algorithm can be used to find a minimum edge covering of an input bipartite graph without isolated vertices.

16.5.5 Find minimum coverings, maximum stable sets, and minimum edge coverings in the graphs of Figures 16.12 and 16.16a.

16.5.6 For any positive integer k, show that the complete k-partite graph $G := K_{2,2,\ldots,2}$ is k-choosable.   (P. ERDŐS, A.L. RUBIN, AND H. TAYLOR)

16.5.7 Apply Edmonds' Algorithm to find a maximum matching in the graph of Figure 16.16b, starting with the matching M indicated. Determine the barrier output by the algorithm.

⋆**16.5.8** Prove Proposition 16.22.

⋆**16.5.9**

a) Show that the set B of blue vertices returned by Edmonds' Algorithm constitutes a barrier of G.

b) Give a proof of Theorem 16.24.

⋆**16.5.10**

a) Let T be an APS-tree returned by $\text{APS}^+(G, M, u)$. Show that every M-
alternating u-path in G that terminates in a blue vertex is of odd length.

b) Deduce that $\text{APS}^+(G, M, u)$ finds all vertices of G that can be reached by
M-alternating u-paths.

16.5.11 Show that Edmonds' Algorithm is a polynomial-time algorithm.

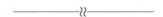

16.5.12 Deduce from Exercise 16.3.10 that the flowers created during the execution of APS^+ are hypomatchable.

16.5.13 Let B be the barrier obtained by applying Edmonds' Algorithm to an
input graph G. Show that:

a) every even component of $G - B$ has a perfect matching,
b) every odd component of $G - B$ is hypomatchable,
c) a vertex v of G is inessential if and only if it belongs to an odd component of
   $G - B$,
d) B is the set of all essential vertices that have some inessential neighbour.

(Gallai (1964a) was the first to show that every graph has a barrier satisfying the
above conditions.)

16.5.14 Shortest Even and Odd Paths
Let $G := G(x, y)$ be a graph, and let H be the graph obtained from $G \,\square\, K_2$ by
deleting the copies of x and y in one of the two copies of G.

a) Find a bijection between the xy-paths of even length in G and the perfect
matchings in H.

b) By assigning weights 0 and 1 to the edges of H in an appropriate way and
applying the weighted version of Edmonds' Algorithm, show how to find, in
polynomial time, a shortest xy-path of even length in G.

c) By means of a similar construction, show how to find, in polynomial time, a
shortest xy-path of odd length in G.   (J. Edmonds)

16.5.15 By using minimum-weight matchings, refine the 2-approximation algorithm for the Metric Travelling Salesman Problem presented in Section 8.4,
so as to obtain a polynomial-time $\frac{3}{2}$-approximation algorithm for this problem.   (N. Christofides)

16.6 Related Reading

STABLE SETS IN CLAW-FREE GRAPHS

Maximum stable sets in line graphs can be determined in polynomial time, by virtue of Edmonds' Algorithm (described in Section 16.5), because a stable set in a line graph $L(G)$ corresponds to a matching in G. More generally, there exist polynomial-time algorithms for finding maximum stable sets in claw-free graphs, a class which includes all line graphs (see Minty (1980), Sbihi (1980), or Lovász and Plummer (1986)).

TRANSVERSAL MATROIDS

Let $G := G[X, Y]$ be a bipartite graph. A subset S of X is *matchable* with a subset of T of Y if there is a matching in G which covers $S \cup T$ and no other vertices. A subset of X is *matchable* if it is matchable with some subset of Y. Edmonds and Fulkerson (1965) showed that the matchable subsets of X are the independent sets of a matroid on X; matroids that arise in this manner are called *transversal matroids*. Various results described in Section 16.2 may be seen as properties of transversal matroids. For example, the König–Ore Formula (Exercise 16.2.9) is an expression for the rank of this matroid.

RADO'S THEOREM

Let $G := G[X, Y]$ be a bipartite graph, and let M be a matroid defined on Y with rank function r. As a far-reaching generalization of Hall's Theorem (16.4), Rado (1942) showed that X is matchable with a subset of Y which is independent in the matroid M if and only if $r(N(S)) \geq |S|$, for all $S \subseteq X$. Many variants and applications of Rado's Theorem can be found in Welsh (1976).

PFAFFIANS

Let $D := (V, A)$ be a strict digraph, and let $\{x_a : a \in A\}$ be a set of variables associated with the arcs of D. The *Tutte matrix* of D is the $n \times n$ skew-symmetric matrix $\mathbf{T} = (t_{uv})$ defined by:

$$
t_{uv} := \begin{cases}
0 & \text{if } u \text{ and } v \text{ are not adjacent in } D, \\
x_a & \text{if } a = (u, v), \\
-x_a & \text{if } a = (v, u).
\end{cases}
$$

Because $\mathbf{T}$ is skew-symmetric, its determinant is zero when n is odd. But when n is even, say $n = 2k$, the determinant of $\mathbf{T}$ is the square of a certain polynomial, called the Pfaffian of $\mathbf{T}$, which may be defined as follows.

For any perfect matching $M := \{a_1, a_2, \ldots, a_k\}$ of D, where $a_i := (u_i, v_i)$, $1 \leq i \leq k$, let $\pi(M)$ denote the product $t_{u_1 v_1} t_{u_2 v_2} \ldots t_{u_k v_k}$ and let $\text{sgn}(M)$ denote the sign of the permutation $(u_1v_1u_2v_2 \ldots u_kv_k)$. (Observe that $\mathrm{sgn}(M)$ does not depend on the order in which the elements of M are listed.) The *Pfaffian* of $\mathbf{T}$ is the sum of $\mathrm{sgn}(M)\pi(M)$ taken over all perfect matchings M of D.

Now, a polynomial in indeterminates is zero if and only if it is identically zero. Thus the digraph D has a perfect matching if and only if the Pfaffian of $\mathbf{T}$ is nonzero. Because the determinant of $\mathbf{T}$ is the square of its Pfaffian, it follows that D has a perfect matching if and only if $\det \mathbf{T} \neq 0$. Tutte's original proof of Theorem 16.13 was based on an ingenious exploitation of this fact (see Tutte (1998) for a delightful account of how he was led to this discovery). In more recent times, properties of the Tutte matrix have played surprisingly useful roles both in the theory of graphs and in its algorithmic applications; see, for example, Lovász and Plummer (1986), McCuaig (2000), and Robertson et al. (1999).

17

Edge Colourings

Contents

17.1 Edge Chromatic Number

In Chapter 14 we studied vertex colourings of graphs. We now turn our attention to the analogous concept of edge colouring.

Recall that a *k-edge-colouring* of a graph $G = (V, E)$ is a mapping $c : E \to S$, where S is a set of k *colours*, in other words, an assignment of k colours to the edges of G. Usually, the set of colours S is taken to be $\{1, 2, \ldots, k\}$. A k-edge-colouring can then be thought of as a partition $\{E_1, E_2, \ldots, E_k\}$ of E, where E_i denotes the (possibly empty) set of edges assigned colour i.

An edge colouring is *proper* if adjacent edges receive distinct colours. Thus a proper k-edge-colouring is a k-edge-colouring $\{M_1, M_2, \ldots, M_k\}$ in which each subset M_i is a matching. (Because loops are self-adjacent, only loopless graphs admit proper edge colourings.) As we are concerned here only with proper edge colourings, all graphs are assumed to be loopless, and we refer to a proper edge colouring simply as an 'edge colouring'. The graph of Figure 17.1 has the 4-edge-colouring $\{\{a, g\}, \{b, e\}, \{c, f\}, \{d\}\}$.

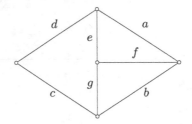

Fig. 17.1. A 4-edge-chromatic graph

A graph is *k-edge-colourable* if it has a *k*-edge-colouring. Clearly, if G is *k*-edge-colourable, G is also ℓ-edge colourable for every $\ell > k$; moreover, every graph G is *m*-edge-colourable. The *edge chromatic number*, $\chi'(G)$, of a graph G is the minimum k for which G is *k*-edge colourable, and G is *k-edge-chromatic* if $\chi'(G) = k$. It is straightforward to verify that the graph in Figure 17.1 is not 3-edge-colourable (see Exercise 17.1.3). This graph is therefore 4-edge-chromatic.

In an edge colouring, the edges incident with any one vertex must evidently be assigned different colours. This observation yields the lower bound

$$\chi' \geq \Delta \tag{17.1}$$

Edge colouring problems arise in practice in much the same way as do vertex colouring problems. Here is a typical example.

Example 17.1 THE TIMETABLING PROBLEM
In a school, there are m teachers $x_1, x_2, \ldots, x_m$, and n classes $y_1, y_2, \ldots, y_n$. Given that teacher x_i is required to teach class y_j for p_{ij} periods, schedule a complete timetable in the minimum number of periods.

To solve this problem, we represent the teaching requirements by a bipartite graph $H[X, Y]$, where $X = \{x_1, x_2, \ldots, x_m\}$, $Y = \{y_1, y_2, \ldots, y_n\}$, and vertices x_i and y_j are joined by p_{ij} edges. It is easy to see (Exercise 17.1.10a) that the problem posed amounts to finding an edge colouring of H in as few colours as possible. This can be solved by means of a polynomial-time algorithm, as indicated below. For a more detailed discussion of this Timetabling Problem, see Bondy and Murty (1976).

EDGE COLOURINGS OF BIPARTITE GRAPHS

Referring to the example of Figure 17.1, we see that inequality (17.1) can be strict. However, as we prove shortly, equality always holds in (17.1) when G is bipartite. In Section 17.2, we derive upper bounds on χ' for other classes of graphs. We show, in particular, that if G is any simple graph, then $\chi' \leq \Delta + 1$. The proofs we present are constructive, and demonstrate how, under suitable conditions, a *k*-edge-colouring of a graph G may be obtained by colouring the edges one by one, adjusting the colouring along the way if necessary. We assume that we have already obtained a k-edge-colouring of a certain subgraph H of G and describe how to extend it to a k-edge-colouring of G. The following notions are basic to our approach.

Let H be a spanning subgraph of a graph G and let $\mathcal{C} := \{M_1, M_2, \ldots, M_k\}$ be a k-edge-colouring of H. A colour i is *represented* at a vertex v if it is assigned to some edge of H incident with v; otherwise it is *available* at v. A colour is *available* for an edge of $E(G) \setminus E(H)$ if it is available at both ends of the edge. Thus, if an edge e is uncoloured, any colour available for e may be assigned to it to extend $\mathcal{C}$ to a k-edge-colouring of $H + e$.

Let i and j be any two distinct colours, and set $H_{ij} := H[M_i \cup M_j]$. Because M_i and M_j are disjoint matchings, each component of H_{ij} is either an even cycle or a path (see the proof of Theorem 16.3); we refer to the path-components of H_{ij} as ij-*paths*. These are akin to Kempe chains (see Section 15.2), and one of the main tools used in our proofs consists of selecting suitable colours i and j and swapping the colours on an appropriately chosen ij-path so as to obtain a new k-edge-colouring with respect to which there is an available colour for some edge of $E(G) \setminus E(H)$. The proof of the following theorem provides a simple illustration of this technique.

Theorem 17.2 *If G is bipartite, then $\chi' = \Delta$.*

Proof By induction on m. Let $e = uv$ be an edge of G. We assume that $H = G \setminus e$ has a Δ-edge-colouring $\{M_1, M_2, \ldots, M_\Delta\}$. If some colour is available for e, that colour can be assigned to e to yield a Δ-edge-colouring of G. So we may assume that each of the Δ colours is represented either at u or at v. Because the degree of u in $G \setminus e$ is at most $\Delta - 1$, at least one colour i is available at u, hence represented at v. Likewise, at least one colour j is available at v and represented at u. Consider the subgraph H_{ij}. Because u has degree one in this subgraph, the component containing u is an ij-path P. This path does not terminate at v. For if it did, it would be of even length, starting with an edge coloured i and ending with an edge coloured j, and $P + e$ would be a cycle of odd length in G, contradicting the hypothesis that G is bipartite. Interchanging the colours on P, we obtain a new Δ-edge-colouring of H with respect to which the colour i is available at both u and v. Assigning colour i to e, we obtain a Δ-edge-colouring of G. $\square$

One may easily extract from the above proof a polynomial-time algorithm for finding a Δ-edge-colouring of a bipartite graph G. An alternative proof of Theorem 17.2, using Exercise 16.4.16, is outlined in Exercise 17.1.11.

Exercises

17.1.1 Show that a d-regular graph G is d-edge-colourable if and only if its edge set can be partitioned into perfect matchings.

17.1.2 By exhibiting an appropriate edge colouring, show that $\chi'(K_{m,n}) = \Delta(K_{m,n})\ (= \max\{m, n\})$.

⋆17.1.3

a) Show that every graph G satisfies the inequality $\chi' \geq m/\lfloor n/2 \rfloor$.
b) Deduce that the graph in Figure 17.1 is not 3-edge-colourable.

17.1.4 Deduce from Exercise 16.1.7 that a cubic graph with a cut edge is not 3-edge-colourable.

17.1.5 Let G be a 2-edge-connected cubic graph with a 2-edge cut $\{e_1, e_2\}$. For $i = 1, 2$, let $e_i := u_i v_i$, where u_i and v_i belong to the component H_i of $G \setminus \{e_1, e_2\}$. Show that:

a) the graph G_i obtained from H_i by joining u_i and v_i by a new edge f_i is a 2-edge-connected cubic graph, $i = 1, 2$,
b) G is 3-edge-colourable if and only if both G_1 and G_2 are 3-edge-colourable.

17.1.6 Let $\partial(X)$ be a 3-edge cut of a cubic graph G. Show that G is 3-edge-colourable if and only if both G / X and $G / \overline{X}$ are 3-edge-colourable, where $\overline{X} := V \setminus X$.

17.1.7

a) Show that the Petersen graph is not 3-edge-colourable (either directly, by considering a hypothetical 3-edge-colouring of one of its 5-cycles, or by appealing to Exercises 16.1.7 and 17.1.1).
b) Deduce that the Petersen graph is 4-edge-chromatic.

17.1.8

a) Show that every hamiltonian cubic graph is 3-edge-colourable.
b) Deduce from Exercise 17.1.7b that the Petersen graph is not hamiltonian.

17.1.9 Let $G = C_5[\overline{K_2}]$.

a) Let M and M' be two perfect matchings of G. Show that there is an automorphism of G mapping M to M'.
b) Deduce that $\chi' = 4$.
c) Deduce further that G does not contain the Petersen graph.

⋆17.1.10

a) Show that the Timetabling Problem (Example 17.1) amounts to finding an edge colouring of a bipartite graph in the minimum possible number of colours.
b) Let p denote the minimum possible number of periods and ℓ the total number of lessons to be given. Show that there exists a timetable in which no more than $\lceil \ell/p \rceil$ classrooms are required in any one period.

17.1.11

a) Show that if G is bipartite, then G has a Δ-regular bipartite supergraph.
b) Using (a) and Exercise 16.4.16, give an alternative proof of Theorem 17.2.

17.1.12 Let G be a graph with $\Delta \leq 3$. Show that G is 4-edge-colourable

a) by appealing to Exercise 2.2.2 and Theorem 17.2,
b) by applying Brooks' Theorem (14.4).

17.1.13 Describe a polynomial-time algorithm for finding a proper Δ-edge-colouring of a bipartite graph G. What is the complexity of your algorithm?

17.1.14 Eight schoolchildren go for a walk in pairs every day. Show how they can arrange their outings so that each child has a different companion on different days of the week?

17.1.15 Deduce from Theorem 17.2 that line graphs of bipartite graphs are perfect.

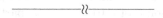

17.1.16 KIRKMAN'S SCHOOLGIRL PROBLEM
A k-*edge-colouring* of a hypergraph $(V, \mathcal{F})$ is a partition $\{\mathcal{F}_1, \mathcal{F}_2, \ldots, \mathcal{F}_k\}$ of $\mathcal{F}$ such that, for $1 \leq i \leq k$, no two edges in $\mathcal{F}_i$ have a vertex in common. Formulate as a hypergraph edge colouring problem, and solve, the following puzzle, posed by the Reverend T. P. Kirkman in 1847. Fifteen schoolgirls in an English boarding school go for a walk every day in groups of three abreast. Can their walks be arranged so that no two shall walk abreast more than once during a week?

17.1.17

a) By exhibiting an appropriate edge colouring, show that $\chi'(K_{2n}) = 2n - 1$, $n \geq 1$.
b) Deduce that $\chi'(K_{2n-1}) = 2n - 1$, $n \geq 2$.

$\star$**17.1.18** GUPTA'S THEOREM
Let G be a bipartite graph with no isolated vertices. Show that G has a (not necessarily proper) δ-edge-colouring in which all δ colours are represented at each vertex.   (R. P. GUPTA)

17.1.19 Consider a (not necessarily proper) 3-edge-colouring of a complete graph in which each colour class induces a connected spanning subgraph. Show that there is a triangle which has an edge of each colour.   (T. GALLAI)

17.2 Vizing's Theorem

As has already been noted, if G is not bipartite one cannot necessarily conclude that $\chi' = \Delta$. An important theorem due to Vizing (1964), and independently Gupta (1966), asserts that for any simple graph G, either $\chi' = \Delta$ or $\chi' = \Delta + 1$.

In proving Vizing's Theorem by induction on m, one may assume (as in the proof of Theorem 17.2) that there is a $(\Delta + 1)$-edge-colouring of $G \setminus e$, where $e \in E$. To complete the proof it suffices to show how a $(\Delta + 1)$-edge-colouring of G itself can be obtained from this $(\Delta+1)$-edge-colouring of $G\backslash e$. With further applications in mind (see, for example, Exercise 17.2.9), we consider the more general problem of deriving a k-edge-colouring of G from a k-edge-colouring of $G \setminus e$, where k is any integer greater than or equal to Δ.

Lemma 17.3 *Let G be a simple graph, v a vertex of G, e an edge of G incident to v, and k an integer, $k \geq \Delta$. Suppose that $G \setminus e$ has a k-edge-colouring c with respect to which every neighbour of v has at least one available colour. Then G is k-edge-colourable.*

Proof Consider the k-edge-colouring c of $G \setminus e$. In seeking a k-edge-colouring of G, it is convenient to study the bipartite graph $H[X,Y]$, where $X := N_G(v)$ and $Y := \{1,2,\ldots,k\}$, vertices $x \in X$ and $i \in Y$ being adjacent if colour i is available at vertex x in the restriction $\tilde{c}$ of c to $G - v$. In particular, for all $x \in X \setminus \{u\}$, where u denotes the other end of e, the colour of the edge xv is available at x in $G - v$, so H contains the matching

$$M := \{(x, c(xv)) : x \in X \setminus \{u\}\}$$

Conversely, every matching in H corresponds to a partial colouring of $\partial(v)$ that is compatible with $\tilde{c}$. In particular, any matching in H which saturates X corresponds to a full colouring of $\partial(v)$ and thus yields, together with $\tilde{c}$, a k-edge-colouring of G. We may suppose that H has no such colouring. Our goal is to modify the colouring c to a colouring c' so that the corresponding bipartite graph H' does contain a matching saturating X.

By hypothesis, each vertex of $X \setminus \{u\}$ is incident with at least one edge of $H \setminus M$, and the vertex u is incident with at least one such edge as well, because

$$d_{G\backslash e}(u) = d_G(u) - 1 \leq \Delta(G) - 1 \leq k - 1$$

Therefore each vertex of X is incident with at least one edge of $H \setminus M$.

Denote by Z the set of all vertices of H reachable from u by M-alternating paths, and set $R := X \cap Z$ and $B := Y \cap Z$. As in the proof of Hall's Theorem (16.4), $N_H(R) = B$ and B is matched under M with $R \setminus \{u\}$, so $|B| = |R| - 1$. Because each vertex of R is incident with at least one edge of $H \setminus M$, some two vertices x, y of R are adjacent in $H \setminus M$ to a common colour $i \in B$, by the pigeonhole principle; this colour i is therefore available at both x and y. Note that every colour in B is represented at v, because B is matched under M with $R \setminus \{u\}$. In particular, colour i is represented at v. On the other hand, because the degree of v in $G \setminus e$ is at most $k - 1$, some colour j is available at v. Observe that $j \notin B$ because every colour in B is represented at v. Thus j is represented at every vertex of R, in particular at both x and y.

Let us return now to the graph $G \setminus e$. By the above observations, each of the vertices v, x, and y is an end of an ij-path in $G \setminus e$. Consider the ij-path starting at v. Evidently, this path cannot terminate at both x and y. We may suppose that the path starting at v does not terminate at y, and let z be the terminal vertex of the ij-path P starting at y. Interchanging the colours i and j on P, we obtain a new colouring c' of $G \setminus e$.

Let $H'[X,Y]$ be the bipartite graph corresponding to c'. The only differences in the edge sets of H and H' occur at y and possibly z (if $z \in X$). Moreover, because v does not lie on P, the matching M is still a matching in H'. Consider the M-alternating uy-path Q in H. If z lies on Q, then $z \in R$ and the alternating path uQz is still an M-alternating path in H', as it terminates with an edge of M. Also, because $j \notin B$, the path P must have originally terminated at z in an edge of colour j, and now terminates in an edge of colour i. With respect to the colouring c', the colour j is therefore available at z, and $Q' := uQzj$ is an M-augmenting path in H'. On the other hand, if z does not lie on Q, then $Q' := uQyj$ is an M-augmenting path in H'.

Set $M' := M \triangle E(Q')$. Then M' is a matching in H' which covers every vertex in X, and this matching corresponds to a full colouring of $\partial(v)$. Combining this colouring with the restriction of c' to $G - v$, we obtain a proper k-edge-colouring of G. □

Figure 17.2 illustrates the steps in this proof as applied to the Petersen graph, with the initial 4-edge-colouring c shown in Figure 17.2a. The bipartite graph H is shown in Figure 17.2b, with $X := \{s,t,u\}$, $Y := \{1,2,3,4\}$, and the matching M indicated by heavy lines. We may take $i = 1$, $j = 3$, $u = x$, and $t = y$. The ij-path from v (to u) and the ij-path P from t to z are shown in Figure 17.2c, and the 4-edge-colouring c' in Figure 17.2d. The corresponding bipartite graph H' is shown in Figure 17.2e, with an M-augmenting u-path Q indicated. Figure 17.2f shows the augmented matching M', and Figure 17.2g the resulting 4-edge-colouring of the Petersen graph.

The above proof yields a polynomial-time algorithm for finding a k-edge-colouring of a simple graph G, given a k-edge-colouring of $G \setminus e$ satisfying the hypothesis of Lemma 17.3. Because the hypothesis of Lemma 17.3 is satisfied when $k = \Delta + 1$, Vizing's Theorem follows directly by induction on m. Moreover, a $(\Delta + 1)$-edge-colouring of any simple graph G can be found, by adding one edge at a time, in polynomial time.

Theorem 17.4 Vizing's Theorem
For any simple graph G, $\chi' \leq \Delta + 1$. □

The observant reader will have noticed that the bound on the edge chromatic number in Vizing's Theorem (17.4) bears a striking formal resemblance to the bound (14.3) on the chromatic number.

There is a natural generalization of Theorem 17.4 to all loopless graphs. Consider such a graph G. For vertices u and v of G, we denote by $\mu(u,v)$ the number of parallel edges joining u and v. The *multiplicity* of G, denoted by $\mu(G)$, is the maximum value of μ, taken over all pairs of vertices of G. Vizing (1964) extended his theorem as follows.

Theorem 17.5 *For any graph G, $\chi' \leq \Delta + \mu$.* □

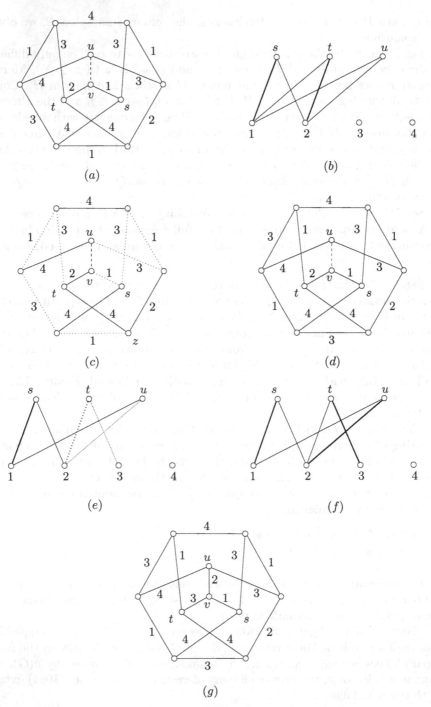

Fig. 17.2. Finding a 4-edge-colouring of the Petersen graph by means of Lemma 17.3

This more general theorem can be established by adapting the proof of Theorem 17.4 (Exercise 17.2.8). The graph G depicted in Figure 17.3 shows that the theorem is best possible for any value of μ. Here $\Delta = 2\mu$ and, the edges being pairwise adjacent, $\chi' = m = 3\mu = \Delta + \mu$.

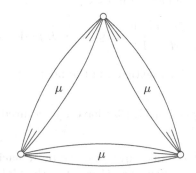

Fig. 17.3. A graph G with $\chi' = \Delta + \mu$

Returning to simple graphs, Theorem 17.4 implies that the edge chromatic number of a simple graph G is equal to either Δ or $\Delta + 1$. Those simple graphs G for which $\chi' = \Delta$ are said to belong to *Class* 1, and the others to *Class* 2. The problem of deciding to which class a graph belongs is $\mathcal{NP}$-hard (Holyer (1981) and Leven and Galil (1983)). It is therefore useful to have simple criteria for a graph to belong to Class 1 or Class 2. For example, with the aid of Lemma 17.3 one can show that a graph G belongs to Class 1 if its vertices of degree Δ induce a forest (Exercise 17.2.9). Also, by virtue of the Four-Colour Theorem (11.2), Tait's Theorem (11.4), and Exercise 17.1.5, every 2-edge-connected cubic planar graph belongs to Class 1. Moreover, the proof of the Four-Colour Theorem outlined in Chapter 15 yields (via Tait's Theorem) a polynomial-time algorithm for 3-edge-colouring such a graph. A simple condition for a graph to belong to Class 2 is described in Exercise 17.2.1.

Exercises

17.2.1 OVERFULL GRAPH
A simple graph G is *overfull* if $m > \lfloor n/2 \rfloor \Delta$.

a) Show that every overfull graph:
    i) is of odd order,
    ii) belongs to Class 2.
b) Show that a nonempty simple graph is overfull if either:
    i) it is obtained from a regular graph of even order by subdividing one edge,
      or

  ii) it is obtained from a simple k-regular graph of odd order by deleting fewer
    than $k/2$ edges.   (L. W. BEINEKE AND R. J. WILSON)

17.2.2

a) Show that, for every loopless graph G,

$$\chi' \geq \max \left\{ \left\lceil \frac{2e(H)}{v(H) - 1} \right\rceil : H \subseteq G, \ v(H) \text{ odd}, \ v(H) \geq 3 \right\} \qquad (17.2)$$

b) Find a graph G for which neither the bound $\chi' \geq \Delta$ nor the bound (17.2) is
   tight.

(Goldberg (1974) and Seymour (1979a) have conjectured that if neither bound is
tight, then $\chi' = \Delta + 1$.)

17.2.3 Let G be a graph obtained from a cycle by replacing each edge by a set of
one or more parallel edges. Show that G satisfies the Goldberg bound (17.2) with
equality.

17.2.4 Let G be a simple graph.

a) Using Vizing's Theorem (17.4), show that $G \square K_2$ belongs to Class 1.
b) Deduce that if H is a nontrivial simple graph which belongs to Class 1, then
   $G \square H$ belongs to Class 1 also.

17.2.5 Let P be the Petersen graph. Show that $P \square K_3$ belongs to Class 1.
(J.D. Horton and W.D. Wallis have shown that, for any 3-connected cubic graph
G, the cartesian product $G \square K_3$ admits a decomposition into two Hamilton cycles
and a perfect matching, hence belongs to Class 1.)

17.2.6 Describe a polynomial-time algorithm for finding a proper $(\Delta + 1)$-edge
colouring of a simple graph G. What is the complexity of your algorithm?

17.2.7 Using Exercise 16.4.16b, show that if Δ is even, then $\chi' \leq 3\Delta/2$.

17.2.8

a) Let G be a graph, let $e = uv$ be an edge of G, and let $k \geq \Delta + \mu - 1$ be
   an integer, where $\mu := \mu(G)$. Suppose that $G \setminus e$ has a k-edge-colouring with
   respect to which every neighbour of v has at least μ available colours. Show
   that G is k-edge-colourable.
b) Deduce that $\chi' \leq \Delta + \mu$.   (V.G. VIZING)
c) By considering separately the cases $\mu \leq \Delta/2$ and $\mu > \Delta/2$, and using induction
   on m, deduce from (b) that $\chi' \leq 3\Delta/2$.   (C. SHANNON)

17.2.9 Let G be a simple graph whose vertices of maximum degree Δ induce a
forest. Show that $\chi' = \Delta$.

17.2.10 UNIQUELY EDGE-COLOURABLE GRAPH
A k-edge-chromatic graph which has exactly one proper k-edge-colouring is said to be *uniquely k-edge-colourable*.

a) Let G be a uniquely k-edge-colourable k-regular graph with k-edge-colouring $\{M_1, M_2, \ldots, M_k\}$. Show that $G[M_i \cup M_j]$ is hamiltonian, $1 \leq i < j \leq k$.
(D. L. GREENWELL AND H. V. KRONK)

b) Let G be a cubic graph with a triangle T. Show that G is uniquely 3-edge-colourable if and only if $H := G / T$ is uniquely 3-edge-colourable.

c) For every even $n \geq 4$, construct a uniquely 3-edge-colourable cubic graph on n vertices. (T. Fowler and R. Thomas have shown that every uniquely 3-edge-colourable cubic planar graph on four or more vertices contains a triangle, and thus can be obtained from K_4 by recursively expanding vertices into triangles.)

d) Show that the generalized Petersen graph $P_{2,9}$ is triangle-free and uniquely 3-edge-colourable.
(S. FIORINI)

17.2.11 Show that every uniquely 3-edge-colourable cubic graph has exactly three Hamilton cycles.

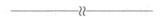

17.2.12 Show that a self-complementary graph belongs to Class 2 if and only if it is regular.
(A.P. WOJDA)

17.2.13

a) Show that if G is simple with $\delta > 1$, then G has a $(\delta - 1)$-edge-colouring (necessarily improper) in which all $\delta - 1$ colours are represented at each vertex.
(R. P. GUPTA)

b) Describe a polynomial-time algorithm for finding such a colouring.

17.2.14 VIZING'S ADJACENCY LEMMA
Let G be a minimal simple graph that is not Δ-edge-colourable, and let u and v be adjacent vertices of G, where $d(u) = k$.

a) Show that v is adjacent to at least $\Delta - k + 1$ vertices of degree Δ different from u.
(V.G. VIZING)

b) Deduce that each vertex of G has at least two neighbours of degree Δ.

17.3 Snarks

The edge chromatic number of a cubic graph G (whether simple or not) always equals either three or four (Exercise 17.1.12). However, as was mentioned in Section 17.2, the problem of deciding between these two values is $\mathcal{NP}$-complete. Thus, unless $\mathcal{P} = \text{co-}\mathcal{NP}$, there is no hope of obtaining any useful characterization of the cubic graphs which are 3-edge-colourable. Nevertheless, edge colourings of cubic graphs have attracted much attention, mainly on account of their relevance to the Four-Colour Problem, discussed in Chapters 11 and 15, and the Cycle Double Cover Conjecture, introduced in Section 3.5.

We have seen that it suffices to prove the Cycle Double Cover Conjecture for essentially 4-edge-connected cubic graphs (Theorem 5.5, Exercises 9.3.9 and 9.4.2). Furthermore, if such a graph is 3-edge-colourable, then it admits a covering by two even subgraphs (Exercise 17.3.4a) and hence has a cycle double cover (Exercise 3.5.4a). Thus it suffices to establish the Cycle Double Cover Conjecture for essentially 4-edge-connected cubic graphs that are not 3-edge-colourable.

The Petersen graph is the smallest such graph (Exercise 17.3.1). For a long time, apart from the Petersen graph, only a few sporadic examples of 4-edge-chromatic essentially 4-edge-connected cubic graphs were known. Because of the elusive nature of such graphs, they were named *snarks* by Descartes (1948), after the Lewis Carroll poem, 'The Hunting of the Snark'. A snark on eighteen vertices, discovered by Blanuša (1946), is shown in Figure 17.4b. That it is indeed a snark can be deduced from the fact the Petersen graph is one (see Exercise 17.1.6).

It can be seen that this Blanuša snark (one of two) has a Petersen graph minor. Tutte (1966b) conjectured that every snark has such a minor. Because the Petersen graph is nonplanar, Tutte's conjecture implies the Four-Colour Theorem, via Tait's Theorem (11.4). Tutte's conjecture was confirmed by N. Robertson, D. Sanders, P. D. Seymour, and R. Thomas (unpublished, see Robertson et al. (1997b)) using the same sorts of techniques as were successful in proving the Four-Colour Theorem.

Isaacs (1975) was the first to succeed in constructing infinite families of snarks (see Exercise 17.3.3). Examples with many interesting properties were found by Kochol (1996), but the general structure of snarks remains a mystery.

Exercises

17.3.1 Verify that the Petersen graph is the smallest snark.

17.3.2 BLANUŠA SNARK
Let G_1 and G_2 be two disjoint snarks, and let $u_ix_iy_iv_i$ be a path of length three in G_i, $i = 1, 2$. Delete the edges u_ix_i and v_iy_i from G_i, $i = 1, 2$, and identify x_1 with x_2 and y_1 with y_2 (and the edge x_1x_2 with the edge y_1y_2). Now add edges joining u_1 and u_2, and v_1 and v_2. Show that the resulting graph is again a snark. (This construction is illustrated in Figure 17.4 when both G_1 and G_2 are Petersen graphs. The resulting snark, shown in Figure 17.4b is known as the *Blanuša snark*.)

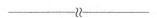

17.3.3 FLOWER SNARK

a) Show that the graph shown in Figure 17.5 is a snark.
b) Find a Petersen graph minor of this snark.

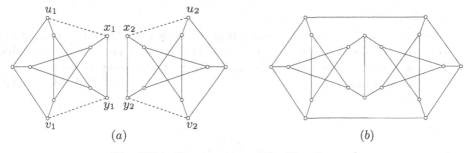

Fig. 17.4. Construction of the Blanuša snark

c) Explain how one may obtain an infinite sequence of snarks by generalizing the above construction. (These snarks are known as the *flower snarks*.)

(R. ISAACS)

d) Show that all flower snarks have Petersen graph minors.

★**17.3.4**

a) Show that a cubic graph admits a covering by two even subgraphs if and only if it is 3-edge-colourable.

b) Deduce that the problem of deciding whether a graph admits a covering by two even subgraphs is $\mathcal{NP}$-complete.

17.3.5 MEREDITH GRAPH

a) Let M be a perfect matching of the Petersen graph. For $k \geq 3$, let G_k denote the graph obtained from the Petersen graph by replacing each edge of M by $k - 2$ parallel edges. Show that:

   i) G_k is k-edge-connected,

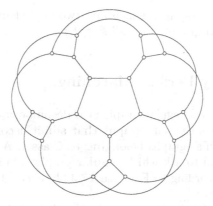

Fig. 17.5. A flower snark on twenty vertices

ii) G_k is not k-edge-colourable.

b) The simple 4-edge-connected 4-regular graph shown in Figure 17.6 is known as
the *Meredith graph*. Deduce from (a) that this graph is not 4-edge-colourable.

c) Explain how to obtain a simple k-edge-connected non-k-edge-colourable k-reg-
ular graph for all $k \geq 4$.   (G.H.J. MEREDITH)

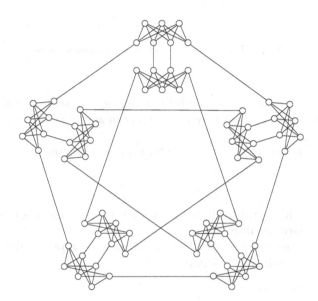

Fig. 17.6. The Meredith graph

17.3.6 Show that the Meredith graph (Figure 17.6) has two disjoint perfect matchings.

(Examples of k-regular graphs in which any two perfect matchings intersect have
been constructed by Rizzi (1999) for all $k \geq 4$.)

17.4 Coverings by Perfect Matchings

A k-edge-colouring of a k-regular graph is a decomposition of the graph into k
perfect matchings. Thus the only graphs that admit decompositions into perfect
matchings are the regular graphs belonging to Class 1. A decomposition being a
1-covering, it is natural to ask which regular graphs admit uniform coverings by
perfect matchings. According to Exercise 17.4.6b, every 2-connected cubic graph
admits such a covering.

Motivated by certain questions concerning the polyhedra defined by the incidence vectors of perfect matchings, Fulkerson (1971) was led to formulate the

following conjecture on uniform coverings by perfect matchings, a conjecture reminiscent of the Cycle Double Cover Conjecture (3.9).

FULKERSON'S CONJECTURE

Conjecture 17.6 *Every 2-connected cubic graph admits a double cover by six perfect matchings.*

Fulkerson's Conjecture certainly holds for cubic graphs of Class 1: it suffices to consider a 3-edge-colouring $\{M_1, M_2, M_3\}$ and take two copies of each perfect matching M_i. The conjecture also holds for the Petersen graph, whose six perfect matchings constitute a double cover (Exercise 17.4.1).

If Fulkerson's Conjecture were true, then deleting one of the perfect matchings from the double cover would result in a covering of the graph by five perfect matchings. This weaker conjecture was proposed by C. Berge (see Seymour (1979a)).

Conjecture 17.7 *Every 2-connected cubic graph admits a covering by five perfect matchings.*

The bound of five in this conjecture cannot be reduced: the Petersen graph admits no covering by fewer than five perfect matchings; with four perfect matchings one may cover fourteen of its fifteen edges, but not all of them (Exercise 17.4.4a).

Exercises

17.4.1 Show that the six perfect matchings of the Petersen graph constitute a double cover of the graph.

17.4.2 Let G be a graph obtained from a 2-connected cubic graph H by duplicating each edge. Show that $\chi' \leq 7$.

17.4.3 Find a double cover of the Blanuša snark (Figure 17.4b) by six perfect matchings.

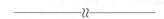

17.4.4 For a graph G and a positive integer k, denote by $m_k(G)$ the largest proportion of the edge set of G that can be covered by k perfect matchings.

a) Let P denote the Petersen graph. Show that $m_k(P) = 1 - \binom{4}{k}/\binom{6}{k}$, $1 \leq k \leq 5$.
b) Let G be a 2-connected cubic graph which has a double cover by six perfect matchings, and let k be an integer, $1 \leq k \leq 5$.

   i) Show that if k of these six perfect matchings are selected uniformly at random, then any given edge of G belongs to at least one of the chosen matchings with probability $1 - \binom{4}{k}/\binom{6}{k}$.

   ii) Deduce that $m_k(G) \geq m_k(P)$.   (V. PATEL)

17.4.5 PERFECT MATCHING POLYTOPE

a) Let G be a graph with at least one perfect matching. The convex hull of the set of incidence vectors of the perfect matchings of G is called the *perfect matching polytope* of G, denoted $PM(G)$. Let $\mathbf{x} := (x(e) : e \in E)$ be an element of $PM(G)$. Show that:

   i) $x(e) \geq 0$ for all $e \in E$,

   ii) $x(\partial(v)) = 1$ for all $v \in V$,

   iii) $x(\partial(S)) \geq 1$ for all odd subsets S of V.

   (Using the result of Exercise 8.6.8, Edmonds (1965b) showed that, conversely, every vector $\mathbf{x} \in \mathbb{R}^E$ which satisfies these constraints belongs to $PM(G)$.)

b) Show that when G is bipartite, a vector $\mathbf{x} \in \mathbb{R}^E$ belongs to $PM(G)$ if and only if it satisfies constraints (i) and (ii). (This assertion is equivalent to the Birkhoff–von Neumann Theorem, see Exercise 16.2.19.)

c) By considering the triangular prism, show that when G is nonbipartite, a vector $\mathbf{x}$ satisfying constraints (i) and (ii) does not necessarily satisfy constraint (iii).

$\star$**17.4.6** Let G be a k-regular graph with $k \geq 1$, such that $d(S) \geq k$ for every odd subset S of V.

a) Using Exercise 16.4.7b, show that each edge of G belongs to a perfect matching.

b) Apply Edmonds' characterization of the perfect matching polytope $PM(G)$, as described in Exercise 17.4.5a, to show that the vector $(\frac{1}{k}, \frac{1}{k}, \ldots, \frac{1}{k})$ belongs to $PM(G)$.

c) Deduce that G admits a uniform covering by perfect matchings.

                                                    (J. EDMONDS)

17.4.7 MATCHING POLYTOPE

The *matching polytope* $MP(G)$ of a graph G is the convex hull of the set of incidence vectors of the matchings of G. Let $\mathbf{x} := (x(e) : e \in E)$ be an element of $MP(G)$. Show that:

   i) $x(e) \geq 0$ for all $e \in E$,

   ii) $x(\partial(v)) \leq 1$ for all $v \in V$,

   iii) $x(E(S)) \leq \lfloor \frac{1}{2}|S| \rfloor$ for all odd subsets S of V.

(Edmonds (1965b) proved that, conversely, every vector $\mathbf{x} \in \mathbb{R}^E$ which satisfies the above constraints belongs to $MP(G)$. This may be deduced from Edmonds' characterization of $PM(G)$, described in Exercise 17.4.5a.)

17.5 List Edge Colourings

The definitions concerning list colourings given in Section 14.5 have obvious analogues for edge colouring: *list edge colouring, k-list-edge-colourable* and *list edge chromatic number*, denoted $\chi'_L(G)$. As already observed, $\chi_L(G) \geq \chi(G)$ for any graph G. Likewise, $\chi'_L(G) \geq \chi'(G)$. Although the former inequality is strict for some graphs, such as $K_{3,3}$, it is conjectured that the latter inequality is always satisfied with equality.

THE LIST EDGE COLOURING CONJECTURE

Conjecture 17.8 *For every loopless graph G, $\chi'_L(G) = \chi'(G)$*

GALVIN'S THEOREM

Conjecture 17.8 was proposed independently by several authors, including V. G. Vizing, R. P. Gupta, and M. O. Albertson and K. L. Collins. It first appeared in print in an article by Bollobás and Harris (1985) (see Häggkvist and Chetwynd (1992) for a brief history). Galvin (1995) showed that the conjecture is true for bipartite graphs. His proof relies on the relationship between kernels and list colourings described in Theorem 14.20. Because colouring the edges of a graph amounts to colouring the vertices of its line graph, the key step in the proof is to show that line graphs of bipartite graphs can be oriented in such a way that (i) the maximum outdegree is not too high, and (ii) every induced subgraph has a kernel.

We present a proof of Galvin's theorem for simple bipartite graphs. Let $G := G[X, Y]$ be such a graph. In the line graph $L(G)$, there is a clique K_v for each vertex v of G, the vertices of K_v corresponding to the edges of G incident to v. Each edge xy of G gives rise to a vertex of $L(G)$ which lies in exactly two of these cliques, namely K_x and K_y. We refer to K_v as an *X-clique* if $v \in X$, and a *Y-clique* if $v \in Y$.

There is a convenient way of visualising this line graph $L(G)$. Because each edge of G is a pair xy, the vertex set of $L(G)$ is a subset of the cartesian product $X \times Y$. Therefore, in a drawing of $L(G)$, we can place its vertices at appropriate lattice points of the $m \times n$ grid, where $m = |X|$ and $n = |Y|$, the rows of the grid being indexed by X and the columns by Y. Any two vertices which lie in the same row or column of the grid are adjacent in $L(G)$, and so the sets of vertices in the same row or column are cliques of $L(G)$, namely its X-cliques and Y-cliques, respectively (see Figure 17.7).

Theorem 17.9 *Let $G[X, Y]$ be a simple bipartite graph, and let D be an orientation of its line graph $L(G)$ in which each X-clique and each Y-clique induces a transitive tournament. Then D has a kernel.*

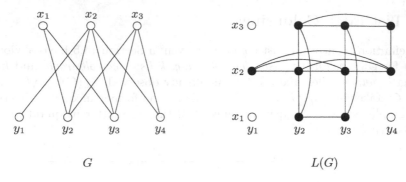

$$G \qquad\qquad\qquad L(G)$$

Fig. 17.7. Representing the line graph $L(G)$ of a bipartite graph G on a grid

Proof By induction on $e(G)$, the case $e(G) = 1$ being trivial. For $v \in V(G)$, denote by T_v the transitive tournament in D corresponding to v, and for $x \in X$, denote by t_x the sink of T_x. Set $K := \{t_x : x \in X\}$. Every vertex of $D - K$ lies in some T_x, and so dominates some vertex of K. Thus if the vertices of K lie in distinct Y-cliques, then K is a kernel of D.

Suppose, then, that the Y-clique T_y contains two vertices of K. One of these, say t_x, is not the source s_y of T_y, so $s_y \to t_x$. Set $D' := D - s_y$. Then D' is an orientation of the line graph $L(G \backslash e)$, where e is the edge of G corresponding to the vertex s_y of $L(G)$. Moreover, each clique of D' induces a transitive tournament. By induction, D' has a kernel K'. We show that K' is also a kernel of D. For this, it suffices to verify that s_y dominates some vertex of K'.

If $t_x \in K'$, then $s_y \to t_x$. On the other hand, if $t_x \notin K'$, then $t_x \to v$, for some $v \in K'$. Because t_x is the sink of its X-clique, v must lie in the Y-clique $T_y \backslash \{s_y\}$. But then s_y, being the source of T_y, dominates v. Thus K' is indeed a kernel of D. $\square$

Theorem 17.10 *Every simple bipartite graph G is Δ-list-edge-colourable.*

Proof Let $G := G[X, Y]$ be a simple bipartite graph with maximum degree k, and let $c : E(G) \to \{1, 2, \ldots, k\}$ be a k-edge-colouring of G. The colouring c induces a k-colouring of $L(G)$. We orient each edge of $L(G)$ joining two vertices of an X-clique from lower to higher colour, and each edge of $L(G)$ joining two vertices of a Y-clique from higher to lower colour, as in Figure 17.8 (where the colour $c(x_i y_j)$ of the edge $x_i y_j$ is indicated inside the corresponding vertex of $L(G)$). This orientation D clearly satisfies the hypotheses of Theorem 17.9; indeed, every induced subgraph of D satisfies these hypotheses. Moreover, $\Delta^+(D) = k - 1$. By Theorem 14.20, $L(G)$ is k-list-colourable, so G is k-list-edge-colourable. $\square$

As mentioned earlier, Galvin (1995) showed that if $G[X, Y]$ is any (not necessarily simple) bipartite graph, then it is Δ-list-edge-colourable. We leave the proof of this more general theorem, which is essentially the same as that of the special case presented above, as Exercise 17.5.3.

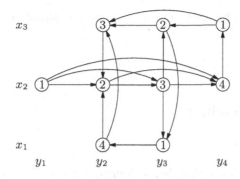

Fig. 17.8. Orienting the line graph of a bipartite graph

Exercises

17.5.1

a) An $n \times n$ array $\mathbf{A} = (a_{ij})$ whose entries are taken from some set S of n symbols is called a *Latin square* of *order* n if each symbol appears precisely once in each row and precisely once in each column of $\mathbf{A}$. Show that there is a one-to-one correspondence between n-edge-colourings of $K_{n,n}$ in colours $1, 2, \ldots, n$ and Latin squares of order n in symbols $1, 2, \ldots, n$.

b) Deduce from Theorem 17.10 the following assertion, a special case of a conjecture due to J. Dinitz (see, for example Galvin (1995)).

For $1 \le i \le j \le n$, let S_{ij} be a set of n elements. Then there exists a Latin square $\mathbf{A} = (a_{ij})$ of order n using a set S of n symbols such that $a_{ij} \in S_{ij} \cap S$, $1 \le i \le j \le n$.

17.5.2 Consider the bipartite graph G shown in Figure 17.9, with the given 5-edge-colouring. Find an orientation of its line graph $L(G)$ each of whose induced subdigraphs has a kernel.

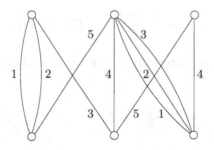

Fig. 17.9. Find an appropriate orientation of $L(G)$ (Exercise 17.5.2)

17.5.3 Prove that a bipartite graph G (whether simple or not) is Δ-list-edge-colourable.   (F. GALVIN)

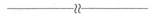

17.6 Related Reading

TOTAL COLOURINGS

A *total colouring* of a graph G is a colouring $c : V \cup E \to S$, where S is a set of *colours*. The colouring c is *proper* if its restriction to V is a proper vertex colouring of G, its restriction to E is a proper edge colouring of G, and no edge receives the same colour as either of its ends. The *total chromatic number* of G, denoted by $\chi''(G)$, is the minimum number of colours in a proper total colouring of G. Vizing (1964) and, independently, Behzad (1965) conjectured that the total chromatic number of a simple graph G never exceeds $\Delta + 2$ (and thus is equal to either $\Delta + 1$ or $\Delta + 2$). Using probabilistic methods, Molloy and Reed (1998) showed that the total chromatic number of a simple graph G is at most $\Delta + 10^{26}$, provided that Δ is sufficiently large. Apart from this result, not much progress has been made on this conjecture, known as the *Total Colouring Conjecture*. A number of interesting problems involving list colouring variants of the total chromatic number have been studied; see, for example, Woodall (2001).

FRACTIONAL EDGE COLOURINGS

Fractional edge colourings can be defined in an analogous manner to fractional vertex colourings, with matchings playing the role of stable sets. The analogue of the fractional chromatic number is called the *fractional edge chromatic number*, denoted χ'^*.

The Goldberg bound (17.2) on the edge chromatic number (without the ceiling function) is satisfied with equality by the fractional edge chromatic number:

$$(\chi')^* = \max \left\{ \frac{2e(H)}{v(H) - 1} : H \subseteq G, \ v(H) \text{ odd} \right\}$$

This follows from the characterization by Edmonds (1965b) of the matching polytope (see Exercise 17.4.7).

Hamilton Cycles

Contents

18.1 Hamiltonian and Nonhamiltonian Graphs

Recall that a path or cycle which contains every vertex of a graph is called a *Hamilton path* or *Hamilton cycle* of the graph. Such paths and cycles are named after Sir William Rowan Hamilton, who described, in a letter to his friend Graves in 1856, a mathematical game on the dodecahedron (Figure 18.1a) in which one person sticks pins in any five consecutive vertices and the other is required to complete the path so formed to a spanning cycle (see Biggs et al. (1986) or Hamilton (1931)). Hamilton was prompted to consider such cycles in his early investigations into group theory, the three edges incident to a vertex corresponding to three generators of a group.

A graph is *traceable* if it contains a Hamilton path, and *hamiltonian* if it contains a Hamilton cycle. The dodecahedron is hamiltonian; a Hamilton cycle is indicated in Figure 18.1a. On the other hand, the *Herschel graph* of Figure 18.1b is nonhamiltonian, because it is bipartite and has an odd number of vertices. This graph is, however, traceable.

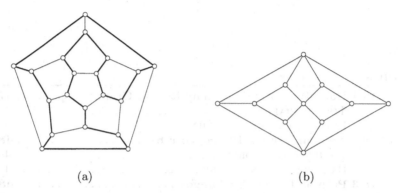

(a)   (b)

Fig. 18.1. Hamiltonian and nonhamiltonian graphs: (a) the dodecahedron, (b) the Herschel graph

Tough Graphs

As we saw in Section 8.3, the problem of deciding whether a given graph is hamiltonian is $\mathcal{NP}$-complete. It is therefore natural to look for reasonable necessary or sufficient conditions for the existence of Hamilton cycles. The following simple necessary condition turns out to be surprisingly useful.

Theorem 18.1 *Let S be a set of vertices of a hamiltonian graph G. Then*

$$c(G - S) \leq |S| \qquad (18.1)$$

Moreover, if equality holds in (18.1), then each of the $|S|$ components of $G - S$ is traceable, and every Hamilton cycle of G includes a Hamilton path in each of these components.

Proof Let C be a Hamilton cycle of G. Then $C - S$ clearly has at most $|S|$ components. But this implies that $G - S$ also has at most $|S|$ components, because C is a spanning subgraph of G.

If $G - S$ has exactly $|S|$ components, $C - S$ also has exactly $|S|$ components, and the components of $C - S$ are spanning subgraphs of the components of $G - S$. In other words, C includes a Hamilton path in each component of $G - S$.   □

A graph G is called *tough* if (18.1) holds for every nonempty proper subset S of V. By Theorem 18.1, a graph which is not tough cannot be hamiltonian. As an illustration, consider the graph G of Figure 18.2a. This graph has nine vertices. On deleting the set S of three vertices indicated, four components remain. This shows that the graph is not tough, and we infer from Theorem 18.1 that it is nonhamiltonian.

Although condition (18.1) has a simple form, it is not always easy to apply. In fact, as was shown by Bauer et al. (1990), recognizing tough graphs is $\mathcal{NP}$-hard.

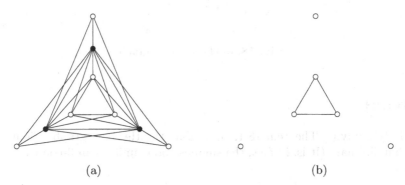

(a)   (b)

Fig. 18.2. (a) A nontough graph G, (b) the components of $G - S$

HYPOHAMILTONIAN GRAPHS

As the above example shows, Theorem 18.1 can sometimes be applied to deduce that a graph is nonhamiltonian. Such an approach does not always work. The Petersen graph is nonhamiltonian (Exercises 2.2.6, 17.1.8), but one cannot deduce this fact from Theorem 18.1. Indeed, the Petersen graph has a very special property: not only is it nonhamiltonian, but the deletion of any one vertex results in a hamiltonian graph (Exercise 18.1.16a). Such graphs are called *hypohamiltonian*. Deleting a single vertex from a hypohamiltonian graph results in a subgraph with just one component, and deleting a set S of at least two vertices produces no more than $|S| - 1$ components, because each vertex-deleted subgraph is hamiltonian, hence tough. The Petersen graph is an example of a vertex-transitive hypohamiltonian graph. Such graphs appear to be extremely rare. Another example is the *Coxeter graph* (see Exercises 18.1.14 and 18.1.16c); the attractive drawing of this graph shown in Figure 18.3 is due to Randić (1981). Its geometric origins and many of its interesting properties are described in Coxeter (1983).

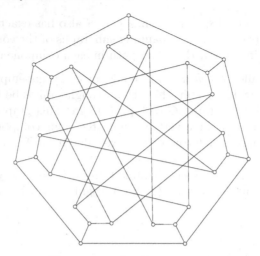

Fig. 18.3. The Coxeter graph

Exercises

18.1.1 By applying Theorem 18.1, show that the Herschel graph (Figure 18.1b) is nonhamiltonian. (It is, in fact, the smallest nonhamiltonian 3-connected planar graph.)

18.1.2 Let G be a cubic graph, and let H be the cubic graph obtained from G by expanding a vertex to a triangle. Exhibit a bijection between the Hamilton cycles of G and those of H.

18.1.3 Show that the Meredith graph (Figure 17.6) is nonhamiltonian.

18.1.4

a) Let G be a graph and let X be a nonempty proper subset of V. If G / X is a nonhamiltonian cubic graph, show that any path of G either misses a vertex of $V \setminus X$ or has an end in $V \setminus X$.

b) Construct a nontraceable 3-connected cubic graph.

18.1.5 Find a 3-connected planar bipartite graph on fourteen vertices which is not traceable.

18.1.6 A graph is *traceable from a vertex* x if it has a Hamilton x-path, *Hamilton-connected* if any two vertices are connected by a Hamilton path, and 1-*hamiltonian* if it and all its vertex-deleted subgraphs are hamiltonian.

Let G be a graph and let H be the graph obtained from G by adding a new vertex and joining it to every vertex of G. Show that:

a) H is hamiltonian if and only if G is traceable,

b) H is traceable from every vertex if and only if G is traceable,

c) H is Hamilton-connected if and only if G is traceable from every vertex,

d) H is 1-hamiltonian if and only if G is hamiltonian.

18.1.7

a) Show that the graph of Figure 18.4 is 1-hamiltonian but not Hamilton-connected.   (T. ZAMFIRESCU)

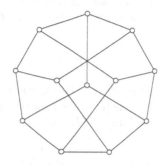

Fig. 18.4. A 1-hamiltonian graph which is not Hamilton-connected (Exercise 18.1.7)

b) Find a Hamilton-connected graph which is not 1-hamiltonian.

18.1.8 Find a 2-diregular hypohamiltonian digraph on six vertices.

(J.-L. FOUQUET AND J.-L. JOLIVET)

18.1.9 k-WALK

A k-*walk* in a graph is a spanning closed walk which visits each vertex at most k times. (Thus a 1-walk is a Hamilton cycle.) If G has a k-walk, show that G is $(1/k)$-tough.

(Jackson and Wormald (1990) showed that, for $k \geq 3$, every $(1/(k-2))$-tough graph has a k-walk.)

18.1.10 PATH PARTITION NUMBER

A *path partition* of a graph is a partition of its vertex set into paths. The *path partition number* of a graph G, denoted $\pi(G)$, is the minimum number of paths into which its vertex set V can be partitioned. (Thus traceable graphs are those whose path partition number is one.) Let G be a graph containing an edge e which lies in no Hamilton cycle, and let H be the graph formed by taking m disjoint copies of G and adding all possible edges between the ends of the m copies of e, so as to form a clique of $2m$ vertices. Show that the path partition number of H is at least $m/2$.

18.1.11 A graph G is *path-tough* if $\pi(G - S) \leq |S|$ for every nonempty proper subset S of V.

a) Show that:
  i) every hamiltonian graph is path-tough,
  ii) every path-tough graph is tough.
b) Give an example of a nonhamiltonian path-tough graph.

18.1.12 Let G be a vertex-transitive graph of prime order. Show that G is hamiltonian.

18.1.13 Let G be the graph whose vertices are the thirty-five 3-element subsets of $\{1,2,3,4,5,6,7\}$, two vertices being joined if the subsets are disjoint. Let X be a set of vertices of G forming a Fano plane. Show that $G - X$ is isomorphic to the Coxeter graph.

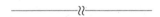

18.1.14   Show that the Petersen graph, the Coxeter graph, and the two graphs derived from them by expanding each vertex to a triangle, are all vertex-transitive nonhamiltonian graphs. (These four graphs are the only examples of such graphs known.)

18.1.15 A graph is *maximally nonhamiltonian* if it is nonhamiltonian but every pair of nonadjacent vertices are connected by a Hamilton path.

a) Show that:
  i) the Petersen graph and the Coxeter graph are maximally nonhamiltonian,
  ii) the Herschel graph is not maximally nonhamiltonian.
b) Find a maximally nonhamiltonian spanning supergraph of the Herschel graph.

18.1.16 Show that:

a) the Petersen graph is hypohamiltonian,
b) there is no smaller hypohamiltonian graph,
                    (J.C. Herz, J.J. Duby, and F. Vigué)
c) the Coxeter graph is hypohamiltonian.

18.1.17 Hypotraceable Graph
A graph is *hypotraceable* if it is not traceable but each of its vertex-deleted subgraphs is traceable. Show that the graph in Figure 18.5 is hypotraceable.
                    (C. Thomassen)

18.1.18 Pancyclic Graph
A simple graph on n vertices is *pancyclic* if it contains at least one cycle of each length l, $3 \leq l \leq n$.

a) Let G be a simple graph and v a vertex of G. Suppose that $G - v$ is hamiltonian and that $d(v) \geq n/2$. Show that G is pancyclic.
b) Prove, by induction on n, that if G is a simple hamiltonian graph with more than $n^2/4$ edges, then G is pancyclic.   (J.A. Bondy; C. Thomassen)

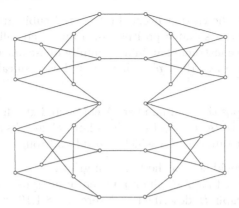

Fig. 18.5. A hypotraceable graph

c) For all $n \geq 4$, give an example of a simple graph G with $n^2/4$ edges which is hamiltonian but not pancyclic.

18.1.19 A simple graph on n vertices is *uniquely pancyclic* if it contains precisely one cycle of each length l, $3 \leq l \leq n$.

a) For $n = 3, 5, 8, 14$, find a uniquely pancyclic graph on n vertices.
b) Let G be a uniquely pancyclic graph. Show that:
   i) $m \geq n + \log_2(n-1) - 1$,
   ii) if $n \leq \binom{r+2}{2} + 1$, then $m \leq n + r - 1$.

18.1.20 Let D be a 2-strong tournament, and let (x, y) be an arc of D such that $d^-(x) + d^+(y) \geq n - 1$. Show that (x, y) lies in a directed cycle of each length l, $3 \leq l \leq n$.   (A. YEO)

18.1.21 Let H be a vertex-deleted subgraph of the Petersen graph P. Define a sequence G_i, $i \geq 0$, of 3-connected cubic graphs, as follows.

▷ $G_0 = P$.
▷ G_{i+1} is obtained from G_i and $v(G_i)$ copies $(H_v : v \in V(G_i))$ of H, by splitting each vertex v of G_i into three vertices of degree one and identifying these vertices with the three vertices of degree two in H_v.

Set $G := G_k$.

a) Show that:
   i) $n = 10 \cdot 9^k$,
   ii) the circumference of G is $9 \cdot 8^k = cn^\gamma$, where $\gamma = \log 8 / \log 9$, and c is a suitable positive constant.
b) By appealing to Exercise 9.1.14, deduce that G has no path of length more than $c'n^\gamma$, where c' is a suitable positive constant.

(Jackson (1986) has shown that every 3-connected cubic graph G has a cycle of length at least n^γ for a suitable positive constant γ. For all $d \geq 3$, Jackson and Parsons (1982) have constructed an infinite family of d-connected d-regular graphs G with circumference less than n^γ, where $\gamma < 1$ is a suitable positive constant depending on d.)

18.1.22 Let t be a positive real number. A connected graph G is *t-tough* if $c(G - S) \leq |S|/t$ for every vertex cut S of V. (Thus 1-tough graphs are the same as tough graphs.) The largest value of t for which a graph is t-tough is called its *toughness*.

a) Determine the toughness of the Petersen graph.
b) Show that every 1-tough graph on an even number of vertices has a 1-factor.
c) Consider the graph H described in Exercise 18.1.10, where G is the graph shown in Figure 18.6 and $m = 5$. Show that $H \vee K_2$ is 2-tough but not path-tough (hence not hamiltonian).

(D. BAUER, H.J. BROERSMA, AND H.J. VELDMAN)

(Chvátal (1973) has conjectured the existence of a constant t such that every t-tough graph is hamiltonian.)

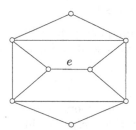

Fig. 18.6. An element in the construction of a nonhamiltonian 2-tough graph (Exercise 18.1.22)

18.2 Nonhamiltonian Planar Graphs

GRINBERG'S THEOREM

Recall (from Section 11.1) that Tait (1880) showed the Four-Colour Conjecture to be equivalent to the statement that every 3-connected cubic planar graph is 3-edge-colourable. Tait thought that he had thereby proved Four-Colour Conjecture, because he believed that every such graph was hamiltonian, and hence 3-edge-colourable. However, Tutte (1946) showed this to be false by constructing a nonhamiltonian 3-connected cubic planar graph (depicted in Figure 18.7) using ingenious *ad hoc* arguments (Exercise 18.2.1). For many years, the *Tutte graph* was

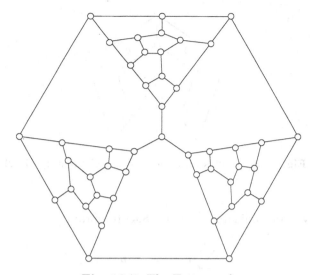

Fig. 18.7. The Tutte graph

the only known example of a nonhamiltonian 3-connected cubic planar graph. But then Grinberg (1968) discovered a simple necessary condition for a plane graph to be hamiltonian. His discovery led to the construction of many nonhamiltonian planar graphs.

Theorem 18.2 GRINBERG'S THEOREM
Let G be a plane graph with a Hamilton cycle C. Then

$$\sum_{i=1}^{n}(i-2)(\phi_i' - \phi_i'') = 0 \qquad (18.2)$$

where ϕ_i' and ϕ_i'' are the numbers of faces of degree i contained in Int C and Ext C, respectively.

Proof Denote by E' the subset of $E(G)\backslash E(C)$ contained in Int C, and set $m' := |E'|$. Then Int C contains exactly $m' + 1$ faces (see Figure 18.8, where $m' = 3$ and the four faces all have degree four).
Therefore

$$\sum_{i=1}^{n}\phi_i' = m' + 1 \qquad (18.3)$$

Now each edge in E' lies on the boundary of two faces in Int C, and each edge of C lies on the boundary of exactly one face in Int C. Therefore

$$\sum_{i=1}^{n}i\phi_i' = 2m' + n \qquad (18.4)$$

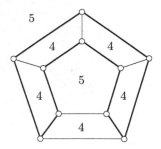

Fig. 18.8. An illustration for Grinberg's Identity (18.2)

Using (18.3), we can eliminate m' from (18.4) to obtain

$$\sum_{i=1}^{n}(i-2)\phi_i' = n-2 \tag{18.5}$$

Likewise,

$$\sum_{i=1}^{n}(i-2)\phi_i'' = n-2 \tag{18.6}$$

Equations (18.5) and (18.6) now yield (18.2).   □

Equation (18.2) is known as *Grinberg's Identity*. With the aid of this identity, it is a simple matter to show, for example, that the *Grinberg graph*, depicted in Figure 18.9, is nonhamiltonian. Suppose that this graph is hamiltonian. Noting

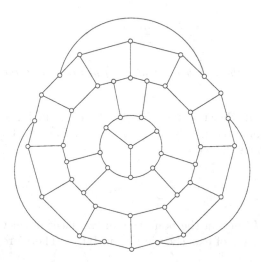

Fig. 18.9. The Grinberg graph

that it only has faces of degrees five, eight, and nine, Grinberg's Identity (18.2) yields

$$3(\phi_5' - \phi_5'') + 6(\phi_8' - \phi_8'') + 7(\phi_9' - \phi_9'') = 0$$

We deduce that

$$7(\phi_9' - \phi_9'') \equiv 0 \pmod{3}$$

But this is clearly impossible, because the value of the left-hand side is 7 or -7, depending on whether the unique face of degree nine lies in Int C or in Ext C. Therefore the graph cannot be hamiltonian.

The Grinberg graph is an example of a nonhamiltonian 3-connected essentially 4-edge-connected cubic planar graph. Tutte (1956) showed, on the other hand, that every 4-connected planar graph is hamiltonian. (Thomassen (1983b) found a shorter proof of this theorem, but his proof is still too complicated to be presented here. The basic idea is discussed in Section 18.6.)

In applying Grinberg's Identity, the parities of the face degrees play a crucial role. This approach fails to provide examples of bipartite nonhamiltonian 3-connected cubic planar graphs. Indeed, Barnette (1969), and independently Kelmans and Lomonosov (1975), conjectured that there are no such graphs.

BARNETTE'S CONJECTURE

Conjecture 18.3 *Every 3-connected cubic planar bipartite graph is hamiltonian.*

Planarity is essential here. An example of a nonhamiltonian 3-connected cubic bipartite graph was constructed by J. D. Horton (see Bondy and Murty (1976), p.240). The smallest example known of such a graph was found independently by Kelmans (1986, 1994) and Georges (1989). It is shown in Figure 18.10.

Historical note. Interestingly, and ironically, Grinberg's Identity (18.2) was known already to Kirkman (1881), some ninety years earlier. But Kirkman, convinced that every 3-connected cubic planar graph was hamiltonian, used it as a tool in searching for Hamilton cycles in particular examples of such graphs.

In the next section, we derive various sufficient conditions for hamiltonicity.

Exercises

18.2.1

a) Show that no Hamilton cycle in the pentagonal prism (the graph G_1 in Figure 18.11) can contain both of the edges e and e'.

b) Using (a), show that no Hamilton cycle in the graph G_2 can contain both of the edges e and e'.

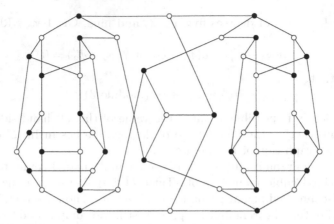

Fig. 18.10. The Kelmans–Georges graph: a nonhamiltonian 3-connected cubic bipartite graph

c) Using (b), show that every Hamilton cycle in the graph G_3 must contain the edge e.
d) Deduce that the Tutte graph (Figure 18.7) is nonhamiltonian.

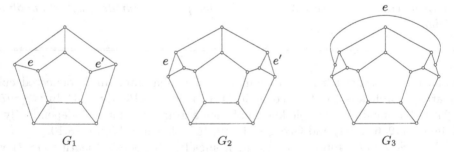

Fig. 18.11. Three steps in the construction of the Tutte graph

18.2.2 Give an example of a simple nonhamiltonian cubic planar graph of connectivity two.

18.2.3 Let $G = G_k$, where G_0 is the plane graph K_4, and G_i the plane triangulation obtained from G_{i-1}, $i \geq 1$, by inserting a vertex in each face and joining it to the three vertices on the boundary of the face. Let l be the circumference of G.

a) Show that $n = 2 \cdot (3^k + 1)$ and $l \leq 2^{k+2}$.
b) Deduce that $l < cn^{\log 2 / \log 3}$ for a suitable constant c.

(J.W. Moon and L. Moser)

———————— ⅔ ————————

18.2.4 Let G be a cubic plane graph which admits a Hamilton double cover (a double cover by three Hamilton cycles).

a) Show that each of these Hamilton cycles induces the same 4-face-colouring of G (see Exercise 11.1.5).
b) For $i \geq 1$ and $1 \leq j \leq 4$, let ϕ_{ij} be the number of faces of degree i assigned colour j. Show that $\sum_{i=1}^{n}(i-2)\phi_{ij} = (n-2)/2$, $1 \leq j \leq 4$ (and thus is independent of the colour j).
c) Deduce that no cubic planar bipartite graph on $0 \pmod 4$ vertices admits a Hamilton double cover.   (H. FETTER)
d) Find an example of such a graph.

18.2.5

a) Using the fact that the Petersen graph, drawn in Figure 18.12a, has no Hamilton cycle, show that the graph in Figure 18.12b has no Hamilton cycle through both of the edges e and f.
b) The graph shown in Figure 18.12c is a redrawing of the one in Figure 18.12b. The *Kelmans–Georges graph*, depicted in Figure 18.10, is obtained from the Petersen graph of Figure 18.12a by substituting two copies of the graph of Figure 18.12c for the two edges e and f. Deduce from (a) that this graph has no Hamilton cycle.   (A.K. KELMANS)

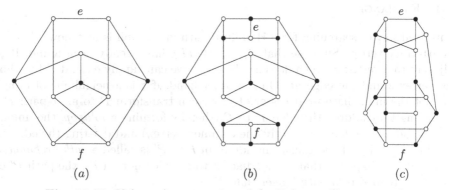

(a) (b) (c)

Fig. 18.12. Kelmans' construction of the Kelmans–Georges graph

18.3 Path and Cycle Exchanges

In this section, we describe how paths and cycles can be transformed into other paths and cycles by means of simple operations. These operations turn out to be very helpful in searching for long paths and cycles (in particular, Hamilton paths and cycles).

First, a little notation. If v is a vertex of a path P or cycle C with a specified sense of traversal, we denote by v^- and v^+ the predecessor and successor of v, respectively, on P or C (provided that these vertices exist) (see Figure 18.13).

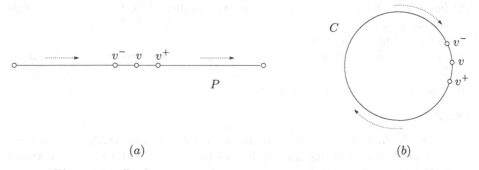

(a)

(b)

Fig. 18.13. Predecessors and successors on a path P and a cycle C

This notation is extended to subsets S of $V(P)$ or $V(C)$ as one would expect:

$$S^- := \{v^- : v \in S\} \quad \text{and} \quad S^+ := \{v^+ : v \in S\}$$

PATH EXCHANGES

A natural way of searching for a Hamilton path in a graph is as follows. Let x be an arbitrary vertex. Suppose that an x-path xPy has already been found. If y is adjacent to a vertex z which does not lie on P, we can simply extend P by adding the vertex z and the edge yz. On the other hand, if z is a neighbour of y on P, but not the immediate predecessor of y, we can transform P to an x-path P' of equal length by adding the edge yz to P (thereby forming a *lollipop*, the union of a path and a cycle having exactly one common vertex) and deleting the edge zz^+ (see Figure 18.14). This transformation from P to P' is called a *path exchange*. Of course, if it so happens that z^+ is adjacent to a vertex not on P', the path P' can then be extended to a path longer than P.

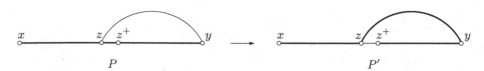

Fig. 18.14. A path exchange

CYCLE EXCHANGES

There is also a simple way of transforming one cycle C into another cycle C' of the same length. If there are nonconsecutive vertices x and y on C such that both xy and x^+y^+ are edges of the graph, the cycle C' obtained by adding these two edges to C, and deleting the edges xx^+ and yy^+ from C, is said to be derived from C by means of a *cycle exchange* (see Figure 18.15).

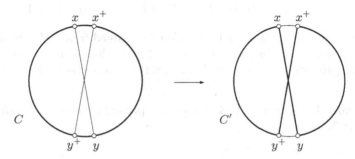

Fig. 18.15. A cycle exchange

Let us remark in passing that cycle exchanges lead to an obvious heuristic for the Travelling Salesman Problem (2.6). If C is a Hamilton cycle in a weighted graph (G, w), and

$$w(xy) + w(x^+y^+) < w(xx^+) + w(yy^+)$$

then the cycle C' will be an improvement on C.

Path and cycle exchanges are the key to establishing the existence of Hamilton paths and cycles in many classes of graphs.

DIRAC'S THEOREM

Every complete graph on at least three vertices is evidently hamiltonian; indeed, the vertices of a Hamilton cycle can be selected one by one, in an arbitrary order. But suppose that our graph has considerably fewer edges. In particular, we may ask how large the minimum degree must be in order to guarantee the existence of a Hamilton cycle. The following theorem of Dirac (1952b) answers this question.

Theorem 18.4   DIRAC'S THEOREM
Let G be a simple graph of minimum degree δ, where $\delta \geq n/2$ and $n \geq 3$. Then G is hamiltonian.

Proof   Form a 2-edge-coloured complete graph K with vertex set V by colouring the edges of G blue and the edges of its complement $\overline{G}$ red. Let C be a Hamilton cycle of K with as many blue edges as possible. We show that every edge of C is blue, in other words, that C is a Hamilton cycle of G.

Suppose not, and let xx^+ be a red edge of C, where x^+ is the successor of x on C. If $S := N_G(x)$ is the set of vertices joined to x by blue edges and $T := N_G(x^+)$ is the set of vertices joined to x^+ by blue edges, then

$$|S^+| + |T| = |S| + |T| = d_G(x) + d_G(x^+) \geq 2\delta \geq n$$

Because $x^+ \notin S^+$ and $x^+ \notin T$, we have $S^+ \cup T \subseteq V \setminus \{x^+\}$, so

$$|S^+ \cap T| = |S^+| + |T| - |S^+ \cup T| \geq n - (n-1) = 1$$

Let $y^+ \in S^+ \cap T$. Then the Hamilton cycle C' obtained from C by exchanging the edges xx^+ and yy^+ for the blue edges xy and x^+y^+ has more blue edges than C, contradicting the choice of C (see Figure 18.15). Thus every edge of C is indeed blue. $\square$

We remark that Theorem 18.4 can also be proved by means of path exchanges (Exercise 18.3.1).

The Closure of a Graph

Observe that the proof of Theorem 18.4 does not make full use of the hypothesis that $\delta \geq n/2$, but only of the weaker condition that the sum of the degrees of the two nonadjacent vertices x and x^+ is at least n. The same method of proof can therefore be used to establish the following lemma.

Lemma 18.5 *Let G be a simple graph and let u and v be nonadjacent vertices in G whose degree sum is at least n. Then G is hamiltonian if and only if $G + uv$ is hamiltonian.*

Proof If G is hamiltonian, so too is $G + uv$. Conversely, suppose that $G + uv$ has a Hamilton cycle C. Then, as in the proof of Theorem 18.4 (with $x := u$ and $x^+ := v$), there is a cycle exchange transforming C to a Hamilton cycle C' of G.
                                                                                    $\square$

Lemma 18.5 motivates the following definition. The *closure* of a graph G is the graph obtained from G by recursively joining pairs of nonadjacent vertices whose degree sum is at least n until no such pair remains. The order in which edges are added to G in forming the closure has no effect on the final result (Exercise 18.3.2).

Lemma 18.6 *The closure of G is well-defined.* $\square$

Figure 18.16 illustrates the formation of the closure G' of a graph G on six vertices. In this example, it so happens that the closure is complete. The pertinence of the closure operation to the study of Hamilton cycles resides in the following observation, due to Bondy and Chvátal (1976).

Theorem 18.7 *A simple graph is hamiltonian if and only if its closure is hamiltonian.*

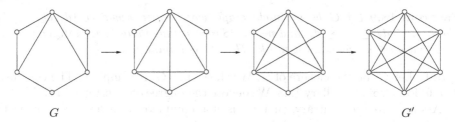

Fig. 18.16. The closure of a graph

Proof Apply Lemma 18.5 each time an edge is added in the formation of the closure. □

Theorem 18.7 has a number of interesting consequences. First, because all complete graphs on at least three vertices are evidently hamiltonian, we obtain the following result.

Corollary 18.8 *Let G be a simple graph on at least three vertices whose closure is complete. Then G is hamiltonian.* □

Consider, for example, the graph of Figure 18.17. One readily checks that its closure is complete. By Corollary 18.8, this graph is therefore hamiltonian. It is perhaps interesting to note that the graph of Figure 18.17 can be obtained from the graph of Figure 18.2 by altering just one end of one edge, and yet we have results (Corollary 18.8 and Theorem 18.1) which tell us that this graph is hamiltonian whereas the other is not.

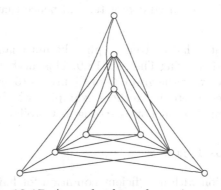

Fig. 18.17. A graph whose closure is complete

Corollary 18.8 can be used to derive various sufficient conditions for a graph to be hamiltonian in terms of its vertex degrees. For example, because the closure is clearly complete when $\delta \geq n/2$, Dirac's Theorem (18.4) is an immediate corollary. Chvátal (1972) extended Dirac's Theorem to a wider class of graphs.

Theorem 18.9 *Let G be a simple graph with degree sequence $(d_1, d_2, \ldots, d_n)$, where $d_1 \leq d_2 \leq \cdots \leq d_n$ and $n \geq 3$. Suppose that there is no integer $k < n/2$ such that $d_k \leq k$ and $d_{n-k} < n - k$. Then G is hamiltonian.*

Proof Let G' be the closure of G. We show that G' is complete. The conclusion then follows from Corollary 18.8. We denote the degree of a vertex v in G' by $d'(v)$.

Assume, to the contrary, that G' is not complete, and let u and v be two nonadjacent vertices in G' with

$$d'(u) \leq d'(v) \tag{18.7}$$

and $d'(u) + d'(v)$ as large as possible. Because no two nonadjacent vertices in G' can have degree sum n or more, we have

$$d'(u) + d'(v) < n \tag{18.8}$$

Now denote by S the set of vertices in $V \setminus \{v\}$ which are nonadjacent to v in G', and by T the set of vertices in $V \setminus \{u\}$ which are nonadjacent to u in G'. Clearly

$$|S| = n - 1 - d'(v), \quad \text{and} \quad |T| = n - 1 - d'(u) \tag{18.9}$$

Furthermore, by the choice of u and v, each vertex of S has degree at most $d'(u)$ and each vertex of $T \cup \{u\}$ has degree at most $d'(v)$. Setting $k := d'(u)$ and using (18.8) and (18.9), we find that G' has at least k vertices of degree not exceeding k and at least $n-k$ vertices of degree strictly less than $n-k$. Because G is a spanning subgraph of G', the same is true of G; that is, $d_k \leq k$ and $d_{n-k} < n - k$. But this is contrary to the hypothesis, because $k < n/2$ by (18.7) and (18.8). We conclude that the closure G' of G is indeed complete, and hence that G is hamiltonian, by Corollary 18.8. $\square$

One can often deduce that a given graph is hamiltonian simply by computing its degree sequence and applying Theorem 18.9. This method works with the graph of Figure 18.17, but not with the graph G of Figure 18.16, even though the closure of the latter graph is complete. From these examples, we see that Theorem 18.9 is stronger than Theorem 18.4 but not as strong as Corollary 18.8.

THE CHVÁTAL–ERDŐS THEOREM

We conclude this section with a sufficient condition for hamiltonicity involving a remarkably simple relationship between the stability number and the connectivity, due to Chvátal and Erdős (1972). Its proof makes use of bridges, introduced in Section 10.4.

Theorem 18.10   THE CHVÁTAL–ERDŐS THEOREM
Let G be a graph on at least three vertices with stability number α and connectivity κ, where $\alpha \leq \kappa$. Then G is hamiltonian.

Proof Let C be a longest cycle in G. Suppose that C is not a Hamilton cycle. Let B be a proper bridge of C in G, and denote by S its set of vertices of attachment to C. For any two vertices x and y of S, there is a path xPy in B; this path is internally disjoint from C and of length at least two (see Figure 18.18). Because C is a longest cycle, it follows that x and y are not consecutive vertices of C. For the same reason, x^+ and y^+ are nonadjacent; otherwise, the cycle obtained by exchanging the edges xx^+ and yy^+ of C for the path xPy and the edge x^+y^+ would be longer than C. Thus S^+ is a stable set of G, and is disjoint from S.

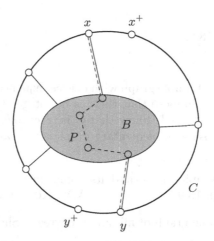

Fig. 18.18. Proof of the Chvátal–Erdős Theorem (18.10)

Let z be an internal vertex of B. Every neighbour of z on C belongs to S. Because S and S^+ are disjoint, $S^+ \cup \{z\}$ is a stable set. This implies that $|S^+| < \alpha$. On the other hand, S is a vertex cut of G because $B - S$ is a component of $G - S$, so $|S| \geq \kappa$. Hence

$$\kappa \leq |S| = |S^+| < \alpha$$

But this contradicts the hypothesis. Therefore C is indeed a Hamilton cycle of G. $\square$

Theorem 18.10 was generalized in a very nice way by Kouider (1994). A Hamilton cycle is a spanning cycle, that is, one which covers the entire vertex set of the graph. Kouider proved that the vertex set of any 2-connected graph G can be covered by a family of at most $\lceil \alpha/\kappa \rceil$ cycles. When $\alpha \leq \kappa$, this is just Theorem 18.10. In Section 19.2, we discuss an analogous theorem for digraphs.

Exercises

18.3.1 Let G be a simple graph on at least three vertices in which the degree sum of any two nonadjacent vertices is at least n. Consider a vertex x of G and an x-path P.

a) Show that P can be transformed to a Hamilton path Q of G, by means of path extensions and path exchanges.
b) By considering the degree sum of the two ends of Q, deduce that G contains a Hamilton cycle.

18.3.2 Prove Lemma 18.6.

18.3.3

a) Let G be a nontrivial simple graph with degree sequence $(d_1, d_2, \ldots, d_n)$, where $d_1 \le d_2 \le \ldots \le d_n$. Suppose that there is no integer $k < (n+1)/2$ such that $d_k < k$ and $d_{n-k+1} < n - k$. Show that G is traceable.
b) Deduce that every self-complementary graph is traceable.

(C.R.J. CLAPHAM)

18.3.4 Show that a simple graph and its complement cannot both satisfy the hypotheses of Theorem 18.9.   (A.V. KOSTOCHKA AND D.B. WEST)

18.3.5 Let G be a simple graph of minimum degree δ. Show that:

a) G contains a path of length 2δ if G is connected and $\delta \le (n-1)/2$,
b) G is traceable if $\delta \ge (n-1)/2$.

18.3.6 Let G be a graph with stability number α and connectivity κ, where $\alpha \le \kappa + 1$. Show that G is traceable.

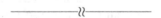

18.3.7 Let G be a simple graph, and let X be the set of vertices of G of degree at least $n/2$. If $|X| \ge 3$, show that G has a cycle which includes X.   (R. SHI)

18.3.8 A graph G is *degree-majorized* by a graph H if $v(G) = v(H)$ and the degree sequence of G (in nondecreasing order) is majorized by that of H (see Exercise 12.2.5).

a) Let m and n be positive integers with $m < n/2$ and $n \ge 3$. Show that the graph $K_m \vee (K_{n-2m} + \overline{K_m})$ is nonhamiltonian.
b) Show that every nonhamiltonian simple graph on n vertices, where $n \ge 3$, is degree-majorized by $K_m \vee (K_{n-2m} + \overline{K_m})$ for some $m < n/2$. (V. CHVÁTAL)

18.3.9 Let $G := G[X, Y]$ be a simple bipartite graph, where $|X| = |Y| \ge 2$, with degree sequence $(d_1, d_2, \ldots, d_n)$, where $d_1 \le d_2 \le \cdots \le d_n$. Suppose that there is no integer $k \le n/4$ such that $d_k \le k$ and $d_{n/2} \le n/2 - k$. Show that G is hamiltonian.

18.3.10

a) Let G be a simple graph with $m > \binom{n-1}{2} + 1$ and $n \geq 3$. Show that G is hamiltonian.   (O. ORE)

b) Show that the only nonhamiltonian simple graphs with n vertices and $\binom{n-1}{2}+1$ edges are the graphs $K_1 \vee (K_{n-2} + K_1)$ and, for $n = 5$, the graph $K_2 \vee \overline{K_3}$.

18.3.11

a) Let G be a Hamilton-connected graph with $n \geq 4$. Show that $m \geq 3n/2$.

b) For all even $n \geq 4$, construct a Hamilton-connected graph G with $m = 3n/2$.

c) For all odd $n \geq 5$, construct a Hamilton-connected graph G with $m = (3n + 1)/2$.   (J.W. MOON)

18.3.12 Deduce from Theorem 18.10 the following two results.

a) Let G be a simple graph with $n \geq 3$ in which the degree sum of any two nonadjacent vertices is at least n. Then G is hamiltonian. (This is also a direct consequence of Corollary 18.8.)   (O. ORE)

b) Let G be a simple k-regular graph on $2k + 1$ vertices, where $k \geq 2$. Then G is hamiltonian.   (C.ST.J.A. NASH-WILLIAMS)

18.3.13 A vertex v is *insertable* into a path P if v is adjacent to two consecutive vertices z, z^+ of P. Let xPy and uQv be disjoint paths. The path Q is *absorbable* into P if there is a path with vertex set $V(P) \cup V(Q)$. Suppose that each vertex of Q is insertable into P. Show that Q is absorbable into P.   (A. AINOUCHE)

18.3.14 Let G be a connected graph which contains a path of length k, and in which every such path is contained in a cycle of length at least l.

a) Show that each path in G of length less than k is contained in a path of length k.

b) Deduce that each path in G of length less than k is contained in a cycle of length at least l.   (T.D. PARSONS)

18.3.15

a) Let G be a simple 2-connected graph, and let xPy be a longest path in G. For a vine $(x_i Q_i y_i : 1 \leq i \leq r)$ on P, consider the cycles $C_i := P_i \cup Q_i$, where $P_i := x_i P y_i$, $1 \leq i \leq r$, and the cycle $C := \triangle_{i=1}^r C_i$. Suppose the vine chosen so that:

  ▷ r is as small as possible,
  ▷   subject to this condition, $|V(C) \cap V(P)|$ is as large as possible.
  Show that:
   i) $\{x\} \cup \{y\} \cup N(x) \cup N(y) \subseteq V(C)$,
   ii) C is either a Hamilton cycle of G or a cycle of length at least $d(x) + d(y)$.

b) Deduce the following statements.
   i) If G is a simple 2-connected graph with $\delta \leq n/2$, then G contains a cycle of length at least 2δ.   (G.A. DIRAC)

ii) If $G := G(x, y)$ is a simple 2-connected graph such that $d(v) \geq k$ for all
$v \neq x, y$, then G contains an xy-path of length at least k.

(P. ERDŐS AND T. GALLAI)

18.3.16 Let G be a claw-free graph.

a) Show that the subgraph of G induced by the neighbours of any vertex is either
connected (Type 1) or has exactly two components, both of which are complete
(Type 2).
b) Let v be a vertex of Type 1 in G whose neighbourhood is not a clique, and
let G' be a graph obtained from G by adding an edge joining two nonadjacent
neighbours of v. Show that G' is hamiltonian if and only if G is hamiltonian.
c) The *Ryjáček closure* of G is the graph H obtained by recursively applying the
operation described in (b) until the neighbourhood of every vertex of Type 1
is a clique. Show that:
  i) H is the line graph of a triangle-free graph,
  ii) H is hamiltonian if and only if G is hamiltonian.   (Z. RYJÁČEK)

18.4 Path Exchanges and Parity

In the previous section, we saw that a simple graph each vertex of which is adjacent
to more than half of the other vertices contains a Hamilton cycle. More surprising,
perhaps, is the fact that one can say something about Hamilton cycles in cubic
graphs. Such graphs need not have any Hamilton cycles at all (the Petersen and
Coxeter graphs being two familiar examples). However, as C. A. B. Smith proved,
each edge of a cubic graph lies in an even number of Hamilton cycles (see Tutte
(1946)). From this one can deduce that if a cubic graph has one Hamilton cycle,
it has at least three of them (Exercise 18.4.1a). Smith's Theorem was extended
by Thomason (1978) using the exchange operation between x-paths introduced in
Section 18.1. We now describe his idea.

THE LOLLIPOP LEMMA

Let G be a graph (not necessarily simple). The *x-path graph* of G is the graph
whose vertices are the longest x-paths of G, two such paths being adjacent if and
only if they are related by a path exchange (see Figure 18.19). Thomason (1978)
made use of this concept to establish a basic property of x-paths.

Theorem 18.11   THE LOLLIPOP LEMMA
*Let G be a connected graph on at least two vertices, and let x be a vertex of G.
Then the number of longest x-paths of G that terminate in a vertex of even degree
is even.*

Proof   Let H denote the x-path graph of G and let P be a longest x-path of G.
If P terminates in y,

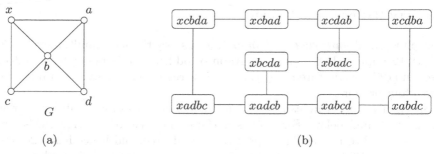

Fig. 18.19. (a) A graph G, (b) the x-path graph of G

$$d_H(P) = d_G(y) - 1$$

Thus y is of even degree in G if and only if P is of odd degree in H. Because the number of vertices of odd degree in H is even by Corollary 1.2, the number of longest x-paths in G that terminate in a vertex of even degree is also even.   □

Corollary 18.12 *Let G be a graph on at least three vertices, and let x and y be two vertices of G. Suppose that each vertex of G other than x and y is of odd degree. Then the number of Hamilton xy-paths in G is even. In particular, if G is a graph in which all vertices are of odd degree, then each edge of G lies in an even number of Hamilton cycles.*

Proof   We may assume that G has at least one Hamilton xy-path, otherwise the conclusion is trivial. Set $G' := G - y$. The longest x-paths of G' are Hamilton paths of G', and each Hamilton xy-path of G is an extension of such a path. Let $xP'z$ be a Hamilton path of G'. If z is of odd degree in G', the number of edges between y and z is even, because z is of odd degree in G. Thus P' gives rise to an even number of Hamilton xy-paths of G in this case. On the other hand, if z is of even degree in G', then P' gives rise to an odd number of Hamilton xy-paths of G. But, by Theorem 18.11, the number of Hamilton x-paths of G' ending in a vertex of even degree is even. Hence the total number of Hamilton xy-paths in G is even.   □

A special case of Corollary 18.12 is the theorem of C. A. B. Smith referred to earlier.

Theorem 18.13   SMITH'S THEOREM
In any cubic graph, each edge lies in an even number of Hamilton cycles.   □

The results of this section give rise to intriguing algorithmic questions. Smith's Theorem implies that every cubic graph with one Hamilton cycle has a second Hamilton cycle. Chrobak and Poljak (1988) asked how hard it is to find a second Hamilton cycle in such a graph when supplied with one of them. The answer is unknown. In particular, no polynomial-time algorithm has been devised for solving this problem.

UNIQUELY HAMILTONIAN GRAPHS

A graph is called *uniquely hamiltonian* if it has exactly one Hamilton cycle. Corollary 18.12 implies that no regular graph of odd degree is uniquely hamiltonian. Sheehan (1975) conjectured that the same is true of regular simple graphs of even degree four or more.

Sheehan's Conjecture can be restricted without loss of generality to 4-regular graphs, as stated below. For if C is a Hamilton cycle of G, then the spanning subgraph $G \setminus E(C)$ is regular of positive even degree, and hence has a 2-factor F (Exercise 16.4.16b). The graph $H := F \cup C$ is a 4-regular spanning subgraph with a Hamilton cycle C. If we could prove that H had a second Hamilton cycle, then G would also have this second Hamilton cycle.

SHEEHAN'S CONJECTURE

Conjecture 18.14 *Every hamiltonian 4-regular simple graph has at least two Hamilton cycles.*

Note that the conditions of simplicity and regularity here are essential. Examples of uniquely hamiltonian 4-regular graphs with multiple edges and of uniquely hamiltonian simple graphs of minimum degree four were constructed by Fleischner (1994, 2007).

By applying the methods described in this chapter, Thomassen (1998) obtained a general sufficient condition for the existence of at least two Hamilton cycles in a hamiltonian graph. Thomassen's argument is based on the following concepts.

Consider a (not necessarily proper) 2-edge-colouring of a graph G in red and blue. A set S of vertices of G is called *red-stable* if no two vertices of S are joined by a red edge, and *blue-dominating* if every vertex of $V \setminus S$ is adjacent by a blue edge to at least one vertex of S.

Theorem 18.15 *Let G be a graph and let C be a Hamilton cycle of G. Colour the edges of C red and the remaining edges of G blue. Suppose that there is a red-stable blue-dominating set S in G. Then G has a second Hamilton cycle.*

Proof Let S be a red-stable blue-dominating set, and let $T := S^- \cup S^+$. Consider the spanning subgraph H of G the edge set of which consists of all red edges and, for each vertex $y \in T$, one blue edge joining y to a vertex of S. In H, each vertex of T has degree three, whereas every other vertex of $V \setminus S$ has degree two. Because S is red-stable, it follows that $H - S$ has exactly $|S|$ components, each of which is a path whose ends are in T. By Theorem 18.1, every Hamilton cycle of H includes all of these paths. Let $e = xy \in E(C)$, where $x \in S$ and $y \in T$. By Theorem 18.11, the number of longest x-paths of $H \setminus e$ that terminate in a vertex of even degree is even. But the Hamilton path $C \setminus e$ is one such path, because y has degree two in $H \setminus e$. Let P be another such path. Then $P + e$ is a second Hamilton cycle of H, hence also of G. □

One easy consequence of Theorem 18.15 is the following result for bipartite graphs.

Corollary 18.16 *Let $G[X, Y]$ be a simple hamiltonian bipartite graph in which each vertex of Y has degree three or more. Then G has at least two Hamilton cycles.*

Proof Let C be a Hamilton cycle of G. Colour the edges of C red and the remaining edges of G blue. Each vertex of Y is then incident to at least one blue edge, so X is a red-stable blue-dominating set in G. By Theorem 18.15, G has a second Hamilton cycle. □

Thomassen (1998) applied the Local Lemma (Theorem 13.12) to show that if k is sufficiently large, then every simple hamiltonian k-regular graph fulfills the requirements of Theorem 18.15 and hence has at least two Hamilton cycles.

Theorem 18.17 *For $k \geq 73$, every simple hamiltonian k-regular graph has at least two Hamilton cycles.*

Proof Let G be a simple hamiltonian k-regular graph, and let C be a Hamilton cycle of G. As in Theorem 18.15, we colour the edges of C red and the remaining edges of G blue. We now select each vertex of G independently, each with probability p, so as to obtain a random subset S of V. We show that, for an appropriate choice of p, this set S is, with positive probability, a red-stable blue-dominating set. The theorem then follows on applying Theorem 18.15.

For each element of $E(C) \cup V(G)$, we define a 'bad' event, as follows.

▷ A_e: *both ends of edge e of C belong to S.*
▷ B_v: *neither vertex v of G nor any vertex joined to v by a blue edge belongs to S.*

We have $p(A_e) = p^2$ and $P(B_v) = (1 - p)^{k-1}$, because each vertex v has blue degree $k - 2$. We define a dependency graph H for these events, with vertex set $E(C) \cup V(G)$, by declaring two vertices to be adjacent in H if the sets of vertices involved in the corresponding events intersect. The vertices involved in A_e, namely the two ends of e, are each involved in one other event A_f and $k - 1$ events B_v. Thus e has degree at most $2 + (2k - 2)$ in the dependency graph H. The $k - 1$ vertices involved in B_v are each involved in two events A_e, and are together involved in a total of at most $(k-2)^2$ other events B_w. Thus v has degree at most $(2k - 2) + (k - 2)^2$ in H. In order to apply the Local Lemma, we must therefore select a value for p and numbers x (associated with each event A_e) and y (associated with each event B_v) such that:

$$p^2 \leq x(1 - x)^2(1 - y)^{2k-2} \quad \text{and} \quad (1 - p)^{k-1} \leq y(1 - x)^{2k-2}(1 - y)^{(k-2)^2}$$

We may simplify these expressions by setting $x := a^2$ and $y := b^{k-1}$:

$$p \le a(1 - a^2)(1 - b^{k-1})^{k-1} \quad \text{and} \quad 1 - p \le b(1 - a^2)^2(1 - b^{k-1})^{k-3}$$

Thus

$$1 \le a(1 - a^2)(1 - b^{k-1})^{k-1} + b(1 - a^2)^2(1 - b^{k-1})^{k-3}$$

For $k \ge 73$, a solution to this inequality is obtained by setting $a = .25$ and $b = .89$, resulting in a value of .2305 for p. $\qquad \square$

The bound $k \ge 73$ in Theorem 18.17 was reduced to $k \ge 23$ by Haxell et al. (2007). However, Sheehan's Conjecture remains open.

To conclude this discussion of Sheehan's Conjecture, let us note a more recent conjecture, due to Fleischner (2007), which bears a close resemblance to it.

Conjecture 18.18   FLEISCHNER'S CONJECTURE
Every hamiltonian 4-connected graph has at least two Hamilton cycles.

This conjecture holds for planar graphs (see Section 18.6).

Exercises

18.4.1

a) Show that:
    i) every hamiltonian cubic graph has at least three Hamilton cycles,
    ii) both the complete graph K_4 and the generalized Petersen graph $P_{2,9}$ have exactly three Hamilton cycles.
b) For each integer $n \ge 2$, construct a simple cubic graph on $2n$ vertices with exactly three Hamilton cycles.

18.4.2 Let G be a cubic hamiltonian graph and let e be an edge of G. Form a bipartite graph $H[C, F]$, where C is the set of proper 3-edge colourings of G and F is the set of even 2-factors of G (that is, 2-factors all components of which are even cycles) that include the edge e, with $c \in C$ joined to $F \in F$ if and only if F is induced by the union of two of the colours in the 3-edge-colouring c.

a) Show that:
    i) every vertex in C has degree two,
    ii) a vertex $F \in F$ with k components has degree 2^{k-1}.
b) Deduce Smith's Theorem (18.13).

18.4.3

a) Show that the following algorithm, when supplied with a Hamilton cycle C of a cubic graph G, returns a second Hamilton cycle C' of G.
    1: colour the chords of C with colour 0 and the edges of C alternately with colours 1 and 2
    2: fix an edge e of colour 2
    3: set $i := 0$

4: **while** $i \neq 2$ **do**
5:     **if** the subgraph F induced by the edges of colours i and 2 is connected
    then
6:         set $C' := F$ and $i := 2$
7:     **else**
8:         swap the colours i and 2 on those components of F not containing e
9:         replace i by $1 - i$
10:    **end if**
11: **end while**
12: return C'

b) Show that this algorithm runs in exponential time when the input graph G consists of a cycle C of length $0 \pmod 4$ with chords joining antipodal vertices.
(T. JENSEN)

18.4.4 Consider the graph G of Figure 18.20.

a) Show that:
    i) the edge e lies in no Hamilton cycle and no Hamilton xy-path of G,
    ii) G has exactly one Hamilton cycle and exactly one Hamilton xy-path.
b) Deduce that:
    i) the graph $G + xy$, in which one vertex is of degree four and the rest are of degree three, has exactly two Hamilton cycles,
    ii) the graph H of Figure 18.20, in which two vertices are of degree four and the rest are of degree three, has a unique Hamilton cycle.
(R.C. ENTRINGER AND H. SWART)

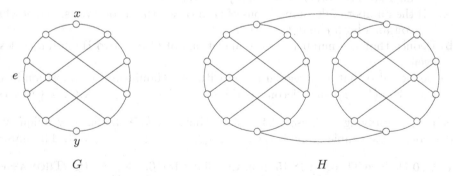

Fig. 18.20. Construction of a uniquely hamiltonian almost cubic graph

18.4.5 Show that every cubic bipartite graph on at least four vertices has an even number of Hamilton cycles.
(A. KOTZIG)

18.4.6

a) Let G be a simple cubic graph. Show that the line graph of G admits a Hamilton decomposition if and only if G is 3-edge-colourable.   (A. KOTZIG)

b) Deduce that the line graph of the Petersen graph is a 4-connected 4-regular graph which admits no Hamilton decomposition.

18.4.7

a) Let G be a 4-regular plane graph that admits a decomposition into two Hamilton cycles, C and D. Denote by F_{11}, F_{12}, F_{21}, and F_{22} the sets of faces of G interior to both C and D, interior to C but exterior to D, interior to D but exterior to C, and exterior to both C and D, respectively. Show that $g(F_{11}) = g(F_{22})$ and $g(F_{12}) = g(F_{21})$, where $g(F_{ij}) := \sum_{f \in F_{ij}} (d(f) - 2)$.

b) i) Draw the medial graph of the Herschel graph (Figure 18.1b).

   ii) Deduce from (a) that if G is a plane graph and if either G or G^* fails to satisfy Grinberg's Identity, then its medial graph has no Hamilton decomposition.

   iii) Conclude that the medial graph of the Herschel graph is a 4-connected 4-regular plane graph which admits no Hamilton decomposition.

(J.A. BONDY AND R. HÄGGKVIST)

18.4.8

a) Let G be a 4-regular graph on at least three vertices, and let e and f be two edges of G. Show that:

   i) the number of decompositions of G into two Hamilton cycles, one containing e and the other containing f, is even,

   ii) the number of decompositions of G into two Hamilton cycles, one of which contains both e and f, is even.

b) Deduce that the number of decompositions of G into two Hamilton cycles is even.

c) Deduce also that a $2k$-regular graph with one Hamilton decomposition has at least $3^{k-1}(k-1)!$ such decompositions.   (A.G. THOMASON)

18.4.9 By applying Exercise 18.4.8, show that, for $k \geq 4$, the only uniquely k-edge-colourable simple graph is the star $K_{1,k}$.   (A.G. THOMASON)

18.4.10 Deduce Theorem 18.15 from Corollary 18.16.   (C. THOMASSEN)

18.4.11 Let G be a graph and x a vertex of G. For $k \geq 1$, denote by p_k the number of x-paths in G of length k and by q_k the number of x-paths in G of length k ending in a vertex of even degree.

a) Show that $p_k \equiv q_{k-1} \pmod 2$.

b) Deduce that:

   i) if G is even, then p_k is even for all $k \geq 1$,

   ii) if G is odd, then p_k is even for all $k \geq 2$.

(J.A. BONDY AND F. HALBERSTAM)

18.4.12 Let G be a graph and let f be a nonnegative integer-valued function on V. An f-*tree* of G is an f-factor which is a spanning tree of G.

a) Suppose that $d(v) - f(v)$ is odd for all $v \in V$. Show that G has an even number of f-trees. (K.A. BERMAN)

b) Let g be a nonnegative integer-valued function on V. Show that the number of decompositions of G into an f-tree and a g-tree is even. (K.A. BERMAN)

18.4.13

a) Let G be a path or cycle of length n. Show that the composition $G[2K_1]$ is decomposable into two cycles of length $2n$.

b) Deduce that if G admits a decomposition into Hamilton cycles, then $G[2K_1]$ admits a decomposition into any list of even 2-factors $F_1, F_2, \ldots, F_k$ such that:
   ▷ $\sum_{i=1}^{k} e(F_i) = 4e(G)$,
   ▷ each F_i is isomorphic to a subgraph of $G[2K_1]$,
   ▷ each subgraph of $G[2K_1]$ occurs as an F_i an even number of times (possibly zero).

c) Show that $K_{4n+2} \setminus M \cong K_{2n+1}[2K_1]$, where M is a perfect matching of K_{4n+2}, and that $K_{4n,4n} \cong K_{2n,2n}[2K_1]$.

d) Deduce that each of the graphs in (c) admits a decomposition into any list $F_1, F_2, \ldots, F_k$ of even 2-factors satisfying the three conditions listed in (b).
(R. HÄGGKVIST)

18.5 Hamilton Cycles in Random Graphs

The graph $K_{k,k+1}$ shows that the bound on the minimum degree required by Dirac's Theorem cannot be reduced. This graph is, however, far from typical. For example, its stability number is $k+1$, roughly half its order, whereas the stability number of a random graph $G \in \mathcal{G}_{n,1/2}$ is almost surely close to $2\log_2 n$, just twice the logarithm of its order (see Exercise 13.2.11 and Theorem 13.9). We show here that a random graph need only have a very small average degree to be almost surely hamiltonian. The proof is due to Pósa (1976), and makes clever use of path exchanges.

PÓSA'S LEMMA

As we have seen, the naïve approach to finding a long path in a graph is to grow a maximal x-path P and consider the x-paths obtainable from P by means of path exchanges. If one of these is not maximal, then it can be extended to a longer x-path. The procedure may then be repeated. Although this approach fails badly on certain graphs, Pósa (1976) proved that it works remarkably well on most graphs. His argument hinges on the following result.

Theorem 18.19   PÓSA'S LEMMA

Let xPy be a longest path in a graph G. Denote by $\mathcal{P}$ the set of all x-paths of G obtainable from P by path exchanges, by T the set of all terminal vertices of paths in $\mathcal{P}$, and by T^- and T^+, respectively, the sets of vertices immediately preceding and following the vertices of T on P. Define $S := V \setminus (T \cup T^- \cup T^+)$. Then $e(S, T) = 0$.

Proof   Let $u \in S$ and $v \in T$. By the definition of T, there exists a path xQv in $\mathcal{P}$. If $u \in V(G) \setminus V(P)$, then u and v cannot be adjacent because Q is a longest path in G. So suppose that $u \in V(P)$. Then u has the same neighbours on each path in $\mathcal{P}$, because an elementary exchange that removed one of these neighbours would, at the same time, establish either it or u as an element of T, contradicting the definition of S. If u and v were adjacent, an elementary exchange applied to Q would yield a x-path terminating in a neighbour of u on P, a contradiction. Thus, in both cases, u and v are nonadjacent.   □

We now apply Pósa's Lemma to show that the set T defined there is almost surely large when G is a random graph whose edge probability is sufficiently high.

Corollary 18.20 *Let $G \in \mathcal{G}_{n,p}$, where $p = 9 \log n / n$, and let T be as defined in the statement of Pósa's Lemma. Then $P(|T| < \lfloor n/4 \rfloor) \ll n^{-1}$.*

Proof   Suppose that $|T| = k$. Then $|T^-| \leq k$ and $|T^+| \leq k$. By Theorem 18.19, there is therefore a subset S of V, disjoint from T, such that $|S| \geq n - 1 - 3k$ and $e(S, T) = 0$. The probability that G has a set of k vertices all of which are nonadjacent to a set of $n - 1 - 3k$ vertices is at most $\binom{n}{k}(1-p)^{k(n-1-3k)}$. Thus the probability that T has at most $l := \lfloor n/4 \rfloor - 1$ vertices is at most

$$\sum_{k=1}^{l} \binom{n}{k}(1-p)^{k(n-1-3k)} < \sum_{k=1}^{l} n^k e^{-pk(n-1-3k)} < \sum_{k=1}^{l} (ne^{-pn/4})^k$$

$$= \sum_{k=1}^{l} (n^{-5/4})^k = n^{-5/4}\left(\frac{1 - n^{-5l/4}}{1 - n^{-5/4}}\right) \ll n^{-1} \quad □$$

Our goal is to show that a random graph in $\mathcal{G}_{n,p}$, where $p = 10 \log n / n$, is almost surely hamiltonian. We do so in two stages, establishing first of all that such a graph is almost surely traceable.

Theorem 18.21 *Let $G \in \mathcal{G}_{n,p}$, where $p = 9 \log n / n$. Then G is almost surely traceable.*

Proof   For $v \in V$, the vertex-deleted subgraph $G - v$ is a random graph on $n - 1$ vertices with independent edge probability p. Let T_v be the set T as defined in the statement of Pósa's Lemma, applied to this random graph $G - v$. We consider the following two events.

▷ A_v: $|T_v| < \lfloor (n-1)/4 \rfloor$.

▷ B_v: $|T_v| \geq \lfloor (n-1)/4 \rfloor$ and there exists a longest path in G which does not include v.

By Corollary 18.20, $P(A_v) \ll n^{-1}$. On the other hand, if $|T_v| \geq \lfloor (n-1)/4 \rfloor$ and there is a longest path in G which does not include v, none of the longest paths in $G - v$ which terminate in a vertex of T_v can be extended to include v, so v is nonadjacent to each vertex of T_v. Thus

$$P(B_v) \leq (1-p)^{\lfloor (n-1)/4 \rfloor} < e^{-p\lfloor (n-1)/4 \rfloor} \ll n^{-2}$$

and

$$\sum_{v \in V} (P(A_v) + P(B_v)) \leq n \left(P(A_v) + P(B_v) \right) \to 0$$

We conclude that almost surely every vertex of G lies on every longest path. But this implies that almost surely G is traceable. □

Theorem 18.22 *Let* $G \in \mathcal{G}_{n,p}$, *where* $p = 10 \log n / n$. *Then* G *is almost surely hamiltonian.*

Proof Let $H = G_1 \cup G_2$, where $G_i \in \mathcal{G}_{n,p_i}$, $i = 1, 2$, with $p_1 = 9 \log n / n$ and $p_2 = \log n / n$. Then $H \in \mathcal{G}_{n,p}$, where $p = 10 \left(\log n / n \right) - 9 \left(\log n / n \right)^2$. It suffices to show that H is almost surely hamiltonian.

By Theorem 18.21, the random graph G_1 is almost surely traceable. Let T_1 be the set T as defined in the statement of Pósa's Lemma, applied to G_1. By Corollary 18.20, almost surely $|T_1| \geq \lfloor n/4 \rfloor$. Let u be the initial vertex of the Hamilton paths in G_1 terminating in T_1. The probability that u is joined in G_2 to no vertex of T_1 is at most $(1 - p_2)^{\lfloor n/4 \rfloor}$. Since

$$(1 - p_2)^{\lfloor n/4 \rfloor} < e^{-p_2 \lfloor n/4 \rfloor} \to 0 \quad \text{as } n \to \infty$$

almost surely u is joined in G_2 to at least one vertex of T_1, resulting in a Hamilton cycle in $H = G_1 \cup G_2$. □

Exercises

18.5.1 Let G be a simple graph and let k and l be integers, where $k \geq -1$, $l \geq 1$ and $k \leq l$. Suppose that $d(X) \geq 2|X| + k$ for every nonempty set X of vertices such that $|X| \leq \lceil (l - k + 1)/3 \rceil$. Show that G contains a path of length l.

──────────── ⁂ ────────────

18.6 Related Reading

THE BRIDGE LEMMA

We remarked in Section 18.2 that every 4-connected planar graph is hamiltonian. Tutte (1956) proved this theorem by establishing inductively the following stronger assertion concerning bridges in plane graphs.

THE BRIDGE LEMMA
Let G be a 2-connected plane graph, e an edge of G, C' and C'' the facial cycles of G which include e, and e' any edge of C'. Then there is a cycle C in G including both e and e' such that:

i) each bridge of C in G has either two or three vertices of attachment,
ii) each bridge of C in G that includes an edge of C' or C'' has exactly two vertices of attachment.

A cycle satisfying the properties described in the Bridge Lemma is called a *Tutte cycle*. The Bridge Lemma implies not only Tutte's theorem but also several other interesting results on cycles in planar graphs, including the truth of Fleischner's Conjecture (18.18) for planar graphs. Refinements and variants of the Bridge Lemma were employed by Thomassen (1983b) to show that every 4-connected planar graph is Hamilton-connected (see also Sanders (1997)), by Thomas and Yu (1994) to extend Tutte's theorem to graphs embeddable on the projective plane, and by Chen and Yu (2002) to prove that every 3-connected planar graph on n vertices has a cycle of length at least cn^γ for some positive constant c, where $\gamma = \log 2/\log 3$. (This bound is best possible in view of the constructive upper bound obtained by Moon and Moser (1963), see Exercise 18.2.3).

THE HOPPING LEMMA

As Pósa's Lemma (Theorem 18.19) illustrates, the approach of iterating path exchanges can be highly effective. This technique was first employed, in the framework of cycles, by Woodall (1973), who proved the following theorem.

THE HOPPING LEMMA
Let G be a graph and C a longest cycle of G such that $H := G - V(C)$ has as few components as possible. Suppose that some component of H is an isolated vertex x. Set $X_0 := \{x\}$ and $Y_0 := N(x)$, and define recursively sets X_i, Y_i, $i \geq 1$, by

$$X_i := X_{i-1} \cup (Y_{i-1}^- \cap Y_{i-1}^+) \quad and \quad Y_i := N(X_i)$$

Then $X := \cup_{i=0}^\infty X_i$ is a stable set and $N(X) \subseteq V(C)$.

As with the Bridge Lemma, there exist a number of variants and extensions of the Hopping Lemma. It was developed by Woodall (1973) in order to prove that a simple 2-connected graph G of minimum degree at least $(n + 2)/3$ is hamiltonian if $d(X) \geq (n + |X| - 1)/3$ for all subsets X of V with $2 \leq |X| \leq (n + 1)/2$. Jackson

(1980) applied it to show that every 2-connected d-regular graph G with $d \geq n/3$ is hamiltonian; see also Broersma et al. (1996). A variant of the Hopping Lemma was devised by Häggkvist and Thomassen (1982) in order to establish a theorem on cycles containing specified edges in k-connected graphs; see also Kawarabayashi (2002).

LONG PATHS AND CYCLES

As we have seen throughout this chapter, investigations into Hamilton cycles inevitably lead to the study of long paths and cycles. For further information on this topic, we refer the reader to the surveys by Bondy (1995a), Broersma (2002), Ellingham (1996), and Gould (2003).

19

Coverings and Packings in Directed Graphs

Contents

19.1 Coverings and Packings in Hypergraphs

In Chapter 3, we introduced the notion of a covering of the edge set of a graph by subgraphs. One may of course consider other notions of covering, for instance covering the vertex set by subgraphs (see, for example, the extension of Theorem 18.10 noted at the end of Section 18.3). And, naturally, the same ideas apply equally well to directed graphs. The language of hypergraphs provides a convenient general setting in which to discuss these and related notions.

COVERINGS AND DECOMPOSITIONS

Let $H := (V, \mathcal{F})$ be a hypergraph, with vertex set V and edge set $\mathcal{F}$. A *covering* of H is a family of edges of H whose union is V. Let us look at some examples of coverings in this language. If E is the edge set of a graph G and $\mathcal{C}$ is the family of edge sets of its cycles, a covering of the hypergraph $(E, \mathcal{C})$ is a simply a cycle covering of G. Likewise, if V is the vertex set of a digraph D and $\mathcal{P}$ is the family of vertex sets of its directed paths, a covering of the hypergraph $(V, \mathcal{P})$ is a covering of the vertices of G by directed paths.

Usually, the existence of a covering is evident, and the goal is to find a minimum covering, that is, one using as few edges as possible. For example, any 2-edge-connected graph G clearly has a covering of its edge set by cycles. The question of interest is how few cycles are needed to achieve such a covering? In terms of n, this number is $\lfloor \frac{3}{4}(n - 1) \rfloor$, as was shown by Fan (2002).

Exact coverings are called decompositions. More precisely, a *decomposition* of a hypergraph is a covering by pairwise disjoint edges. In contrast to coverings, the existence of a decomposition often requires much stronger assumptions. For instance, although every graph without cut edges has a cycle covering, only even graphs admit cycle decompositions. Such questions can be very challenging. For instance, the problem of deciding whether a 3-uniform hypergraph admits a decomposition is $\mathcal{NP}$-hard (see Garey and Johnson (1979)).

PACKINGS AND TRANSVERSALS

Let us now turn to the related notion of packing. A *packing* of a hypergraph $H := (V, \mathcal{F})$ is a set of pairwise disjoint edges of H. (Note that decompositions are both packings and coverings.) Whereas every hypergraph clearly has a packing (by the empty set), the objects of interest here are maximum packings, that is, packings with as many edges as possible. The Maximum Matching Problem (16.1) is perhaps the most basic nontrivial instance of such a problem, the hypergraph H being just the graph G.

Hand in hand with the concept of a packing goes that of a transversal. A *transversal* of a hypergraph $H := (V, \mathcal{F})$ is a subset X of V which intersects each edge of H. For example, if the vertex set of H is the edge set of a graph G and its edges are the edge sets of the cycles of G, a transversal of H is a subset S of E such that $G \setminus S$ is acyclic; in a digraph, such a subset is called a *feedback arc set*. (Transversals are sometimes referred to as *vertex coverings* — indeed, this is standard practice in matching theory — but we prefer to use the term 'transversal' here in order to avoid confusion with the notion of hypergraph covering.) Every hypergraph $H := (V, \mathcal{F})$ clearly has a transversal, namely V; it is the minimum transversals that are of primary interest.

MIN–MAX THEOREMS

We denote by $\nu(H)$ the number of edges in a maximum packing of H, and by $\tau(H)$ the number of vertices in a minimum transversal. These two parameters are related by a simple inequality (Exercise 19.1.1):

$$\nu(H) \leq \tau(H)$$

A family $\mathcal{H}$ of hypergraphs is said to satisfy the *Min–Max Property* if equality holds in the above inequality for each member H of $\mathcal{H}$, and the assertion of such an equality is called a *min–max theorem*. Menger's Theorems (7.16, 9.1, 9.7, and 9.8) are important examples of min–max theorems. Such theorems are particularly interesting from an algorithmic standpoint, because they provide succinct certificates of the optimality of the relevant packings and transversals.

THE ERDŐS–PÓSA PROPERTY

Most families of hypergraphs do not satisfy the Min–Max Property. Nevertheless, it is still of interest in such cases to find an upper bound on τ in terms of ν, if there is one. A prototypical example is provided by the following theorem of Erdős and Pósa (1965).

Theorem 19.1   THE ERDŐS–PÓSA THEOREM
For any positive integer k, a graph either contains k disjoint cycles or else has a set of at most $4k \log k$ vertices whose deletion destroys all cycles.   □

Here, the vertex set of the hypergraph H is the vertex set of the graph and its edges are the vertex sets of the cycles of the graph. (The case $k = 2$ is the topic of Exercise 9.4.9). In terms of the parameters ν and τ, the Erdős–Pósa Theorem states that

$$\tau(H) \leq f(\nu(H)) \tag{19.1}$$

where $f(k) := 4k \log k$.

It was shown by Robertson and Seymour (1986) that, for fixed k, this theorem yields a linear-time algorithm for deciding whether an input graph G has a family of k disjoint cycles. For if G does not contain k disjoint cycles, there is a set of at most $4k \log k$ vertices whose deletion results in a forest. Now forests have tree-width at most two (according to the definition given in Section 9.8), and it follows easily that G has tree-width at most $4k \log k + 2$, a constant. As noted in Section 9.8, many $\mathcal{NP}$-hard problems become easy when restricted to graphs of bounded tree-width, and this is the case with the search for k disjoint cycles.

In general, a family $\mathcal{H}$ of hypergraphs is said to satisfy the *Erdős–Pósa Property* if there is a function f satisfying (19.1) for all $H \in \mathcal{H}$. Erdős and Pósa showed that the hypergraphs whose edges are the edge sets of cycles of graphs have this property also. An example of a simple family of hypergraphs which fails to satisfy the Erdős–Pósa Property is given in Exercise 19.1.3.

Recall that a *clutter* is a hypergraph no edge of which is contained in another. Let $H = (V, \mathcal{F})$ be a clutter, and let $\mathcal{T}$ denote the set of all minimal transversals of H. The hypergraph $(V, \mathcal{T})$ is denoted by $H^{\perp}$ and called the *blocker* of H. (Blockers are also known as *transversal hypergraphs* or *Menger duals*.) It may be shown that $(H^\perp)^\perp = H$ for every clutter H (Exercise 19.1.4).

The blocker of a hypergraph H satisfies the Min–Max Property if the size of a maximum packing of transversals of H is equal to the size of a smallest edge of H. If a hypergraph satisfies the Min–Max Property, it is not necessarily true that its blocker also satisfies this property (Exercise 19.1.5). However, there are many interesting hypergraphs associated with graphs for which this is so. Consider, for example, the hypergraph whose edges are the arc sets of the directed (x, y)-paths in a digraph $D(x, y)$. Not only does this hypergraph satisfy the Min–Max Property, as is shown by Menger's Theorem, but its blocker does too (Exercise 19.1.6). A number of other such examples are described by Woodall (1978).

In this chapter, we discuss three theorems on coverings and packings in digraphs, all of which are min–max theorems. These theorems concern directed cycles, branchings, and directed bonds.

Exercises

$\star$**19.1.1** Show that $\nu(H) \le \tau(H)$ for any hypergraph H.

19.1.2 Show that the family of all graphs has the Erdős-Pósa Property.

19.1.3 Let G be a graph which can be embedded on the projective plane so that each face is bounded by an even cycle. (One such graph is obtained from a planar grid by identifying antipodal points on its boundary.)

a) Show that any two odd cycles of G intersect.
b) Construct a family of hypergraphs (based on graphs of the type defined above) which does not have the Erdős-Pósa Property.

19.1.4 Let $H = (V, \mathcal{F})$ be a clutter. Show that $(H^\perp)^\perp = H$.

19.1.5 Consider the graph G shown in Figure 19.1. Let H be the hypergraph whose vertex set is $E(G)$ and whose edges are the edge sets of the x_1y_1- and x_2y_2-paths in G. Show that H has the Min–Max Property, but not its blocker. (More generally, it was shown by Rothschild and Whinston (1966) that if G is any eulerian graph with four distinguished vertices x_1, y_1, x_2, and y_2, then the hypergraph H defined above has the Min–Max Property.)

19.1.6 Let $D := D(x, y)$ be a digraph. Show that the length of a shortest directed (x, y)-path in D is equal to the size of a maximum packing of outcuts separating y from x.

19.1.7 Let G be a bipartite graph, and let H be the hypergraph whose vertex set is the edge set of G and whose edges are the trivial edge cuts of G. By quoting appropriate results from Chapter 16, show that both H and $H^\perp$ have the Min–Max Property.

———————⟨⟨———————

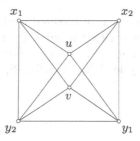

Fig. 19.1. The x_1y_1- and x_2y_2-paths define a hypergraph with the Min–Max Property (Exercise 19.1.5)

19.2 Coverings by Directed Cycles

Just as the Gallai–Milgram Theorem extends Rédei's Theorem to all digraphs, there is a natural extension of Camion's Theorem to all strong digraphs. Let us first recall the statement of this theorem (see the inset on Induction in Section 2.2, and Exercise 3.4.12a).

Theorem 19.2 CAMION'S THEOREM
Every nontrivial strong tournament has a directed Hamilton cycle.

A moment's reflection shows that one cannot hope to partition the vertices of every strong digraph into directed cycles. On the other hand, because each vertex of a strong digraph lies in some directed cycle, the entire vertex set can certainly be covered by directed cycles. Gallai (1964b) made a conjecture regarding the number of cycles needed for such a covering.

Conjecture 19.3 GALLAI'S CONJECTURE
The vertex set of any nontrivial strong digraph D can be covered by α directed cycles.

Gallai's Conjecture remained unresolved for several decades, but was eventually confirmed by Bessy and Thomassé (2004). A key idea in their proof is the concept of a coherent cyclic order of the vertices of a digraph.

COHERENT CYCLIC ORDERS

Let $D = (V, A)$ be a digraph. By a *cyclic order* of D we mean a cyclic order $O := (v_1, v_2, \ldots, v_n, v_1)$ of its vertex set V. Given such an order O, each directed cycle of D can be thought of as winding around O a certain number of times. In order to make this notion precise, we define the *length* of an arc (v_i, v_j) of D (with respect to O) to be $j - i$ if $i < j$ and $n + j - i$ if $i > j$. Informally, the length of an arc is just the length of the segment of O 'jumped' by the arc. If C is a directed cycle of D, the sum of the lengths of its arcs is a certain multiple of n. This multiple is called the *index* of C (with respect to O), and denoted $i(C)$. By extension, the *index* of a family $\mathcal{C}$ of directed cycles, denoted $i(\mathcal{C})$, is the sum of the indices of its constituent cycles. Consider, for example, the orientation of the Petersen graph shown in Figure 19.2a. With respect to the cyclic order shown in Figure 19.2b, the directed cycle $(1, 2, 3, 4, 5, 6, 7, 8, 9, 1)$ is of index two.

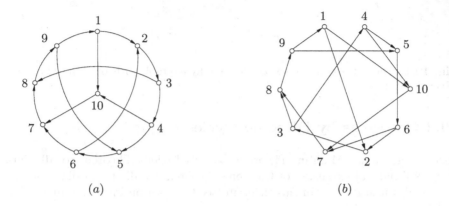

Fig. 19.2. (a) An orientation of the Petersen graph, and (b) a cyclic order of this digraph

A directed cycle of index one is called a *simple cycle*. If every arc lies in such a cycle, the cyclic order is *coherent*. The cyclic order shown in Figure 19.2b is not coherent because the arc $(4, 10)$ lies in no simple cycle; on the other hand, the cyclic order of Figure 19.3a is coherent.

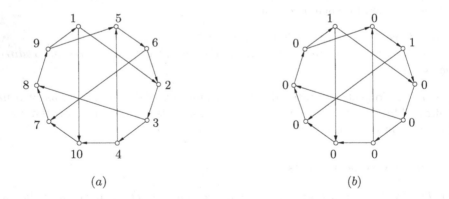

Fig. 19.3. (a) A coherent cyclic order, and (b) an index-bounded weighting

Bessy and Thomassé (2004) showed that every strong digraph admits a coherent cyclic order. It follows that a digraph admits a coherent cyclic order if and only if each of its arcs lies in a directed cycle or, equivalently, if and only if each of its connected components is strong. We deduce this fact from a result of Knuth

(1974), established in Chapter 5 — the two theorems are, in fact, equivalent. They have several significant implications.

Theorem 19.4 *Every strong digraph admits a coherent cyclic order.*

Proof Let D be a strong digraph. By Theorem 5.14, D has a coherent feedback arc set S. Since $D \setminus S$ is acyclic, the vertices of D may be ordered as $v_1, v_2, \ldots, v_n$, so that every arc not in S joins some vertex v_i to a vertex v_j with $j > i$ (Exercise 2.1.11). Every fundamental cycle of D with respect to S contains exactly one arc (v_i, v_j) with $i > j$, namely its unique arc of S. With respect to the cyclic order $(v_1, v_2, \ldots, v_n, v_1)$, these fundamental cycles are therefore simple. Because S is coherent, we deduce that the cyclic order $(v_1, v_2, \ldots, v_n, v_1)$ is coherent. $\square$

Observe that in a coherent cyclic order of a strong tournament, each vertex is adjacent to its successor, and thus dominates it. The cyclic order therefore determines a directed Hamilton cycle, and we obtain Camion's Theorem as an immediate corollary of Theorem 19.4.

THE BESSY–THOMASSÉ THEOREM

Bessy and Thomassé (2004) established Gallai's Conjecture by proving a stronger result, namely a min–max theorem relating a cyclic analogue of the stability number to the minimum index of a cycle covering. They did so by using Theorem 19.4 in conjunction with Dilworth's Theorem (12.5). Here, we present a closely related min–max theorem, establishing it by applying the linear programming proof technique described in Section 8.6.

A *weighting* of the vertices of a digraph D is a function $w : V \to \mathbb{N}$. We refer to $w(v)$ as the *weight* of vertex v. By extension, the *weight* $w(H)$ of a subgraph H of D is the sum of the weights of its vertices. If D is equipped with a cyclic order O, and if $w(C) \leq i(C)$ for every directed cycle C of D, we say that the weighting w is *index-bounded* (with respect to O). It can be checked that the $(0,1)$-weighting indicated in Figure 19.3b is index-bounded. Observe that for any cycle covering $\mathcal{C}$ of D and any index-bounded weighting w,

$$i(\mathcal{C}) \geq \sum_{C \in \mathcal{C}} w(C) \geq w(D) \tag{19.2}$$

Bondy and Charbit (2004) (see, also, Charbit (2005)) showed that equality holds in (19.2) for some cycle covering $\mathcal{C}$ and some index-bounded weighting w.

Theorem 19.5 *Let D be a digraph each of whose vertices lies in a directed cycle, and let O be a cyclic order of D. Then:*

$$\min i(\mathcal{C}) = \max w(D) \tag{19.3}$$

where the minimum is taken over all cycle coverings $\mathcal{C}$ of D and the maximum over all index-bounded weightings w of D.

In order to deduce Gallai's Conjecture from Theorem 19.5, it suffices to apply it to a coherent cyclic order O of D and observe that:

▷   for every family $\mathcal{C}$ of directed cycles of D, we have $|\mathcal{C}| \leq i(\mathcal{C})$,
▷   because each vertex lies in a directed cycle and O is coherent, each vertex lies in a simple cycle, so an index-bounded weighting of D is necessarily $(0, 1)$-valued,
▷   because each arc lies in a simple cycle, in an index-bounded weighting w no arc can join two vertices of weight one, so the support of w is a stable set, and $w(D) \leq \alpha(D)$.

These observations yield the theorem of Bessy and Thomassé (2004).

Theorem 19.6 *The vertex set of any nontrivial strong digraph D can be covered by α directed cycles.*

Proof of Theorem 19.5 Let D be a strict digraph, with vertex set $V = \{v_1, \ldots, v_n\}$ and arc set $A = \{a_1, \ldots, a_m\}$. It suffices to show that equality holds in (19.2) for some cycle covering $\mathcal{C}$ and some index-bounded weighting w.

An arc (v_i, v_j) is called a *forward arc* of D if $i < j$, and a *reverse arc* if $j < i$. Consider the matrix

$$\mathbf{Q} := \begin{bmatrix} \mathbf{M} \\ \mathbf{N} \end{bmatrix}$$

where $\mathbf{M} = (m_{ij})$ is the incidence matrix of D and $\mathbf{N} = (n_{ij})$ is the $n \times m$ matrix defined by:

$$n_{ij} = \begin{cases} 1 & \text{if } v_i \text{ is the tail of } a_j \\ 0 & \text{otherwise} \end{cases}$$

We know that $\mathbf{M}$ is totally unimodular (Exercise 1.5.7). Let us show that $\mathbf{Q}$ is totally unimodular also. Consider the matrix $\widetilde{\mathbf{Q}}$ obtained from $\mathbf{Q}$ by subtracting each row of $\mathbf{N}$ from the corresponding row of $\mathbf{M}$. Each column of $\widetilde{\mathbf{Q}}$ contains one 1 and one -1, the remaining entries being 0. It follows easily by induction on k that every $k \times k$ submatrix of $\widetilde{\mathbf{Q}}$ has determinant 1, -1 or 0 (Exercise 19.2.2). Thus $\widetilde{\mathbf{Q}}$ is totally unimodular. Because $\widetilde{\mathbf{Q}}$ was derived from $\mathbf{Q}$ by elementary row operations, the matrix $\mathbf{Q}$ is totally unimodular too.

We now define vectors $\mathbf{b} = (b_1, \ldots, b_{2n})$ and $\mathbf{c} = (c_1, \ldots, c_m)$ as follows.

$$b_i := \begin{cases} 0 & \text{if } 1 \leq i \leq n \\ 1 & \text{otherwise} \end{cases}$$

$$c_j := \begin{cases} 1 & \text{if } a_j \text{ is a reverse arc} \\ 0 & \text{otherwise} \end{cases}$$

Before proceeding with the proof, let us make two observations:

▷   If $\mathbf{x} := f_C$ is the circulation associated with a directed cycle C, then $\mathbf{cx} = i(C)$, the index of C.
▷   If $\mathbf{Nx} \geq \mathbf{1}$, where $\mathbf{x} := \sum\{\gamma_C f_C : C \in \mathcal{C}\}$ is a linear combination of circulations associated with a family $\mathcal{C}$ of directed cycles of D, then $\mathcal{C}$ is a covering of D.

Consider the LP:

$$minimize \quad \mathbf{cx}$$
$$subject\ to \quad \mathbf{Qx} \geq \mathbf{b} \tag{19.4}$$
$$\mathbf{x} \geq \mathbf{0}$$

The system of constraints $\mathbf{Qx} \geq \mathbf{b}$ is equivalent to the two systems $\mathbf{Mx} \geq \mathbf{0}$ and $\mathbf{Nx} \geq \mathbf{1}$, where $\mathbf{0}$ and $\mathbf{1}$ are vectors of 0s and 1s, respectively. Because the rows of $\mathbf{M}$ sum to $\mathbf{0}$, the rows of $\mathbf{Mx}$ sum to 0, which implies that $\mathbf{Mx} = \mathbf{0}$. Thus every feasible solution to (19.4) is a nonnegative circulation in D, hence, by Proposition 7.14, a nonnegative linear combination $\sum \gamma_C f_C$ of circulations associated with directed cycles of D. Moreover, because $\mathbf{Nx} \geq \mathbf{1}$, the cycles of positive weight in this sum form a covering of D. Conversely, every cycle covering of D yields a feasible solution to (19.4). The LP (19.4) is feasible because, by assumption, D has at least one cycle covering, and it is bounded because $\mathbf{c}$ is nonnegative. Thus (19.4) has an optimal solution. Indeed, by Theorem 8.28, (19.4) has an integral optimal solution, because $\mathbf{Q}$ is totally unimodular and the constraints are integral. This solution corresponds to a cycle covering $\mathcal{C}$ of minimum index, the optimal value being its index $i(\mathcal{C})$.

We now study the dual of (19.4):

$$maximize \quad \mathbf{yb}$$
$$subject\ to \quad \mathbf{yQ} \leq \mathbf{c} \tag{19.5}$$
$$\mathbf{y} \geq \mathbf{0}$$

Let us write $\mathbf{y} := (z_1, \ldots, z_n, w_1, \ldots, w_n)$. Then (19.5) is the problem of maximizing $\sum_{i=1}^{n} w_i$ subject to the constraints:

$$z_i - z_k + w_i \leq \begin{cases} 1 & \text{if } a_j := (v_i, v_k) \text{ is a reverse arc} \\ 0 & \text{if } a_j \text{ is a forward arc} \end{cases}$$

Consider an integral optimal solution to (19.5). If we sum the above constraints over the arc set of a directed cycle C of D, we obtain the inequality

$$\sum_{v_i \in V(C)} w_i \leq i(C)$$

In other words, the function w defined by $w(v_i) := w_i$, $1 \leq i \leq n$, is an index-bounded weighting, and the optimal value is the weight $w(D)$ of D. By linear programming duality (Theorem 8.27), we have $i(\mathcal{C}) = w(D)$. □

An alternative proof of Theorem 19.6 based on network flows, as well as a variety of extensions and generalizations, are described in Sebő (2007). A fast algorithm for finding coherent cyclic orders can be found in Iwata and Matsuda (2007).

CYCLE COVERINGS AND EAR DECOMPOSITIONS

The Bessy–Thomassé Theorem shows that any nontrivial strong digraph D has a spanning subdigraph which is the union of α directed cycles. However, the structure of this subdigraph might be rather complicated. This leads one to ask whether there always exists a spanning subdigraph whose structure is relatively simple, one which is easily seen to be the union of α directed cycles. A natural candidate would be a spanning subdigraph built from a directed cycle by adding $\alpha - 1$ directed ears. Such a digraph is readily seen to be the union of α directed cycles (Exercise 19.2.3). Unfortunately, not every strong digraph has such a spanning subdigraph, as is illustrated by the example shown in Figure 19.4. This digraph has stability number two but has no strong ear decomposition with fewer than two directed ears. It can be extended in a straightforward manner to an example with stability number α requiring at least $2\alpha - 2$ directed ears, for any $\alpha \geq 2$ (Exercise 19.2.4).

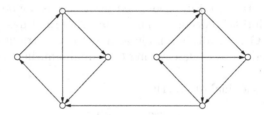

Fig. 19.4. A digraph with stability number two, but which has no strong ear decomposition with just one ear

A possible way around this problem is to allow spanning subdigraphs which are disconnected but whose components are strong. Each component then has a strong ear decomposition, and the number of directed cycles needed to cover the subdigraph is simply the sum of the numbers required for its components. We call such a spanning subdigraph *cyclic*, because each arc lies on a directed cycle. The following conjecture was formulated by Bondy (1995b), based on a remark of Chen and Manalastas (1983).

Conjecture 19.7 *Let D be a digraph all of whose strong components are nontrivial. Then D contains a cyclic spanning subdigraph with cyclomatic number at most α.*

Conjecture 19.7 holds for $\alpha = 1$ by Camion's Theorem (19.2) and also for $\alpha = 2$ and $\alpha = 3$ by theorems of Chen and Manalastas (1983) and S. Thomassé (unpublished), respectively. In addition, it implies not only the Bessy–Thomassé Theorem, but also a result of Thomassé (2001), that the vertex set of any strong digraph D with $\alpha \geq 2$ can be partitioned into $\alpha - 1$ directed paths, as well as another theorem of Bessy and Thomassé (2003), that every strong digraph D has a strong spanning subdigraph with at most $n + 2\alpha - 2$ arcs (Exercise 19.2.5).

Exercises

19.2.1 Describe a polynomial-time algorithm for finding a coherent cyclic order of a strong digraph.

$\star$**19.2.2** Prove that the matrix $\mathbf{Q}$ defined in the proof of Theorem 19.5 is totally unimodular.

19.2.3 Let D be a digraph obtained from a directed cycle by adding $k - 1$ directed ears. Show that D is a union of k directed cycles.

19.2.4 Consider the digraph of Figure 19.4.

a) Show that this digraph has no strong spanning subdigraph of cyclomatic number less than three.
b) For all $\alpha \geq 2$, extend this example to a strong digraph D having no strong spanning subdigraph of cyclomatic number less than $2\alpha - 1$.   (O. FAVARON)

19.2.5 Let D be a strong digraph which has a cyclic spanning subdigraph of cyclomatic number α. Show that:

a) if $\alpha \geq 2$, the vertex set of D can be partitioned into $\alpha - 1$ directed paths,
b) D has a strong spanning subdigraph with at most $n + 2\alpha - 2$ arcs.

(S. BESSY AND S. THOMASSÉ)

19.3 Packings of Branchings

EDMONDS' BRANCHING THEOREM

Recall that an x-*branching* is an oriented tree in which each vertex has indegree one, apart from the root x, which is a source. We consider here the problem of packing x-branchings in a digraph $D := D(x)$. A necessary condition for the existence of k arc-disjoint spanning x-branchings in D is that $d^+(X) \geq k$ for every proper subset X of V that includes x, because each of the k branchings must include a separate arc of $\partial^+(X)$. This condition is sufficient when $k = 1$ (Exercise 4.2.9). Indeed, as Edmonds (1973) showed, it is sufficient for all k. The proof of Edmonds' theorem given below is due to Lovász (1976). It leads to a polynomial-time algorithm for finding such a family of branchings.

Theorem 19.8 EDMONDS' BRANCHING THEOREM
A digraph $D := D(x)$ has k arc-disjoint spanning x-branchings if and only if

$$d^+(X) \geq k, \quad \text{for all } X \subset V \text{ with } x \in X \tag{19.6}$$

Proof As remarked above, the necessity of condition (19.6) is clear. We establish its sufficiency by induction on k. Having already noted its validity for $k = 1$, we may suppose that $k \geq 2$. Let us call an x-branching with arc set S *removable* if

$$d^+_{D \backslash S}(X) \geq k - 1, \quad \text{for all } X \subset V \text{ with } x \in X \tag{19.7}$$

Our aim is to show that D has a removable spanning x-branching. The theorem will then follow by induction. We construct such a branching by means of a tree-search algorithm, ensuring that at each stage the current x-branching is removable. Observe that the trivial x-branching, consisting only of the root x, is clearly removable.

Assume that we have already constructed a removable x-branching B, with vertex set Z and arc set S. If B is spanning, there is nothing more to prove. If not, consider an arc $a := (u, v)$ in the outcut $\partial^+(Z)$. Suppose that the branching obtained by adjoining the vertex v and arc a to B is not removable, in other words that $d^+_{D \backslash (S \cup \{a\})}(X) < k - 1$ for some proper subset X of V that contains x. Because B is removable, $d^+_{D \backslash S}(X) \geq k - 1$. We deduce that $a \in \partial^+(X)$ and $d^+_{D \backslash S}(X) = k - 1$ (see Figure 19.5a).

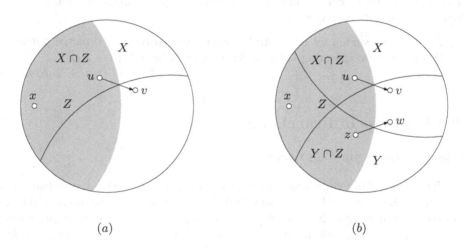

(a)   (b)

Fig. 19.5.

Noting that $v \notin X \cup Z$, we see that the set X has the following properties.

$$x \in X, \quad X \cup Z \subset V, \quad d^+_{D \backslash S}(X) = k - 1$$

Let us call such a set X a *critical set*, and its outcut $\partial^+(X)$ a *critical cut*. In order to grow B, it suffices to find an arc in $\partial^+(Z)$ which lies in no critical cut.

Consider a maximal critical set X. Because X is critical, $d^+_{D \backslash S}(X) = k - 1$. On the other hand, by (19.6), $d^+(X \cup Z) \geq k$. Because no arc of B lies in the outcut

$\partial^+(X \cup Z)$, we have $d^+_{D\backslash S}(X \cup Z) \geq k$, so the outcut $\partial^+(Z \setminus X)$ contains an arc $b := (z, w)$ (see Figure 19.5b). We claim that this arc b belongs to no critical cut. Suppose, to the contrary, that b lies in a critical cut $\partial^+(Y)$. By Exercise 2.5.4c,

$$(k - 1) + (k - 1) = d^+_{D\backslash S}(X) + d^+_{D\backslash S}(Y)$$
$$\geq d^+_{D\backslash S}(X \cup Y) + d^+_{D\backslash S}(X \cap Y) \geq (k - 1) + (k - 1)$$

implying, in particular, that $d^+_{D\backslash S}(X \cup Y) = k - 1$. Noting now that $w \notin X \cup Y \cup Z$, we see that the set $X \cup Y$ is critical. But this contradicts the maximality of the critical set X. We conclude that the removable x-branching B can be grown by adjoining the vertex w and the arc b, as claimed. □

An arc by which a removable x-branching B can be grown may be found in polynomial time by means of the Max-Flow Min-Cut Algorithm. This observation leads to a polynomial-time algorithm for finding a maximum family of arc-disjoint x-branchings in a digraph $D(x)$ (Exercise 19.3.2). An undirected analogue of Theorem 19.8 is discussed in Section 21.4.

Exercises

19.3.1

a) Let $D := D(x, y)$ be a digraph. Construct a new digraph D' from D by adding t arcs from y to each $v \in V \setminus \{x, y\}$. Show that D' has k arc-disjoint spanning x-branchings if and only if D has k arc-disjoint directed (x, y)-paths, provided that t is sufficiently large.

b) Deduce the arc version of Menger's Theorem (Theorem 7.16) from Edmonds' Branching Theorem.

19.3.2 Describe a polynomial-time algorithm for finding a maximum collection of arc-disjoint spanning x-branchings in a digraph $D(x)$.

19.3.3 Suppose that we have k roots, $x_1, x_2, \ldots, x_k$, and we wish to find k pairwise arc-disjoint spanning branchings $B_1, B_2, \ldots, B_k$, such that B_i is rooted at x_i, $1 \leq i \leq k$. Solve this problem by reducing it to the arc-disjoint spanning branching problem (with a single root).

———————— ⁂ ————————

$\star$**19.3.4** A *branching forest* in a digraph D is a subdigraph of D each component of which is a branching. Show that the arc set of a digraph D can be decomposed into k branching forests if and only if:

i) $d^-(v) \leq k$ for all $v \in V$,
ii) $a(X) \leq k(|X| - 1)$ for all nonempty subsets X of V.   (A. FRANK)

19.4 Packings of Directed Cycles and Directed Bonds

We noted in Section 19.1 that whereas packings of cycles in undirected graphs do not satisfy the Min–Max Property, they do satisfy the weaker Erdős–Pósa Property (both for vertex packings and for edge packings). A similar situation pertains to digraphs. It was conjectured by Gallai (1968c) and Younger (1973), and proved by Reed et al. (1996), that directed cycles in digraphs enjoy the Erdős–Pósa Property (again with regard to both vertices and arcs). But again, as noted below, they do not satisfy the Min–Max Property. Even when one restricts attention to planar digraphs, the Min–Max Property fails to hold for vertex packings of cycles (Exercise 19.4.9a). On the other hand, as we now prove, there does exist a min–max theorem for arc packings of directed cycles in planar digraphs.

Theorem 19.9 *In any planar digraph, the maximum number of arc-disjoint directed cycles is equal to the minimum number of arcs which meet all directed cycles.*

Planarity is crucial here: Theorem 19.9 does not extend to all digraphs. Consider, for example, the orientation of $K_{3,3}$ shown in Figure 19.6. In this digraph, any two directed cycles have an arc in common, but no single arc lies in every directed cycle.

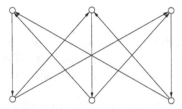

Fig. 19.6. A digraph with $\nu = 1$ and $\tau = 2$

DIRECTED BONDS AND CUTS

Recall that a *directed bond* is a bond $\partial(X)$ such that $\partial^-(X) = \emptyset$ (so that $\partial(X)$ is an outcut $\partial^+(X)$) and that a connected digraph has no directed bond if and only if it is strong (Exercise 2.5.7b). The more general notion of a *directed cut* is defined similarly. Note that every directed cut can be decomposed into directed bonds.

Corresponding to any statement concerning directed cycles in a plane digraph there is one concerning directed bonds in its dual (see Section 10.2). Thus Theorem 19.9 can be restated as follows.

Theorem 19.10 *In any planar directed graph, the maximum number of arc-disjoint directed bonds is equal to the minimum number of arcs which meet all directed bonds.*

We have seen that the hypothesis of planarity in Theorem 19.9 cannot be dropped. Remarkably, its dual, Theorem 19.10, does extend to all directed graphs, as was conjectured by N. Robertson (unpublished) and Younger (1965), and proved by Lucchesi and Younger (1978). The proof that we give here is due to Lovász (1976).

THE LUCCHESI–YOUNGER THEOREM

The notion of a 2-packing of directed bonds plays a key role in Lovász's proof. A family $\mathcal{C}$ of directed bonds in a digraph D is a 2-*packing* if no arc appears in more than two of its members. One way of obtaining a 2-packing is simply to take two copies of each member of a packing. This observation shows that the size of a maximum 2-packing of directed bonds is at least twice the size of a maximum packing. Rather surprisingly, one cannot do better.

Proposition 19.11 *In any digraph, the size of a maximum 2-packing of directed bonds is equal to twice the size of a maximum packing of directed bonds.*

Proof To establish the required equality, it is enough to show that any 2-packing $\mathcal{B}$ of directed bonds contains a packing consisting of at least half its members.

Recall that subsets X and Y of V *cross* if $X \cap Y$, $X \setminus Y$, $Y \setminus X$, and $V \setminus (X \cup Y)$ are all nonempty. As with edge cuts (see Section 9.3), we say that two directed cuts $\partial^+(X)$ and $\partial^+(Y)$ *cross* if the sets X and Y cross. The first step is to show that the members of $\mathcal{B}$ may be assumed to be pairwise noncrossing.

Suppose that $\mathcal{B}$ contains two directed bonds, $\partial^+(X)$ and $\partial^+(Y)$, which cross. Then (see Figure 19.7):

▷ both $\partial^+(X \cap Y)$ and $\partial^+(X \cup Y)$ are directed cuts,
▷ every arc of the digraph is covered exactly the same number of times (once, twice, or not at all) by these two directed cuts as it is by the two directed bonds $\partial^+(X)$ and $\partial^+(Y)$.

The family obtained from $\mathcal{B}$ by replacing the two directed bonds $\partial^+(X)$ and $\partial^+(Y)$ by two new directed bonds, one contained in $\partial^+(X \cap Y)$ and one contained in $\partial^+(X \cup Y)$, is therefore also a 2-packing, and moreover one of the same size as $\mathcal{B}$. By repeatedly uncrossing in this fashion, one obtains a pairwise noncrossing 2-packing of the same size as $\mathcal{B}$ (see Exercise 19.4.5). We may thus assume that the 2-packing $\mathcal{B}$ already has this property.

Some directed bonds might occur twice in the family $\mathcal{B}$. Define a graph G in which each vertex represents a directed bond that appears just once in $\mathcal{B}$, joining two such bonds $\partial^+(X)$ and $\partial^+(Y)$ by an edge if they have an arc in common. Because $\partial^+(X)$ and $\partial^+(Y)$ do not cross, they can be adjacent in G only if either $X \subset Y$ or $Y \subset X$. It follows that the graph G is bipartite (Exercise 19.4.6). The subfamily of $\mathcal{B}$ consisting of the larger of the two parts of G, together with one copy of each directed bond that appears twice in $\mathcal{B}$, is a packing of directed bonds whose size is at least half that of $\mathcal{B}$.   □

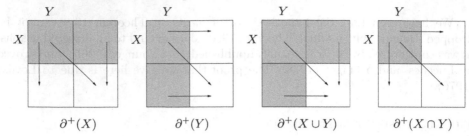

$$\partial^+(X) \qquad \partial^+(Y) \qquad \partial^+(X \cup Y) \qquad \partial^+(X \cap Y)$$

Fig. 19.7. Uncrossing two crossing directed cuts

For convenience, we denote by $\nu(D)$ the size of a maximum packing of directed bonds in a digraph D and by $\tau(D)$ the size of a minimum transversal of the directed bonds of D. In this notation, the Lucchesi–Younger Theorem states that $\nu(D) = \tau(D)$ for every digraph D. Before proceeding with its proof, let us note two simple facts about directed bonds (Exercise 19.4.4).

Proposition 19.12 *Let D be a digraph. Then:*

i) the digraph $\widetilde{D}$ obtained by subdividing each arc of D satisfies the identity
$\nu(\widetilde{D}) = 2\nu(D)$,
ii) the digraph D' obtained by subdividing a single arc of D satisfies the inequalities
$\nu(D) \leq \nu(D') \leq \nu(D) + 1$,
iii) for any arc a, the directed bonds of D / a are precisely the directed bonds of D which do not include the arc a. □

Theorem 19.13 THE LUCCHESI–YOUNGER THEOREM
In any digraph, the maximum number of arc-disjoint directed bonds is equal to the minimum number of arcs which meet all directed bonds.

Proof When $\nu(D) = 0$, the digraph D has no directed bonds, and equality clearly holds. Suppose that the theorem is false. Then there is a smallest positive integer k, and a digraph D, such that:

$$\nu(D) = k \quad \text{and} \quad \tau(D) > k \qquad (19.8)$$

Denote by $\mathcal{D}$ the set of all digraphs D for which (19.8) holds, and let D_0 be any member of $\mathcal{D}$. We subdivide the arcs of D_0 one by one, thereby obtaining a sequence of digraphs $D_0, D_1, D_2, \ldots, D_m$. By Proposition 19.12(i), $\nu(D_m) = 2\nu(D_0) > \nu(D_0)$. Hence, by Proposition 19.12(ii), there is an index i such that $\nu(D_i) = \nu(D_0) = k$ and $\nu(D_{i+1}) = k + 1$. Because $D_0 \in \mathcal{D}$ and $\nu(D_i) = \nu(D_0)$, we have $D_i \in \mathcal{D}$ (Exercise 19.4.7). Now set $D := D_i$, and let a be the arc of D_i that was subdivided to obtain D_{i+1}.

Because $\nu(D_{i+1}) = k + 1$ and $\nu(D) = k$, in any packing of D_{i+1} by $k + 1$ directed bonds, the two arcs resulting from the subdivision of a both belong to members of the packing. These $k + 1$ directed bonds therefore correspond to a family $\mathcal{B}'$ of $k+1$ directed bonds of D which together cover a twice and each other arc of D at most once.

Let us show, now, that D has a packing $\mathcal{B}''$ of k directed bonds, none of which includes the arc a. By Proposition 19.12(iii), this amounts to showing that $\nu(D \,/\, a) = k$. Suppose, to the contrary, that $\nu(D \,/\, a) = k-1$. Then $\tau(D \,/\, a) = k-1$ also, by the minimality of k. Let T be a minimum transversal of the directed bonds of $D \,/\, a$. Then $T \cup \{a\}$ is a transversal of the directed bonds of D, implying that $\tau(D) = k = \nu(D)$ and contradicting (19.8). Therefore $\nu(D \,/\, a) = k$, and D indeed has such a packing $\mathcal{B}''$.

The family $\mathcal{B} := \mathcal{B}' \cup \mathcal{B}''$ is thus a 2-packing of D by $2k+1 = 2\nu(D)+1$ directed bonds. But this contradicts Proposition 19.11. We conclude that $\mathcal{D} = \emptyset$.   □

Lucchesi (1976) gave a polynomial-time algorithm for finding a maximum packing of directed cuts in a digraph. We refer the reader to Schrijver (2003) for a comprehensive survey of results related to the Lucchesi–Younger Theorem, including its algorithmic aspects.

The Lucchesi–Younger Theorem shows that the hypergraph defined by the arc sets of directed bonds of any digraph satisfies the Min–Max Property. Woodall (1978) conjectured that the blocker of this hypergraph also satisfies the Min–Max Property.

WOODALL'S CONJECTURE

Conjecture 19.14 *In any digraph, the maximum number of arc-disjoint transversals of directed bonds is equal to the minimum number of arcs in a directed bond.*

It can be shown that any digraph which does not contain a directed bond of size one has two disjoint transversals of directed bonds (Exercise 19.4.3). Schrijver (1982) and Feofiloff and Younger (1987) verified Woodall's Conjecture for all digraphs in which each source is connected to each sink by a directed path. Little else is known about this conjecture. For planar digraphs, by duality, Woodall's Conjecture has the following equivalent formulation.

Conjecture 19.15 *In any planar digraph, the maximum number of arc-disjoint transversals of directed cycles is equal to the length of a shortest directed cycle.*

Even in this special case, it is not even known if there is some constant k such that every planar digraph of directed girth k or more has three disjoint transversals of directed cycles.

Exercises

19.4.1 Show that any two directed cycles in the digraph of Figure 19.8 have a vertex in common, but that at least three vertices must be deleted in order to destroy all directed cycles.

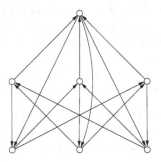

Fig. 19.8. A digraph with $\nu = 1$ and $\tau = 3$

19.4.2 By splitting vertices of the digraph in Figure 19.8, construct a digraph in which any two directed cycles have an arc in common, but from which at least three arcs must be deleted in order to destroy all directed cycles.

19.4.3 Let D be a 2-edge-connected digraph. Show that D has two disjoint transversals of directed bonds.

★**19.4.4** Prove Proposition 19.12.

★**19.4.5**

a) Let D be a digraph, and let $\mathcal{B} = \{\partial^+(X) : X \in \mathcal{F}\}$ be a 2-packing of directed bonds of D for which $\sum\{|X \setminus Y||Y \setminus X| : X, Y \in \mathcal{F}\}$ is minimized. Show that $\mathcal{B}$ is a pairwise noncrossing family of directed bonds.

b) Deduce that, given any 2-packing of directed bonds, one can obtain a pairwise noncrossing 2-packing by repeatedly uncrossing pairs of crossing directed bonds, as described in the proof of Proposition 19.11.

★**19.4.6** Show that the graph G defined in the proof of Proposition 19.11 is bipartite.

★**19.4.7** Consider the family $\mathcal{D}$ of digraphs D satisfying (19.8).

a) Let $D \in \mathcal{D}$, and let D' be any digraph obtained from D by subdividing one arc of D. Show that if $\nu(D') = \nu(D)$, then $D' \in \mathcal{D}$.

b) Deduce that $D_i \in \mathcal{D}$, where D_i is the digraph defined in the proof of Theorem 19.13.

————————⟨⟨————————

19.4.8 Let D be a strong plane digraph. Show that:

a) D has a facial directed cycle,
b) if D has two vertex-disjoint directed cycles, then it has two such cycles which are facial.

19.4.9 Consider vertex-disjoint packings of directed cycles in digraphs.

a) Find a 2-regular orientation of the octahedron for which $\nu = 1$ and $\tau = 2$.
b) Let D be a planar digraph with $\nu = 1$. Show that:
   i) $\tau \leq 2$,
   ii) if $\tau = 2$ and D is simple, with $\delta^- \geq 2$ and $\delta^+ \geq 2$, then the underlying graph of D is either the octahedron or a graph obtained from a wheel W_n, $n \geq 2$, by replacing each of its spokes by two parallel edges.

(A. METZLAR AND U.S.R. MURTY)

19.4.10 Let D be a plane digraph. Two directed cycles C_1 and C_2 of D *cross* if they have a common vertex v and their arcs that are incident with v alternate around v. A family $\mathcal{C}$ of directed cycles in D is *laminar* if no two members of $\mathcal{C}$ cross. Show that, given any family $\mathcal{C}$ of arc-disjoint directed cycles in D, there exists a laminar family $\mathcal{C}'$ of arc-disjoint directed cycles in D with $|\mathcal{C}'| = |\mathcal{C}|$.

19.4.11 Let D be a 2-diregular planar digraph in which a maximum packing of vertex-disjoint directed cycles has size k. Show that D contains a set of at most $4k$ vertices whose deletion destroys all directed cycles.

(A. METZLAR AND U.S.R. MURTY)

19.5 Related Reading

PACKING T-CUTS

The analogue for undirected graphs of the Lucchesi–Younger Theorem (19.13) fails already for a triangle. However, it does hold when restricted to T-cuts in bipartite graphs.

Let G be a graph and let T be an even subset of V. Recall that an edge cut of the form $\partial(X)$, where $|X \cap T|$ is odd, is called a T-*cut* of G. (In the special case where T is the set of vertices of odd degree in G, the T-cuts are the odd cuts of G.) Recall also that a subset F of E is called a T-*join* if the vertices of odd degree in the subgraph $G[F]$ are precisely the vertices in T. It can be seen that the minimal T-joins (with respect to inclusion) are precisely the minimal transversals of T-cuts. Seymour (1981c) showed that for any bipartite graph G and any even subset T of V, the minimum number of edges in a T-join is equal to the maximum number of T-cuts in a packing. He used his theorem to solve an interesting special case of the multicommodity flow problem for planar graphs. (This problem was discussed in Section 7.4.)

One may obtain a bipartite graph from an arbitrary graph by the simple device of subdividing each edge exactly once. This transformation, together with the above theorem of Seymour, yields a result on 2-packings due to Lovász (1975a): for any graph G and any even subset T of V, the minimum number of edges in a T-join is equal to one-half the maximum number of T-cuts in a 2-packing.

20

Electrical Networks

Contents

20.1 Circulations and Tensions

We saw in Section 2.6 that the even subgraphs and edge cuts of a graph form vector spaces over $GF(2)$, namely the cycle and bond spaces of the graph. Here, we consider analogous vector spaces over the reals, and more generally over any field. Throughout this section, D denotes a connected (though not necessarily strongly connected) digraph and T a spanning tree of D.

THE CIRCULATION AND TENSION SPACES

In Section 7.3, we defined a *circulation* in a digraph D as a function $f : A \rightarrow \mathbb{R}$ which satisfies the conservation condition at every vertex:

$$f^+(v) = f^-(v), \quad \text{for all } v \in V \tag{20.1}$$

If one thinks of D as an electrical network, such a function defines a circulation of currents in D. We noted in Section 7.3 that (20.1) may be expressed in matrix notation as:

$$\mathbf{M}\mathbf{f} = \mathbf{0} \tag{20.2}$$

where $\mathbf{M}$ is the $n \times m$ incidence matrix of D and $\mathbf{0}$ the $n \times 1$ zero-vector. The set of all circulations in D is thus a vector space. We denote this space by $\mathcal{C} := \mathcal{C}(D)$, and refer to it as the *circulation space* of D. As a consequence of (20.2), we have:

Proposition 20.1 *The circulation space $\mathcal{C}$ of a digraph D is the orthogonal complement of the row space of its incidence matrix $\mathbf{M}$.*   □

We now turn our attention to the row space of $\mathbf{M}$. Let $\mathbf{g}$ be an element of the row space, so that $\mathbf{g} = \mathbf{p}\mathbf{M}$ for some vector $\mathbf{p} \in \mathbb{R}^V$. Consider an arc $a := (x, y)$. In the column a of $\mathbf{M}$, there are just two nonzero entries: $+1$ in row x and -1 in row y. Thus

$$g(a) = p(x) - p(y) \tag{20.3}$$

If D is thought of as an electrical network, with potential $p(v)$ at vertex v, then by (20.3) g represents the potential difference, or tension, in the wires of the network. For this reason, the row space of $\mathbf{M}$ is called the *tension space* of D, denoted $\mathcal{B}(D)$ and its elements are referred to as *tensions*. Figure 20.1a shows a digraph with an assignment of potentials to its vertices and the corresponding tension.

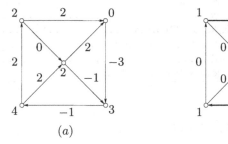

$$(a) \qquad\qquad\qquad (b)$$

Fig. 20.1. (a) A tension in a digraph, (b) the tension associated with a bond

Recall from Section 7.3 that with each cycle C one may associate a circulation $\mathbf{f}_C$. Analogously, with each bond $B := \partial(X)$, one may associate the tension $\mathbf{g}_B$ defined by:

$$g_B(a) := \begin{cases} 1 & \text{if } a \in \partial^+(X) \\ -1 & \text{if } a \in \partial^-(X) \\ 0 & \text{if } a \notin B \end{cases}$$

It can be verified that $\mathbf{g}_B = \mathbf{pM}$, where

$$p(v) := \begin{cases} 1 & \text{if } v \in X \\ 0 & \text{if } v \in V \setminus X \end{cases}$$

so $\mathbf{g}_B$ is indeed a tension. This definition extends naturally to edge cuts. In the case of a trivial edge cut $B := \partial(v)$, the vector $\mathbf{g}_B$ is simply the row $\mathbf{m}(v)$ of the incidence matrix $\mathbf{M}$ of D. Figure 20.1b depicts the tension associated with a bond. In the remainder of this chapter, we find it convenient to identify a set of arcs in a digraph with the subdigraph induced by that set.

Recall that the support of a nonzero circulation contains a cycle (Lemma 7.12). Analogously, we have:

Lemma 20.2 *Let g be a nonzero tension in a digraph D. Then the support of g contains a bond. Moreover, if g is nonnegative, then the support of g contains a directed bond.*

Proof Let $\mathbf{g} := \mathbf{pM}$ be a nonzero tension in a digraph D, with support S, and let $(x, y) \in S$. Set $X := \{v \in V : p(v) = p(x)\}$. Then $(x, y) \in \partial(X)$ and $\partial(X) \subseteq S$. Thus S contains the edge cut $\partial(X)$ which, being nonempty, contains a bond. If g is nonnegative, this bond is a directed bond. □

The following two propositions are the direct analogues for tensions of Propositions 7.13 and 7.14. We leave their proofs as an exercise (20.1.3).

Proposition 20.3 *Every tension in a digraph is a linear combination of the tensions associated with its bonds.* □

Proposition 20.4 *Every nonnegative tension in a digraph is a nonnegative linear combination of the tensions associated with its directed bonds. Moreover, if the tension is integer-valued, the coefficients of the linear combination may be chosen to be nonnegative integers.* □

CIRCULATIONS AND TENSIONS IN PLANE DIGRAPHS

In Section 10.2, we studied the relationship between cycles and bonds in plane graphs and digraphs. Theorem 10.18 may be extended to the spaces of circulations and tensions as follows. For a function f on the arc set $A(D)$ of a plane digraph D, let f^* denote the function on $A(D^*)$ defined by $f^*(a^*) := f(a)$, for all $a \in A(D)$. Applying Theorem 10.18, one may deduce the following theorem.

Theorem 20.5 *Let D be a plane digraph. A function f on $A(D)$ is a circulation in D if and only if the function f^* is a tension in D^*. Thus the circulation space of D is isomorphic to the tension space of D^*.* □

We leave the proof of Theorem 20.5 to the reader (Exercise 20.1.6). The relationship discussed here between circulations and tensions in plane digraphs is the foundation for the theory of integer-valued flows in graphs. We explore this topic in Chapter 21.

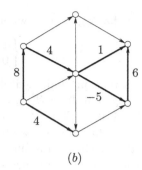

Fig. 20.2. Circulations and tensions are determined by their values on trees and cotrees (Exercise 20.1.4)

Exercises

$\star$**20.1.1** Let $D = (V, A)$ be a digraph and g a real-valued function on A. Show that g is a tension in D if and only if $g^+(C) = g^-(C)$ for each cycle C of D and each sense of traversal of C.

$\star$**20.1.2**

a) Let f be a circulation in a digraph D. Show that the function obtained by negating the value of f on an arc a of D is a circulation in the digraph derived from D by reversing the orientation of a.
b) State and prove an analogous result for tensions.

$\star$**20.1.3** Prove Propositions 20.3 and 20.4.

20.1.4

a) Figure 20.2a displays a function defined on a spanning tree of a digraph, and Figure 20.2b a function defined on its cotree. Extend the function shown in Figure 20.2a to a tension and the function shown in Figure 20.2b to a circulation.
b) Let f be a circulation and g a tension in a digraph D, and let T be a spanning tree of D. Show that f is uniquely determined by $f|\overline{T}$, and that g is uniquely determined by $g|T$.

$\star$**20.1.5** Let D be a digraph, and let $\mathbf{b}$ be a real-valued function defined on A. Show that:

a) if $\mathbf{f}$ is a nonnegative circulation such that $\mathbf{bf} > 0$, then there is a directed cycle C in D such that $\mathbf{bf}_C > 0$,
b) if $\mathbf{g}$ is a nonnegative tension such that $\mathbf{gb} > 0$, then there is a directed bond B in D such that $\mathbf{g}_B\mathbf{b} > 0$.

$\star$**20.1.6** Deduce Theorem 20.5 from Theorem 10.18.

———————— ℓℓ ————————

20.2 Basis Matrices

A matrix $\mathbf{B}$ is called a *basis matrix* of the tension space $\mathcal{B}$ of a digraph D if the rows of $\mathbf{B}$ form a basis for $\mathcal{B}$; basis matrices of the circulation space $\mathcal{C}$ are defined analogously. In this section, all basis matrices are understood to be with respect to a fixed digraph D. We find the following notation convenient. If $\mathbf{R}$ is a matrix whose columns are labelled with the elements of A, and S is a subset of A, we denote by $\mathbf{R}|_S$ the submatrix of $\mathbf{R}$ consisting of the columns of $\mathbf{R}$ that are labelled by the elements of S. If $\mathbf{R}$ has a single row, our notation is the same as the usual notation for the restriction of a function to a subset of its domain.

Theorem 20.6 *Let $\mathbf{B}$ and $\mathbf{C}$ be basis matrices of $\mathcal{B}$ and $\mathcal{C}$, respectively, and let $S \subseteq A$. Then:*

i) the columns of $\mathbf{B}|_S$ are linearly independent if and only if S contains no cycle,
ii) the columns of $\mathbf{C}|_S$ are linearly independent if and only if S contains no bond.

Proof Denote the column of $\mathbf{B}$ corresponding to arc a by $\mathbf{b}(a)$. The columns of $\mathbf{B}|_S$ are linearly dependent if and only if there exists a function f on A such that $\sum_{a \in A} f(a)\mathbf{b}(a) = \mathbf{0}$, with $f(a) \neq 0$ for some $a \in S$ and $f(a) = 0$ for all $a \notin S$, that is, if and only if there exists a nonzero circulation $\mathbf{f}$ whose support is contained in S. Now if there is such an $\mathbf{f}$, then S contains a cycle, by Lemma 7.12. Conversely, if S contains a cycle C, then $\mathbf{f}_C$ is a nonzero circulation whose support C is contained in S. It follows that the columns of $\mathbf{B}|_S$ are linearly independent if and only if S is acyclic. A similar argument, using Lemma 20.2, yields a proof of (ii). $\qquad \square$

Theorem 20.7 *The dimensions of the tension and circulation spaces of a connected digraph D are given by the formulae:*

$$dim\ \mathcal{B} = n - 1 \qquad\qquad (20.4)$$
$$dim\ \mathcal{C} = m - n + 1 \qquad\qquad (20.5)$$

Proof Consider a basis matrix $\mathbf{B}$ of $\mathcal{B}$. By Theorem 20.6,

$$\text{rank } \mathbf{B} = \max\{|S| : S \subseteq A, S \text{ acyclic}\}$$

The above maximum is attained when S is a spanning tree of D, and is therefore equal to $n - 1$. Because $dim\ \mathcal{B} = \text{rank } \mathbf{B}$, this establishes (20.4). Now (20.5) follows directly, because $\mathcal{C}$ is the orthogonal complement of $\mathcal{B}$. $\qquad \square$

Note that the formulae (20.4) and (20.5) for the dimensions of the tension and circulation spaces of a digraph D depend only on the underlying graph G of D. This is a common feature of most properties of circulations and tensions that are of interest to us, the reason being that, for any circulation $\mathbf{f}$ or tension $\mathbf{g}$, the function obtained by negating the value of $\mathbf{f}$ or $\mathbf{g}$ on an arc of D is a circulation or tension in the digraph obtained from D by reversing the orientation of that arc (Exercise 20.1.2).

Let T be a spanning tree of a digraph D. We may associate with T a special basis matrix of $\mathcal{B}$. Consider an arc a of the tree T, and the corresponding fundamental bond B_a. We have seen that to each bond there is an associated tension. We denote this tension by $\mathbf{g}_a$, defined so that $g_a(a) = 1$. The $(n-1) \times m$ matrix $\mathbf{B}$ whose rows are the vectors $\mathbf{g}_a$, $a \in T$, is then a basis matrix of $\mathcal{B}$. This follows from the fact that each row is a tension and that rank $\mathbf{B} = n - 1$ (because $\mathbf{B}|_T$ is an identity matrix). We refer to $\mathbf{B}$ as the *basis matrix* of $\mathcal{B}$ *corresponding to* T. Figure 20.3b shows the basis matrix of $\mathcal{B}$ corresponding to the spanning tree $\{1, 2, 4, 5\}$ indicated in Figure 20.3a.

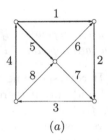

(a)

	1	2	3	4	5	6	7	8
g_1	1	0	-1	0	0	1	1	0
g_2	0	1	-1	0	0	0	1	0
g_4	0	0	-1	1	0	0	0	1
g_5	0	0	0	0	1	-1	-1	1

(b)

	1	2	3	4	5	6	7	8
f_3	1	1	1	1	0	0	0	0
f_6	-1	0	0	0	1	1	0	0
f_7	-1	-1	0	0	1	0	1	0
f_8	0	0	0	-1	-1	0	0	1

(c)

Fig. 20.3. Basis matrices of $\mathcal{B}$ and $\mathcal{C}$ corresponding to a spanning tree

Analogously, if a is an arc of $\overline{T}$, the fundamental cycle C_a corresponding to a has an associated circulation $\mathbf{f}_a$, defined so that $f_a(a) = 1$. The $(m - n + 1) \times m$ matrix $\mathbf{C}$ whose rows are the vectors $\mathbf{f}_a$, $a \in \overline{T}$, is a basis matrix of $\mathcal{C}$, the *basis matrix* of $\mathcal{C}$ *corresponding to* T. Figure 20.3c gives an example of such a matrix.

In light of the above observations, we now have the following fundamental theorem.

Theorem 20.8 *Let D be a connected digraph and T a spanning tree of D. Then:*

i) *the set $\{\mathbf{g}_a : a \in T\}$ of tensions associated with the fundamental bonds of D with respect to T is a basis for $\mathcal{B}$,*

ii) *the set $\{\mathbf{f}_a : a \in \overline{T}\}$ of circulations associated with the fundamental cycles of D with respect to T is a basis for $\mathcal{C}$.*   □

The defining conditions for circulations and tensions involve only addition and subtraction, so these notions may be defined over any (additive) abelian group Γ.

We denote by $\mathcal{B}_\Gamma$ the set of all tensions, and by $\mathcal{C}_\Gamma$ the set of all circulations, over Γ. When Γ is the additive group of a field F, these sets are vector spaces over F, and Theorems 20.6 and 20.7 remain valid (Exercise 20.2.3a).

In the case of fields of characteristic two, and more generally groups in which each element is its own additive inverse, circulations and tensions in a digraph D depend only on the underlying graph G; orientations of arcs play no role. For example, over $GF(2)$, a function f on A satisfies the conservation condition (20.1) if and only if $\sum\{f(a) : a \in \partial(v)\} = 0$, for all $v \in V$. This simply means that f is a circulation over $GF(2)$ if and only if its support is an even subgraph of G. Similarly, a function g on A over $GF(2)$ is a tension if and only if its support is an edge cut of G. Thus, when F is the field $GF(2)$, the space $\mathcal{C}_F$ is simply the cycle space of G and $\mathcal{B}_F$ its bond space, as defined in Section 2.6.

Exercises

20.2.1

a) Let $\mathbf{B}$ and $\mathbf{C}$ be basis matrices of $\mathcal{B}$ and $\mathcal{C}$ and let T be a spanning tree of D. Show that $\mathbf{B}$ is uniquely determined by $\mathbf{B}|_T$ and that $\mathbf{C}$ is uniquely determined by $\mathbf{C}|_{\overline{T}}$.

b) Let T and T_1 be two spanning trees of a connected digraph D. Denote by $\mathbf{B}$ and $\mathbf{B}_1$ the basis matrices of $\mathcal{B}$, and by $\mathbf{C}$ and $\mathbf{C}_1$ the basis matrices of $\mathcal{C}$, corresponding to the trees T and T_1, respectively. Show that $\mathbf{B} = (\mathbf{B}|_{T_1})\mathbf{B}_1$ and that $\mathbf{C} = (\mathbf{C}|_{\overline{T}_1})\mathbf{C}_1$.

20.2.2 KIRCHHOFF MATRIX
A *Kirchhoff matrix* of a loopless connected digraph D is a matrix $\mathbf{K} := \mathbf{M}_x$ obtained from the incidence matrix $\mathbf{M}$ of D by deleting the row $\mathbf{m}(x)$. Show that $\mathbf{K}$ is a basis matrix of $\mathcal{B}$.

20.2.3 Let F be a field and let $\mathcal{C}_F$ and $\mathcal{B}_F$ denote the circulation and tension spaces of D over F. Show that Theorems 20.6 and 20.7 remain valid if $\mathcal{B}$ and $\mathcal{C}$ are replaced by $\mathcal{B}_F$ and $\mathcal{C}_F$, respectively.

20.2.4 Show that a function $f : A \to \Gamma$ is a circulation in a digraph D over an additive abelian group Γ if and only if

$$f^+(X) - f^-(X) = 0$$

for every subset X of V.

20.2.5 Show that a function $g : A \to \Gamma$ is a tension in a digraph D over an additive abelian group Γ if and only if

$$g^+(C) - g^-(C) = 0$$

for every cycle C of D with a given sense of traversal.

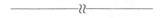

20.3 Feasible Circulations and Tensions

Frequently, in both theoretical and practical situations, one seeks circulations or tensions whose values on arcs satisfy prescribed bounds. In this section, we describe necessary and sufficient conditions for the existence of such circulations and tensions.

Let $D := (V, A)$ be a directed graph. Suppose that, with each arc a of D, are associated two real numbers, $b(a)$ and $c(a)$, such that $b(a) \leq c(a)$. A circulation f in D is *feasible* (with respect to the functions b and c) if $b(a) \leq f(a) \leq c(a)$ for all $a \in A$. The functions b and c are called *lower* and *upper bounds*, respectively. A *feasible tension* is defined similarly.

Let f be a feasible circulation in D, and let X be a subset of V. Because f is a circulation, we have $f^+(X) = f^-(X)$ (Exercise 7.3.1). On the other hand, because f is feasible, $c^+(X) \geq f^+(X)$ and $f^-(X) \geq b^-(X)$. It follows that

$$c^+(X) \geq b^-(X), \quad \text{for all subsets } X \text{ of } V \tag{20.6}$$

Hoffman (1960) showed that this necessary condition for the existence of a feasible circulation is sufficient.

Theorem 20.9   HOFFMAN'S CIRCULATION THEOREM
A digraph D has a feasible circulation with respect to bounds b and c if and only if these bounds satisfy inequality (20.6). Furthermore, if both b and c are integer-valued and satisfy this inequality, then D has an integer-valued feasible circulation.

Now let g be a feasible tension in D with respect to b and c. Consider any cycle C of D, and a sense of traversal of C. Because g is a tension, $g(C^+) = g(C^-)$ (Exercise 20.1.1). Moreover, because g is feasible, $c(C^+) \geq g(C^+)$ and $g(C^-) \geq b(C^-)$. Therefore,

$$c(C^+) \geq b(C^-), \quad \text{for each cycle } C \text{ of } D \text{ and each sense of traversal of } C \tag{20.7}$$

Ghouila-Houri (1960) showed that this necessary condition for the existence of a feasible tension is sufficient.

Theorem 20.10   GHOUILA-HOURI'S THEOREM
A digraph D has a feasible tension with respect to bounds b and c if and only if these bounds satisfy inequality (20.7). Furthermore, if both b and c are integer-valued and satisfy this inequality, then D has an integer-valued feasible tension.

Hoffman's Circulation Theorem and Ghouila-Houri's Theorem may both be proved with the aid of a fundamental tool in linear algebra known as Farkas' Lemma (see inset).

Proof Technique: Farkas' Lemma

A system $\mathbf{Ax} = \mathbf{0}$ of linear equations always has at least one solution. However, in many practical applications, one would like to find a solution which satisfies certain additional constraints, such as one of the form $\mathbf{x} \geq \mathbf{b}$, where $\mathbf{b}$ is a given vector of lower bounds. This may or may not be possible. Let us suppose that there is, indeed, a vector $\mathbf{x}$ such that

$$\mathbf{Ax} = \mathbf{0}, \quad \mathbf{x} \geq \mathbf{b} \qquad (20.8)$$

Consider any vector $\mathbf{y}$ such that $\mathbf{yA} \geq \mathbf{0}$. Because $\mathbf{b} \leq \mathbf{x}$ and $\mathbf{yA} \geq \mathbf{0}$, we have $\mathbf{yAb} \leq \mathbf{yAx} = \mathbf{y0} = 0$. Thus, if (20.8) has a solution, the linear system

$$\mathbf{yA} \geq \mathbf{0}, \quad \mathbf{yAb} > 0 \qquad (20.9)$$

can have no solution. Farkas (1902) showed that this necessary condition for the feasibility of the system (20.8) is also sufficient.

Lemma 20.11   Farkas' Lemma
For any real matrix $\mathbf{A}$ and any real vector $\mathbf{b}$, precisely one of the two linear systems (20.8) and (20.9) has a solution. □

Farkas' Lemma can be deduced from the Duality Theorem of Linear Programming, discussed in Chapter 8 (see Exercise 8.6.9). A number of feasibility theorems in graph theory may be derived from it very easily. In most of these applications, the matrix $\mathbf{A}$ is the incidence matrix $\mathbf{M}$ of the digraph D under consideration, and thus is totally unimodular (Exercise 1.5.7). Consequently, if $\mathbf{b}$ is integer-valued, there is either an integer-valued solution $\mathbf{x}$ to (20.8) or else an integer-valued solution $\mathbf{y}$ to (20.9).
As an illustration, we describe how Hoffman's Circulation Theorem (20.9) may be derived from Farkas' Lemma. We first obtain a condition for the existence of a circulation subject only to lower bounds.

Proposition 20.12 *Let D be a digraph, and let $\mathbf{b}$ be a real-valued function defined on A. Then either there is a circulation $\mathbf{f}$ in D such that $\mathbf{f} \geq \mathbf{b}$, or there is a nonnegative tension $\mathbf{g}$ in D such that $\mathbf{gb} > 0$.*

Proof Consider the incidence matrix $\mathbf{M}$ of D, and the two linear systems:

$$\mathbf{Mf} = \mathbf{0}, \qquad \mathbf{f} \geq \mathbf{b} \qquad (20.10)$$

$$\mathbf{pM} \geq \mathbf{0}, \qquad \mathbf{pMb} > 0 \qquad (20.11)$$

By Farkas' Lemma, exactly one of these two systems has a solution. The proposition follows on observing that a solution $\mathbf{f}$ to the system $\mathbf{Mf} = \mathbf{0}$ is a circulation in D, and a vector of the form $\mathbf{pM}$ is a tension $\mathbf{g}$ in D. □

FARKAS' LEMMA (CONTINUED)

Applying Exercise 20.1.5b, we now have the following corollary.

Corollary 20.13 *Let D be a digraph, and let $\mathbf{b}$ be a real-valued function defined on A. Then either there is a circulation $\mathbf{f}$ in D such that $\mathbf{f} \geq \mathbf{b}$, or there is a directed bond B in D such that $\mathbf{g}_B \mathbf{b} > 0$.* □

Hoffman's Circulation Theorem (with both lower and upper bounds) can be deduced from Corollary 20.13 by means of a simple transformation (see Exercise 20.3.5). Ghouila-Houri's Theorem (20.10) can be derived in an analogous fashion from a variant of Farkas' Lemma asserting that exactly one of the two linear systems $\mathbf{yA} \geq \mathbf{b}$, and $\mathbf{Ax} = \mathbf{0}$, $\mathbf{x} \geq \mathbf{0}$, $\mathbf{bx} > 0$, has a solution (Exercises 8.6.10 and 20.3.6). A constructive proof of Hoffman's Circulation Theorem is given below.

FINDING A FEASIBLE CIRCULATION

An algebraic proof of Hoffman's Circulation Theorem (20.9) is given in the inset. Here, we describe a constructive proof of this theorem based on network flows.

Proof We show how to find a feasible circulation in any network satisfying the necessary condition (20.6). For clarity of presentation, we assume that both b and c are integer-valued functions on A. Our proof may be adapted to the general case in a straightforward manner (Exercise 20.3.1).

Let f be a real-valued function on A satisfying the lower and upper bounds $b(a) \leq f(a) \leq c(a)$ for all $a \in A$. We define a vertex v to be *positive, balanced,* or *negative* according as the net flow $f^+(v) - f^-(v)$ out of v is positive, zero, or negative. Because $\sum_{v \in V}(f^+(v) - f^-(v)) = 0$ (Exercise 7.1.1a), either all vertices are balanced, in which case f is a feasible circulation, or there are both positive and negative vertices.

Let us call the quantity $\sum_{v \in V}|f^+(v) - f^-(v)|$ the *excess* of f. If not all vertices are balanced, the excess is positive. In this case, we show how f can be modified to a flow f' which also satisfies the lower and upper bounds, but which has a smaller excess. Repeating this procedure results in a feasible circulation.

Consider a negative vertex x. By analogy with the notion of an f-augmenting path in the Max-Flow Min-Cut Algorithm (7.9), call an x-path P in D an *f-improving path* if $f(a) < c(a)$ for each forward arc a of P, and $f(a) > b(a)$ for each reverse arc a of P. Let X be the set of all vertices reachable from x by f-improving paths. Then $f(a) = c(a)$ for each arc $a \in \partial^+(X)$ and $f(a) = b(a)$ for each arc $a \in \partial^-(X)$. Therefore, applying Exercise 7.1.2,

$$\sum_{v \in X}(f^+(v) - f^-(v)) = f^+(X) - f^-(X) = c^+(X) - b^-(X) \geq 0$$

Because X includes a negative vertex, namely x, we deduce that X also includes a positive vertex y, so there exists an f-improving path xPy. Let f' be the function on A defined by:

$$f'(a) = \begin{cases} f(a) + 1 & \text{if } a \text{ is a forward arc of } P \\ f(a) - 1 & \text{if } a \text{ is a reverse arc of } P \\ f(a) & \text{otherwise} \end{cases}$$

It is straightforward to verify that $b(a) \le f'(a) \le c(a)$ for all $a \in A$, and that the excess of f' is two less than the excess of f. $\square$

We remark that the above constructive proof can be adapted easily to an algorithm which accepts as input a network N with lower and upper bounds b and c, and returns either a feasible circulation in N or a set X violating condition (20.6).

Exercises

⋆**20.3.1** Give a constructive proof of Theorem 20.9, without the assumption that b and c are integer-valued.

20.3.2 Give an algorithm, based on the constructive proof of Theorem 20.9, for finding either a feasible circulation in a digraph D or a subset X of V which violates (20.6).

20.3.3 Given a digraph $D := D(x, y)$ and two real-valued functions b and c on the arc set A of D, a function f on A is a *feasible flow* in D (relative to b and c) if (i) $b(a) \le f(a) \le c(a)$, for all $a \in A$, and (ii) $f^+(v) - f^-(v) = 0$, for all $v \in V \setminus \{x, y\}$. (This is just a slight generalization of our earlier definition of feasibility; when $b = 0$, condition (i) reduces to (7.2).) Show how the Max-Flow Min-Cut Algorithm (7.9) can be modified to find a feasible flow of maximum value starting from an initial feasible flow.

20.3.4 Deduce the Max-Flow Min-Cut Theorem (7.7) from Hoffman's Circulation Theorem.

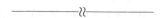

20.3.5

a) Let D be a digraph with lower and upper bound functions b and c defined on its arc set A. Define a digraph D' with vertex set $V(D) \cup \{v(a) : a \in A\}$ and arc set

$$A' := \{(t(a), v(a)) : a \in A\} \cup \{(h(a), v(a)) : a \in A\}$$

where $t(a)$ and $h(a)$ are the tail and head of a, respectively. (Equivalently, subdivide each arc a of D by a single vertex $v(a)$, thereby transforming it into a directed path of length two, and then reverse the second arc of each such path.) Define a lower bound function b' on the arc set A' of D' by:

$$b'(t(a), v(a)) := b(a), \quad b'(h(a), v(a)) := -c(a), \quad a \in A$$

Show that:
  i) if f is a circulation in D, then the function f' defined on A' by:

$$f'(t(a), v(a)) := f(a), \quad f'(h(a), v(a)) := -f(a), \quad a \in A$$

  is a circulation in D',
  ii) $b \le f \le c$ if and only if $f' \ge b'$.
b) Deduce Hoffman's Circulation Theorem (20.9) from Corollary 20.13.

20.3.6 A variant of Farkas' Lemma says that exactly one of the two linear systems $\mathbf{yA} \ge \mathbf{b}$, and $\mathbf{Ax} = \mathbf{0}$, $\mathbf{x} \ge \mathbf{0}$, $\mathbf{bx} > 0$, has a solution (see Exercise 8.6.10). Let D be a digraph, and let $\mathbf{b} \in \mathbb{R}^A$.

a) Using this variant of Farkas' Lemma, show that either there is a tension $\mathbf{g}$ in D such that $\mathbf{g} \ge \mathbf{b}$, or there is a nonnegative circulation $\mathbf{f}$ in D such that $\mathbf{bf} > 0$.
b) Deduce that either there is a tension $\mathbf{g}$ in D such that $\mathbf{g} \ge \mathbf{b}$, or there is a directed cycle C in D such that $\mathbf{bf}_C > 0$.
c) Deduce Ghouila-Houri's Theorem from (b).

20.3.7 Give a constructive proof of the statement in Exercise 20.3.6b based on the Bellman–Ford Algorithm (described in Exercise 6.3.11).

20.3.8 MIN-COST CIRCULATION
Let $D = (V, A)$ be a weighted directed graph, with weight (or *cost*) function w. The *cost* of a circulation f is the quantity $\sum_{a \in A} w(a)f(a)$. Suppose that with each arc a of D are associated lower and upper bounds $b(a)$ and $c(a)$, where $b(a) \le c(a)$. The *Min-Cost Circulation Problem* consists of finding a feasible circulation of minimum cost.

a) Define a cycle C to be *cost-reducing* with respect to a feasible circulation f if:
  ▷ $f(a) < c(a)$, for each $a \in C^+$,
  ▷ $f(a) > b(a)$, for each $a \in C^-$,
  ▷ $\sum_{a \in C^+} w(a) - \sum_{a \in C^-} w(a)$ is negative.
  i) Let f^* be a min-cost circulation. Show that there can exist no cost-reducing cycle with respect to f^*.
  ii) Let f be a feasible circulation that is not one of minimum cost. By considering the circulation $f^* - f$, and using the transformation described in Exercise 20.1.2 and Proposition 7.14, show that there exists a cost-reducing cycle with respect to f.
b) Starting with any feasible circulation, describe how the Bellman–Ford Algorithm (Exercise 6.3.11) may be applied to find a min-cost circulation, assuming that the functions b, c and w are rational-valued.

(There exist polynomial-time algorithms for solving the Min-Cost Circulation Problem; see Schrijver (2003).)

20.4 The Matrix–Tree Theorem

In Section 4.2, we gave a proof of Cayley's Formula for the number of spanning trees in a complete graph. Here, we derive several expressions for the number of spanning trees in any connected graph G. Recall that this parameter is denoted $t(G)$.

A matrix is said to be *unimodular* if the determinants of all its full square submatrices have values 0, $+1$ or -1; in particular, any totally unimodular matrix is unimodular. Kirchhoff matrices are examples of unimodular matrices (Exercise 20.4.1a). Other examples are provided by basis matrices corresponding to spanning trees (Exercise 20.4.3).

The proof of the following theorem is due to Tutte (1965a).

Theorem 20.14 *Let D be a connected digraph and $\mathbf{B}$ a unimodular basis matrix of its tension space $\mathcal{B}$. Then*

$$t(G) = \det \mathbf{B}\mathbf{B}^t$$

Proof Using the Cauchy–Binet Formula[1] for the determinant of the product of two rectangular matrices, we obtain

$$\det \mathbf{B}\mathbf{B}^t = \sum \{ (\det(\mathbf{B}|_S))^2 \; : \; S \subseteq A, \; |S| = n - 1 \}$$

By Theorem 20.6(i), the number of nonzero terms in this sum is equal to $t(G)$. Moreover, because $\mathbf{B}$ is unimodular, each such term has value 1. $\square$

This observation, together with the fact, noted above, that Kirchhoff matrices are unimodular, yields the following formula for the number of spanning trees in a graph, implicit in the work of Kirchhoff (1847).

Theorem 20.15 THE MATRIX–TREE THEOREM
Let G be a loopless connected graph, D an orientation of G, and $\mathbf{K}$ a Kirchhoff matrix of D. Then

$$t(G) = \det \mathbf{K}\mathbf{K}^t \qquad\qquad \square$$

The *conductance matrix* or *Laplacian* of a loopless graph G with adjacency matrix $\mathbf{A} = (a_{ij})$ is the $n \times n$ matrix $\mathbf{C} = (c_{ij})$, where

$$c_{ij} := \begin{cases} \displaystyle\sum_k a_{ki}, & \text{if } i = j \\ -a_{ij}, & \text{if } i \neq j \end{cases}$$

Figure 20.4 shows a graph together with its conductance matrix.

We leave the proof of the following corollary as Exercise 20.4.2.

---

[1] Let $\mathbf{A}$ and $\mathbf{B}$ be two $k \times m$ matrices ($k \leq m$), whose columns are indexed by the elements of a set E. The *Cauchy–Binet Formula* states that

$$\det \mathbf{A}\mathbf{B}^t = \sum_{S \subseteq E, \; |S| = k} \det(\mathbf{A}|_S) \det(\mathbf{B}|_S)$$

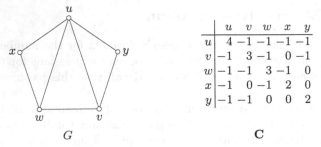

	u	v	w	x	y
u	4	−1	−1	−1	−1
v	−1	3	−1	0	−1
w	−1	−1	3	−1	0
x	−1	0	−1	2	0
y	−1	−1	0	0	2

$$G \qquad\qquad\qquad\qquad C$$

Fig. 20.4. The conductance matrix of a graph

Corollary 20.16 *Let G be a loopless connected graph, $\mathbf{C}$ its conductance matrix, and D an orientation of G. Then:*

i) $\mathbf{C} = \mathbf{MM}^t$, *where $\mathbf{M}$ is the incidence matrix of D,*
ii) every cofactor of $\mathbf{C}$ has value $t(G)$. $\square$

The proof of the following theorem is analogous to that of Theorem 20.14.

Theorem 20.17 *Let D be a connected digraph and $\mathbf{C}$ a unimodular basis matrix of its circulation space $\mathcal{C}$. Then*

$$t(G) = \det \mathbf{CC}^t$$

Corollary 20.18 *Let D be a connected digraph and $\mathbf{B}$ and $\mathbf{C}$ unimodular basis matrices of $\mathcal{B}$ and $\mathcal{C}$, respectively. Then*

$$t(G) = \pm \det \begin{bmatrix} \mathbf{B} \\ \mathbf{C} \end{bmatrix}$$

Proof By Theorems 20.14 and 20.17,

$$(t(G))^2 = \det \mathbf{BB}^t \det \mathbf{CC}^t = \det \begin{bmatrix} \mathbf{BB}^t & \mathbf{0} \\ \mathbf{0} & \mathbf{CC}^t \end{bmatrix}$$

Because $\mathcal{B}$ and $\mathcal{C}$ are orthogonal, $\mathbf{BC}^t = \mathbf{CB}^t = \mathbf{0}$. Thus

$$(t(G))^2 = \det \begin{bmatrix} \mathbf{BB}^t & \mathbf{BC}^t \\ \mathbf{CB}^t & \mathbf{CC}^t \end{bmatrix} = \det \left(\begin{bmatrix} \mathbf{B} \\ \mathbf{C} \end{bmatrix} [\mathbf{B}^t | \mathbf{C}^t] \right)$$

$$= \det \begin{bmatrix} \mathbf{B} \\ \mathbf{C} \end{bmatrix} \det [\mathbf{B}^t | \mathbf{C}^t] = \left(\det \begin{bmatrix} \mathbf{B} \\ \mathbf{C} \end{bmatrix} \right)^2$$

The corollary follows on taking square roots. $\square$

Exercises

★**20.4.1** Let $\mathbf{K}$ be a Kirchhoff matrix of a loopless connected digraph D, and let G be the underlying graph of D. Show that:

  a) $\mathbf{K}$ is unimodular,

  b) $t(G) = \pm \det \begin{bmatrix} \mathbf{K} \\ \mathbf{C} \end{bmatrix}$, where $\mathbf{C}$ is a basis matrix of $\mathcal{C}$ associated with a spanning tree of G.

★**20.4.2** Prove Corollary 20.16.

★**20.4.3** Let D be a connected digraph and $\mathbf{B}$ and $\mathbf{C}$ basis matrices of $\mathcal{B}$ and $\mathcal{C}$, respectively, corresponding to a spanning tree T. Show that $\mathbf{B}$ and $\mathbf{C}$ are totally unimodular.

★**20.4.4** Use the Matrix–Tree Theorem (20.15) to compute the number of spanning trees in the graph shown in Figure 20.5.

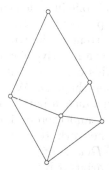

Fig. 20.5. How many spanning trees are there in this graph? (Exercise 20.4.4)

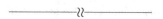

20.4.5 Let F be a finite field of characteristic p, let $\mathbf{B}$ and $\mathbf{C}$ be basis matrices of $\mathcal{B}_F$ and $\mathcal{C}_F$, respectively, corresponding to a spanning tree of a digraph D, and let G be the underlying graph of D. Show that:

  a)
$$\det \begin{bmatrix} \mathbf{B} \\ \mathbf{C} \end{bmatrix} \equiv \pm t(G) (\mathrm{mod}\ p)$$

  b) $\dim(\mathcal{B}_F \cap \mathcal{C}_F) > 0$ if and only if $p | t(G)$.   (H. Shank)

20.4.6 The *Laplacian* of a loopless digraph D with vertex set $\{v_1, v_2, \ldots, v_n\}$ and adjacency matrix $\mathbf{A} = (a_{ij})$ is the $n \times n$ matrix $\mathbf{C} = (c_{ij})$, where:

$$c_{ij} := \begin{cases} \sum_k a_{ki}, & \text{if } i = j \\ -a_{ij}, & \text{if } i \neq j \end{cases}$$

Let C_{ii} denote the matrix obtained from $\mathbf{C}$ by deleting its ith row and ith column. Show that the determinant of C_{ii} is equal to the number of spanning branchings of D rooted at v_i.   (W.T. TUTTE)

20.5 Resistive Electrical Networks

We have already remarked that a circulation in a digraph may be viewed as a flow of currents in an electrical network, and that tensions represent voltage drops along wires. In this section, we make precise these relationships, and show how to compute the currents in a network using the matrix equations developed in the previous section.

A *resistive electrical network* is one in which each wire has a specific resistance. By *Ohm's Law*, the voltage drop v between the ends of the wire is given by the equation v = ir, where i is the current in the wire and r its resistance.

A graph G can be regarded as a resistive electrical network in which each edge is a wire of unit resistance. In this case, the voltage drop between the ends of an edge is equal to the current along the edge. We take one edge of G, say $e := xy$, to be a *current generator* (for instance, a battery). This current generator creates a voltage drop between x and y, and thereby induces in $G \setminus e$ a flow of current from x, the *negative pole* of the network, to y, its *positive pole*. This flow, in turn, determines an orientation $D := D(x, y)$ of $G \setminus e$ (except for edges which carry no current, which may be oriented arbitrarily).

KIRCHHOFF'S LAWS

Kirchhoff (1847) formulated two basic laws of resistive electrical networks. When each wire is of unit resistance, they can be stated in terms of the above notation as follows.

▷ *Kirchhoff's Current Law*: the currents in $G \setminus e$ constitute an (x, y)-flow in D.
▷ *Kirchhoff's Voltage Law*: the currents in $G \setminus e$ constitute a tension in D.

A function on the arc set A of a digraph $D := D(x, y)$ which is both an (x, y)-flow and a tension is called a *current flow* in D from x to y. (Thus what distinguishes a current flow from an arbitrary (x, y)-flow is that it is also a tension.) The *value* of a current flow is its value as an (x, y)-flow.

Theorem 20.19 *Let* $D := D(x, y)$ *be a connected directed graph. For any real number* i, *there exists a unique current flow in* D *of value* i *from* x *to* y.

Proof Let $\mathbf{K} = \mathbf{M}_y$ be a Kirchhoff matrix of D. We assume that the first row of $\mathbf{K}$ is indexed by x. By definition, a function $f : A \to \mathbb{R}$ is a current flow of value i from x to y if it satisfies the two systems of equations:

$$\mathbf{K}f = \begin{bmatrix} i \\ 0 \end{bmatrix} \quad \text{and} \quad \mathbf{C}f = 0$$

where $\mathbf{C}$ is any basis matrix of $\mathcal{C}$. Because $\mathbf{K}$ has $n - 1$ rows, the first system consists of $n - 1$ equations. Similarly, because $\mathbf{C}$ has $m - n + 1$ rows, the second system consists of $m - n + 1$ equations. Combining these two systems, we obtain the following system of m equations in m variables.

$$\begin{bmatrix} \mathbf{K} \\ \mathbf{C} \end{bmatrix} f = \begin{bmatrix} i \\ 0 \end{bmatrix} \tag{20.12}$$

Because the rows of $\mathbf{K}$ are a basis for $\mathcal{B}$ and the rows of $\mathbf{C}$ are a basis for its orthogonal complement $\mathcal{C}$, the matrix $\begin{bmatrix} \mathbf{K} \\ \mathbf{C} \end{bmatrix}$ is nonsingular. Therefore the system (20.12) has a unique solution, yielding a unique current flow f of value i from x to y. $\qquad\square$

For an arbitrary positive integer i, the values of the currents obtained by solving (20.12) might well not be integral. However, by Exercise 20.4.1b,

$$\det \begin{bmatrix} \mathbf{K} \\ \mathbf{C} \end{bmatrix} = \pm t(D)$$

so, by Cramér's Rule, we can guarantee a solution in integers by taking $i = t(D)$. Thus, in computing currents, it is convenient to take the total current leaving x to be equal to the number of spanning trees of D.

Example 20.20 Consider the planar graph G in Figure 20.6a. On deleting the edge xy and orienting each remaining edge, as shown, we obtain the digraph D of Figure 20.6b. It can be checked that $t(D) = 66$ (Exercise 20.4.4). By considering the tree $T := \{a_1, a_2, a_3, a_4, a_5\}$ we obtain the following nine equations, as in (20.12), (with $f(a_i)$ written simply as f_i).

$$
\begin{aligned}
f_1 + f_2 & & & & = 66 \\
f_1 & & & -f_8 - f_9 & = 0 \\
f_2 - f_3 - f_4 & & & & = 0 \\
f_3 & -f_5 - f_6 & & +f_9 & = 0 \\
f_4 & +f_6 - f_7 & & & = 0 \\
f_3 - f_4 & +f_6 & & & = 0 \\
-f_3 + f_4 - f_5 & & +f_7 & & = 0 \\
f_1 - f_2 - f_3 & -f_5 & & +f_8 & = 0 \\
f_1 - f_2 - f_3 & & & +f_9 & = 0
\end{aligned}
$$

The solution to this system of equations is given by

$$(f_1, f_2, f_3, f_4, f_5, f_6, f_7, f_8, f_9) = (36, 30, 14, 16, 20, 2, 18, 28, 8) \tag{20.13}$$

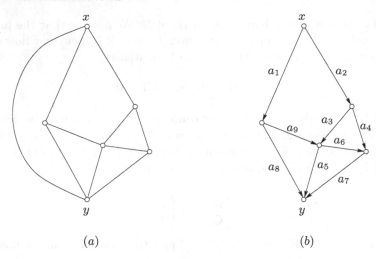

Fig. 20.6. (a) A planar graph G, (b) an orientation D of $G \setminus xy$

We now derive a second expression for the current flow, due to Thomassen (1990).

Given an xy-path P in a digraph $D := D(x, y)$, the *signed incidence vector* of P is the function $f_P : A \to \mathbb{R}$ defined by

$$f_P(a) := \begin{cases} 1 & \text{if } a \in P^+ \\ -1 & \text{if } a \in P^- \\ 0 & \text{if } a \notin P \end{cases}$$

For each spanning tree T of D, set $f_T := f_P$, where $P := xTy$. Observe that f_T is an (x, y)-flow in D of value one. Consequently, the function f defined by

$$f := \sum_T f_T \tag{20.14}$$

where the sum is taken over all spanning trees T of D, is an (x, y)-flow in D of value $t(D)$.

Now consider the digraph D' obtained from D by adding a new arc a' from y to x. For each spanning tree T' of D' containing a', denote by T'_x the component of $T' \setminus a'$ containing x, and by $g_{T'}$ the tension in D associated with the bond $\partial(T'_x)$. Then the function g defined by

$$g := \sum_{T'} g_{T'} \tag{20.15}$$

where the sum is taken over all spanning trees T' of D' containing a', is a tension in D.

Thomassen (1990) showed that $f = g$, and hence (by virtue of Theorem 20.19) that this function is the unique current flow in D of value $t(D)$.

Theorem 20.21 *In any directed graph $D := D(x, y)$, the functions f and g defined by (20.14) and (20.15) are equal. This function is therefore the unique current flow in D of value $t(D)$.*

Proof Let T be a spanning tree of D, and a an arc of the path $P := xTy$. Consider the spanning tree $T' := (T \setminus a) + a'$ of D'. Then the arc a is a forward arc of P if it belongs to $\partial^+(T'_x)$, and a reverse arc of P if it belongs to $\partial^-(T'_x)$.

Conversely, let T' be a spanning tree of D' containing the arc a', and let a be an arc of D such that $T := (T' \setminus a') + a$ is a spanning tree of D. Then the arc a belongs to $\partial^+(T'_x)$ if it is a forward arc of the path $P := xTy$, and to $\partial^-(T'_x)$ if it is a reverse arc of P. It follows that $f = g$. □

EFFECTIVE RESISTANCE

Given a current flow from x to y in a digraph $D := D(x, y)$, Kirchhoff's Voltage Law implies that the voltage drop along each xy-path (that is, the sum of the tensions on its edges) is the same. When the current flow is of value one, this common voltage drop is called the *effective resistance* between x and y, and is denoted r_{xy}. (It is easily seen, by reversing the flow, that $r_{xy} = r_{yx}$, so the terminology and notation adopted here are unambiguous.)

As an example, consider the current flow of value one in the digraph $D(x, y)$ of Figure 20.7. This flow shows that the effective resistance between x and y is $8/7$.

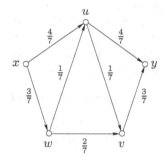

Fig. 20.7. A current flow of value one in a digraph $D(x, y)$

Denote by $\mathbf{K}$ the Kirchhoff matrix $\mathbf{M}_x$ and by $\mathbf{L}$ the matrix $\mathbf{M}_{xy}$, where $\mathbf{M}$ is the incidence matrix of D. Thomassen (1990) gave a simple formula for the effective resistance.

Theorem 20.22 *The effective resistance between x and y in an electrical network $D(x, y)$ is given by the formula*

$$r_{xy} = \frac{\det \mathbf{LL^t}}{\det \mathbf{KK^t}}$$

Proof Add a new arc $a' := (y, x)$ to D, so as to obtain a digraph D'. By Theorem 20.21, $g := \sum_{T'} g_{T'}$ is the unique current flow of value $t(D)$. For any xy-path in D, each tension $g_{T'}$ contributes a voltage drop of one along the path. Thus the total voltage drop along any path is the number of spanning trees of D' that contain a'. By Exercise 4.2.1a, this number is $t(D' / a') = t(D / \{x, y\})$ (the digraph obtained from D by identifying x and y). Therefore,

$$r_{xy}(D) = \frac{t(D / \{x, y\})}{t(D)}$$

By the Matrix–Tree Theorem (20.15), $t(D) = \det \mathbf{KK^t}$, so it remains to show that $t(D / \{x, y\}) = \det \mathbf{LL^t}$.

The incidence matrix of $D / \{x, y\}$ is obtained from the incidence matrix $\mathbf{M}$ of D by 'merging' the rows $\mathbf{m}(x)$ and $\mathbf{m}(y)$, the new row corresponding to the vertex resulting from the identification of x and y. The principal submatrix of $\mathbf{MM^t}$ obtained by deleting this row and the corresponding column is precisely $\mathbf{LL^t}$, so $t(D / \{x, y\}) = \det \mathbf{LL^t}$, by the Matrix–Tree Theorem (20.15). $\qquad \square$

As an illustration, consider the digraph of Figure 20.7. This digraph has 21 spanning trees (see Exercise 4.2.5), whereas the digraph obtained from it by identifying x and y has 24 spanning trees. Theorem 20.22 asserts that

$$r_{xy} = \frac{\det \mathbf{LL^t}}{\det \mathbf{KK^t}} = \frac{24}{21} = \frac{8}{7}$$

confirming our earlier computation.

If xy is an edge of a connected graph G, we denote by $t_{xy}(G)$ the number of spanning trees of G containing xy. Note that $t_{xy}(G) = t(G / xy)$, by Exercise 4.2.1a. The following expression for the effective resistance between adjacent vertices is due to Thomassen (1990).

Corollary 20.23 *If x and y are adjacent vertices of a digraph D,*

$$r_{xy} = \frac{t_{xy}(D)}{t(D)}$$

Proof As in the proof of Theorem 20.22, one has $t_{xy}(D) = \det \mathbf{LL^t}$. Also, $t(D) = \det \mathbf{KK^t}$. $\qquad \square$

In the remaining sections of this chapter, we present two surprising, quite different applications of the results established above.

Exercises

20.5.1 THOMSON'S PRINCIPLE
Let $D := D(x, y)$ be a connected digraph.

a) For any real number i, show that:
    i) there is an (x, y)-flow in D of value i,
    ii) the set of all such flows is a closed subset of $\mathbb{R}^A$.
b) The *power* of an (x, y)-flow f in D is the quantity $\sum\{(f(a))^2 : a \in A(D)\}$.
Show that:
    i) there is an (x, y)-flow which minimizes the power,
    ii) this flow is a current flow in D,
    iii) there is only one such flow.

(The fact that the unique flow of minimum power is a current flow is known as *Thomson's Principle*.)

20.5.2 Let G be an edge-transitive graph, and let $xy \in E$. Express r_{xy} in terms of the numbers of vertices and edges of G.

20.5.3 Compute the effective resistances between all pairs of vertices in the digraph of Figure 20.7 in two ways:

a) by determining current flows of value one,
b) by applying Theorem 20.22 or Corollary 20.23.

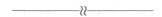

20.6 Perfect Squares

A *squared rectangle* is a rectangle dissected into a finite number (at least two) of squares. If no two of the squares in the dissection have the same size, the squared rectangle is said to be *perfect*. The *order* of a squared rectangle is the number of squares into which it is dissected. Figure 20.8 shows a perfect rectangle of order nine, found by Moron (1925). Note that this squared rectangle does not properly contain another one. Such squared rectangles are called *simple*. Clearly, every squared rectangle is composed of ones that are simple.

For a long time, no simple perfect squared squares were known. It was even conjectured that such squares do not exist. The first person to describe one was Sprague (1939); it was of order 55. At about the same time, Brooks et al. (1940) developed systematic methods for constructing simple perfect squared squares by using the theory of electrical networks. In this section, we outline their approach.

We first show how a current flow in a digraph can be associated with a given squared rectangle R. The union of the horizontal sides of the constituent squares of R consists of horizontal line segments; each such segment is called a *horizontal dissector* of R. In Figure 20.9a the horizontal dissectors are indicated by solid lines. To each horizontal dissector H_i of R there corresponds a vertex v_i of the digraph D associated with R. Two vertices v_i and v_j of D are joined by an arc (v_i, v_j) if and only if the horizontal dissectors H_i and H_j flank some square of the dissection and H_i lies above H_j in R. Figure 20.9b shows the digraph associated with the

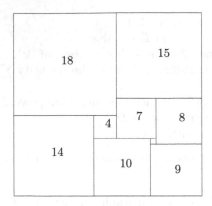

Fig. 20.8. A simple perfect squared rectangle

squared rectangle in Figure 20.9a. The vertices corresponding to the upper and lower sides of R are called the *poles* of D and are denoted by x and y, respectively.

We assign to each vertex v of D a potential $p(v)$, equal to the height (above the lower side of R) of the corresponding horizontal dissector (see Figures 20.9a and 20.9b). If we regard D as an electrical network in which each wire has unit resistance, the tension f determined by this potential can be seen to satisfy both of Kirchhoff's Laws, and hence is a current flow from x to y in D (see Figure 20.9c).

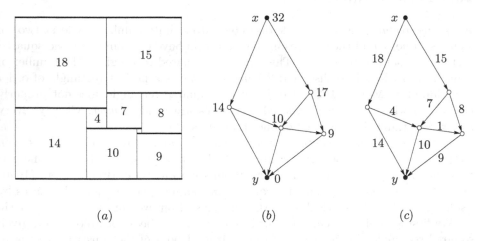

Fig. 20.9. (a) A perfect squared rectangle, (b) its associated electrical network, (c) the resulting current flow

Let D be the digraph corresponding to a squared rectangle R, with poles x and y, and let G' be the underlying graph of D. Then the graph $G := G' + xy$ is called the *horizontal graph* of R. This graph is easily seen to be connected and planar. Brooks et al. (1940) showed that when the squared rectangle R is simple, its horizontal graph is 3-connected. Conversely, they showed that if G is a simple 3-connected planar graph and xy is any edge of G, then a current flow from x to y in $G \setminus xy$ determines a simple squared rectangle. Thus one way of searching for squared rectangles of order k is as follows.

i) List all simple 3-connected planar graphs with $k + 1$ edges.

ii) For each such graph G and each edge xy of G, determine a current flow from x to y in $G \setminus xy$ by solving the system of equations (20.12).

For example, if we consider the graph G shown in Figure 20.6a, the current flow from x to y in the digraph D of Figure 20.6b is given by (20.13). The squared rectangle based on this current flow is just the one displayed in Figure 20.9a, but with all dimensions doubled. Brooks et al. (1940) examined by hand many 3-connected planar graphs, and eventually succeeded in finding a simple perfect squared square. Much later, Duijvestijn (1978) applied this same strategy systematically using a computer, and found several other examples, including the one of order 21 shown in Figure 20.10; this is the unique simple perfect squared square of least order (up to reflections and rotations).

Tutte (1948b) generalized the above theory to dissections of equilateral triangles into equilateral triangles. Further results on perfect squares can be found in the survey article by Tutte (1965b).

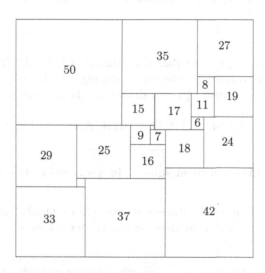

Fig. 20.10. The unique perfect squared square of smallest order

Exercises

20.6.1

a) Determine the unique current flow of value 69 from x to y in the graph of Figure 20.11a, and the unique current flow of value 65 from x to y in the graph of Figure 20.11b.

b) Based on these flows, construct simple squared rectangles of orders nine and ten.

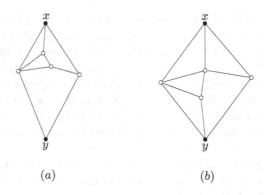

(a) (b)

Fig. 20.11. Construct squared rectangles of orders nine and ten from these graphs (Exercise 20.6.1)

20.6.2 The *vertical graph* of a squared rectangle R is the horizontal graph of the squared rectangle obtained by rotating R through 90 degrees. If no point of R is the corner of four constituent squares, show that the horizontal and vertical graphs of R are planar duals.

(R.L. Brooks, C.A.B. Smith, A.H. Stone, and W.T. Tutte)

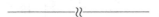

20.6.3 Show that the constituent squares in a squared rectangle have commensurable sides.

20.6.4 A *perfect cube* is a cube dissected into a finite number of smaller cubes, no two of the same size. Show that there exists no perfect cube.

20.6.5

a) The *Fibonacci numbers* F_i, $i \geq 1$, are the integers satisfying the initial condition $F_1 = F_2 = 1$ and the recurrence relation $F_i = F_{i-1} + F_{i-2}$, $i \geq 3$. A dissection of the plane into squares which uses exactly one square of side F_i for each $i \geq 1$ is called a *Fibonacci tiling*. Find such a tiling.

b) Using (a), find a dissection of the plane into squares, no two of the same size. (Henle and Henle (2006) have described a dissection of the plane into squares using exactly one square of each integral dimension.)

20.7 Random Walks on Graphs

A *random walk* on a simple undirected graph G is a walk on G whereby, when it reaches a vertex v, the edge along which to travel next is chosen at random, the $d(v)$ edges incident with v being equiprobable. If its initial vertex is x, the walk is called a *random x-walk*. A classical example of such a walk, on the n-dimensional *integer lattice* $\mathbb{Z}^n$ (the cartesian product of n two-way infinite paths), is the *Drunkard's Walk*.

Example 20.24 THE DRUNKARD'S WALK
The two-dimensional integer lattice $\mathbb{Z}^2$ (depicted in Figure 1.27) represents an infinite system of streets. A drunkard sets out from home, one of the vertices. What is the probability that he eventually returns home, assuming that he takes a random walk? Pólya (1921) proved that this probability is one, in other words, that the drunkard is sure to get back home eventually (despite his inebriated state). On the other hand, in a random walk on the three-dimensional lattice $\mathbb{Z}^3$, this probability is strictly less than one; with positive probability, the drunkard will stray farther and farther from home.

HITTING, COMMUTE, AND COVER TIMES

Doyle and Snell (1984) discovered that certain basic properties of random walks on graphs depend principally on the effective resistances of the graphs, regarded as electrical networks. This phenomenon, which at first sight seems quite surprising, can perhaps be understood by imagining the trajectory of an electron in a network as a random walk. It is illustrated by the following theorem. A random x-walk is said to *hit* a vertex y when it reaches y, and to *return* to x when it reaches x after hitting at least one other vertex.

Theorem 20.25 *Let x and y be distinct vertices of a simple connected graph G. The probability P_x that a random x-walk on G hits y before returning to x is given by*

$$P_x = \frac{1}{d(x)\, r_{xy}}$$

Proof For $v \in V \setminus \{x\}$, denote by P_v the probability that a random v-walk on G hits y before hitting x. Then $P_y = 1$, and

$$P_v = \frac{1}{d(v)} \sum_w P_w, \quad v \in V \setminus \{y\}$$

that is,

$$d(v)P_v - \sum_w P_w = 0, \quad v \in V \setminus \{y\}$$

where the summations are over all vertices w in $N(v) \setminus \{x\}$.

Let $\mathbf{N}$ be the matrix derived from the conductance matrix $\mathbf{C} := \mathbf{MM^t}$ by setting to zero all nondiagonal entries in the column corresponding to x and replacing the row corresponding to y by the unit vector consisting of a one on the diagonal and zeros elsewhere. Let $\mathbf{P}$ be the vector $(P_v : v \in V)$. We assume that the first and last vertices in the indexing of $\mathbf{MM^t}$, $\mathbf{N}$ and $\mathbf{P}$ are x and y, respectively. Then

$$\mathbf{NP} = \begin{bmatrix} 0 \\ 0 \\ \cdot \\ \cdot \\ 0 \\ 1 \end{bmatrix} \quad \text{where} \quad \mathbf{N} = \begin{bmatrix} d(x) & * & \cdot & \cdot & \cdot & * & * \\ 0 & & & & & & * \\ \cdot & & & & & & \cdot \\ \cdot & & & \mathbf{LL^t} & & & \cdot \\ \cdot & & & & & & \cdot \\ 0 & & & & & & * \\ 0 & 0 & \cdot & \cdot & \cdot & 0 & 1 \end{bmatrix}$$

where $\mathbf{L} = \mathbf{M}_{xy}$ (the asterisks denoting unspecified values). By Cramér's Rule,

$$P_x = \frac{-\det \mathbf{N}_y^x}{\det \mathbf{N}}$$

where $\mathbf{N}_y^x$ is the matrix obtained from $\mathbf{N}$ by deleting the first column (corresponding to x) and the last row (corresponding to y). But this is identical to the submatrix of $\mathbf{MM^t}$ obtained by deleting its first column and last row. Therefore $-\det \mathbf{N}_y^x$ is a cofactor of $\mathbf{MM^t}$ and hence is equal to $t(G) = \det \mathbf{KK^t}$ by Corollary 20.16. Because $\det \mathbf{N} = d(x) \det \mathbf{LL^t}$, we have

$$P_x = \frac{\det \mathbf{KK^t}}{d(x) \det \mathbf{LL^t}} = \frac{1}{d(x)\, r_{xy}} \qquad \square$$

We illustrate the proof of Theorem 20.25 with the graph of Figure 20.4. This graph and its associated matrix $\mathbf{N}$ are shown in Figure 20.12.

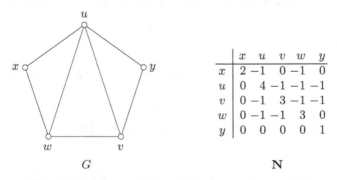

	x	u	v	w	y
x	2	-1	0	-1	0
u	0	4	-1	-1	-1
v	0	-1	3	-1	-1
w	0	-1	-1	3	0
y	0	0	0	0	1

G $\qquad\qquad\qquad\qquad\qquad$ $\mathbf{N}$

Fig. 20.12. A graph G and its associated matrix $\mathbf{N}$

The corresponding system of equations is:

$$
\begin{aligned}
2P_x - P_u \quad - P_w \quad &= 0 \\
4P_u - P_v - P_w - P_y &= 0 \\
- P_u + 3P_v - P_w - P_y &= 0 \\
- P_u - P_v + 3P_w \quad &= 0 \\
P_y &= 1
\end{aligned}
$$

with solution $P_x = 7/16$, $P_u = 1/2$, $P_v = 5/8$, $P_w = 3/8$, $P_y = 1$. From a previous computation, $r_{xy} = 8/7$. Because $d(x) = 2$, this confirms the formula $P_x = 1/(d(x)\, r_{xy})$ of Theorem 20.25.

Another connection between random walks and electrical networks was discovered by Nash-Williams (1959). The *hitting time* H_{xy} is the expected number of steps taken by a random x-walk before hitting vertex y. The *commute time* C_{xy} between x and y is defined by

$$
C_{xy} := H_{xy} + H_{yx} \tag{20.16}
$$

(Note that, in general, $H_{xy} \neq H_{yx}$; see Exercise 20.7.3.)

Theorem 20.26 *Let x and y be distinct vertices of a simple connected graph G. The commute time between x and y is given by*

$$
C_{xy} = 2m r_{xy}
$$

Proof The hitting time satisfies the linear system

$$
H_{vy} = \sum_{w \in N(v)} \frac{1}{d(v)} (1 + H_{wy}), \quad v \in V \setminus \{y\} \tag{20.17}
$$

For $z \in \{x, y\}$ and $v \in V \setminus \{z\}$, denote by f_{vz} the current flow in G from v to z of value $d(v)$, and set

$$
f_z := \sum_{v \in V \setminus \{z\}} f_{vz}
$$

The net flow out of v in f_{vy} is $d(v)$, whereas the net flow out of v in f_{uy}, where $u \neq v$, is zero. Therefore, the net flow out of v in f_y is $d(v)$, $v \in V \setminus \{y\}$, and the net flow into y is $\sum_{v \in V \setminus \{y\}} d(v) = 2m - d(y)$. Denote by V_{vy} the voltage drop from v to y in f_y. The current in f_y between adjacent vertices v and w is equal to the voltage drop between v and w, that is $V_{vy} - V_{wy}$, and the sum of these quantities over all neighbours w of v is the net flow out of v, namely $d(v)$. Thus the voltage drops V_{vy}, $v \in V \setminus \{y\}$, satisfy the linear system

$$
\sum_{w \in N(v)} (V_{vy} - V_{wy}) = d(v), \quad v \in V \setminus \{y\} \tag{20.18}
$$

System (20.17) can be rewritten as

$$d(v)H_{vy} - \sum_w H_{wy} = d(v), \quad v \in V \setminus \{y\}$$

and system (20.18) can be rewritten as

$$d(v)V_{vy} - \sum_w V_{wy} = d(v), \quad v \in V \setminus \{y\}$$

where the summations are over all vertices w in $N(v) \setminus \{y\}$. Thus the H_{vy} and the V_{vy} satisfy the same system of equations. Moreover, this system has a unique solution, because its matrix $\mathbf{M}_y \mathbf{M}_y^t$ is nonsingular ($\mathbf{M}_y$ is a Kirchhoff matrix, so has rank $n-1$). It follows that $H_{vy} = V_{vy}$ for all $v \in V \setminus \{y\}$.

Consider, now, the current flow $f_y - f_x$. Here, the net flow out of each vertex $v \in V \setminus \{x, y\}$ is $d(v) - d(v) = 0$, the net flow out of x is $2m$, and the net flow into y is $2m$. Note that the voltage drop between x and y in $-f_x$ is the same as the voltage drop between y and x in f_x, namely V_{yx}, so the voltage drop between x and y in the current flow $f_y - f_x$ is $V_{xy} + V_{yx}$. By the definition of effective resistance,

$$r_{xy} = \frac{V_{xy} + V_{yx}}{2m}$$

and so

$$C_{xy} = H_{xy} + H_{yx} = V_{xy} + V_{yx} = 2m r_{xy} \qquad \square$$

These computations are illustrated in Figure 20.13, where the functions f_y and f_x, and the current flow $f_y - f_x$, are displayed for the graph of Figure 20.12. (Observe that the symmetry between x and y in this example is reflected in the functions f_x and f_y.) We have:

$$H_{uy} = \frac{19}{3}, \quad H_{vy} = \frac{17}{3}, \quad H_{wy} = \frac{23}{3}, \quad H_{xy} = 8$$

$$H_{ux} = \frac{19}{3}, \quad H_{vx} = \frac{23}{3}, \quad H_{wx} = \frac{17}{3}, \quad H_{yx} = 8$$

The commute time between x and y is therefore given by $C_{xy} = H_{xy} + H_{yx} = 8 + 8 = 16$. Because $m = 7$ and the effective resistance between x and y, as calculated previously, is $r_{xy} = 8/7$, the above computation agrees with the formula $C_{xy} = 2m r_{xy}$ of Theorem 20.26.

The *cover time* of G is defined by $C := \max \{C_v : v \in V\}$, where C_v is the expected number of steps taken by a random v-walk on G to hit every vertex of G. A bound on the cover time was obtained by Aleliunas et al. (1979). We derive it here from our earlier results.

Corollary 20.27 *The cover time C of a graph G is at most $2m(n-1)$.*

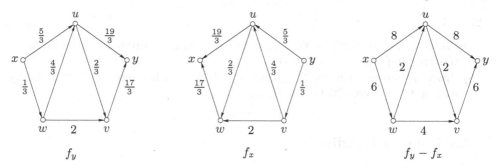

Fig. 20.13. Computing hitting times: the functions f_y and f_x, and the current flow $f_y - f_x$

Proof Let T be any spanning tree of G, and let $(v = v_0, v_1, \ldots, v_{2n-2} = v)$ be the sequence of vertices encountered in a walk around T (not a random walk) which starts at an arbitrary vertex v and traverses each edge of T once in each direction. Consider now a random v-walk on G. By Theorem 20.26 and Corollary 20.23, the expected time taken to visit the vertices $v_1, \ldots, v_{2n-2}$ in this order is

$$\sum_{i=1}^{2n-2} H_{v_{i-1}v_i} = \sum_{xy \in E(T)} (H_{xy} + H_{yx})$$

$$= \sum_{xy \in E(T)} C_{xy} = 2m \sum_{xy \in E(T)} r_{xy} = \frac{2m}{t(G)} \sum_{xy \in E(T)} t_{xy}(G)$$

This is clearly an upper bound for C_v, and is independent of v. It follows that

$$C \le \frac{2m}{t(G)} \sum_{xy \in E(T)} t_{xy}(G) \le 2m(v-1) \qquad \square$$

Exercises

20.7.1 Compute the hitting times and commute times between all remaining pairs of vertices in the example of Figure 20.12.

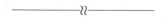

20.7.2 Determine an upper bound on the cover time of an edge-transitive graph G in terms of n.

20.7.3 Let $G = P \cup Q$, where P is an xy-path of length k and Q is a complete graph on $2k$ vertices such that $P \cap Q = \{y\}$. Determine H_{xy}, H_{yx} and r_{xy}.

20.7.4

a) Prove Theorem 20.25 by the proof technique of Theorem 20.26 (that is, by constructing appropriate flows).

b) Prove Theorem 20.26 by the proof technique of Theorem 20.25 (that is, by solving the system (20.17) for H_{xy}).

20.8 Related Reading

RANDOM WALKS ON INFINITE GRAPHS

The notion of a random walk introduced in Section 20.7 applies equally well to all connected locally finite graphs. Let G be such a graph. A basic result in the theory of Markov chains implies that either a random x-walk in G returns to its origin x with probability one, regardless of the choice of x, in which case G is called *recurrent*, or the probability of return is strictly less than one, in which case G is called *transient* (see Feller (1968)). Pólya (1921) showed that the integer lattice $\mathbb{Z}^n$ is recurrent for $n = 1, 2$, and transient for $n \geq 3$. (The case $n = 2$ is the Drunkard's Walk mentioned in Section 20.7.) Nash-Williams (1959) established a far-reaching generalization of Pólya's result by providing a characterization of recurrent locally finite graphs. In intuitive terms, his theorem says that a connected locally finite graph which 'widens out' rapidly from a vertex is transient, and one which does not is recurrent.

21

Integer Flows and Coverings

Contents

21.1 Circulations and Colourings

In Chapter 20, we studied circulations and tensions over fields. In this concluding chapter, we return to the same notions, but regarded here as functions defined over arbitrary abelian groups. Studying them in this more general setting leads to interesting applications on colourings and coverings, and to some of the most intriguing unsolved problems of the subject.

NOWHERE-ZERO CIRCULATIONS AND TENSIONS

A function f on the arc set A of a digraph D is *nowhere-zero* if $f(a) \neq 0$ for each arc $a \in A$ (that is, the support of f is the entire arc set A). Here our interest is in nowhere-zero circulations and tensions. As circulations take value zero on cut edges and tensions take value zero on loops, we assume that all graphs and digraphs considered in this chapter are 2-edge-connected and loopless. We start by observing a simple correspondence between nowhere-zero tensions and vertex colourings.

Proposition 21.1 *A digraph D is k-vertex-colourable if and only if it admits a nowhere-zero tension over $\mathbb{Z}_k$.*

Proof Firstly, suppose that D has a proper k-vertex-colouring $c : V \to \mathbb{Z}_k$. Consider the tension $g : A \to \mathbb{Z}_k$ defined by $g(a) := c(u) - c(v)$ for each arc $a := (u, v)$. This tension is nowhere-zero because c is a proper colouring. Conversely, let g be a nowhere-zero tension in D over $\mathbb{Z}_k$. Obtain a colouring $c : V \to \mathbb{Z}_k$ recursively, as follows.

▷   Select an arbitrary vertex x and assign it the colour $c(x) := 0$.
▷   Subsequently, if an arc a links a coloured vertex u and an uncoloured vertex v, assign to v the colour

$$c(v) := \begin{cases} c(u) - g(a) & \text{if } a = (u, v) \\ c(u) + g(a) & \text{if } a = (v, u) \end{cases}$$

Using the fact that g is a tension, it can be shown that the resulting colouring c is well-defined (Exercise 21.1.1). Furthermore, this colouring is proper because g is nowhere-zero. $\qquad\square$

In the case of plane digraphs, as Tutte (1954a) observed, one has the following dual version of Proposition 21.1.

Theorem 21.2 *A plane digraph D is k-face-colourable if and only if it admits a nowhere-zero circulation over $\mathbb{Z}_k$.*

Proof By the analogue of Theorem 20.5 for circulations and tensions over $\mathbb{Z}_k$, a function $f : A(D) \to \mathbb{Z}_k$ is a circulation in D over $\mathbb{Z}_k$ if and only if the corresponding function $f^* : A(D^*) \to \mathbb{Z}_k$ is a tension in D^*. Face colourings of D correspond to vertex colourings of D^* (and conversely), thus the assertion follows. $\qquad\square$

Figure 21.1a shows a face colouring c of an oriented 3-prism by the elements of $\mathbb{Z}_4$, and Figure 21.1b the resulting nowhere-zero circulation f over $\mathbb{Z}_4$, where $f(a) := c(l_a) - c(r_a)$ for each arc a (l_a and r_a being the faces to the left and right of a, respectively).

Observe that whether a digraph has a nowhere-zero circulation over a given additive abelian group Γ depends only on its underlying graph (Exercise 21.1.4). Thus we may simply speak of nowhere-zero circulations in undirected graphs, without reference to a specific orientation. In the same vein, we often find it convenient

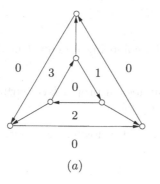
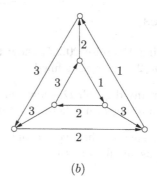

$$(a) \qquad\qquad (b)$$

Fig. 21.1. (a) a 4-face-colouring of the 3-prism, and (b) the resulting nowhere-zero circulation over $\mathbb{Z}_4$

to refer to arcs as 'edges'. For example, by virtue of Theorem 21.2, the Four-Colour Theorem may be restated as: *every planar graph admits a nowhere-zero circulation over* $\mathbb{Z}_4$.

What can one say about circulations in graphs that are not planar? Not every graph has a nowhere-zero circulation over $\mathbb{Z}_4$. For example, as we show in Theorem 21.11, the Petersen graph has no such circulation. It does, however, have nowhere-zero circulations over $\mathbb{Z}_5$, for example, the ones shown in Figure 21.2. This leads one to ask whether every graph has a nowhere-zero circulation over $\mathbb{Z}_k$ for some k. We show that this is indeed so. In fact, all graphs have nowhere-zero circulations over $\mathbb{Z}_6$. We prove this theorem by exploiting a close connection between circulations and coverings of graphs, as explained in the next section.

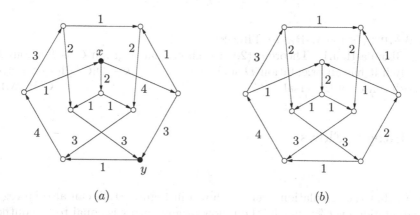

Fig. 21.2. Two nowhere-zero circulations over $\mathbb{Z}_5$ in the Petersen graph

Exercises

21.1.1 Using Exercise 20.2.5, show that the colouring c defined in the proof of Proposition 21.1 is well-defined.

21.1.2 For any digraph D, show that the number of proper k-colourings of D is k times the number of nowhere-zero tensions over $\mathbb{Z}_k$.

21.1.3 For each of the five platonic graphs, find a nowhere-zero circulation over $\mathbb{Z}_k$ with k as small as possible.

⋆**21.1.4**

a) Let D be a digraph, and let D' be the digraph obtained by reversing an arc a. Show that there is a bijection between the nowhere-zero circulations in D and D' over any additive abelian group Γ.

b) Deduce, more generally, that if D and D' are any two orientations of a graph G, there is a bijection between the nowhere-zero circulations in D and D' over any additive abelian group Γ.

21.1.5

a) Let D be a digraph embedded on an orientable surface. Show that if D is k-face-colourable, then it admits a nowhere-zero circulation over $\mathbb{Z}_k$.

b) Using the embedding of the Petersen graph on the torus shown in Figure 3.9b, find a nowhere-zero circulation over $\mathbb{Z}_5$ in an orientation of that graph.

c) Give an example to show that the converse of the statement in part (a) is not true in general. (Compare with Theorem 21.2.)

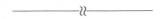

21.1.6 MINTY'S FLOW-RATIO THEOREM
Using Ghouila-Houri's Theorem (20.10), show that a graph G is k-colourable if and only if it has an orientation D such that $|C^-| \leq (k-1)|C^+|$ for each cycle C and both senses of traversal of C.   (G.H. MINTY)

21.2 Integer Flows

K-FLOWS

Observe that the circulation over $\mathbb{Z}_5$ shown in Figure 21.2b can also be regarded as a circulation over $\mathbb{Z}$; that is, the inflow at any vertex is equal to the outflow at the vertex, not just equal to it modulo 5. Furthermore, the values taken by the circulation all lie between one and four. Such a circulation is called a 5-flow. More generally, a nowhere-zero circulation f over $\mathbb{Z}$ in a digraph D is a k-flow if

$$-(k-1) \leq f(a) \leq k-1, \quad \text{for all } a \in A$$

(In this context, the term 'flow' is commonly used, rather than 'circulation'; we adopt this terminology.)

If a graph has a k-flow, it clearly also has a nowhere-zero circulation over $\mathbb{Z}_k$; one may simply regard the flow values as elements of $\mathbb{Z}_k$. Surprisingly, the converse also holds, as was shown by Tutte (1954a). The proof that we give here closely resembles the constructive proof of Hoffman's Circulation Theorem (20.9) in Section 20.3.

Theorem 21.3 *A graph admits a nowhere-zero circulation over $\mathbb{Z}_k$ if and only if it admits a k-flow.*

Proof As observed above, if a graph has a k-flow, then it has a nowhere-zero circulation over $\mathbb{Z}_k$. It remains to prove the converse statement.

Let G be a graph which has a nowhere-zero circulation over $\mathbb{Z}_k$. Consider an orientation D of G, and let f be a nowhere-zero circulation in D over $\mathbb{Z}_k$. If we regard the elements of $\mathbb{Z}_k$ as elements of $\mathbb{Z}$, the function f, although not necessarily a flow, has the following properties.

i) For all $a \in A$, $f(a) \in \{1, 2, \ldots, k-1\}$.
ii) For all $v \in V$, $f^+(v) - f^-(v) \equiv 0 \pmod{k}$,

where $f^+(v) - f^-(v)$ is the net flow out of v. Define a vertex v to be *positive, balanced*, or *negative* according as this net flow is positive, zero, or negative. Because $\sum_{v \in V}(f^+(v) - f^-(v)) = 0$ (Exercise 7.1.1a), either all vertices are balanced, in which case f is a k-flow, or there are both positive and negative vertices.

Let us call the quantity $\sum_{v \in V} |f^+(v) - f^-(v)|$ the *excess* of f. If not all vertices are balanced, the excess is positive. In this case, we show how f can be modified to a nowhere-zero circulation f' over $\mathbb{Z}_k$ which has a smaller excess. Repeating this procedure results in a k-flow.

Consider a positive vertex x, and let X denote the set of all vertices of D reachable from x by directed paths. Then $\partial^+(X) = \emptyset$, and so

$$\sum_{v \in X}(f^+(v) - f^-(v)) = f^+(X) - f^-(X) \leq 0$$

implying that there exists a negative vertex y in X. By the definition of X, there is a directed path xPy in D. Let D' be the digraph obtained from D by reversing the direction of all arcs of P, replacing each arc a of P by $\overleftarrow{a}$, and let f' be the function derived from f by setting $f'(\overleftarrow{a}) := k - f(a)$ if $a \in A(P)$, and $f'(a) := f(a)$ otherwise. Then f' also satisfies the properties (i) and (ii), but the excess of f' is $2k$ less than the excess of f. Iterating this procedure results in a k-flow in G. $\square$

The technique of modifying a circulation f whose excess is positive to a circulation f' with smaller excess is illustrated in Figure 21.2. With respect to the nowhere-zero circulation over $\mathbb{Z}_5$ shown in Figure 21.2a, there is just one positive vertex, namely x, and just one negative vertex, namely y, the remaining vertices being balanced. The excess of this circulation is ten. Taking P to be the directed (x, y)-path of length two, we obtain the digraph D' and 5-flow f' shown in Figure 21.2b.

FLOW NUMBER

By definition, a k-flow in a graph is also a k'-flow for all $k' \geq k$. Tutte (1954a) conjectured that there is a positive integer k such that every graph has a k-flow. This is true for planar graphs because, by virtue of Theorem 21.3, Theorem 21.2 asserts that a 2-edge-connected plane graph G has a k-flow if and only if its dual G^* is k-colourable, and every planar graph is 4-colourable by the Four-Colour Theorem. Jaeger (1976) and Kilpatrick (1975) independently confirmed Tutte's conjecture by showing that every graph has an 8-flow. Shortly thereafter, Seymour (1981b) improved upon their bound by showing that every graph has a 6-flow. We give proofs of these two theorems in Sections 21.4 and 21.6, respectively.

The *flow number* of a graph is defined to be the smallest positive integer k for which it has a k-flow. The flow number of a (2-edge-connected) planar graph lies between two and four. The Petersen graph has the 5-flow shown in Figure 21.2b, but (as already mentioned) no 4-flow, so its flow number is equal to five.

A k-flow f in a digraph D is *positive* if $f(a) > 0$, for all $a \in A$. If a graph has a k-flow, then it has a positive k-flow (Exercise 21.2.1). In particular, if G is an even graph and D is an even orientation of G, then the flow f in D with $f(a) = 1$ for all $a \in A$ is a positive 2-flow in D. Conversely, any graph which has a 2-flow is necessarily even (Exercise 21.2.2). We thus have:

Theorem 21.4 *A graph admits a 2-flow if and only if it is even.*   □

For any k, the problem of deciding whether a graph G has a k-flow is clearly in $\mathcal{NP}$. For $k = 3$, this problem is in fact $\mathcal{NP}$-complete, even when G is planar. This follows from Theorem 21.3 and the fact, noted in Chapter 14, that the problem of deciding whether a given planar graph is 3-vertex-colourable is $\mathcal{NP}$-complete. The following theorem characterizes those cubic graphs which have 3-flows.

Theorem 21.5 *A 2-edge-connected cubic graph admits a 3-flow if and only if it is bipartite.*

Proof Let $G := G[X, Y]$ be a bipartite cubic graph. By Theorem 17.2, G is 3-edge-colourable, so there exist three disjoint perfect matchings M_1, M_2, and M_3 in G such that $E = M_1 \cup M_2 \cup M_3$. Orient the edges of M_1 from X to Y and the edges of $M_2 \cup M_3$ from Y to X. The function $f : A \to \{1, 2\}$ defined by

$$f(a) := \begin{cases} 2 & \text{if } a \in M_1 \\ 1 & \text{if } a \in M_2 \cup M_3 \end{cases}$$

is a 3-flow in G.

Conversely, let G be a cubic graph with a 3-flow f. Reversing orientations of arcs if necessary, we may assume that f is a positive 3-flow. The conservation condition now implies that, at any vertex v, either (i) there are two incoming arcs carrying a flow of one unit, and one outgoing arc carrying two units, or (ii) there is one incoming arc carrying a flow of two units, and two outgoing arcs each carrying one unit. Denote by X the set of vertices at which the flow pattern is as in (i), and by Y its complement. Then (X, Y) is a bipartition of G.   □

THE FLOW POLYNOMIAL

We now turn briefly to another striking analogy between colourings and nowhere-zero circulations in graphs. Consider any graph G, and any finite additive abelian group Γ. Let $F(G, \Gamma)$ denote the number of nowhere-zero circulations in G over Γ (that is, the number of nowhere-zero circulations over Γ in some fixed orientation of G; by Exercise 21.1.4, this number is independent of the orientation).

For any link e of G, each circulation f' in G / e is the restriction to $E \setminus e$ of a unique circulation f in G. Therefore, $F(G / e, \Gamma)$ is the number of circulations f in G which have nonzero values on all edges, except possibly on e. On the other hand, each circulation f' in $G \setminus e$ is the restriction to $E \setminus e$ of a unique circulation f in G with $f(e) = 0$, so $F(G \setminus e, \Gamma)$ is the number of circulations f in G which take nonzero values on all edges in $E \setminus \{e\}$, and value zero on e. Therefore, as Tutte (1954a) observed, the function $F(G, \Gamma)$ satisfies the following recursion formula, reminiscent of the recursion formulae for the number of spanning trees (Proposition 4.9) and the chromatic polynomial (14.6).

Theorem 21.6 *For any graph G, any link e of G, and any finite additive abelian group Γ,*

$$F(G, \Gamma) = F(G / e, \Gamma) - F(G \setminus e, \Gamma)$$

$\square$

By a simple inductive argument similar to the one used to demonstrate that the number of k-colourings is a polynomial in k, one can derive the following implication of Theorem 21.6 (Exercise 21.2.4). What is striking here is that $F(G, \Gamma)$ depends not on the structure of the group Γ, but only on its order.

Theorem 21.7 *For any graph G without cut edges, there exists a polynomial $Q(G, x)$ such that $F(G, \Gamma) = Q(G, k)$ for every additive abelian group Γ of order k. Moreover, if G is simple and e is any edge of G, then $Q(G, x)$ satisfies the recursion formula:*

$$Q(G, x) = Q(G / e, x) - Q(G \setminus e, x)$$

$\square$

The polynomial $Q(G, x)$ is known as the *flow polynomial* of G. An explicit expression for the flow polynomial, analogous to Whitney's expansion formula for the chromatic polynomial (see Exercise 14.7.12) is given in Exercise 21.2.13.

INTEGER FLOWS AND COVERS BY EVEN SUBGRAPHS

A particularly interesting instance of Theorem 21.7 arises when the order of the group Γ is a nontrivial product $k_1 k_2$. In this case, the group Γ may be chosen to be either $\mathbb{Z}_{k_1 k_2}$ or $\mathbb{Z}_{k_1} \times \mathbb{Z}_{k_2}$, resulting in the following corollary.

Corollary 21.8 *Let G be a graph, and let k_1 and k_2 be integers, where $k_i \geq 2$, $i = 1, 2$. Then the number of nowhere-zero circulations in G over $\mathbb{Z}_{k_1 k_2}$ is equal to the number of nowhere-zero circulations in G over $\mathbb{Z}_{k_1} \times \mathbb{Z}_{k_2}$.*

$\square$

When combined with Theorem 21.3, Corollary 21.8 provides an important link between flows and structural properties of graphs.

Theorem 21.9 *Let G be a graph and let k_1 and k_2 be integers, where $k_i \geq 2$, $i = 1, 2$. Then G admits a $k_1 k_2$-flow if and only if $G = G_1 \cup G_2$, where G_i admits a k_i-flow, $i = 1, 2$.*

Proof If G has a $k_1 k_2$-flow, then it has a nowhere-zero circulation over $\mathbb{Z}_{k_1 k_2}$, by virtue of Theorem 21.3. By Corollary 21.8, this implies that G has a nowhere-zero circulation $f := (f_1, f_2)$ over $\mathbb{Z}_{k_1} \times \mathbb{Z}_{k_2}$. Let $G_i := G[E_i]$, where E_i is the support of f_i, $i = 1, 2$. Then $G = G_1 \cup G_2$, because f is nowhere zero. Moreover, f_i is a nowhere-zero circulation in G_i over $\mathbb{Z}_{k_i}$, $i = 1, 2$. Appealing once more to Theorem 21.3, we conclude that $G = G_1 \cup G_2$, where G_i has a k_i-flow, $i = 1, 2$. The converse is proved by reversing the above argument. $\square$

The following consequence of Theorem 21.9 is implicit in the work of Jaeger (1976) and Kilpatrick (1975).

Corollary 21.10 *A graph admits a 2^k-flow if and only if it admits a covering by k even subgraphs.*

Proof Apply Theorem 21.9 recursively, with $k_i = 2$, $1 \leq i \leq k$, and invoke Theorem 21.4. $\square$

Matthews (1978) gave a proof of Corollary 21.10 by describing a bijection between 2^k-flows and coverings by k-tuples of even subgraphs.

Corollary 21.10 implies, in particular, that a graph has a 4-flow if and only if it has a covering by two even subgraphs. In the case of cubic graphs, this condition may be expressed in terms of edge colourings because a cubic graph has a covering by two even subgraphs if and only if it is 3-edge-colourable (Exercise 17.3.4a). Thus, as a consequence of the case $k = 2$ of Corollary 21.10, we have:

Theorem 21.11 *A cubic graph admits a 4-flow if and only if it is 3-edge-colourable.* $\square$

The Petersen graph, being cubic but not 3-edge-colourable, therefore has no 4-flow. As noted earlier, this implies that its flow number is equal to five.

The foregoing discussion reveals a close connection between cycle covers and integer flows in graphs. Jaeger (1985) found a still stronger link, between orientable cycle double covers and integer flows. Recall that a cycle double cover of a graph is *orientable* if its members can be oriented as directed cycles so that, taken together, they traverse each edge once in each direction. For instance, the cycle double cover of the cube given in Figure 3.8 is orientable, as shown in Figure 21.3. A double cover by even subgraphs is likewise said to be *orientable* if its members can be decomposed to form an orientable cycle double cover. Jaeger (1985) proved the following theorem.

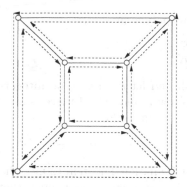

Fig. 21.3. An oriented cycle double cover of the cube

Theorem 21.12 *Every graph which admits an orientable double cover by k even subgraphs admits a k-flow.*

Proof Let $\{C_i : 1 \leq i \leq k\}$ be an orientable double cover of G by k even subgraphs and let f_i be the positive 2-flow on an even orientation D_i of C_i. Now consider a fixed orientation D of G and, for $1 \leq i \leq k$, let g_i denote the function on $A(D)$ such that

$$g_i(a) := \begin{cases} f_i(a), & \text{if } a \in A(D_i) \\ -f_i(a), & \text{if } \overleftarrow{a} \in A(D_i) \\ 0, & \text{otherwise} \end{cases}$$

where $\overleftarrow{a}$ denotes the arc obtained by reversing the orientation of a. Then $g := \sum_{i=1}^{k} i g_i$ is a k-flow on G. $\qquad\square$

Exercises

$\star$**21.2.1** Show that:

  a) if a graph G has a k-flow, then some orientation of G has a positive k-flow,
  b) a connected digraph has a positive k-flow for some $k \geq 1$ if and only if it is strongly connected.

$\star$**21.2.2** Show that any graph which admits a 2-flow is even.

21.2.3 Let G be a 2-edge-connected graph, and let e be an edge of G. If $G \setminus e$ has a k-flow, show that G has a $(k+1)$-flow.   (C.Q.ZHANG)

$\star$**21.2.4** Deduce from Theorem 21.6 that, for any graph G, the function $F(G, \Gamma)$ is a polynomial in the order of Γ.

21.2.5 Show that every hamiltonian graph has a 4-flow.

21.2.6 Determine the flow polynomials of:

a) the wheel W_n, $n \geq 3$,
b) the complete graph K_5.

$\star$**21.2.7** Let G be a graph and let k_1 and k_2 be integers, where $k_i \geq 2$, $i = 1, 2$. Prove that G has a $k_1 k_2$-flow if and only if there is a subgraph F of G such that F has a k_1-flow and G / F has a k_2-flow.

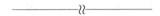

21.2.8 Using Hoffman's Circulation Theorem (20.9), show that a graph G admits a k-flow if and only if it has an orientation D such that $d^-(X) \leq (k - 1)d^+(X)$ for all subsets X of V.

21.2.9 A $(k + l)$-regular graph is (k, l)-*orientable* if it has an orientation in which each indegree (and each outdegree) is either k or l.

a) Suppose that $k \geq l$. Show that a loopless $(k + l)$-regular graph G is (k, l)-orientable if and only if there is a partition (X, Y) of V such that, for every subset S of V,
$$d(S) \geq (k - l)(|S \cap X| - |S \cap Y|)$$
b) Deduce that every (k, l)-orientable graph, where $k > l$, is also $(k - 1, l + 1)$-orientable.

21.2.10 Let f be a positive k-flow on a digraph D. Set $S := \{a \in A : f(a) = k-1\}$.

a)   i) Show that there is a $(k-1)$-flow f' on $D \backslash S$ such that $f'(a) \equiv f(a) \pmod{k-1}$ for all $a \in A \backslash S$.
   ii) Deduce that there is a positive 2-flow on D whose support contains S.
b) Deduce from (a) that f can be expressed as a sum of $k - 1$ positive 2-flows on D.   (C.H.C. LITTLE, W.T. TUTTE, AND D.H. YOUNGER)

21.2.11   a) Show that if G and H are disjoint, then $Q(G \cup H, x) = Q(G, x)Q(H, x)$.
b) Deduce that the flow polynomial of a graph is equal to the product of the flow polynomials of its components.

21.2.12

a) Let D be a digraph, T a maximal forest of D, and Γ a finite additive abelian group. Show that any mapping from $A(D) \setminus A(T)$ to Γ extends to a unique circulation in D over Γ.
b) Deduce that the number of circulations in D over Γ is k^{m-n+c}, where k is the order of Γ.

21.2.13

a) Let a be an arc of a digraph D. Show that:
   i) a function on $A \setminus a$ is a circulation in D / a if and only if it is the restriction to $A \setminus a$ of a circulation in D,

ii) if a is a link of D and f' is a circulation in D/a, then there is a unique circulation f in D such that f' is the restriction of f to $A \setminus a$.

b) Let G be a graph and let Γ be an additive abelian group of order k. For a subset S of E, let $c(S)$ denote the number of components of the spanning subgraph of G with edge set S.

    i) Show that the number of nowhere-zero circulations in G over Γ is equal to $\sum_{S \subseteq A} (-1)^{m-|S|} k^{|S|-n+c(S)}$.

    ii) Conclude that $Q(G, x) = \sum_{S \subseteq E} (-1)^{m-|S|} x^{|S|-n+c(S)}$.   (W.T. TUTTE)

21.3 Tutte's Flow Conjectures

Tutte (1954a, 1966b, 1972) proposed three celebrated conjectures on integer flows. They are arguably the most significant problems in the whole of graph theory.

THE FIVE-FLOW CONJECTURE

Conjecture 21.13 *Every 2-edge-connected graph admits a 5-flow.*

If true, then by Theorems 21.2 and 21.3 this conjecture would be a generalization of the Five-Colour Theorem (11.6). There has been essentially no progress on this problem. Theorem 21.12 suggests the following conjecture, formulated independently by Archdeacon (1984) and Jaeger (1988), which is a common generalization of the Five-Flow Conjecture and the Cycle Double Cover Conjecture (3.9).

Conjecture 21.14 *Every 2-edge-connected graph admits an orientable double cover by five even subgraphs.*

The second of Tutte's three conjectures seeks to generalize the Four-Colour Conjecture.

THE FOUR-FLOW CONJECTURE

Conjecture 21.15 *Every 2-edge-connected graph with no Petersen graph minor admits a 4-flow.*

For cubic graphs, this conjecture has been proved by N. Robertson, D. Sanders, P. D. Seymour and R. Thomas[1], using a similar approach to the one employed to establish the Four-Colour Theorem (described in Section 15.2). By Theorem 21.11, this special case is equivalent to yet another conjecture of Tutte (1966b), namely that every 2-edge-connected cubic graph with no Petersen graph minor is 3-edge-colourable.

Tutte's third conjecture, if true, would generalize Grötzsch's Theorem (15.10).

THE THREE-FLOW CONJECTURE

Conjecture 21.16 *Every 2-edge-connected graph without 3-edge cuts admits a 3-flow.*

Although Tutte's flow conjectures are about 2-edge-connected graphs, all three may be restricted to 3-edge-connected graphs (Exercise 21.3.5). Moreover, because the Three-Flow Conjecture only concerns graphs without 3-edge cuts, it can be reformulated as: *every 4-edge-connected graph admits a 3-flow.* It is not even known whether there exists any integer k such that every k-edge-connected graph admits a 3-flow. This weaker version of the Three-Flow Conjecture was proposed by Jaeger (1979). However, as we prove in Section 21.5, every 4-edge-connected graph has a 4-flow.

Exercises

21.3.1 Let G be a graph and let G' be a graph obtained from G by splitting off two edges. Show that if G' admits a k-flow then so does G.

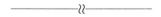

21.3.2

a) Let G be a 2-edge-connected graph with no 3-edge cut, and let v be a vertex of G whose degree is not equal to five. Show that some pair of edges incident with v may be split off so that the resulting graph is 2-edge-connected and has no 3-edge cut.

b) Deduce that the Three-Flow Conjecture (21.16) is equivalent to the statement that every 4-edge-connected 5-regular graph has a 3-flow.

21.3.3 Show that every 2-edge-connected graph with no $K_{3,3}$-minor admits a 4-flow.

---

[1] The proof consists of five papers by various subsets of these authors, of which only one, Robertson et al. (1997b), has so far appeared in print.

21.3.4 Let G be a 2-edge-connected graph and let (X, Y) be a nontrivial partition of V. Show that:

$$Q(G, x) = \frac{Q(G\,/\,X, x) \cdot Q(G\,/\,Y, x)}{x - 1} \quad \text{if } d(X) = 2$$

$$Q(G, x) = \frac{Q(G\,/\,X, x) \cdot Q(G\,/\,Y, x)}{(x - 1)(x - 2)} \quad \text{if } d(X) = 3$$

(K. SEKINE AND C.-Q. ZHANG)

$\star$**21.3.5** Show that:

a) it suffices to prove the Three-Flow, Four-Flow, and Five-Flow Conjectures for 3-edge-connected graphs,

b) for $k \geq 6$, every 2-edge-connected graph admits a k-flow if and only if every 3-edge-connected graph admits a k-flow.

21.3.6 Let k be a positive integer. A digraph D is said to be *(mod k)-balanced* if $d^-(v) \equiv d^+(v) \pmod{k}$. Show that:

a) the Three-Flow Conjecture is true if and only if every 4-edge-connected graph admits a (mod 3)-balanced orientation,

b) the Five-Flow Conjecture is true if and only if every 8-edge-connected graph admits a (mod 5)-balanced orientation.   (F. JAEGER)

21.4 Edge-Disjoint Spanning Trees

In the previous section, we saw that a graph has a 2^k-flow if and only if it has a covering by k even subgraphs (Corollary 21.10). This prompts one to ask which graphs have coverings by few even subgraphs. For instance, a graph which contains a Hamilton cycle or two edge-disjoint spanning trees has such a covering (Exercises 4.3.9 and 4.3.10).

Motivated by the latter observation, we consider here the problem of determining the maximum possible number of edge-disjoint spanning trees in a graph. According to Theorem 4.6, a graph has a spanning tree if and only if it is connected, that is, if and only if $\partial(X) \neq \emptyset$ for every nonempty proper subset X of V. Thus a graph has a spanning tree if and only if, for every partition of its vertex set into two nonempty parts, there is an edge with one end in each part. Generalizing this result, we present a necessary and sufficient condition for a graph to contain k edge-disjoint spanning trees, where k is an arbitrary positive integer. This fundamental structural theorem, found independently by Nash-Williams (1961) and Tutte (1961a), has a number of important applications, in particular to flows. These are described in the next section.

THE NASH-WILLIAMS–TUTTE THEOREM

Recall that the graph obtained from a graph G by shrinking a subset X of the vertex set V of G is denoted G/X. We may extend this shrinking operation to partitions of V, as follows. Given a partition $\mathcal{P} = \{V_1, V_2, \ldots, V_p\}$ of V into nonempty parts, we *shrink* $\mathcal{P}$ by shrinking each set V_i, $1 \leq i \leq p$, and we denote the resulting p-vertex graph by $G/\mathcal{P}$. Note that $G/\mathcal{P}$ might have multiple edges, even if G is simple, but not loops.

Suppose now that G is connected, and consider a spanning tree T of G and a partition $\mathcal{P}$ of V. Because T is connected, so is $T/\mathcal{P}$. Therefore, $e(T/\mathcal{P}) \geq |\mathcal{P}|-1$, where $|\mathcal{P}|$ denotes the number of parts of $\mathcal{P}$. Moreover, if G has k edge-disjoint spanning trees, then this inequality is valid for each of them, and so $e(G/\mathcal{P}) \geq k(|\mathcal{P}| - 1)$. Consequently, a necessary condition for a graph G to have k edge-disjoint spanning trees is that this inequality hold for all partitions $\mathcal{P}$ of V. What Nash-Williams and Tutte proved is that this condition is also sufficient. The proof we give here is due to Frank (1978).

Theorem 21.17 THE NASH-WILLIAMS–TUTTE THEOREM
A graph G has k edge-disjoint spanning trees if and only if, for every partition $\mathcal{P}$ of V into nonempty parts,

$$e(G/\mathcal{P}) \geq k(|\mathcal{P}| - 1) \tag{21.1}$$

Proof We have already shown that (21.1) is a necessary condition for G to have k edge-disjoint spanning trees. It remains to establish its sufficiency. We do so by proving that if condition (21.1) holds, then some orientation of G has k edge-disjoint spanning branchings rooted at some vertex x. By virtue of Theorem 19.8, it suffices to show that G has an orientation D satisfying condition (19.6). On taking complements, we may state this condition in the equivalent form:

$$d^-(X) \geq k, \quad \text{for every nonempty subset } X \text{ of } V \setminus \{x\} \tag{21.2}$$

Because no family of k edge-disjoint spanning trees can use more than k parallel edges, we may assume that each edge of G has multiplicity at most k.

Let G be a graph which satisfies (21.1) but does not satisfy (21.2) for any $x \in V$. Subject to these conditions, let its maximum degree Δ be as large as possible, and consider a vertex x of degree Δ in G. Note that x is joined to some vertex y by fewer than k edges, for if x were joined to every other vertex by k edges, the desired orientation would clearly exist. We add a new edge e to G joining x and y, and set $G' := G + e$.

Observe that G' satisfies (21.1). Therefore, by the choice of G, the graph G' has an orientation D' such that $d^-_{D'}(X) \geq k$ for every nonempty subset X of $V \setminus \{x\}$. We may assume that x is a source of D', because no arc entering x can lie in an x-branching. We denote by a the arc of D' obtained by orienting e, and by Y the set of all vertices reachable from y by means of directed paths in D'. Setting $\overline{Y} := V \setminus Y$, we have $x \in \overline{Y}$, because x is a source of D', and $y \in Y$.

Call a nonempty subset X of $V \setminus \{x\}$ *critical* if $d_{D'}^-(X) = k$. We claim that any critical set X that meets Y is a subset of Y. Suppose, to the contrary, that both $X \cap Y$ and $X \cap \overline{Y}$ are nonempty (see Figure 21.4).

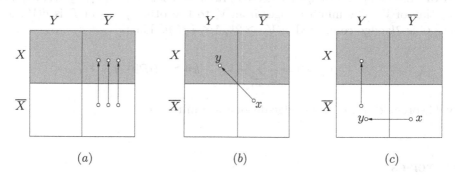

(a)   (b)   (c)

Fig. 21.4. (a) k arcs entering $X \cap \overline{Y}$, (b) the case $y \in X \cap Y$, (c) the case $y \in \overline{X} \cap Y$

By the definition of Y, every arc that enters $X \cap \overline{Y}$ must enter from $\overline{X} \cap \overline{Y}$ (see Figure 21.4a). There are at least k such arcs because $d_{D'}^-(X \cap \overline{Y}) \geq k$. If $y \in X \cap Y$, then the arc $a = (x, y)$ is in $\partial_{D'}^-(X)$ also (Figure 21.4b). And if $y \in \overline{X} \cap Y$, at least one arc enters the nonempty set $X \cap Y$ from $\overline{X} \cap Y$, by the definition of Y (Figure 21.4(c)). In either case, $d_{D'}^-(X) > k$, contradicting the assumption that X is critical. We conclude that $X \subseteq Y$.

To complete the proof, we consider two cases.

Case 1: There exists a vertex z in Y which belongs to no critical set. (This means that $d_{D'}^-(X)) > k$ for every subset X of $V \setminus \{x\}$ which contains z.) Because $y \in Y$, there exists a directed (y, z)-path P in D'. The digraph D obtained from D' by reversing the arcs of P and deleting the arc a is an orientation of G satisfying condition (21.2) (Exercise 21.4.2).

Case 2: Every vertex of Y is contained in a critical set. As in the proof of Edmonds' Branching Theorem (19.8), one may show that the union of intersecting critical sets is critical (Exercise 21.4.1). Because every vertex of Y is contained in a critical set, and every critical set which meets Y is a subset of Y, the maximal critical sets contained in Y form a partition $\{Y_1, Y_2, \ldots, Y_{p-1}\}$ of Y. Without loss of generality, we may assume that $y \in Y_1$.

Now set $Y_p := \overline{Y}$, and consider the partition $\mathcal{P} = \{Y_1, Y_2, \ldots, Y_p\}$ of V. Because $Y_1, Y_2, \ldots, Y_{p-1}$ are critical sets, $d_{D'}^-(Y_i) = k$ for $1 \leq i \leq p-1$. Thus $d_D^-(Y_1) = k-1$, and $d_D^-(Y_i) = k$ for $2 \leq i \leq p - 1$. Finally, by the definition of Y, $d_D^-(Y_p) = 0$. We now have:

$$e(G / \mathcal{P}) = \sum_{i=1}^{p} d_D^-(Y_i) = k(p - 1) - 1 < k(|\mathcal{P}| - 1)$$

contradicting (21.1). $\square$

The following corollary is due to Polesskiĭ (1971).

Corollary 21.18 *Every $2k$-edge-connected graph contains k edge-disjoint spanning trees.*

Proof Let G be a $2k$-edge-connected graph and let $\mathcal{P} := \{V_1, V_2, \ldots, V_p\}$ be a partition of V. The number edges from V_i to the other parts of $\mathcal{P}$ is $d(V_i)$ and, since G is $2k$-edge-connected, $d(V_i) \geq 2k$, $1 \leq i \leq p$. Thus

$$e(G \,/\, \mathcal{P}) = \frac{1}{2} \sum_{i=1}^{p} d(V_i) \geq kp > k(|\mathcal{P}| - 1)$$

By Theorem 21.17, G has k edge-disjoint spanning trees. $\square$

Exercises

$\star$**21.4.1** Show that the union of two intersecting critical sets is critical.

$\star$**21.4.2** In Case 1 of the proof of Theorem 21.17, verify that the orientation D of G satisfies condition (21.2).

21.4.3

a) According to Theorem 21.17, a graph G has a spanning tree if and only if $e(G \,/\, \mathcal{P}) \geq |\mathcal{P}| - 1$ for every partition $\mathcal{P}$ of V into nonempty parts. Show that this condition holds if and only if it holds for all partitions $\mathcal{P}$ of V into two nonempty parts.

b) For some integer $k \geq 2$, give an example of a graph G which satisfies condition (21.1) for all partitions $\mathcal{P}$ of V into two nonempty parts, but which does not have k edge-disjoint spanning trees.

21.4.4 Describe a polynomial-time algorithm which accepts any graph G as input and returns either a family of k edge-disjoint spanning trees of G or a partition $\mathcal{P}$ of G which fails condition (21.1). Determine the complexity of your algorithm.

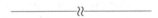

21.4.5

a) Let $G = (V, E)$ be a graph, and let $f : V \to \mathbb{N}$ be a mapping of its vertices. Show that G admits an orientation D with $d^+(v) \leq f(v)$ for all $v \in V$ if and only if $e(F) \leq \sum_{v \in X} f(v)$ for all induced subgraphs $F = G[X]$.

(S.L. HAKIMI)

b) Deduce that every simple plane triangulation on more than four vertices admits a $K_{1,3}$-decomposition.

(M. DEHN)

21.4.6 ARBORICITY
The *arboricity* of a graph is the least number of forests into which it can be decomposed.

a) By appealing to Exercises 19.3.4 and 21.4.5, show that a graph G has arboricity at most k if and only if $e(X) \leq k(|X| - 1)$ for all nonempty subsets X of V.

(C.St.J.A. Nash-Williams)

b) Deduce that every simple planar graph can be decomposed into three forests.

21.4.7 Let G be a 4-edge-connected graph.

a) Show that G has an eulerian spanning subgraph.   (C. Thomassen)

b) Deduce from (a) and Exercise 3.3.8 that the line graph of G is hamiltonian.

c) Let $m_1, m_2, \ldots, m_k$ be positive integers whose sum is m. Deduce from (b) that G admits a decomposition into k connected subgraphs, of sizes $m_1, m_2, \ldots, m_k$.

21.5 The Four-Flow and Eight-Flow Theorems

By exploiting the relationship between spanning trees and even subgraphs described in Corollary 4.12, and applying the results of the preceding section, Jaeger (1976) and Kilpatrick (1975) derived two fundamental theorems on integer flows, the *Four-Flow Theorem* and the *Eight-Flow Theorem*. For notational clarity, we identify trees with their edge sets.

Theorem 21.19 *Every 4-edge-connected graph admits a covering by two even subgraphs.*

Proof Let G be a 4-edge-connected graph. By Corollary 21.18, G has two edge-disjoint spanning trees, hence (Exercise 4.3.10) a covering by two even subgraphs. □

Corollary 21.10 and Theorem 21.19 now yield:

Theorem 21.20   The Four-Flow Theorem
Every 4-edge-connected graph admits a 4-flow.   □

The proof of the Eight-Flow Theorem proceeds along similar lines, but requires a more subtle argument.

Theorem 21.21 *Every 2-edge-connected graph admits a covering by three even subgraphs.*

Proof It suffices to prove the assertion for 3-edge-connected graphs (Exercise 21.5.1). Thus let G be a 3-edge-connected graph. Denote by H the graph obtained by duplicating each edge of G. Being 6-edge-connected, H has three edge-disjoint spanning trees, by Corollary 21.18. These trees correspond to three spanning trees of G, T_1, T_2, and T_3, such that $T_1 \cap T_2 \cap T_3 = \emptyset$. By Corollary 4.12, there exist even subgraphs C_1, C_2, and C_3 such that $C_i \supseteq E \setminus T_i$, $i = 1, 2, 3$. Thus

$$C_1 \cup C_2 \cup C_3 \supseteq (E \setminus T_1) \cup (E \setminus T_2) \cup (E \setminus T_3) = E \setminus (T_1 \cap T_2 \cap T_3) = E$$

in other words, $\{C_1, C_2, C_3\}$ is a covering of G.   □

Theorem 21.21 combined with Corollary 21.10 yields:

Theorem 21.22 THE EIGHT-FLOW THEOREM
Every 2-edge-connected graph admits an 8-flow.   □

UNIFORM COVERS BY EVEN SUBGRAPHS

Theorems 21.19 and 21.21 also yield conditions for the existence of uniform covers by even subgraphs. The link between coverings and uniform covers is provided by the following proposition.

Proposition 21.23 *If a graph admits a covering by k even subgraphs, then it admits a 2^{k-1}-cover by $2^k - 1$ even subgraphs.*

Proof Let $\{C_1, C_2 \ldots, C_k\}$ be a covering of a graph G by k even subgraphs, and let e be an edge of G that belongs to j of these subgraphs, without loss of generality $C_1, C_2, \ldots, C_j$. Then e belongs to all the even subgraphs $\triangle\{C_i : i \in S\}$, such that $S \subseteq \{1, 2, \ldots, k\}$ and $|S \cap \{1, 2, \ldots, j\}|$ is odd. There are 2^{k-1} such subgraphs. Therefore

$$\{\triangle\{C_i : i \in S\} : S \subseteq \{1, 2, \ldots, k\}, S \neq \emptyset\}$$

is a 2^{k-1}-cover of G by $2^k - 1$ even subgraphs.   □

The case $k = 2$ of Proposition 21.23 (or Exercise 3.5.4a), combined with Theorem 21.19, implies the truth of the Cycle Double Cover Conjecture for 4-edge-connected graphs.

Theorem 21.24 *Every 4-edge-connected graph admits a double cover by three even subgraphs.*   □

The same approach, using Theorem 21.21, yields the following result of Bermond et al. (1983) (see, also, Exercise 3.5.4b).

Theorem 21.25 *Every 2-edge-connected graph admits a quadruple cover by seven even subgraphs.*   □

Exercises

⋆**21.5.1** Suppose that every 3-edge-connected graph has a covering by three even subgraphs. Show that every 2-edge-connected graph also has such a covering.

21.5.2 Let G be a 2-edge-connected graph which has no k-flow. Show that there exists an essentially 4-edge-connected cubic graph H with $v(H) + e(H) \leq v(G) + e(G)$ which has no k-flow.

21.5.3

a) Let G be an essentially 4-edge-connected cubic graph and let C be a 2-factor of G which is not a Hamilton cycle. Show that the graph obtained from G by contracting the edges of C is 4-edge-connected.

b) Using (a) and Exercise 21.5.2, derive the Eight-Flow Theorem (21.22) from the Four-Flow Theorem (21.20).

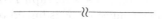

21.6 The Six-Flow Theorem

We present here a proof of Seymour's Six-Flow Theorem (Seymour (1981b)), and also an elegant application of this theorem to uniform covers by even subgraphs, due to Fan (1992). For notational clarity, we identify even subgraphs with their edge sets.

The Six-Flow Theorem is proved by showing that every 3-edge-connected graph G contains an even subgraph C such that the graph G / C has a nowhere-zero circulation over $\mathbb{Z}_3$; equivalently, G has a circulation over $\mathbb{Z}_3$ whose support includes $E \setminus C$ (see Exercise 21.2.7). The notion of the 2-closure of a subgraph plays a key role in establishing this property.

Let S be a set of edges of a graph G. The *2-closure* of S is the (unique) maximal subset of E obtained from S by recursively adding edges, one or two at a time, subject to the condition that the edge or edges added at each stage form a cycle with some of the edges of the subset constructed up to that stage. For example, in the Petersen graph, the 2-closure of a 5-cycle is just the edge set of the cycle itself, whereas the 2-closure of a 9-cycle is the entire edge set.

Lemma 21.26 *Let S be a set of edges of a graph G whose 2-closure is the entire set E. Then there exists a circulation in G over $\mathbb{Z}_3$ whose support includes $E \setminus S$.*

Proof The proof is by induction on $|E \setminus S|$, the result being trivial when $S = E$. Assume that S is a proper subset of E. By hypothesis, there is a cycle C in G such that $1 \leq |C \setminus S| \leq 2$. Set $S' := S \cup C$. Then $|E \setminus S'| < |E \setminus S|$ and the 2-closure of S' is E. Let $D = (V, A)$ be an orientation of G in which C is a directed cycle. By the induction hypothesis, there is a circulation f in D over $\mathbb{Z}_3$ whose support includes $A \setminus S'$. Let f_C be a circulation in D over $\mathbb{Z}_3$ such that $f_C(e) := 1$ if $e \in C$, and $f_C(e) := 0$ otherwise. Then one of f, $f + f_C$, $f - f_C$ is a circulation in G over $\mathbb{Z}_3$ whose support includes $E \setminus S$ (Exercise 21.6.3). $\square$

Lemma 21.27 *Every 3-edge-connected graph G contains an even subgraph C whose 2-closure is E.*

Proof Let C be an even subgraph of G such that:

i) the subgraph H of G induced by the 2-closure of C is connected,

ii) subject to (i), C is as large as possible.

We may assume that H is not a spanning subgraph of G. Otherwise, by the definition of 2-closure, $H = G$ and the assertion holds. Let K be a component of $G - V(H)$. Note that no vertex of K is adjacent to two vertices of H (again by the definition of 2-closure). Because $\delta(G) \geq 3$, we therefore have $\delta(K) \geq 2$, so the endblocks of K are 2-connected.

Consider a maximal 2-edge-connected subgraph F of K containing one of these endblocks. Because G is 3-edge-connected, F is linked to H by two (or more) edges. The ends of these edges in F are distinct because no vertex of K is adjacent to two vertices of H, and these ends are connected by two edge-disjoint paths in F by the edge version of Menger's Theorem (7.17). Let S be the union of the edge sets of these paths. Then $C \cup S$ contradicts the choice of C. $\square$

The above proof of Seymour's Lemma 21.27 is due to Younger (1983).

Theorem 21.28   THE SIX-FLOW THEOREM
Every 2-edge-connected graph admits a 6-flow.

Proof   By Exercise 21.3.5b, it suffices to prove the theorem for 3-edge-connected graphs. Let G be such a graph. By Lemma 21.27, G contains an even subgraph C whose 2-closure is E. Let D be an orientation of G, f_1 a circulation in D over $\mathbb{Z}_2$ whose support is C, and f_2 a circulation in D over $\mathbb{Z}_3$ whose support includes $E \setminus C$; such a circulation f_2 exists by Lemma 21.26. Then (f_1, f_2) is a nowhere-zero circulation in D over $\mathbb{Z}_2 \times \mathbb{Z}_3$. By Corollary 21.8 with $k_1 = 2$ and $k_2 = 3$, G has a nowhere-zero circulation over $\mathbb{Z}_6$. Theorem 21.3 now implies that G has a 6-flow. $\square$

SEXTUPLE COVERS BY EVEN SUBGRAPHS

We now give the application by Fan (1992) of the Six-Flow Theorem to uniform covers by even subgraphs.

Theorem 21.29 *Every 2-edge-connected graph admits a sextuple cover by ten even subgraphs.*

Proof   It suffices to prove the theorem for 3-edge-connected graphs (Exercise 21.6.4). Let G be such a graph. By Lemma 21.27, G contains an even subgraph C whose edge set has 2-closure E. Consider an orientation D of G whose restriction to C is even, and let f' be the circulation in D over $\mathbb{Z}_3$ defined by $f'(e) := 1$ for $e \in C$, and $f'(e) := 0$ for $e \in E \setminus C$. By Lemma 21.26, D has a circulation f over $\mathbb{Z}_3$ whose support contains $E \setminus C$. Set

$$S_i := \{e \in C : f(e) \equiv i \,(\mathrm{mod}\ 3)\}$$

so that (S_0, S_1, S_2) is a partition of C. Then $f - if'$ is a nowhere-zero circulation in $G_i := G \setminus S_i$ over $\mathbb{Z}_3$, so there is a 3-flow f_i in G_i, $i = 0, 1, 2$, by Theorem 21.3. Note that each edge of C lies in two of the three subgraphs G_i and each edge of $E \setminus C$ lies in all three of them. We may regard f_i as a 4-flow in G_i. Theorem 21.21 then implies that G_i has a double cover by three even subgraphs $\mathcal{C}_i := \{C_{i1}, C_{i2}, C_{i3}\}$. Thus $\mathcal{C} := \mathcal{C}_1 \cup \mathcal{C}_2 \cup \mathcal{C}_3$ is a collection of nine even subgraphs of G which together cover each edge of C four times and each edge of $E \setminus C$ six times. It follows that $\mathcal{D} := \{C \bigtriangleup C' : C' \in \mathcal{C}\}$ is a collection of nine even subgraphs of G which together cover each edge of C five times and each edge of $E \setminus C$ six times. Now $\mathcal{C} \cup \mathcal{D}$ is a sextuple cover of G by ten even subgraphs. $\square$

The complement of a perfect matching in a cubic graph is an even subgraph. Using this relationship between perfect matchings and even subgraphs, Fulkerson's Conjecture (17.6) may be restated as: *every 2-edge-connected cubic graph has a quadruple cover by six even subgraphs.* Jaeger (1988) observed that the following more general statement is equivalent to Fulkerson's Conjecture (Exercise 21.6.5).

JAEGER'S CONJECTURE

Conjecture 21.30 *Every 2-edge-connected graph admits a quadruple cover by six even subgraphs.*

By Theorem 21.25, this assertion is true if six is replaced by seven. On the other hand, it is not even known whether there is any integer k such that every 2-edge-connected graph has a $2k$-cover by $3k$ even subgraphs. Such an integer must be even, because the Petersen graph has no $2k$-cover by $3k$ even subgraphs when k is odd (Exercise 21.6.6). In particular, the Petersen graph has no sextuple cover by nine even subgraphs. In this sense, Theorem 21.29 is sharp.

Exercises

⋆**21.6.1** Show that the 2-closure is well-defined.

21.6.2 Find an example of a 2-edge-connected graph G which contains no even subgraph with 2-closure E.

⋆**21.6.3** . In the proof of Lemma 21.26, show that one of the functions f, $f + f_C$, $f - f_C$ is a circulation in G over $\mathbb{Z}_3$ whose support includes $E \setminus S$

⋆**21.6.4** Let k and l be positive integers. Suppose that every 3-edge-connected graph has an l-cover by k even subgraphs. Show that every 2-edge-connected graph has such a cover.

————————⁊⁊————————

21.6.5 Show that Fulkerson's Conjecture (17.6) and Jaeger's Conjecture (21.30) are equivalent.

21.6.6 Show that the Petersen graph has no $2k$-cover by $3k$ even subgraphs if k is odd.

21.7 The Tutte Polynomial

A function on the class of all graphs is called a *graphical invariant* if it takes the same value on isomorphic graphs. The connectivity and stability number are examples of graphical invariants, as are the chromatic polynomial and flow polynomial.

Tutte (1947c) showed that several important graphical invariants satisfy two simple and natural recursion rules. Consider, for example, the number of maximal spanning forests of a graph G. Let us denote this number by $f(G)$. Now a subgraph of a graph G is a maximal spanning forest if and only if it is a union of spanning trees of the components of G. A consequence of the recursion formula for the number of spanning trees of a graph (Proposition 4.9) is that f satisfies the following identities.

$$f(G) = f(G \setminus e) + f(G / e) \quad \text{if } e \text{ is a link of } G \qquad (21.3)$$

$$f(G \cup H) = f(G)f(H) \quad \text{if } G \text{ and } H \text{ are disjoint} \qquad (21.4)$$

The chromatic and flow polynomials also satisfy (21.4) (Exercises 14.7.7 and 21.2.11). Moreover, even though they do not quite satisfy (21.3), they can be made to do so by a minor adjustment of signs. Indeed, the polynomials $(-1)^n P(G, x)$ and $(-1)^{m-n} Q(G, x)$ satisfy both (21.3) and (21.4) (Exercise 21.7.1).

Tutte (1947c) showed that there is a two-variable polynomial with the remarkable property that every invariant which satisfies (21.3) and (21.4), and appropriate initial conditions, is an evaluation of that polynomial. This fundamental result has far-reaching ramifications, with applications to diverse areas of mathematics.

We denote by L_m the graph consisting of one vertex and m incident loops; thus $L_0 = K_1$ and L_1 is the loop graph. Tutte (1947c) proved that any graphical invariant which satisfies (21.3) and (21.4), and takes its values in a commutative ring, is determined by the values it takes on the graphs L_m, $m \geq 0$. As an example of such an invariant, he introduced a two-variable polynomial obtained by combining the expression for the chromatic polynomial due to Whitney (1932b) with its analogue for nowhere-zero flows (see Exercises 14.7.12 and 21.2.13). This polynomial differs by a multiplicative factor of $(-1)^{c(G)}$ from a polynomial now commonly known as the Whitney (rank) polynomial. We give here a definition of this latter polynomial as it lends itself more easily to generalizations in other areas of mathematics such as matroid theory.

For a given graph G, the *Whitney polynomial* of G, denoted by $W(G; x, y)$, is defined by the expression

$$W(G; x, y) := \sum_{S \subseteq E} x^{c(S)-c(G)} y^{|S|-v(G)+c(S)}$$

where $c(S)$ is the number of components of the spanning subgraph of G with edge set S.

As an example, consider the m-bond B_m, the graph consisting of two vertices joined by m links, $m \geq 1$; for instance, $B_1 = K_2$ and B_2 is the 2-cycle. The term of $W(B_m; x, y)$ corresponding to $S = \emptyset$ is $x^{2-1} y^{0-2+2} = x$. If S is any subset of $E(B_m)$ of cardinality i, $1 \leq i \leq m$, the spanning subgraph of B_m with edge set S is connected, and so the term of $W(B_m; x, y)$ corresponding to S is $x^{1-1} y^{i-2+1} = y^{i-1}$. Thus the Whitney polynomial of B_m is given by

$$W(B_m; x, y) = x + \sum_{i=1}^{m} \binom{m}{i} y^{i-1}$$

Tutte (1954a) introduced a polynomial which is obtained by a simple modification of the Whitney polynomial. He called it the *dichromate*. This polynomial, now known as the *Tutte polynomial* and denoted $T(G; x, y)$, is defined by $T(G; x, y) := W(G; x-1, y-1)$. Thus

$$T(G; x, y) = \sum_{S \subseteq E} (x-1)^{c(S)-c(G)} (y-1)^{|S|-v(G)+c(S)}$$

Although the two polynomials $W(G; x, y)$ and $T(G; x, y)$ encode the same information about the graph G, the Tutte polynomial tends to have smaller coefficients. For example, it may be verified (Exercise 21.7.2a) that

$$T(B_m; x, y) = y^{m-1} + y^{m-2} + \cdots + y^2 + x + y$$

The following theorem lists various basic properties of the Tutte polynomial.

Theorem 21.31 *The Tutte polynomial $T(G; x, y)$ has the following properties.*

▷ $T(L_0; x, y) = 1$, $T(B_1; x, y) = x$, $T(L_1; x, y) = y$.

▷ *If e is a loop of G,*
$$T(G; x, y) = y \cdot T(G \setminus e; x, y)$$

▷ *If e is a cut edge of G,*
$$T(G; x, y) = x \cdot T(G / e; x, y)$$

▷ *If e is neither a loop nor a cut edge of G,*
$$T(G; x, y) = T(G \setminus e; x, y) + T(G / e; x, y)$$

▷ *If $G_1, G_2, \ldots, G_k$ are the blocks of G,*
$$T(G; x, y) = \prod_{i=1}^{k} T(G_i; x, y) \qquad \square$$

What makes the Tutte polynomial so special is that it is a universal invariant in the sense of the following theorem, due to Tutte (1948a) and rediscovered by Brylawski (1972).

Theorem 21.32 *Let f be a graphical invariant which takes values in a commutative ring R, is multiplicative on the blocks of the graph, and satisfies $f(L_0) = 1$, the unit element of R. Suppose, furthermore, that $f(B_1) = x$ and $f(L_1) = y$, and that:*

▷ $f(G) = y \cdot f(G \setminus e)$ *if e is a loop of G,*
▷ $f(G) = x \cdot f(G / e)$ *if e is a cut edge of G,*
▷ $f(G \setminus e) + f(G / e)$ *otherwise.*

Then $f(G) = T(G; x, y)$.   □

We leave the proofs of Theorems 21.31 and 21.32 as an exercise (21.7.6).

Evaluations of the Tutte polynomial at certain special values of x and y are given in the table of Figure 21.5. For convenience, the graph G is assumed to be connected; there are analogous results for disconnected graphs.

$T(G; 1, 1)$	number of spanning trees
$T(G; 1, 2)$	number of connected spanning subgraphs
$T(G; 2, 1)$	number of spanning forests
$T(G; 0, 2)$	number of strong orientations
$T(G; 2, 0)$	number of acyclic orientations
$T(G; -1, -1)$	$(-1)^{n-1}(-2)^{\dim(\mathcal{B} \cap \mathcal{C})}$
$T(G; 1 - x, 0)$	$(-1)^{n-1}P(G, x)$
$T(G; 0, 1 - x)$	$(-1)^{m-n+1}Q(G, x)$

Fig. 21.5. Special evaluations of the Tutte polynomial of a connected graph G

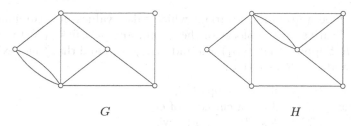

$$G \qquad\qquad\qquad H$$

Fig. 21.6. Nonisomorphic graphs with the same Tutte polynomial

The formula for the number of acyclic orientations is due to Winder (1966) and, independently, Stanley (1973), the formula for the number of strong orientations is due to Las Vergnas (1977/78), and the formula for the dimension of the intersection of the cycle and bond spaces is due to Rosenstiehl and Read (1978). We leave the verification of these evaluations to the reader (Exercise 21.7.5).

As mentioned earlier, the importance of the Tutte polynomial extends to many other areas of mathematics. For example, a well-known polynomial in knot theory, the *Jones polynomial*, is closely related to the Tutte polynomial. The article by Brylawski and Oxley (1992) provides an excellent survey of applications of the Tutte polynomial. For an account of the computational aspects of the Tutte polynomial, we refer the reader to Welsh (1993).

Exercises

21.7.1 Verify that $(-1)^n P(G, x)$ and $(-1)^{m-n} Q(G, x)$ satisfy (21.3) and (21.4).

21.7.2 Verify the following results.

a) For the m-bond B_m, $T(B_m; x, y) = y^{m-1} + y^{m-2} + \ldots + y^2 + x + y$.
b) For the n-cycle C_n, $T(C_n; x, y) = x^{n-1} + x^{n-2} + \ldots + x^2 + x + y$.
c) For any tree G on n vertices, $T(G; x, y) = x^{n-1}$.

21.7.3 For the two nonisomorphic graphs G and H in Figure 21.6, show that $T(G; x, y) = T(H; x, y)$.

21.7.4 Let G be a plane graph, and let G^* be its dual. Show that $T(G; x, y) = T(G^*; y, x)$.

21.7.5 Verify the entries in the table of Figure 21.5.

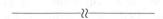

21.7.6 Supply proofs of Theorems 21.31 and 21.32.

21.7.7 Show that the Tutte polynomial is reconstructible.

21.7.8 Let f be a graphical invariant which takes values in a commutative ring R, is multiplicative on the blocks of the graph, and satisfies $f(K_1) = 1$, the unit element of R. Suppose that $f(B_1) = x$ and $f(L_1) = y$, and that there exist nonzero elements r and s of R such that:

▷ $f(G) = y \cdot f(G \setminus e)$ if e is a loop of G,
▷ $f(G) = x \cdot f(G / e)$ if e is a cut edge of G,
▷ $f(G) = r \cdot f(G \setminus e) + s \cdot f(G / e)$ otherwise.

Show that $f(G) = r^{m-n+1} s^{n-1} T(G; x/s, y/r)$.   (J. OXLEY AND D.J.A. WELSH)

21.7.9 THE RELIABILITY POLYNOMIAL
 Let G be a connected graph and let G_p be a random spanning subgraph of G obtained by retaining each edge independently with probability $p \in [0, 1)$ (or deleting it with probability $q := 1 - p$). Denote by $R(G, p)$ the probability that G_p is connected. Show that $R(G, p) = p^{n-1} q^{m-n+1} T(G; 1, 1/q)$.
(If G represents a communication network and q denotes the probability of failure of a link, then $R(G, p)$ is the probability of communications remaining possible between all centres. For this reason, $R(G, p)$ is known as the *reliability polynomial* of G.)

21.8 Related Reading

Several of the concepts and theorems covered in this chapter have natural generalizations to matroids.

PACKING BASES IN MATROIDS

Edmonds (1965a) showed that a matroid M defined on a set E has k disjoint bases if and only if

$$k\, r(S) + |E \setminus S| \geq k\, r(E) \quad \text{for all } S \subseteq E$$

When this theorem is applied to the cycle matroid of a graph, one obtains the Nash-Williams–Tutte Theorem (21.17). It can be derived from Rado's generalization of Hall's Theorem mentioned in Section 16.6, but Edmonds' proof has the advantage that it gives rise to a polynomial-time algorithm.

THE TUTTE POLYNOMIAL FOR MATROIDS

Crapo (1969) observed that with any matroid one may associate a two-variable polynomial analogous to the Tutte polynomial for graphs. As with graphs, it is convenient to define first the *Whitney polynomial* of a matroid M on a set E:

$$W(M; x, y) := \sum_{S \subseteq E} x^{r(E)-r(S)} y^{r^*(E)-r^*(E \setminus S)}$$

The Tutte polynomial of M can then be obtained from the Whitney polynomial by a change of variables, as in the case of graphs. The Tutte polynomial of the cycle matroid of a graph is simply the Tutte polynomial of the graph. For a survey of applications of Tutte polynomials of matroids, see Brylawski and Oxley (1992).

Appendix A

Unsolved Problems

Our first book *Graph Theory with Applications* included a list of fifty unsolved problems.[1] Many have since been settled, notably the Four-Colour and Perfect Graph Conjectures. But several other basic problems, such as the Reconstruction and Five-Flow Conjectures, remain unresolved.

We present here an updated selection of interesting unsolved problems and conjectures; statements of conjectures are set in italics. A number of them have been discussed in detail in the text. In such cases, a reference to the problem or conjecture is included. In other instances, where appropriate, pertinent exercises or theorems from the book are indicated. A more detailed discussion of the problems and conjectures listed here may be found on the book's web page.

Proving and conjecturing, as Paul Erdős was fond of saying, is an integral part mathematical activity. He did plenty of both in his life! The article by Chung (1997) contains many of Erdős's favourite problems. Another excellent source of unsolved problems is Jensen and Toft (1995).

Reconstruction

1. (Conjecture 2.19) *Every simple graph on at least three vertices is reconstructible.*
KELLY (1942)

2. (Conjecture 2.23) *Every simple graph on at least four edges is edge-reconstructible.*
HARARY (1964)

3. *If two infinite graphs are hypomorphic, then each is isomorphic to an induced subgraph of the other.* [Exercise 4.2.10]
HALIN (1970)

4. *Every simple graph on five or more vertices is reconstructible from its vertex-switchings.* [Exercise 2.7.19]
STANLEY (1985)

---

[1] S. C. Locke of Florida Atlantic University maintains a web site which contains information on the status of these fifty problems.

Subgraphs

5. *Every simple graph G with at least $p\binom{n}{2}$ edges has at least $cn^{v(F)}p^{e(F)}$ labelled copies of every fixed bipartite graph F, where $c \sim 1$ (that is, at least as many copies as the random graph with the same number of vertices and edges as G.)* [Exercise 13.2.6]   SIDORENKO (1991)

6. Is there an integer d such that the vertices of any strict digraph with minimum outdegree d can be partitioned into two classes so that the minimum outdegree of the subgraph induced by each class is at least two?   ALON (1996)

Coverings, Decompositions, and Packings

7. *Every simple connected graph on n vertices can be decomposed into at most $\frac{1}{2}(n+1)$ paths.* [Exercise 2.4.6b]   T. GALLAI, SEE LOVÁSZ (1968B)

8. *Every simple even graph on n vertices can be decomposed into at most $\frac{1}{2}(n-1)$ cycles.* [Exercise 2.4.6a]   G. HAJÓS, SEE LOVÁSZ (1968B)

9. *Every regular tournament can be decomposed into directed Hamilton cycles.*
                    P.J. KELLY, SEE MOON (1968)

10. (Conjecture 3.9) *Every graph without cut edges has a cycle double cover.*
         FOLKLORE (ROBERTSON (2007)); SZEKERES (1973); SEYMOUR (1979B)

11. (Conjecture 3.11) *Every simple graph on n vertices without cut edges has a double cover by at most $n-1$ cycles.*   BONDY (1990)

12. *Every 2-edge-connected graph has a double cover by at most five even subgraphs* [Exercise 3.5.3].   PREISSMANN (1981)

13. *Let G be a loopless even graph of minimum degree four, and let W be an Euler tour of G. Then G admits a decomposition into cycles none of which contains two consecutive edges of W.* [Exercise 3.3.9]
                    G. SABIDUSSI 1975, SEE FLEISCHNER (1990)

14. *Every simple k-regular graph has linear arboricity $\lceil (k+1)/2 \rceil$.* [Theorem 13.19]
                    AKIYAMA ET AL. (1981), HILTON (1982)

15. *Let G and H be simple graphs on n vertices such that $(\Delta(G)+1)(\Delta(H)+1) \leq n+1$. Then K_n contains edge-disjoint copies of G and H.*
                    CATLIN (1974); BOLLOBÁS AND ELDRIDGE (1978)

16. Does every 3-connected cubic graph on $3k$ vertices admit a partition into k paths of length two?   KELMANS (2005)

Complexity

17. (Conjecture 8.2) $\mathcal{P} \neq \mathcal{NP}$

J. Edmonds 1966, Cook (1971), Levin (1973)

18. (Conjecture 8.3) $\mathcal{P} = \mathcal{NP} \cap \mathrm{co} - \mathcal{NP}$ Edmonds (1965c)

19. Is the following problem in $\mathcal{P}$?

Given: a cubic graph G and a Hamilton cycle in G,
Find: a second Hamilton cycle in G.
[Exercise 18.4.1]   Chrobak and Poljak (1988)

20. Is the following problem in co-$\mathcal{NP}$?

Given: a graph G and two k-subsets X and Y of V,
Decide: do there exist k internally disjoint (X, Y)-paths of odd length in G?
Thomassen (1980)

Connectivity

21. *There is an integer-valued function $f(k)$ such that if G is any $f(k)$-connected graph and x and y are any two vertices of G, then there exists an induced xy-path P such that $G - V(P)$ is k-connected.*

L. Lovász, see Kawarabayashi et al. (2007)

22. *Let G be a graph and T a set of vertices in G any two of which are connected by $2k$ edge-disjoint paths. Then there are k edge-disjoint trees in G all of which contain T.* [Corollary 21.18]   Kriesell (2003)

23. *Every $2k$-connected digraph contains a spanning k-connected oriented subdigraph.*   Thomassen (1989)

Embeddings

24. (Conjecture 21.14) *Every 2-edge-connected graph has an orientable double cover by five even subgraphs.*   Archdeacon (1984), Jaeger (1988)

25. *Every simple planar graph G has an induced forest on at least $n/2$ vertices.*   Albertson and Berman (1979)

26. *Every planar oriented graph can be partitioned into two induced acyclic subgraphs.*

Neumann-Lara (1985), R. Škrekovski 2001, see Bokal et al. (2004)

27. *Every* 5-*connected nonplanar graph contains a* K_5-*subdivision.* [Exercise 10.5.14]
    P.D. SEYMOUR 1974, SEE SEYMOUR (2007); A.K. KELMANS 1979, SEE KELMANS (1993)

28. *Every* 6-*connected graph with no* K_6-*minor has a vertex whose deletion results in a planar graph.*   JØRGENSEN (1994)

29. *No graph with more edges than vertices has a thrackle embedding.* [Exercise 10.1.11]   J.H. CONWAY C.1968, SEE WOODALL (1971)

30. *Every simple planar graph admits a straight-line embedding with integer edge lengths.*   KEMNITZ AND HARBORTH (2001)

Extremal Problems

31. *If* G *is simple and* $m > n(k-1)/2$, *then* G *contains every tree with* k *edges.* [Exercise 4.1.9]   P. ERDŐS AND V.T. SÓS 1963, SEE ERDŐS (1964)

32. *There exists a positive constant* c *such that* $\mathrm{ex}(n, C_{2k}) \geq cn^{1+1/k}$. [Exercise 12.2.14]   ERDŐS (1971)

33. *If* G *has at most* k *edge-disjoint triangles, then there is a set of* $2k$ *edges whose deletion destroys every triangle.*   Zs. TUZA 1981, SEE TUZA (1990)

34. *If* G *is a simple triangle-free graph, then there is a set of at most* $n^2/25$ *edges whose deletion destroys every odd cycle.*   ERDŐS ET AL. (1988)

Ramsey Numbers

35. Give a constructive proof that $r(k,k) \geq c^k$ for some $c > 1$ and all $k \geq 1$. [Theorem 12.12]   ERDŐS (1969)

36. Does the limit $\lim_{k \to \infty}(r(k,k))^{1/k}$ exist? If so, determine its value.
                                            P. ERDŐS 1947, SEE ERDŐS (1961B)

37. *For every tree* T *on* n *vertices,* $r(T,T) \leq 2n - 2$.   BURR AND ERDŐS (1976)

38. *For every simple graph* F, *there is a positive constant* $\epsilon := \epsilon(F)$ *such that every simple graph* G *which contains no induced copy of* F *has either a stable set or a clique of cardinality* n^ϵ.   ERDŐS AND HAJNAL (1989)

Vertex Colourings

39. (Conjecture 15.11) *Every k-chromatic graph has a K_k-minor.*

HADWIGER (1943)

40. Does a k-chromatic graph necessarily contain a K_k-subdivision when $k = 5$ and $k = 6$? [Exercise 15.4.3] CATLIN (1979)

41. *Every $2k$-chromatic digraph contains a copy of every oriented tree on $k + 1$ vertices.* [Theorem 4.5] BURR (1980)

42. *Every graph which can be decomposed into k complete graphs on k vertices is k-colourable.* P. ERDŐS, V. FABER, AND L. LOVÁSZ 1972, SEE ERDŐS (1976)

43. *Every graph which can be decomposed into $k - 1$ complete bipartite graphs is k-colourable.* [Theorem 2.8]

N. ALON, M. SAKS, AND P.D. SEYMOUR, SEE KAHN (1994)

44. *The chromatic number of the weak product of two graphs is equal to the lesser of their chromatic numbers: $\chi(G \times H) = \min\{\chi(G), \chi(H)\}$.* [Exercise 14.1.18]

A. KOTZIG, AND HEDETNIEMI (1966)

45. *Let G be a k-chromatic graph which contains no k-clique, and let $k_1 + k_2$ be a partition of $k + 1$ with k_1, $k_2 \geq 2$. Then there are disjoint subgraphs G_1 and G_2 of G such that G_i is k_i-chromatic, $i = 1, 2$.* [Exercise 16.3.13]

L. LOVÁSZ 1968, SEE ERDŐS (1968)

46. *The union of a 1-degenerate graph (a forest) and a 2-degenerate graph is 5-colourable.* M. TARSI, SEE KLEIN (1994)

47. *For any graph G, $\chi \leq \lceil (\omega + \Delta + 1)/2 \rceil$.* [Equation (14.2), Theorem 14.4, Exercise 14.1.14] REED (1998)

48. *Every triangle-free graph of infinite chromatic number contains every finite tree as an induced subgraph.* GYÁRFÁS (1975)

49. *Every nonempty graph which contains no induced odd cycle of length five or more admits a 2-vertex-colouring with no monochromatic maximum clique.*

HOÀNG AND MCDIARMID (2002)

50. *If $\chi \geq (n - 1)/2$, then $\chi_L = \chi$.* OHBA (2002)

51. *There is a constant c such that the list chromatic number of any bipartite graph G is at most $c \log \Delta$.* [Exercise 14.5.6] ALON (2000)

52. *The absolute values of the coefficients of a chromatic polynomial form a unimodal sequence.* READ (1968)

Colourings of Embedded Graphs

53. *Every planar graph without 4-cycles or 5-cycles is 3-colourable.* [Discharging, Section 15.2]   R. STEINBERG 1977, SEE STEINBERG (1993)

54. *Every toroidal graph has three vertices whose deletion results in a 4-colourable graph.* [Section 15.5]   ALBERTSON (1981)

55. Determine the chromatic number of the unit distance graph. [Exercise 14.1.20]   E. NELSON 1950, SEE JENSEN AND TOFT (1995), P.150

Matchings

56. *There is a constant $c > 1$ such that every 2-edge-connected cubic graph G has at least c^n perfect matchings.* [Exercise 16.2.27]   LOVÁSZ AND PLUMMER (1986), P.314

57. (Conjecture 17.6) *Every 2-edge-connected cubic graph admits a double covering by six perfect matchings.*   FULKERSON (1971)

Edge Colourings

58. *If G is a loopless graph and $\chi' > \Delta + 1$, then*

$$\chi' = \max \left\{ \left\lceil \frac{2e(H)}{v(H) - 1} \right\rceil : H \subseteq G, \ v(H) \ odd, \ v(H) \geq 3 \right\}$$

[Exercise 17.2.2]   GOLDBERG (1974), SEYMOUR (1979A)

59. *Every d-regular simple graph on n vertices, with n even and $d \geq n/2$, is d-edge-colourable.*   HILTON (1989)

60. (Conjecture 17.8) *Every k-edge-colourable simple graph is k-list-edge-colourable.* [Theorem 17.10] VARIOUS AUTHORS, SEE HÄGGKVIST AND CHETWYND (1992)

61. *Every simple graph G has a total colouring in $\Delta + 2$ colours.*   VIZING (1964), BEHZAD (1965)

62. *Given any proper edge colouring of a graph G, one can obtain a proper edge colouring of G in χ' colours by means of a sequence of alternating path and cycle exchanges.*   VIZING (1965)

63. *Every simple graph G admits a proper $(\Delta + 2)$-edge-colouring such that the union of any two colour classes is acyclic.*   FIAMČÍK (1978)

Paths and Cycles in Graphs

64. *For $k \geq 3$, there exists no graph in which each pair of vertices is connected by a unique path of length k.* [Theorem 3.1]   KOTZIG (1979)

65. *Every longest cycle in a 3-connected graph has a chord.*
C. THOMASSEN 1976, SEE THOMASSEN (1997A)

66. *In a k-connected graph, where $k \geq 2$, any two longest cycles have at least k vertices in common.* [Exercise 5.1.5]   S. SMITH, SEE GRÖTSCHEL (1984)

67. *Every 3-connected cyclically 4-edge-connected cubic graph on n vertices contains a cycle of length at least cn, where c is a positive constant.*
J.A. BONDY, SEE FLEISCHNER AND JACKSON (1989)

68. *Every graph in which any two cycles of length k or more intersect has a set of k vertices which meet all such cycles.*   BIRMELÉ (2003)

69. Do any three longest paths in a connected graph have a vertex in common?
[Exercise 2.2.13a]   GALLAI (1968B)

Paths and Cycles in Digraphs

The digraphs here are assumed to be *strict*, that is, without loops or multiple arcs.

70. *Every oriented graph in which each vertex has outdegree at least k contains a directed path of length $2k$.*   S. THOMASSÉ 2005, SEE SULLIVAN (2006)

71. *Every strong oriented graph in which each vertex has indegree and outdegree at least k contains a directed cycle of length at least $2k + 1$.*   JACKSON (1981)

72. *Every positively weighted strong digraph (D, w) such that $w^-(v) \geq 1$ and $w^+(v) \geq 1$ for all $v \in V$ contains a directed cycle of weight at least one.*
[Exercise 2.5.9]   BOLLOBÁS AND SCOTT (1996)

73. *Every digraph of girth g in which each vertex has outdegree at least k has at least $k(g - 1) + 1$ vertices.* [Exercise 2.2.22]   CACCETTA AND HÄGGKVIST (1978)

74. *Every digraph in which each vertex has outdegree at least k contains k directed cycles $C_1, \ldots, C_k$ such that C_j meets $\cup_{i=1}^{j-1} C_i$ in at most one vertex, $2 \leq j \leq k$.*
HOÀNG AND REED (1987)

75. *Every digraph in which each vertex has outdegree at least $2k - 1$ contains k disjoint directed cycles.*   BERMOND AND THOMASSEN (1981)

76. *Every digraph with at least one directed cycle has an arc whose reversal reduces the number of directed cycles.*   ÁDÁM (1964)

77. *Every digraph has a stable set meeting all longest directed paths.*
[Theorem 14.5, Exercise 12.1.13]   LABORDE ET AL. (1983)

78. *Every oriented graph has a vertex with at least as many second outneighbours as (first) outneighbours.* [Exercise 4.1.18]

P.D. SEYMOUR 1995, SEE BONDY (1997)

Hamilton Paths and Cycles in Graphs

79. *Every 3-connected cubic bipartite planar graph is hamiltonian.* [Theorem 18.2, Exercise 18.2.5]   BARNETTE (1969), A. KELMANS AND M. LOMONOSOV 1975

80. *Every 4-connected claw-free graph is hamiltonian.* [Exercise 18.3.16]

MATTHEWS AND SUMNER (1984)

81. *Every 4-regular 4-polytope is hamiltonian.*

D.W. BARNETTE, SEE GRÜNBAUM (1970), P.1145

82. *Every prism over a 3-connected planar graph is hamiltonian.*

KAISER ET AL. (2007)

83. (Conjecture 18.14) *Every simple 4-regular graph with a Hamilton cycle has a second Hamilton cycle.*   SHEEHAN (1975)

84. (Conjecture 18.18) *Every 4-connected graph with a Hamilton cycle has a second Hamilton cycle.*   FLEISCHNER (2007)

85. *Every prism over a 3-connected cubic planar graph admits a Hamilton decomposition.*   ALSPACH AND ROSENFELD (1986)

86. *Every cubic planar graph with exactly three Hamilton cycles contains a triangle.*   CANTONI (1950), SEE ALSO NINČÁK (1974), TUTTE (1976)

87. Does every connected vertex-transitive graph have a Hamilton path?

LOVÁSZ (1970)

88. *Every Cayley graph is hamiltonian.*

T.D. PARSONS, SEE WITTE AND GALLIAN (1984)

89. *All but a finite number of vertex-transitive connected graphs are hamiltonian.*
[Exercise 18.1.12]   C. THOMASSEN 1976, SEE BERMOND (1978)

90. *There exists a positive integer k such that every k-tough graph is hamiltonian.*
[Exercise 18.1.22]   CHVÁTAL (1973)

91. Is there a hypohamiltonian graph of minimum degree at least four? [Exercise 18.1.16]   THOMASSEN (1978)

92. *There is no bipartite hypotraceable graph.* [Exercise 18.1.17].

GRÖTSCHEL (1978)

Coverings and Packings in Directed Graphs

93. (Conjecture 14.6) *Let D be a digraph and $\mathcal{P}$ a k-optimal path partition of D. Then there is a partial k-colouring of D orthogonal to $\mathcal{P}$.*   BERGE (1982)

94. (Conjecture 19.14) *In every digraph, the maximum number of disjoint transversals of directed cuts is equal to the size of a smallest directed cut.*

WOODALL (1978)

Integer Flows

95. (Conjecture 21.13) *Every 2-edge-connected graph has a 5-flow.*

TUTTE (1954A)

96. (Conjecture 21.15) *Every 2-edge-connected graph with no Petersen minor has a 4-flow.*   TUTTE (1966B)

97. (Conjecture 21.16) *Every 2-edge-connected graph without 3-edge cuts has a 3-flow.*   W.T. TUTTE 1972, SEE BONDY AND MURTY (1976), P.252

98. *Every $4k$-edge-connected graph admits a balanced orientation $\mod(2k+1)$.* [Exercise 21.3.6]   JAEGER (1984)

Hypergraphs

99. *Every simple 3-uniform hypergraph on $3n$ vertices which contains no complete 3-uniform hypergraph on four vertices has at most $\frac{1}{2}n^2(5n-3)$ edges.* [Exercise 12.2.6]   TURÁN (1941)

100. *Every simple 3-uniform hypergraph on $2n$ vertices which contains no complete 3-uniform hypergraph on five vertices has at most $n^2(n-1)$ edges.* [Exercise 12.2.6]   TURÁN (1941)

References

ABBOTT H.L. and MOSER L. (1966). Sum-free sets of integers. *Acta Arith.* **11**, 393–396.

ABBOTT H.L. and ZHOU B. (1991). On small faces in 4-critical planar graphs. *Ars Combin.* **32**, 203–207.

ÁDÁM A. (1964). Problem 2. In *Theory of Graphs and its Applications* (M. Fiedler, ed.), 234. Publishing House of the Czechoslovak Academy of Sciences, Prague.

AGRAWAL M., KAYAL N. and SAXENA N. (2004). PRIMES is in P. *Ann. of Math.* *(2)* **160**, 781–793.

AHARONI R., HARTMAN I.B.A. and HOFFMAN A.J. (1985). Path partitions and packs of acyclic digraphs. *Pacific J. Math.* **118**, 249–259.

AHO A.V., HOPCROFT J.E. and ULLMAN J.D. (1975). *The Design and Analysis of Computer Algorithms.* Addison-Wesley Series in Computer Science and Information Processing, Addison-Wesley, Reading, MA. Second printing.

AHO A.V., HOPCROFT J.E. and ULLMAN J.D. (1983). *Data Structures and Algorithms.* Addison-Wesley Series in Computer Science and Information Processing, Addison-Wesley, Reading, MA.

AIGNER M. (1995). Turán's graph theorem. *Amer. Math. Monthly* **102**, 808–816.

AIGNER M. and ZIEGLER G.M. (2004). *Proofs from The Book.* Third edition. Springer, Berlin. Including illustrations by Karl H. Hofmann.

AJTAI M., KOMLÓS J. and SZEMERÉDI E. (1980). A note on Ramsey numbers. *J. Combin. Theory Ser. A* **29**, 354–360.

AJTAI M., CHVÁTAL V., NEWBORN M.M. and SZEMERÉDI E. (1982). Crossing-free subgraphs. In *Theory and Practice of Combinatorics*, 9–12. North-Holland Math. Stud., Vol. 60, North-Holland, Amsterdam.

AKIYAMA J., EXOO G. and HARARY F. (1981). Covering and packing in graphs. IV. Linear arboricity. *Networks* **11**, 69–72.

ALBERTSON M.O. (1981). Open Problem 2. In *The Theory and Applications of Graphs, Proceedings of the Fourth International Conference, Kalamazoo, Mich., 1980*, 609. Wiley, New York.

ALBERTSON M.O. and BERMAN D.M. (1979). A conjecture on planar graphs. In *Graph Theory and Related Topics* (J.A. Bondy and U.S.R. Murty, eds.), 357. Academic Press, New York.

ALBERTSON M.O. and HUTCHINSON J.P. (1978). On the independence ratio of a graph. *J. Graph Theory* **2**, 1–8.

ALELIUNAS R., KARP R.M., LIPTON R.J., LOVÁSZ L. and RACKOFF C. (1979). Random walks, universal traversal sequences, and the complexity of maze problems. In *20th Annual Symposium on Foundations of Computer Science (San Juan, Puerto Rico, 1979)*, 218–223. IEEE, New York.

ALON N. (1988). The linear arboricity of graphs. *Israel J. Math.* **62**, 311–325.

ALON N. (1996). Disjoint directed cycles. *J. Combin. Theory Ser. B* **68**, 167–178.

ALON N. (1999). Combinatorial Nullstellensatz. *Combin. Probab. Comput.* **8**, 7–29. Recent Trends in Combinatorics (Mátraháza, 1995).

ALON N. (2000). Degrees and choice numbers. *Random Structures Algorithms* **16**, 364–368.

ALON N. and LINIAL N. (1989). Cycles of length 0 modulo k in directed graphs. *J. Combin. Theory Ser. B* **47**, 114–119.

ALON N. and SPENCER J.H. (2000). *The Probabilistic Method.* Second edition. Wiley-Interscience Series in Discrete Mathematics and Optimization, Wiley, New York. With an appendix on the life and work of Paul Erdős.

ALON N., TUZA Z. and VOIGT M. (1997). Choosability and fractional chromatic numbers. *Discrete Math.* **165/166**, 31–38. Graphs and Combinatorics (Marseille, 1995).

ALSPACH B. and ROSENFELD M. (1986). On Hamilton decompositions of prisms over simple 3-polytopes. *Graphs Combin.* **2**, 1–8.

ALSPACH B., GODDYN L. and ZHANG C.Q. (1994). Graphs with the circuit cover property. *Trans. Amer. Math. Soc.* **344**, 131–154.

AMINI O., MAZOIT F., NISSE N. and THOMASSÉ S. (2007). Submodular partition functions. *Discrete Math.* In press.

APPEL K. and HAKEN W. (1977a). Every planar map is four colorable. I. Discharging. *Illinois J. Math.* **21**, 429–490.

APPEL K. and HAKEN W. (1977b). The solution of the four-color-map problem. *Sci. Amer.* **237**, 108–121, 152.

APPEL K., HAKEN W. and KOCH J. (1977). Every planar map is four colorable. II. Reducibility. *Illinois J. Math.* **21**, 491–567.

APPLEGATE D., BIXBY R., CHVÁTAL V. and COOK W. (2007). *The Traveling Salesman Problem: A Computational Study.* Princeton Series in Applied Mathematics, Princeton University Press, Princeton, NJ.

ARCHDEACON D. (1984). Face colorings of embedded graphs. *J. Graph Theory* **8**, 387–398.

ARNBORG S. and PROSKUROWSKI A. (1989). Linear time algorithms for NP-hard problems restricted to partial k-trees. *Discrete Appl. Math.* **23**, 11–24.

ARNBORG S., CORNEIL D.G. and PROSKUROWSKI A. (1987). Complexity of finding embeddings in a k-tree. *SIAM J. Algebraic Discrete Methods* **8**, 277–284.

ARTZY R. (1956). Self-dual configurations and their Levi graphs. *Proc. Amer. Math. Soc.* **7**, 299–303.

BABAI L. (1995). Automorphism groups, isomorphism, reconstruction. In *Handbook of Combinatorics, Vol. 2*, 1447–1540. Elsevier, Amsterdam.

BAER R. (1946). Polarities in finite projective planes. *Bull. Amer. Math. Soc.* **52**, 77–93.

BANG-JENSEN J. and GUTIN G. (2001). *Digraphs*. Springer Monographs in Mathematics, Springer, London.

BÁRÁNY I. (1978). A short proof of Kneser's conjecture. *J. Combin. Theory Ser. A* **25**, 325–326.

BARNETTE D. (1969). Conjecture 5. In *Recent Progress in Combinatorics* (W.T. Tutte, ed.), 343. Academic Press, New York.

BATTLE J., HARARY F. and KODAMA Y. (1962). Every planar graph with nine points has a nonplanar complement. *Bull. Amer. Math. Soc.* **68**, 569–571.

BAUER D., HAKIMI S.L. and SCHMEICHEL E. (1990). Recognizing tough graphs is NP-hard. *Discrete Appl. Math.* **28**, 191–195.

BEHZAD M. (1965). *Graphs and their Chromatic Numbers*. Ph.D. thesis, Michigan State University.

BEHZAD M., CHARTRAND G. and WALL C.E. (1970). On minimal regular digraphs with given girth. *Fund. Math.* **69**, 227–231.

BEINEKE L.W. and HARARY F. (1965). The thickness of the complete graph. *Canad. J. Math.* **17**, 850–859.

BEINEKE L.W., HARARY F. and MOON J.W. (1964). On the thickness of the complete bipartite graph. *Proc. Cambridge Philos. Soc.* **60**, 1–5.

BENSON C.T. (1966). Minimal regular graphs of girths eight and twelve. *Canad. J. Math.* **18**, 1091–1094.

BERGE C. (1957). Two theorems in graph theory. *Proc. Nat. Acad. Sci. U.S.A.* **43**, 842–844.

BERGE C. (1958). Sur le couplage maximum d'un graphe. *C. R. Acad. Sci. Paris* **247**, 258–259.

BERGE C. (1963). Some classes of perfect graphs. In *Six Papers in Graph Theory*, 1–21. Indian Statistical Institute, Calcutta.

BERGE C. (1973). *Graphs and Hypergraphs*. North-Holland Mathematical Library, Vol. 6, North-Holland, Amsterdam. Translated by Edward Minieka.

BERGE C. (1977). L'art subtil du hex. Unpublished manuscript.

BERGE C. (1982). k-optimal partitions of a directed graph. *European J. Combin.* **3**, 97–101.

BERGE C. (1985). *Graphs*. North-Holland Mathematical Library, Vol. 6, North-Holland, Amsterdam. Second revised edition of part 1 of the 1973 English version.

BERGE C. (1995). *Who Killed the Duke of Densmore?* Bibliothèque Oulipienne, Vol. 67, OuLiPo, Paris.

BERGE C. (1996). The history of the perfect graphs. *Southeast Asian Bull. Math.* **20**, 5–10.

602   References

BERGE C. (1997). Motivations and history of some of my conjectures. *Discrete Math.* **165/166**, 61–70. Graphs and Combinatorics (Marseille, 1995).

BERGER E. and HARTMAN I. (2008). Proof of Berge's strong path partition conjecture for $k = 2$. *European J. Combin.* **29**, 179–192.

BERMOND J. (1978). Hamiltonian graphs. In *Selected Topics in Graph Theory* (L.W. Beineke and R.J. Wilson, eds.), xii+451. Academic Press, London.

BERMOND J.C. and THOMASSEN C. (1981). Cycles in digraphs—a survey. *J. Graph Theory* **5**, 1–43.

BERMOND J.C., JACKSON B. and JAEGER F. (1983). Shortest coverings of graphs with cycles. *J. Combin. Theory Ser. B* **35**, 297–308.

BERNHART A. (1947). Six-rings in minimal five-color maps. *Amer. J. Math.* **69**, 391–412.

BESSY S. and THOMASSÉ S. (2003). Every strong digraph has a spanning strong subgraph with at most $n + 2\alpha - 2$ arcs. *J. Combin. Theory Ser. B* **87**, 289–299.

BESSY S. and THOMASSÉ S. (2004). Three min-max theorems concerning cyclic orders of strong digraphs. In *Integer Programming and Combinatorial Optimization*, 132–138. Lecture Notes in Comput. Sci., Vol. 3064, Springer, Berlin.

BIENSTOCK D. and DEAN N. (1993). Bounds for rectilinear crossing numbers. *J. Graph Theory* **17**, 333–348.

BIENSTOCK D. and LANGSTON M.A. (1995). Algorithmic implications of the graph minor theorem. In *Network Models*, 481–502. Handbooks Oper. Res. Management Sci., Vol. 7, North-Holland, Amsterdam.

BIGGS N. (1993). *Algebraic graph theory*. Second edition. Cambridge Mathematical Library, Cambridge University Press, Cambridge.

BIGGS N.L., LLOYD E.K. and WILSON R.J. (1986). *Graph Theory. 1736–1936*. Second edition. Clarendon Press, New York.

BIRKHOFF G.D. (1912/13). A determinant formula for the number of ways of coloring a map. *Ann. of Math. (2)* **14**, 42–46.

BIRKHOFF G.D. (1913). The reducibility of maps. *Amer. J. Math.* **35**, 115–128.

BIRMELÉ E. (2003). *Largeur d'arborescence, q-cliques-mineurs et propriété d'Erdős-Pósa*. Ph.D. thesis, Université Claude Bernard Lyon 1.

BLANUŠA D. (1946). Le problème des quatre couleurs. *Hrvatsko Prirodoslovno Društvo. Glasnik Mat.-Fiz. Astr. Ser. II.* **1**, 31–42.

BLUM A. and KARGER D. (1997). An $O(n^{3/14})$-coloring algorithm for 3-colorable graphs. *Inform. Process. Lett.* **61**, 49–53.

BODLAENDER H.L. (2006). Treewidth: characterizations, applications, and computations. In *Graph-Theoretic Concepts in Computer Science*, 1–14. Lecture Notes in Comput. Sci., Vol. 4271, Springer, Berlin.

BOKAL D., FIJAVŽ G., JUVAN M., KAYLL P.M. and MOHAR B. (2004). The circular chromatic number of a digraph. *J. Graph Theory* **46**, 227–240.

BOLLOBÁS B. (1978). *Extremal Graph Theory*. London Mathematical Society Monographs, Vol. 11, Academic Press, London.

BOLLOBÁS B. (1980). A probabilistic proof of an asymptotic formula for the number of labelled regular graphs. *European J. Combin.* **1**, 311–316.

BOLLOBÁS B. (2001). *Random Graphs.* Second edition. Cambridge Studies in Advanced Mathematics, Vol. 73, Cambridge University Press, Cambridge.

BOLLOBÁS B. and ELDRIDGE S.E. (1978). Packings of graphs and applications to computational complexity. *J. Combin. Theory Ser. B* **25**, 105–124.

BOLLOBÁS B. and ERDŐS P. (1976). Cliques in random graphs. *Math. Proc. Cambridge Philos. Soc.* **80**, 419–427.

BOLLOBÁS B. and HARRIS A.J. (1985). List-colourings of graphs. *Graphs Combin.* **1**, 115–127.

BOLLOBÁS B. and SCOTT A.D. (1996). A proof of a conjecture of Bondy concerning paths in weighted digraphs. *J. Combin. Theory Ser. B* **66**, 283–292.

BOLLOBÁS B., CATLIN P.A. and ERDŐS P. (1980). Hadwiger's conjecture is true for almost every graph. *European J. Combin.* **1**, 195–199.

BONDY J.A. (1978). Hamilton cycles in graphs and digraphs. In *Proceedings of the Ninth Southeastern Conference on Combinatorics, Graph Theory, and Computing*, volume 21, 3–28. Utilitas Math., Winnipeg.

BONDY J.A. (1990). Small cycle double covers of graphs. In *Cycles and Rays (Montreal, PQ, 1987)*, 21–40. NATO Adv. Sci. Inst. Ser. C Math. Phys. Sci., Vol. 301, Kluwer, Dordrecht.

BONDY J.A. (1991). A graph reconstructor's manual. In *Surveys in Combinatorics, 1991 (Guildford, 1991)*, 221–252. London Math. Soc. Lecture Note Ser., Vol. 166, Cambridge Univ. Press, Cambridge.

BONDY J.A. (1995a). Basic graph theory: paths and circuits. In *Handbook of Combinatorics, Vol. 1*, 3–110. Elsevier, Amsterdam.

BONDY J.A. (1995b). A short proof of the Chen-Manalastas theorem. *Discrete Math.* **146**, 289–292.

BONDY J.A. (1997). Counting subgraphs: a new approach to the Caccetta–Häggkvist conjecture. *Discrete Math.* **165/166**, 71–80. Graphs and Combinatorics (Marseille, 1995).

BONDY J.A. and CHARBIT P. (2004). Cyclic orders, circuit coverings and circular chromatic number. *Technical report*, Université Claude-Bernard, Lyon 1.

BONDY J.A. and CHVÁTAL V. (1976). A method in graph theory. *Discrete Math.* **15**, 111–135.

BONDY J.A. and HELL P. (1990). A note on the star chromatic number. *J. Graph Theory* **14**, 479–482.

BONDY J.A. and MURTY U.S.R. (1976). *Graph Theory with Applications.* American Elsevier, New York.

BONNINGTON C.P. and LITTLE C.H.C. (1995). *The Foundations of Topological Graph Theory.* Springer, New York.

BORŮVKA O. (1926a). O jistém problému minimálním. *Práce Moravské Přírodovědecké Společnosti Brno* **3**, 37–58. Czech, with German summary; On a minimal problem.

BORŮVKA O. (1926b). Příspěvek k řešení otázky economické stavby elektrovodných sítí. *Elektrotechnický Obzor* **15:10**, 153–154.

BORODIN O.V., GLEBOV A.N., RASPAUD A. and SALAVATIPOUR M.R. (2005). Planar graphs without cycles of length from 4 to 7 are 3-colorable. *J. Combin. Theory Ser. B* **93**, 303–311.

BROERSMA H.J. (2002). On some intriguing problems in Hamiltonian graph theory—a survey. *Discrete Math.* **251**, 47–69. Cycles and Colourings (Stará Lesná, 1999).

BROERSMA H.J., VAN DEN HEUVEL J., JACKSON B. and VELDMAN H.J. (1996). Hamiltonicity of regular 2-connected graphs. *J. Graph Theory* **22**, 105–124.

BROOKS R.L. (1941). On colouring the nodes of a network. *Proc. Cambridge Philos. Soc.* **37**, 194–197.

BROOKS R.L., SMITH C.A.B., STONE A.H. and TUTTE W.T. (1940). The dissection of rectangles into squares. *Duke Math. J.* **7**, 312–340.

BROWNE C. (2000). *Hex Strategy: Making the Right Connections*. A. K. Peters, Natick, MA.

BRYLAWSKI T. and OXLEY J. (1992). The Tutte polynomial and its applications. In *Matroid Applications*, 123–225. Encyclopedia Math. Appl., Vol. 40, Cambridge Univ. Press, Cambridge.

BRYLAWSKI T.H. (1972). A decomposition for combinatorial geometries. *Trans. Amer. Math. Soc.* **171**, 235–282.

BUCZAK J. (1980). *Finite Group Theory*. Ph.D. thesis, Oxford University.

BURR S.A. (1980). Subtrees of directed graphs and hypergraphs. In *Proceedings of the Eleventh Southeastern Conference on Combinatorics, Graph Theory and Computing*, volume 28, 227–239.

BURR S.A. and ERDŐS P. (1976). Extremal Ramsey theory for graphs. *Utilitas Math.* **9**, 247–258.

CACCETTA L. and HÄGGKVIST R. (1978). On minimal digraphs with given girth. In *Proceedings of the Ninth Southeastern Conference on Combinatorics, Graph Theory, and Computing*, volume 21, 181–187. Utilitas Math., Winnipeg.

CAMERON K. (1986). On k-optimum dipath partitions and partial k-colourings of acyclic digraphs. *European J. Combin.* **7**, 115–118.

CAMERON P.J. (1980). 6-transitive graphs. *J. Combin. Theory Ser. B* **28**, 168–179.

CAMERON P.J. (1983). Automorphism groups of graphs. In *Selected Topics in Graph Theory, 2* (L.W. Beineke and R.J. Wilson, eds.), 89–127. Academic Press, London.

CAMERON P.J. (1997). The random graph. In *The Mathematics of Paul Erdős, II*, 333–351. Algorithms Combin., Vol. 14, Springer, Berlin.

CAMERON P.J. (2001). The random graph revisited. In *European Congress of Mathematics, Vol. I (Barcelona, 2000)*, 267–274. Progr. Math., Vol. 201, Birkhäuser, Basel.

CAMION P. (1959). Chemins et circuits hamiltoniens des graphes complets. *C. R. Acad. Sci. Paris* **249**, 2151–2152.

CANTONI R. (1950). Conseguenze dell'ipotesi del circuito totale pari per le reti con vertici tripli. *Ist. Lombardo Sci. Lett. Rend. Cl. Sci. Mat. Nat. (3)* **14(83)**, 371–387.

CARVALHO M., LUCCHESI C. and MURTY U.S.R. (2002). Optimal ear decompositions of matching covered graphs and bases for the matching lattice. *J. Combin. Theory Ser. B* **85**, 59–93.

CATLIN P.A. (1974). Subgraphs of graphs. I. *Discrete Math.* **10**, 225–233.

CATLIN P.A. (1979). Hajós' graph-coloring conjecture: variations and counterexamples. *J. Combin. Theory Ser. B* **26**, 268–274.

CAYLEY A. (1889). A theorem on trees. *Quart. J. Math.* **23**, 376–378.

CHARBIT P. (2005). *Plongements de graphes et étude des circuits.* Ph.D. thesis, Université Claude Bernard Lyon 1.

CHARTRAND G. and KRONK H.V. (1968). Randomly traceable graphs. *SIAM J. Appl. Math.* **16**, 696–700.

CHEN C.C. and MANALASTAS JR. P. (1983). Every finite strongly connected digraph of stability 2 has a Hamiltonian path. *Discrete Math.* **44**, 243–250.

CHEN G. and YU X. (2002). Long cycles in 3-connected graphs. *J. Combin. Theory Ser. B* **86**, 80–99.

CHRISTOFIDES N. (1976). Worst-case analysis of a new heuristic for the travelling salesman problem. *Management Sciences Research Report 388*, Graduate School of Industrial Administration, Carnegie-Mellon University, Pittsburgh, PA.

CHROBAK M. and POLJAK S. (1988). On common edges in optimal solutions to traveling salesman and other optimization problems. *Discrete Appl. Math.* **20**, 101–111.

CHUDNOVSKY M. and SEYMOUR P. (2005). The structure of claw-free graphs. In *Surveys in Combinatorics 2005*. London Math. Soc. Lecture Note Series, Vol. 327.

CHUDNOVSKY M., ROBERTSON N., SEYMOUR P. and THOMAS R. (2003). Progress on perfect graphs. *Math. Program.* **97**, 405–422. ISMP, 2003 (Copenhagen).

CHUDNOVSKY M., CORNUÉJOLS G., LIU X., SEYMOUR P. and VUŠKOVIĆ K. (2005). Recognizing Berge graphs. *Combinatorica* **25**, 143–186.

CHUDNOVSKY M., ROBERTSON N., SEYMOUR P. and THOMAS R. (2006). The strong perfect graph theorem. *Ann. of Math. (2)* **164**, 51–229.

CHUNG F.R.K. (1997). Open problems of Paul Erdős in graph theory. *J. Graph Theory* **25**, 3–36.

CHVÁTAL V. (1972). On Hamilton's ideals. *J. Combin. Theory Ser. B* **12**, 163–168.

CHVÁTAL V. (1973). Tough graphs and Hamiltonian circuits. *Discrete Math.* **5**, 215–228.

CHVÁTAL V. (1983). *Linear Programming.* A Series of Books in the Mathematical Sciences, Freeman, New York.

CHVÁTAL V. and ERDŐS P. (1972). A note on Hamiltonian circuits. *Discrete Math.* **2**, 111–113.

CHVÁTAL V. and LOVÁSZ L. (1974). Every directed graph has a semi-kernel. In *Hypergraph Seminar (Proc. First Working Sem., Columbus, Ohio, 1972; dedicated to Arnold Ross)*, 175. Lecture Notes in Math., Vol. 411, Springer, Berlin.

CHVÁTAL V., RÖDL V., SZEMERÉDI E. and TROTTER JR. W.T. (1983). The Ramsey number of a graph with bounded maximum degree. *J. Combin. Theory Ser. B* **34**, 239–243.

COOK S.A. (1971). The complexity of theorem-proving procedures. In *Proceedings of the Third Annual ACM Symposium on Theory of Computing*, 151–158. ACM Press, New York.

CORMEN T.H., LEISERSON C.E., RIVEST R.L. and STEIN C. (2001). *Introduction to Algorithms*. Second edition. MIT Press, Cambridge, MA.

CORNEIL D.G. (2004). Lexicographic breadth first search—a survey. In *Graph-Theoretic Concepts in Computer Science*, 1–19. Lecture Notes in Comput. Sci., Vol. 3353, Springer, Berlin.

COXETER H.S.M. (1950). Self-dual configurations and regular graphs. *Bull. Amer. Math. Soc.* **56**, 413–455.

COXETER H.S.M. (1983). My graph. *Proc. London Math. Soc. (3)* **46**, 117–136.

CRAPO H.H. (1969). The Tutte polynomial. *Aequationes Math.* **3**, 211–229.

CROSSLEY M.D. (2005). *Essential Topology*. Springer Undergraduate Mathematics Series, Springer-Verlag London Ltd., London.

CUNNINGHAM W.H. and EDMONDS J. (1980). A combinatorial decomposition theory. *Canad. J. Math.* **32**, 734–765.

DESCARTES B. (1948). Network-colourings. *Mat. Gaz.* **32**, 67–69.

DEVILLERS A. (2002). *Classification of some Homogeneous and Ultrahomogeneous Structures*. Ph.D. thesis, Université Libre de Bruxelles.

DIESTEL R. (2005). *Graph Theory*. Third edition. Graduate Texts in Mathematics, Vol. 173, Springer, Berlin.

DIJKSTRA E.W. (1959). A note on two problems in connexion with graphs. *Numer. Math.* **1**, 269–271.

DILWORTH R.P. (1950). A decomposition theorem for partially ordered sets. *Ann. of Math. (2)* **51**, 161–166.

DINIC E.A. (1970). An algorithm for the solution of the problem of maximal flow in a network with power estimation. *Dokl. Akad. Nauk SSSR* **194**, 754–757.

DIRAC G.A. (1951). Note on the colouring of graphs. *Math. Z.* **54**, 347–353.

DIRAC G.A. (1952a). A property of 4-chromatic graphs and some remarks on critical graphs. *J. London Math. Soc.* **27**, 85–92.

DIRAC G.A. (1952b). Some theorems on abstract graphs. *Proc. London Math. Soc. (3)* **2**, 69–81.

DIRAC G.A. (1953). The structure of k-chromatic graphs. *Fund. Math.* **40**, 42–55.

DIRAC G.A. (1957). A theorem of R. L. Brooks and a conjecture of H. Hadwiger. *Proc. London Math. Soc. (3)* **7**, 161–195.

DIRAC G.A. (1961). On rigid circuit graphs. *Abh. Math. Sem. Univ. Hamburg* **25**, 71–76.

DONALD A. (1980). An upper bound for the path number of a graph. *J. Graph Theory* **4**, 189–201.

DOUSSE O., FRANCESCHETTI M. and THIRAN P. (2006). On the throughput scaling of wireless relay networks. *IEEE Trans. Inform. Theory* **52**, 2756–2761.

DOYLE P.G. and SNELL J.L. (1984). *Random Walks and Electric Networks.* Carus Mathematical Monographs, Vol. 22, Mathematical Association of America, Washington, DC.

DUIJVESTIJN A.J.W. (1978). Simple perfect squared square of lowest order. *J. Combin. Theory Ser. B* **25**, 240–243.

EDMONDS J. (1965a). Lehman's switching game and a theorem of Tutte and Nash-Williams. *J. Res. Nat. Bur. Standards Sect. B* **69B**, 73–77.

EDMONDS J. (1965b). Maximum matching and a polyhedron with 0, 1-vertices. *J. Res. Nat. Bur. Standards Sect. B* **69B**, 125–130.

EDMONDS J. (1965c). Minimum partition of a matroid into independent subsets. *J. Res. Nat. Bur. Standards Sect. B* **69B**, 67–72.

EDMONDS J. (1965d). Paths, trees, and flowers. *Canad. J. Math.* **17**, 449–467.

EDMONDS J. (1973). Edge-disjoint branchings. In *Combinatorial Algorithms*, 91–96. Algorithmics Press, New York.

EDMONDS J. and FULKERSON D.R. (1965). Transversals and matroid partition. *J. Res. Nat. Bur. Standards Sect. B* **69B**, 147–153.

EDMONDS J. and KARP R.M. (1970). Theoretical improvements in algorithmic efficiency for network flow problems. In *Combinatorial Structures and their Applications*, 93–96. Gordon and Breach, New York.

EGERVÁRY E. (1931). On combinatorial properties of matrices. *Mat. Lapok.* **38**, 16–28. Hungarian with German summary.

ELIAS P., FEINSTEIN A. and SHANNON C.E. (1956). A note on the maximum flow through a network. *IRE. Trans. Inf. Theory* **IT-2**, 117–119.

ELLINGHAM M.N. (1988). Recent progress in edge reconstruction. *Congr. Numer.* **62**, 3–20. Seventeenth Manitoba Conference on Numerical Mathematics and Computing.

ELLINGHAM M.N. (1996). Spanning paths, cycles, trees and walks for graphs on surfaces. *Congr. Numer.* **115**, 55–90. Surveys in Graph Theory.

ERDŐS P. (1947). Some remarks on the theory of graphs. *Bull. Amer. Math. Soc.* **53**, 292–294.

ERDŐS P. (1955). Problem 250. *Elemente der Math.* **10**, 114.

ERDŐS P. (1956). Solution to Problem 250 (H. Debrunner). *Elemente der Math.* **11**, 137.

ERDŐS P. (1961a). Graph theory and probability. II. *Canad. J. Math.* **13**, 346–352.

ERDŐS P. (1961b). Some unsolved problems. *Magyar Tud. Akad. Mat. Kutató Int. Közl.* **6**, 221–254.

ERDŐS P. (1964). Extremal problems in graph theory. In *Theory of Graphs and its Applications (Proc. Sympos. Smolenice, 1963)*, 29–36. Publ. House Czechoslovak Acad. Sci., Prague.

ERDŐS P. (1964/1965). On an extremal problem in graph theory. *Colloq. Math.* **13**, 251–254.

ERDŐS P. (1965). On some extremal problems in graph theory. *Israel J. Math.* **3**, 113–116.

ERDŐS P. (1968). Problem 2. In *Theory of Graphs (Proc. Colloq., Tihany, 1966)*, 361. Academic Press, New York.

ERDŐS P. (1969). Problems and results in chromatic graph theory. In *Proof Techniques in Graph Theory*, 27–35. Academic Press, New York.

ERDŐS P. (1971). Some unsolved problems in graph theory and combinatorial analysis. In *Combinatorial Mathematics and its Applications*, 97–109. Academic Press, London.

ERDŐS P. (1976). Problems and results in graph theory and combinatorial analysis. In *Proceedings of the Fifth British Combinatorial Conference*, volume 15, 169–192. Utilitas Math., Winnipeg.

ERDŐS P. and FAJTLOWICZ S. (1981). On the conjecture of Hajós. *Combinatorica* **1**, 141–143.

ERDŐS P. and GALLAI T. (1960). Graphs with prescribed degrees of vertices. *Mat. Lapok* **11**, 264–274. Hungarian.

ERDŐS P. and HAJNAL A. (1989). Ramsey-type theorems. *Discrete Appl. Math.* **25**, 37–52. Combinatorics and Complexity (Chicago, IL, 1987).

ERDŐS P. and KELLY P. (1967). The minimal regular graph containing a given graph. In *A Seminar on Graph Theory*, 65–69. Holt, Rinehart and Winston, New York.

ERDŐS P. and LOVÁSZ L. (1975). Problems and results on 3-chromatic hypergraphs and some related questions. In *Infinite and Finite Sets, Vol. II*, 609–627. Colloq. Math. Soc. János Bolyai, Vol. 10, North-Holland, Amsterdam.

ERDŐS P. and PÓSA L. (1965). On independent circuits contained in a graph. *Canad. J. Math.* **17**, 347–352.

ERDŐS P. and PURDY G. (1995). Extremal problems in combinatorial geometry. In *Handbook of Combinatorics, Vol. 1*, 809–874. Elsevier, Amsterdam.

ERDŐS P. and RÉNYI A. (1959). On random graphs. I. *Publ. Math. Debrecen* **6**, 290–297.

ERDŐS P. and RÉNYI A. (1960). On the evolution of random graphs. *Magyar Tud. Akad. Mat. Kutató Int. Közl.* **5**, 17–61.

ERDŐS P. and RÉNYI A. (1962). On a problem in the theory of graphs. *Magyar Tud. Akad. Mat. Kutató Int. Közl.* **7**, 623–641 (1963).

ERDŐS P. and RÉNYI A. (1963). Asymmetric graphs. *Acta Math. Acad. Sci. Hungar* **14**, 295–315.

ERDŐS P. and SIMONOVITS M. (1966). A limit theorem in graph theory. *Studia Sci. Math. Hungar* **1**, 51–57.

ERDŐS P. and STONE A.H. (1946). On the structure of linear graphs. *Bull. Amer. Math. Soc.* **52**, 1087–1091.

ERDŐS P. and SZEKERES G. (1935). A combinatorial problem in geometry. *Compositio Math.* **2**, 463–470.

ERDŐS P., KO C. and RADO R. (1961). Intersection theorems for systems of finite sets. *Quart. J. Math. Oxford Ser. (2)* **12**, 313–320.

ERDŐS P., RÉNYI A. and SÓS V.T. (1966). On a problem of graph theory. *Studia Sci. Math. Hungar.* **1**, 215–235.

ERDŐS P., MEIR A., SÓS V.T. and TURÁN P. (1971). On some applications of graph theory. II. In *Studies in Pure Mathematics (Presented to Richard Rado)*, 89–99. Academic Press, London.

ERDŐS P., MEIR A., SÓS V.T. and TURÁN P. (1972a). On some applications of graph theory. I. *Discrete Math.* **2**, 207–228.

ERDŐS P., MEIR A., SÓS V.T. and TURÁN P. (1972b). On some applications of graph theory. III. *Canad. Math. Bull.* **15**, 27–32.

ERDŐS P., FAUDREE R., PACH J. and SPENCER J. (1988). How to make a graph bipartite. *J. Combin. Theory Ser. B* **45**, 86–98.

EULER L. (1736). Solutio problematis ad geometriam situs pertinentis. *Comment. Academiae Sci. I. Petropolitanae* **8**, 128–140.

EULER L. (1752). Elementa doctrinae solidorum.— Demonstratio nonnullarum insignium proprietatum, quibus solida hedris planis inclusa sunt praedita. *Novi Comment. Acad. Sc. Imp. Petropol.* **4**, 109–140–160.

FAN G. (1992). Integer flows and cycle covers. *J. Combin. Theory Ser. B* **54**, 113–122.

FAN G. (2002). Subgraph coverings and edge switchings. *J. Combin. Theory Ser. B* **84**, 54–83.

FARKAS J. (1902). Über die Theorie der einfachen Ungleichungen. *J. Reine Angewandte Math.* **124**, 1–27.

FELLER W. (1968). *An Introduction to Probability Theory and its Applications. Vol. I.* Third edition. Wiley, New York.

FEOFILOFF P. and YOUNGER D.H. (1987). Directed cut transversal packing for source-sink connected graphs. *Combinatorica* **7**, 255–263.

FIAMČÍK I. (1978). The acyclic chromatic class of a graph. *Math. Slovaca* **28**, 139–145.

FISK S. (1978). The nonexistence of colorings. *J. Combin. Theory Ser. B* **24**, 247–248.

FLEISCHNER H. (1974). The square of every two-connected graph is Hamiltonian. *J. Combin. Theory Ser. B* **16**, 29–34.

FLEISCHNER H. (1990). *Eulerian Graphs and Related Topics. Part 1. Vol. 1.* Annals of Discrete Mathematics, Vol. 45, North-Holland, Amsterdam.

FLEISCHNER H. (1991). *Eulerian Graphs and Related Topics. Part 1. Vol. 2.* Annals of Discrete Mathematics, Vol. 50, North-Holland, Amsterdam.

FLEISCHNER H. (1992). Spanning Eulerian subgraphs, the splitting lemma, and Petersen's theorem. *Discrete Math.* **101**, 33–37. Special volume to mark the centennial of Julius Petersen's "Die Theorie der regulären Graphs", Part II.

FLEISCHNER H. (1994). Uniqueness of maximal dominating cycles in 3-regular graphs and of Hamiltonian cycles in 4-regular graphs. *J. Graph Theory* **18**, 449–459.

FLEISCHNER H. (2007). Uniquely Hamiltonian graphs of minimum degree four. *J. Graph Theory.* To appear.

FLEISCHNER H. and JACKSON B. (1989). A note concerning some conjectures on cyclically 4-edge connected 3-regular graphs. In *Graph Theory in Memory of G. A. Dirac (Sandbjerg, 1985)*, 171–177. Ann. Discrete Math., Vol. 41, North-Holland, Amsterdam.

FLEISCHNER H. and STIEBITZ M. (1992). A solution to a colouring problem of P. Erdős. *Discrete Math.* **101**, 39–48. Special volume to mark the centennial of Julius Petersen's "Die Theorie der regulären Graphs", Part II.

FLEURY (1883). Deux problèmes de géométrie de situation. *Journal de Mathématiques Elémentaires,* 257–261.

FOLKMAN J. (1970). Graphs with monochromatic complete subgraphs in every edge coloring. *SIAM J. Appl. Math.* **18**, 19–24.

FORD JR. L.R. and FULKERSON D.R. (1956). Maximal flow through a network. *Canad. J. Math.* **8**, 399–404.

FORD JR. L.R. and FULKERSON D.R. (1962). *Flows in Networks.* Princeton University Press, Princeton, NJ.

FORTUNE S., HOPCROFT J. and WYLLIE J. (1980). The directed subgraph homeomorphism problem. *Theoret. Comput. Sci.* **10**, 111–121.

FRANK A. (1978). On disjoint trees and arborescences. In *Algebraic Methods in Graph Theory, Colloquia Mathematica Soc. J. Bolyai,* volume 25, 159–169. North-Holland.

FRANK A. (1992). On a theorem of Mader. *Discrete Math.* **101**, 49–57. Special volume to mark the centennial of Julius Petersen's "Die Theorie der regulären Graphs", Part II.

FRANK A. (1995). Connectivity and network flows. In *Handbook of Combinatorics, Vol. 1,* 111–177. Elsevier, Amsterdam.

FRANKL P. and WILSON R.M. (1981). Intersection theorems with geometric consequences. *Combinatorica* **1**, 357–368.

FRANKLIN P. (1934). A six-color problem. *J. Math. Phys.* **13**, 363–369.

FRÉCHET M. and FAN K. (2003). *Invitation to Combinatorial Topology.* Dover, Mineola. Translated from the French, with notes, by Howard W. Eves, Reprint of the 1967 English translation.

FRUCHT R. (1938). Herstellung von Graphen mit vorgegebener abstrakter Gruppe. *Compositio Math.* **6**, 239–250.

FUJISHIGE S. (2005). *Submodular Functions and Optimization.* Second edition. Annals of Discrete Mathematics, Vol. 58, Elsevier, Amsterdam.

FULKERSON D.R. (1971). Blocking and anti-blocking pairs of polyhedra. *Math. Programming* **1**, 168–194.

FÜREDI Z. (1991). Turán type problems. In *Surveys in Combinatorics, 1991 (Guildford, 1991),* 253–300. London Math. Soc. Lecture Note Ser., Vol. 166, Cambridge Univ. Press, Cambridge.

FÜREDI Z. (1996). On the number of edges of quadrilateral-free graphs. *J. Combin. Theory Ser. B* **68**, 1–6.

FÜREDI Z. and SIMONOVITS M. (2005). Triple systems not containing a Fano configuration. *Combin. Probab. Comput.* **14**, 467–484.

GALLAI T. (1959). Über extreme Punkt- und Kantenmengen. *Ann. Univ. Sci. Budapest. Eötvös Sect. Math.* **2**, 133–138.

GALLAI T. (1964a). Maximale Systeme unabhängiger Kanten. *Magyar Tud. Akad. Mat. Kutató Int. Közl.* **9**, 401–413 (1965).

GALLAI T. (1964b). Problem 15. In *Theory of Graphs and its Applications* (M. Fiedler, ed.), 161. Czech. Acad. Sci. Publ.

GALLAI T. (1968a). On directed paths and circuits. In *Theory of Graphs (Proc. Colloq., Tihany, 1966)*, 115–118. Academic Press, New York.

GALLAI T. (1968b). Problem 4. In *Theory of Graphs (Proc. Colloq., Tihany, 1966)*, 362. Academic Press, New York.

GALLAI T. (1968c). Problem 6. In *Theory of Graphs (Proc. Colloq., Tihany, 1966)*, 362. Academic Press, New York.

GALLAI T. and MILGRAM A.N. (1960). Verallgemeinerung eines graphentheoretischen Satzes von Rédei. *Acta Sci. Math. (Szeged)* **21**, 181–186.

GALVIN F. (1995). The list chromatic index of a bipartite multigraph. *J. Combin. Theory Ser. B* **63**, 153–158.

GARDINER A. (1976). Homogeneous graphs. *J. Combin. Theory Ser. B* **20**, 94–102.

GAREY M.R. and JOHNSON D.S. (1979). *Computers and Intractability: a Guide to the Theory of NP-completeness.* Freeman, San Francisco.

GASPARIAN G.S. (1996). Minimal imperfect graphs: a simple approach. *Combinatorica* **16**, 209–212.

GEORGES J.P. (1989). Non-Hamiltonian bicubic graphs. *J. Combin. Theory Ser. B* **46**, 121–124.

GHOUILA-HOURI A. (1960). Sur l'existence d'un flot ou d'une tension prenant ses valeurs dans un groupe abélien. *C. R. Acad. Sci. Paris* **250**, 3931–3933.

GODSIL C. and ROYLE G. (2001). *Algebraic Graph Theory.* Graduate Texts in Mathematics, Vol. 207, Springer, New York.

GOLDBERG M.K. (1974). A remark on the chromatic class of a multigraph. In *Numerical Mathematics and Computer Technology, No. V (Russian)*, 128–130, 168. Akad. Nauk Ukrain. SSR Fiz.-Tehn. Inst. Nizkih Temperatur, Kharkov.

GOLUMBIC M.C. (2004). *Algorithmic Graph Theory and Perfect Graphs.* Second edition. Annals of Discrete Mathematics, Vol. 57, Elsevier, Amsterdam. With a foreword by Claude Berge.

GOMORY R.E. and HU T.C. (1961). Multi-terminal network flows. *J. Soc. Indust. Appl. Math.* **9**, 551–570.

GÖRING F. (2000). Short proof of Menger's theorem. *Discrete Math.* **219**, 295–296.

GOULD R.J. (2003). Advances on the Hamiltonian problem—a survey. *Graphs Combin.* **19**, 7–52.

GOWERS W.T. (2006). Quasirandomness, counting and regularity for 3-uniform hypergraphs. *Combin. Probab. Comput.* **15**, 143–184.

GRAHAM R.L. and POLLAK H.O. (1971). On the addressing problem for loop switching. *Bell System Tech. J.* **50**, 2495–2519.

GRAHAM R.L., ROTHSCHILD B.L. and SPENCER J.H. (1990). *Ramsey Theory.* Second edition. Wiley-Interscience Series in Discrete Mathematics and Optimization, Wiley, New York.

GRAHAM R.L., GRÖTSCHEL M. and LOVÁSZ L., eds. (1995). *Handbook of Combinatorics. Vol. 1, 2.* Elsevier, Amsterdam.

GREENE J.E. (2002). A new short proof of Kneser's conjecture. *Amer. Math. Monthly* **109**, 918–920.

GREENWOOD R.E. and GLEASON A.M. (1955). Combinatorial relations and chromatic graphs. *Canad. J. Math.* **7**, 1–7.

GRINBERG È.J. (1968). Plane homogeneous graphs of degree three without Hamiltonian circuits. In *Latvian Math. Yearbook, 4 (Russian)*, 51–58. Izdat. "Zinatne", Riga.

GROSS J.L. and TUCKER T.W. (1987). *Topological Graph Theory.* Wiley-Interscience Series in Discrete Mathematics and Optimization, Wiley, New York.

GRÖTSCHEL M. (1978). Hypo-Hamiltonian facets of the symmetric travelling salesman polytype. *Z. Angew. Math. Mech.* **58**, T469–T471.

GRÖTSCHEL M. (1984). On intersections of longest cycles. In *Graph Theory and Combinatorics (Cambridge, 1983)*, 171–189. Academic Press, London.

GRÖTSCHEL M., LOVÁSZ L. and SCHRIJVER A. (1988). *Geometric Algorithms and Combinatorial Optimization.* Algorithms and Combinatorics: Study and Research Texts, Vol. 2, Springer, Berlin.

GRÖTZSCH H. (1958/1959). Zur Theorie der diskreten Gebilde. VII. Ein Dreifarbensatz für dreikreisfreie Netze auf der Kugel. *Wiss. Z. Martin-Luther-Univ. Halle-Wittenberg. Math.-Nat. Reihe* **8**, 109–120.

GRÜNBAUM B. (1970). Polytopes, graphs, and complexes. *Bull. Amer. Math. Soc.* **76**, 1131–1201.

GUPTA R.P. (1966). The chromatic index and the degree of a graph. *Notices Amer. Math. Soc.* **13**. Abstract 66T-429.

GYÁRFÁS A. (1975). On Ramsey covering-numbers. In *Infinite and Finite Sets, Vol. II*, 801–816. Colloq. Math. Soc. János Bolyai, Vol. 10, North-Holland, Amsterdam.

GYÁRFÁS A., JENSEN T. and STIEBITZ M. (2004). On graphs with strongly independent color-classes. *J. Graph Theory* **46**, 1–14.

HADWIGER H. (1943). Über eine Klassifikation der Streckenkomplexe. *Vierteljschr. Naturforsch. Ges. Zürich* **88**, 133–142.

HÄGGKVIST R. and CHETWYND A. (1992). Some upper bounds on the total and list chromatic numbers of multigraphs. *J. Graph Theory* **16**, 503–516.

HÄGGKVIST R. and THOMASSEN C. (1982). Circuits through specified edges. *Discrete Math.* **41**, 29–34.

HAHN G. and JACKSON B. (1990). A note concerning paths and independence number in digraphs. *Discrete Math.* **82**, 327–329.

HAJNAL A. and SZEMERÉDI E. (1970). Proof of a conjecture of P. Erdős. In *Combinatorial Theory and its Applications, II (Proc. Colloq., Balatonfüred, 1969)*, 601–623. North-Holland, Amsterdam.

HALIN R. (1969). A theorem on *n*-connected graphs. *J. Combin. Theory* **7**, 150–154.

HALIN R. (1970). Unpublished.

HALL P. (1935). On representatives of subsets. *J. London Math. Soc.* **10**, 26–30.

HAMILTON W.R. (1931). Letter to John Graves on the Icosian, 17 oct., 1856. In *The Mathematical Papers of Sir William Rowan Hamilton*, volume 3 (H. Halberstam and R. Ingram, eds.), 612–625. Cambridge University Press, Cambridge.

HARARY F. (1964). On the reconstruction of a graph from a collection of subgraphs. In *Theory of Graphs and its Applications (Proc. Sympos. Smolenice, 1963)*, 47–52. Publ. House Czechoslovak Acad. Sci., Prague.

HARARY F., PLANTHOLT M. and STATMAN R. (1982). The graph isomorphism problem is polynomially equivalent to the legitimate deck problem for regular graphs. *Caribbean J. Math.* **1**, 15–23.

HARTMAN I.B.A. (2006). Berge's conjecture on path partitions — a survey. In *Creation and Recreation — A Volume in Memory of Claude Berge, Discrete Math.*, volume 306 (J.A. Bondy and V. Chvátal, eds.), 2498–2514. Elsevier, Amsterdam.

HARTSFIELD N. and RINGEL G. (1994). *Pearls in Graph Theory: A Comprehensive Introduction*. Academic Press, Boston. Revised reprint of the 1990 original.

HAVET F. and THOMASSÉ S. (2000). Median orders of tournaments: a tool for the second neighborhood problem and Sumner's conjecture. *J. Graph Theory* **35**, 244–256.

HAXELL P., SEAMONE B. and VERSTRAETE J. (2007). Independent dominating sets and Hamiltonian cycles. *J. Graph Theory* **54**, 233–244.

HEAWOOD P.J. (1890). Map-colour theorem. *Quart. J. Pure Appl. Math.* **24**, 332–338.

HEDETNIEMI S. (1966). Homomorphisms of graphs and automata. *Technical Report 03105-44-T*, University of Michigan.

HEESCH H. (1969). Untersuchungen zum Vierfarbenproblem. *B.I.-Hochschulskripten 810/810a/810b*, Bibliographisches Institut, Mannheim.

HELL P. and NEŠETŘIL J. (2004). *Graphs and Homomorphisms*. Oxford Lecture Series in Mathematics and its Applications, Vol. 28, Oxford University Press, Oxford.

HENLE F. and HENLE J. (2006). Squaring the plane. Unpublished manuscript.

HERSTEIN I.N. (1996). *Abstract Algebra*. Third edition. Prentice Hall, Upper Saddle River, NJ. With a preface by Barbara Cortzen and David J. Winter.

HILTON A.J.W. (1982). Canonical edge-colourings of locally finite graphs. *Combinatorica* **2**, 37–51.

HILTON A.J.W. (1989). Two conjectures on edge-colouring. *Discrete Math.* **74**, 61–64.

HOÀNG C.T. and McDIARMID C. (2002). On the divisibility of graphs. *Discrete Math.* **242**, 145–156.

HOÀNG C.T. and REED B. (1987). A note on short cycles in digraphs. *Discrete Math.* **66**, 103–107.

HOFFMAN A.J. (1960). Some recent applications of the theory of linear inequalities to extremal combinatorial analysis. In *Proc. Sympos. Appl. Math., Vol. 10*, 113–127. American Mathematical Society, Providence, R.I.

HOFFMAN A.J. and SINGLETON R.R. (1960). On Moore graphs with diameters 2 and 3. *IBM J. Res. Develop.* **4**, 497–504.

HOLYER I. (1981). The NP-completeness of edge-coloring. *SIAM J. Comput.* **10**, 718–720.

HOPCROFT J. and TARJAN R. (1974). Efficient planarity testing. *J. Assoc. Comput. Mach.* **21**, 549–568.

HOPCROFT J.E. and WONG J.K. (1974). Linear time algorithm for isomorphism of planar graphs: preliminary report. In *Sixth Annual ACM Symposium on Theory of Computing (Seattle, Wash., 1974)*, 172–184. Assoc. Comput. Mach., New York.

ISAACS R. (1975). Infinite families of nontrivial trivalent graphs which are not Tait colorable. *Amer. Math. Monthly* **82**, 221–239.

IWATA S. and MATSUDA T. (2007). Finding coherent cyclic orders in strong digraphs. *Technical report*, Graduate School of Information Sciences and Technology, University of Tokyo.

JACKSON B. (1980). Hamilton cycles in regular 2-connected graphs. *J. Combin. Theory Ser. B* **29**, 27–46.

JACKSON B. (1981). Long paths and cycles in oriented graphs. *J. Graph Theory* **5**, 145–157.

JACKSON B. (1986). Longest cycles in 3-connected cubic graphs. *J. Combin. Theory Ser. B* **41**, 17–26.

JACKSON B. (1993a). On circuit covers, circuit decompositions and Euler tours of graphs. In *Surveys in Combinatorics, 1993 (Keele)*, 191–210. London Math. Soc. Lecture Note Ser., Vol. 187, Cambridge Univ. Press, Cambridge.

JACKSON B. (1993b). A zero-free interval for chromatic polynomials of graphs. *Combin. Probab. Comput.* **2**, 325–336.

JACKSON B. and PARSONS T.D. (1982). A shortness exponent for r-regular r-connected graphs. *J. Graph Theory* **6**, 169–176.

JACKSON B. and WORMALD N.C. (1990). k-walks of graphs. *Australas. J. Combin.* **2**, 135–146. Combinatorial Mathematics and Combinatorial Computing, Vol. 2 (Brisbane, 1989).

JAEGER F. (1976). On nowhere-zero flows in multigraphs. In *Proceedings of the Fifth British Combinatorial Conference*, volume 15, 373–378. Utilitas Math., Winnipeg.

JAEGER F. (1979). Flows and generalized coloring theorems in graphs. *J. Combin. Theory Ser. B* **26**, 205–216.

JAEGER F. (1984). On circular flows in graphs. In *Finite and Infinite Sets, Vol. I, II (Eger, 1981)*, 391–402. Colloq. Math. Soc. János Bolyai, Vol. 37, North-Holland, Amsterdam.

JAEGER F. (1985). A survey of the cycle double cover conjecture. In *Cycles in Graphs (Burnaby, B.C., 1982)*, 1–12. North-Holland Math. Stud., Vol. 115, North-Holland, Amsterdam.

JAEGER F. (1988). Nowhere-zero flow problems. In *Selected Topics in Graph Theory, 3* (L.W. Beineke and R.J. Wilson, eds.), 71–95. Academic Press, San Diego.

JANSON S., ŁUCZAK T. and RUCINSKI A. (2000). *Random Graphs*. Wiley-Interscience Series in Discrete Mathematics and Optimization, Wiley, New York.

JARNÍK V. (1930). O jistém problému minimálním. *Práce Moravské Přírodovědecké Společnosti Brno* **6**, 57–63.

JENSEN T.R. and TOFT B. (1995). *Graph Coloring Problems.* Wiley-Interscience Series in Discrete Mathematics and Optimization, Wiley, New York.

JØRGENSEN L.K. (1994). Contractions to K_8. *J. Graph Theory* **18**, 431–448.

JÜNGER M., REINELT G. and RINALDI G. (1995). The traveling salesman problem. In *Network Models*, 225–330. Handbooks Oper. Res. Management Sci., Vol. 7, North-Holland, Amsterdam.

KAHN J. (1994). Recent results on some not-so-recent hypergraph matching and covering problems. In *Extremal Problems for Finite Sets (Visegrád, 1991)*, 305–353. Bolyai Soc. Math. Stud., Vol. 3, János Bolyai Math. Soc., Budapest.

KAISER T., KRÁL D., ROSENFELD M., RYJAČEK Z. and VOSS H.J. (2007). Hamilton cycles in prisms over graphs. *J. Graph Theory* .

KARP R.M. (1972). Reducibility among combinatorial problems. In *Complexity of Computer Computations (Proc. Sympos., IBM Thomas J. Watson Res. Center, Yorktown Heights, N.Y., 1972)*, 85–103. Plenum, New York.

KAUFFMAN L.H. (1990). Map coloring and the vector cross product. *J. Combin. Theory Ser. B* **48**, 145–154.

KAWARABAYASHI K. (2002). One or two disjoint circuits cover independent edges. Lovász-Woodall conjecture. *J. Combin. Theory Ser. B* **84**, 1–44.

KAWARABAYASHI K., LEE O., REED B. and WOLLAN P. (2007). Progress on Lovász' path removal conjecture. *Technical report*, McGill University.

KEEVASH P. and SUDAKOV B. (2005). The Turán number of the Fano plane. *Combinatorica* **25**, 561–574.

KELLY P.J. (1942). *On Isometric Transformations.* Ph.D. thesis, University of Wisconsin.

KELLY P.J. (1957). A congruence theorem for trees. *Pacific J. Math.* **7**, 961–968.

KELMANS A.K. (1986). Constructions of cubic bipartite and 3-connected graphs without hamiltonian cycles. In *Analiz Zadach Formirovaniya i Vybora Alternativ, Vol. 10*, 64–72. VNIISI, Moscow.

KELMANS A.K. (1993). Graph planarity and related topics. In *Graph Structure Theory (Seattle, WA, 1991)*, 635–667. Contemp. Math., Vol. 147, Amer. Math. Soc., Providence, RI.

KELMANS A.K. (1994). Constructions of cubic bipartite 3-connected graphs without Hamiltonian cycles. In *Selected Topics in Discrete Mathematics (Moscow, 1972–1990)*, 127–140. Amer. Math. Soc. Transl. Ser. 2, Vol. 158, Amer. Math. Soc., Providence, RI.

KELMANS A.K. (2005). On λ-packings in 3-connected graphs. *RUTCOR Research Report 23-2005*, Rutgers University.

KELMANS A.K. and LOMONOSOV M. (1975). Moscow Discrete Mathematics Seminar. Personal communication, A.K. Kelmans, 1980.

KEMNITZ A. and HARBORTH H. (2001). Plane integral drawings of planar graphs. *Discrete Math.* **236**, 191–195. Graph theory (Kazimierz Dolny, 1997).

KEMPE A.B. (1879). On the geographical problem of the four colours. *Amer. J. Math.* **2**, 193–200.

KIERSTEAD H.A. and KOSTOCHKA A.V. (2008). A short proof of the Hajnal-Szemerédi theorem on equitable colouring. *Combin. Probab. Comput.* **17**, 265–270.

KILPATRICK P.A. (1975). *Tutte's First Colour-Cycle Conjecture*. Ph.D. thesis, University of Cape Town.

KIM J.H. (1995). The Ramsey number $R(3,t)$ has order of magnitude $t^2/\log t$. *Random Structures Algorithms* **7**, 173–207.

KIRCHHOFF G. (1847). Über die Auflösung der Gleichungen, auf welche man bei der Untersuchung der linearen Verteilung galvanischer Ströme geführt wird. *Ann. Phys. Chem.* **72**, 497–508.

KIRKMAN T.P. (1881). Question 6610, solution by the proposer. *Math. Quest. Solut. Educ. Times* **35**, 112–116.

KLEIN R. (1994). On the colorability of m-composed graphs. *Discrete Math.* **133**, 181–190.

KNESER M. (1955). Aufgabe 300. *Jber. Deutsch. Math.-Verein.* **58**, 27.

KNUTH D.E. (1969). *The Art of Computer Programming. Vol. 1: Fundamental Algorithms.* Addison-Wesley, Reading, MA. Second printing.

KNUTH D.E. (1974). Wheels within wheels. *J. Combin. Theory Ser. B* **16**, 42–46.

KOCAY W.L. (1987). A family of nonreconstructible hypergraphs. *J. Combin. Theory Ser. B* **42**, 46–63.

KOCHOL M. (1996). Snarks without small cycles. *J. Combin. Theory Ser. B* **67**, 34–47.

KOHAYAKAWA Y. and RÖDL V. (2003). Szemerédi's regularity lemma and quasi-randomness. In *Recent Advances in Algorithms and Combinatorics*, 289–351. CMS Books Math./Ouvrages Math. SMC, Vol. 11, Springer, New York.

KOMLÓS J. and SIMONOVITS M. (1996). Szemerédi's regularity lemma and its applications in graph theory. In *Combinatorics, Paul Erdős is Eighty, Vol. 2 (Keszthely, 1993)*, 295–352. Bolyai Soc. Math. Stud., Vol. 2, János Bolyai Math. Soc., Budapest.

KOMLÓS J., SHOKOUFANDEH A., SIMONOVITS M. and SZEMERÉDI E. (2002). The regularity lemma and its applications in graph theory. In *Theoretical Aspects of Computer Science (Tehran, 2000)*, 84–112. Lecture Notes in Comput. Sci., Vol. 2292, Springer, Berlin.

KÖNIG D. (1931). Graphs and matrices. *Mat. Fiz. Lapok* **38**, 116–119. Hungarian.

KÖNIG D. (1936). *Theorie der Endlichen und Unendlichen Graphen.* Akademische Verlagsgesellschaft, Leipzig.

KORTE B. and NEŠETŘIL J. (2001). Vojtěch Jarnik's work in combinatorial optimization. *Discrete Math.* **235**, 1–17. Combinatorics (Prague, 1998).

KOSTOCHKA A.V. (1982). A class of constructions for Turán's (3, 4)-problem. *Combinatorica* **2**, 187–192.

KOTZIG A. (1979). Selected open problems in graph theory. In *Graph Theory and Related Topics* (J.A. Bondy and U.S.R. Murty, eds.), xxxii+371. Academic Press, New York.

KOUIDER M. (1994). Cycles in graphs with prescribed stability number and connectivity. *J. Combin. Theory Ser. B* **60**, 315–318.

KRIESELL M. (2003). Edge-disjoint trees containing some given vertices in a graph. *J. Combin. Theory Ser. B* **88**, 53–65.

KŘÍŽ I. (1991). Permutation groups in Euclidean Ramsey theory. *Proc. Amer. Math. Soc.* **112**, 899–907.

KRUSKAL J.B. (1997). A reminiscence about shortest spanning subtrees. *Arch. Math. (Brno)* **33**, 13–14.

KRUSKAL JR. J.B. (1956). On the shortest spanning subtree of a graph and the traveling salesman problem. *Proc. Amer. Math. Soc.* **7**, 48–50.

KUMAR R., RAGHAVAN P., RAJAGOPALAN S., SIVAKUMAR D., TOMKINS A. and UPFAL E. (2000). Stochastic models for the web graph. In *41st Annual Symposium on Foundations of Computer Science (Redondo Beach, CA, 2000)*, 57–65. IEEE Comput. Soc. Press, Los Alamitos, CA.

KÜNDGEN A. and RAMAMURTHI R. (2002). Coloring face-hypergraphs of graphs on surfaces. *J. Combin. Theory Ser. B* **85**, 307–337.

KURATOWSKI C. (1930). Sur le problème des courbes gauches en topologie. *Fund. Math.* **15**, 271–283.

LABORDE J.M., PAYAN C. and XUONG N.H. (1983). Independent sets and longest directed paths in digraphs. In *Graphs and other Combinatorial Topics (Prague, 1982)*, 173–177. Teubner-Texte Math., Vol. 59, Teubner, Leipzig.

LAS VERGNAS M. (1977/78). Acyclic and totally cyclic orientations of combinatorial geometries. *Discrete Math.* **20**, 51–61.

LAURI J. and SCAPELLATO R. (2003). *Topics in Graph Automorphisms and Reconstruction*. London Mathematical Society Student Texts, Vol. 54, Cambridge University Press, Cambridge.

LEIGHTON F.T. (1983). *Complexity Issues in VLSI: Optimal Layouts for the Shuffle-Exchange Graph and Other Networks*. MIT Press, Cambridge, MA.

LEONARDI S. (2004). *Algorithms and Models for the Web-Graph*. Springer, New York.

LEVEN D. and GALIL Z. (1983). NP completeness of finding the chromatic index of regular graphs. *J. Algorithms* **4**, 35–44.

LEVIN L.A. (1973). Universal search problems. *Problemy Peredači Informacii* **9**, 265–266.

LINIAL N. (1978). Covering digraphs by paths. *Discrete Math.* **23**, 257–272.

LINIAL N. (1981). Extending the Greene–Kleitman theorem to directed graphs. *J. Combin. Theory Ser. A* **30**, 331–334.

LOCKE S. (1995). Problem 10447. *Amer. Math. Monthly* **102**, 360.

LOVÁSZ L. (1968a). On chromatic number of finite set-systems. *Acta Math. Acad. Sci. Hungar.* **19**, 59–67.

LOVÁSZ L. (1968b). On covering of graphs. In *Theory of Graphs (Proc. Colloq., Tihany, 1966)*, 231–236. Academic Press, New York.

LOVÁSZ L. (1970). Problem 11. In *Combinatorial Structures and their Applications*, 497. Gordon and Breach, New York.

LOVÁSZ L. (1972a). A characterization of perfect graphs. *J. Combin. Theory Ser. B* **13**, 95–98.

LOVÁSZ L. (1972b). Normal hypergraphs and the perfect graph conjecture. *Discrete Math.* **2**, 253–267.

LOVÁSZ L. (1972c). A note on the line reconstruction problem. *J. Combin. Theory Ser. B* **13**, 309–310.

LOVÁSZ L. (1975a). 2-matchings and 2-covers of hypergraphs. *Acta Math. Acad. Sci. Hungar.* **26**, 433–444.

LOVÁSZ L. (1975b). Three short proofs in graph theory. *J. Combin. Theory Ser. B* **19**, 269–271.

LOVÁSZ L. (1976). On two minimax theorems in graph. *J. Combin. Theory Ser. B* **21**, 96–103.

LOVÁSZ L. (1978). Kneser's conjecture, chromatic number, and homotopy. *J. Combin. Theory Ser. A* **25**, 319–324.

LOVÁSZ L. (1979). On the Shannon capacity of a graph. *IEEE Trans. Inform. Theory* **25**, 1–7.

LOVÁSZ L. (1993). *Combinatorial Problems and Exercises.* Second edition. North-Holland, Amsterdam.

LOVÁSZ L. and PLUMMER M.D. (1986). *Matching Theory.* Annals of Discrete Mathematics, Vol. 29, North-Holland, Amsterdam.

LUCAS E. (1894). *Récréations Mathématiques, Vol. IV.* Gauthier-Villars et fils, Paris. Reprinted by Blanchard, Paris, 1960.

LUCCHESI C. (1976). *A Minimax Equality for Directed Graphs.* Ph.D. thesis, University of Waterloo.

LUCCHESI C.L. and YOUNGER D.H. (1978). A minimax theorem for directed graphs. *J. London Math. Soc. (2)* **17**, 369–374.

LUKS E.M. (1982). Isomorphism of graphs of bounded valence can be tested in polynomial time. *J. Comput. System Sci.* **25**, 42–65.

MACLANE S. (1937). A combinatorial condition for planar graphs. *Fund. Math.* **28**, 22–32.

MADER W. (1967). Homomorphieeigenschaften und mittlere Kantendichte von Graphen. *Math. Ann.* **174**, 265–268.

MADER W. (1971a). Minimale n-fach kantenzusammenhängende Graphen. *Math. Ann.* **191**, 21–28.

MADER W. (1971b). Minimale n-fach zusammenhängende Graphen mit maximaler Kantenzahl. *J. Reine Angew. Math.* **249**, 201–207.

MADER W. (1978). A reduction method for edge-connectivity in graphs. In *Advances in Graph Theory*, volume 3, 145–164. North-Holland, Amsterdam.

MANSFIELD A. (1982). The relationship between the computational complexities of the legitimate deck and isomorphism problems. *Quart. J. Math. Oxford Ser. (2)* **33**, 345–347.

MATTHEWS K.R. (1978). On the Eulericity of a graph. *J. Graph Theory* **2**, 143–148.

MATTHEWS M.M. and SUMNER D.P. (1984). Hamiltonian results in $K_{1,3}$-free graphs. *J. Graph Theory* **8**, 139–146.

MATULA D. (1976). The largest clique size in a random graph. *Technical report*, Dept. Comp. Sci., Southern Methodist University, Dallas, Texas.

McCUAIG W. (1993). Intercyclic digraphs. In *Graph Structure Theory (Seattle, WA, 1991)*, 203–245. Contemp. Math., Vol. 147, Amer. Math. Soc., Providence, RI.

McCUAIG W. (2000). Even dicycles. *J. Graph Theory* **35**, 46–68.

McKAY B.D. (1977). Computer reconstruction of small graphs. *J. Graph Theory* **1**, 281–283.

MENGER K. (1927). Zur allgemeinen Kurventheorie. *Fund. Math.* **10**, 96–115.

MINTY G.J. (1980). On maximal independent sets of vertices in claw-free graphs. *J. Combin. Theory Ser. B* **28**, 284–304.

MOHAR B. and THOMASSEN C. (2001). *Graphs on Surfaces*. Johns Hopkins Studies in the Mathematical Sciences, Johns Hopkins University Press, Baltimore.

MOLLOY M. and REED B. (1998). A bound on the total chromatic number. *Combinatorica* **18**, 241–280.

MOLLOY M. and REED B. (2002). *Graph Colouring and the Probabilistic Method*. Algorithms and Combinatorics, Vol. 23, Springer, Berlin.

MOON J.W. (1967). Various proofs of Cayley's formula for counting trees. In *A Seminar on Graph Theory*, 70–78. Holt, Rinehart and Winston, New York.

MOON J.W. (1968). *Topics on Tournaments*. Holt, Rinehart and Winston, New York.

MOON J.W. and MOSER L. (1963). Simple paths on polyhedra. *Pacific J. Math.* **13**, 629–631.

MORON Z. (1925). O rozkladach prostokatów na kwadraty. *Przeglad Matematyczno-Fizyczny* **3**, 152–153.

MÜLLER V. (1977). The edge reconstruction hypothesis is true for graphs with more than $n \cdot \log_2 n$ edges. *J. Combin. Theory Ser. B* **22**, 281–283.

MYCIELSKI J. (1955). Sur le coloriage des graphs. *Colloq. Math.* **3**, 161–162.

NAGAMOCHI H. and IBARAKI T. (1992). Computing edge-connectivity in multigraphs and capacitated graphs. *SIAM J. Discrete Math.* **5**, 54–66.

NASH-WILLIAMS C. (1978). The reconstruction problem. In *Selected Topics in Graph Theory* (L.W. Beineke and R.J. Wilson, eds.), 205–236. Academic Press, London.

NASH-WILLIAMS C.S.J.A. (1959). Random walk and electric currents in networks. *Proc. Cambridge Philos. Soc.* **55**, 181–194.

NASH-WILLIAMS C.S.J.A. (1960). On orientations, connectivity and odd-vertex-pairings in finite graphs. *Canad. J. Math.* **12**, 555–567.

NASH-WILLIAMS C.S.J.A. (1961). Edge-disjoint spanning trees of finite graphs. *J. London Math. Soc.* **36**, 445–450.

NEŠETŘIL J. and RÖDL V. (1975). Partitions of subgraphs. In *Recent Advances in Graph Theory*, 413–423. Academia, Prague.

NEŠETŘIL J. and RÖDL V. (1979). A short proof of the existence of highly chromatic hypergraphs without short cycles. *J. Combin. Theory Ser. B* **27**, 225–227.

NEŠETŘIL J., MILKOVÁ E. and NEŠETŘILOVÁ H. (2001). Otakar Borůvka on minimum spanning tree problem: translation of both the 1926 papers, comments, history. *Discrete Math.* **233**, 3–36. Graph Theory (Prague, 1998).

NEUMANN-LARA V. (1985). Vertex colourings in digraphs. some problems. *Technical report*, University of Waterloo.

NINČÁK J. (1974). Hamiltonian circuits in cubic graphs. *Comment. Math. Univ. Carolinae* **15**, 627–630.

OHBA K. (2002). On chromatic-choosable graphs. *J. Graph Theory* **40**, 130–135.

OXLEY J.G. (1992). *Matroid Theory.* Oxford Science Publications, Clarendon Press, New York.

PADBERG M.W. and RAO M.R. (1982). Odd minimum cut-sets and *b*-matchings. *Math. Oper. Res.* **7**, 67–80.

PALMER E.M. (1985). *Graphical Evolution: An Introduction to the Theory of Random Graphs.* Wiley-Interscience Series in Discrete Mathematics, Wiley, Chichester.

PAPADIMITRIOU C.H. (1994). *Computational Complexity.* Addison-Wesley, Reading, MA.

PETERSEN J. (1891). Die Theorie der regulären Graphs. *Acta Math.* **15**, 193–220. English translation in: N.L. Biggs, E.K. Lloyd and R.J. Wilson, *Graph Theory 1736–1936.* Clarendon Press, Oxford, 1986, p. 190.

PITMAN J. (1999). Coalescent random forests. *J. Combin. Theory Ser. A* **85**, 165–193.

POLESSKIĬ V.P. (1971). A certain lower bound for the reliability of information networks. *Problemy Peredači Informacii* **7**, 88–96.

PÓLYA G. (1921). Über eine Aufgabe der Wahrscheinlichkeitsrechnung betreffend die Irrfahrt im Strassennetz. *Math. Ann.* **84**, 149–160.

PÓLYA G. (2004). *How to Solve It: A New Aspect of Mathematical Method.* Princeton Science Library, Princeton University Press, Princeton, NJ. Expanded version of the 1988 edition, with a new foreword by John H. Conway.

PÓSA L. (1976). Hamiltonian circuits in random graphs. *Discrete Math.* **14**, 359–364.

PREISSMANN M. (1981). *Sur les colorations des arêtes des graphes cubiques.* Ph.D. thesis, Université de Grenoble.

PRIM R.C. (1957). Shortest connection networks and some generalizations. *Bell Sys. Tech. J.* **36**, 1389–1401.

RADO R. (1942). A theorem on independence relations. *Quart. J. Math., Oxford Ser.* **13**, 83–89.

RADO R. (1957). Note on independence functions. *Proc. London Math. Soc. (3)* **7**, 300–320.

RAMSEY F.P. (1930). On a problem of formal logic. *Proc. London Math. Soc.* **30**, 264–286.

RANDÍC M. (1981). Personal communication. (Letter, January 18, 1981).

RAZBOROV A. (2006). On the minimal density of triangles in graphs. *Technical report*, Mathematical Institute, Russian Academy of Sciences.

READ R.C. (1968). An introduction to chromatic polynomials. *J. Combin. Theory* **4**, 52–71.

READ R.C. and TUTTE W.T. (1988). Chromatic polynomials. In *Selected Topics in Graph Theory, 3* (L.W. Beineke and R.J. Wilson, eds.), 15–42. Academic Press, San Diego.

RÉDEI L. (1934). Ein kombinatorischer Satz. *Acta. Litt. Sci. Szeged* **7**, 39–43.

REED B. (1998). ω, Δ, and χ. *J. Graph Theory* **27**, 177–212.

REED B. and SEYMOUR P. (1998). Fractional colouring and Hadwiger's conjecture. *J. Combin. Theory Ser. B* **74**, 147–152.

REED B., ROBERTSON N., SEYMOUR P. and THOMAS R. (1996). Packing directed circuits. *Combinatorica* **16**, 535–554.

REED B.A. (2003). Algorithmic aspects of tree width. In *Recent Advances in Algorithms and Combinatorics*, 85–107. CMS Books Math./Ouvrages Math. SMC, Vol. 11, Springer, New York.

REIMAN I. (1958). Über ein Problem von K. Zarankiewicz. *Acta. Math. Acad. Sci. Hungar.* **9**, 269–273.

RICHARDSON M. (1953). Solutions of irreflexive relations. *Ann. of Math. (2)* **58**, 573–590; errata 60 (1954), 595.

ŘÍHA S. (1991). A new proof of the theorem by Fleischner. *J. Combin. Theory Ser. B* **52**, 117–123.

RINGEL G. (1974). *Map Color Theorem.* Springer, New York. Foundations of Mathematical Sciences, Vol. 209.

RINGEL G. and YOUNGS J.W.T. (1968). Solution of the Heawood map-coloring problem. *Proc. Nat. Acad. Sci. U.S.A.* **60**, 438–445.

RIZZI R. (1999). Indecomposable r-graphs and some other counterexamples. *J. Graph Theory* **32**, 1–15.

RIZZI R. (2001). On 4-connected graphs without even cycle decompositions. *Discrete Math.* **234**, 181–186.

ROBBINS H.E. (1939). A theorem on graphs, with an application to a problem of traffic control. *Amer. Math. Monthly* **46**, 281–283. Questions, Discussions, and Notes.

ROBERTSON N. (2007). Personal communication.

ROBERTSON N. and SEYMOUR P.D. (1986). Graph minors. II. Algorithmic aspects of tree-width. *J. Algorithms* **7**, 309–322.

ROBERTSON N. and SEYMOUR P.D. (1995). Graph minors. XIII. The disjoint paths problem. *J. Combin. Theory Ser. B* **63**, 65–110.

ROBERTSON N. and SEYMOUR P.D. (2004). Graph minors. XX. Wagner's conjecture. *J. Combin. Theory Ser. B* **92**, 325–357.

ROBERTSON N., SEYMOUR P. and THOMAS R. (1993). Hadwiger's conjecture for K_6-free graphs. *Combinatorica* **13**, 279–361.

ROBERTSON N., SANDERS D., SEYMOUR P. and THOMAS R. (1997a). The four-colour theorem. *J. Combin. Theory Ser. B* **70**, 2–44.

ROBERTSON N., SEYMOUR P. and THOMAS R. (1997b). Tutte's edge-colouring conjecture. *J. Combin. Theory Ser. B* **70**, 166–183.

ROBERTSON N., SEYMOUR P.D. and THOMAS R. (1999). Permanents, Pfaffian orientations, and even directed circuits. *Ann. of Math. (2)* **150**, 929–975.

RÖDL V., NAGLE B., SKOKAN J., SCHACHT M. and KOHAYAKAWA Y. (2005). The hypergraph regularity method and its applications. *Proc. Natl. Acad. Sci. USA* **102**, 8109–8113 (electronic).

ROSE D.J., TARJAN R.E. and LUEKER G.S. (1976). Algorithmic aspects of vertex elimination on graphs. *SIAM J. Comput.* **5**, 266–283.

ROSENFELD M. (1967). On a problem of C. E. Shannon in graph theory. *Proc. Amer. Math. Soc.* **18**, 315–319.

ROSENKRANTZ D.J., STEARNS R.E. and LEWIS P.M. (1974). Approximate algorithms for the traveling salesperson problem. In *15th Annual Symposium on Switching and Automata Theory (1974)*, 33–42. IEEE Comput. Soc., Long Beach, CA.

ROSENSTIEHL P. and READ R.C. (1978). On the principal edge tripartition of a graph. In *Advances in Graph Theory*, volume 3, 195–226. North-Holland, Amsterdam.

ROTA G.C. (1964). On the foundations of combinatorial theory. I. Theory of Möbius functions. *Z. Wahrscheinlichkeitstheorie und Verw. Gebiete* **2**, 340–368 (1964).

ROTHSCHILD B. and WHINSTON A. (1966). Feasiblity of two commodity network flows. *Operations Res.* **14**, 1121–1129.

ROY B. (1967). Nombre chromatique et plus longs chemins d'un graphe. *Rev. Française Informat. Recherche Opérationnelle* **1**, 129–132.

SAATY T.L. (1972). Thirteen colorful variations on Guthrie's four-color conjecture. *Amer. Math. Monthly* **79**, 2–43.

SABIDUSSI G. (1992). Correspondence between Sylvester, Petersen, Hilbert and Klein on invariants and the factorisation of graphs 1889–1891. *Discrete Math.* **100**, 99–155. Special volume to mark the centennial of Julius Petersen's "Die Theorie der regulären Graphs", Part I.

SACHS H. (1993). Elementary proof of the cycle-plus-triangles theorem. In *Combinatorics, Paul Erdős is Eighty, Vol. 1*, 347–359. Bolyai Soc. Math. Stud., János Bolyai Math. Soc., Budapest.

SALAVATIPOUR M. (2003). *Graph Colouring via the Discharging Method*. Ph.D. thesis, University of Toronto.

SANDERS D.P. (1997). On paths in planar graphs. *J. Graph Theory* **24**, 341–345.

SBIHI N. (1980). Algorithme de recherche d'un stable de cardinalité maximum dans un graphe sans étoile. *Discrete Math.* **29**, 53–76.

SCHEINERMAN E.R. and ULLMAN D.H. (1997). *Fractional Graph Theory: a Rational Approach to the Theory of Graphs*. Wiley-Interscience Series in Discrete Mathematics and Optimization, Wiley, New York. With a foreword by Claude Berge.

SCHLÄFLI L. (1858). An attempt to determine the twenty-seven lines upon a surface of the third order and to divide such surfaces into species in reference to the reality of the lines upon the surface. *Quart. J. Math.* **2**, 55–65, 110–121.

SCHRIJVER A. (1978). Vertex-critical subgraphs of Kneser graphs. *Nieuw Arch. Wisk. (3)* **26**, 454–461.

SCHRIJVER A. (1982). Min-max relations for directed graphs. In *Bonn Workshop on Combinatorial Optimization (Bonn, 1980)*, 261–280. Ann. Discrete Math., Vol. 16, North-Holland, Amsterdam.

SCHRIJVER A. (2003). *Combinatorial Optimization: Polyhedra and Efficiency.* Algorithms and Combinatorics, Vol. 24, Springer, Berlin.

SCHUR I. (1916). Über die Kongruenz $x^m + y^m = z^m$ (mod p). *Jber. Deutsch. Math.-Verein.* **25**, 114–117.

SEBÖ A. (2007). Minmax relations for cyclically ordered digraphs. *J. Combin. Theory Ser. B* **97**, 518–552.

SEYMOUR P.D. (1979a). On multicolourings of cubic graphs, and conjectures of Fulkerson and Tutte. *Proc. London Math. Soc. (3)* **38**, 423–460.

SEYMOUR P.D. (1979b). Sums of circuits. In *Graph Theory and Related Topics* (J.A. Bondy and U.S.R. Murty, eds.), 341–355. Academic Press, New York.

SEYMOUR P.D. (1980). Disjoint paths in graphs. *Discrete Math.* **29**, 293–309.

SEYMOUR P.D. (1981a). Even circuits in planar graphs. *J. Combin. Theory Ser. B* **31**, 327–338.

SEYMOUR P.D. (1981b). Nowhere-zero 6-flows. *J. Combin. Theory Ser. B* **30**, 130–135.

SEYMOUR P.D. (1981c). On odd cuts and plane multicommodity flows. *Proc. London Math. Soc. (3)* **42**, 178–192.

SEYMOUR P.D. (2007). Personal communication.

SEYMOUR P.D. and THOMAS R. (1993). Graph searching and a min-max theorem for tree-width. *J. Combin. Theory Ser. B* **58**, 22–33.

SHANNON C.E. (1956). The zero error capacity of a noisy channel. *IRE Trans.Inf. Theory* **IT-2**, 8–19.

SHEEHAN J. (1975). The multiplicity of Hamiltonian circuits in a graph. In *Recent Advances in Graph Theory*, 477–480. Academia, Prague.

SHILOACH Y. (1980). A polynomial solution to the undirected two paths problem. *J. Assoc. Comput. Mach.* **27**, 445–456.

SIDORENKO A. (1995). What we know and what we do not know about Turán numbers. *Graphs Combin.* **11**, 179–199.

SIDORENKO A.F. (1991). Inequalities for functionals generated by bipartite graphs. *Diskret. Mat.* **3**, 50–65.

SINGLETON R. (1966). On minimal graphs of maximum even girth. *J. Combin. Theory* **1**, 306–332.

SIPSER M. (2005). *Introduction to the Theory of Computation.* Second edition. Course Technology, Boston, MA.

SMITH C.A.B. and TUTTE W.T. (1950). A class of self-dual maps. *Canadian J. Math.* **2**, 179–196.

SÓS V.T. (1976). Remarks on the connection of graph theory, finite geometry and block designs. In *Colloquio Internazionale sulle Teorie Combinatorie (Roma, 1973), Tomo II*, 223–233. Atti dei Convegni Lincei, No. 17. Accad. Naz. Lincei, Rome.

624   References

SPENCER J. (1987). *Ten Lectures on the Probabilistic Method.* CBMS-NSF Regional Conference Series in Applied Mathematics, Vol. 52, Society for Industrial and Applied Mathematics (SIAM), Philadelphia.

SPENCER J., SZEMERÉDI E. and TROTTER JR. W. (1984). Unit distances in the Euclidean plane. In *Graph Theory and Combinatorics (Cambridge, 1983)*, 293–303. Academic Press, London.

SPRAGUE R. (1939). Beispiel einer Zerlegung des Quadrats in lauter verschiedene Quadrate. *Math. Z.* **45**, 607–608.

STANLEY R.P. (1973). Acyclic orientations of graphs. *Discrete Math.* **5**, 171–178.

STANLEY R.P. (1985). Reconstruction from vertex-switching. *J. Combin. Theory Ser. B* **38**, 132–138.

STEINBERG R. (1993). The state of the three color problem. In *Quo Vadis, Graph Theory?*, 211–248. Ann. Discrete Math., Vol. 55, North-Holland, Amsterdam.

STEINITZ E. (1922). Polyeder und Raumeinteilungen. In *Enzyklopädie der mathematischen Wissenschaften*, volume 3, 1–139. .

STOCKMEYER P.K. (1981). A census of nonreconstructible digraphs. I. Six related families. *J. Combin. Theory Ser. B* **31**, 232–239.

SULLIVAN B. (2006). A summary of results and problems related to the Caccetta-Häggkvist conjecture. *Technical report*, Princeton University.

SZÉKELY L.A. (1997). Crossing numbers and hard Erdős problems in discrete geometry. *Combin. Probab. Comput.* **6**, 353–358.

SZEKERES G. (1973). Polyhedral decompositions of cubic graphs. *Bull. Austral. Math. Soc.* **8**, 367–387.

SZEMERÉDI E. (1978). Regular partitions of graphs. In *Problèmes Combinatoires et Théorie des Graphes (Colloq. Internat. CNRS, Univ. Orsay, Orsay, 1976)*, 399–401. Colloq. Internat. CNRS, Vol. 260, CNRS, Paris.

SZEMERÉDI E. and TROTTER JR. W.T. (1983). Extremal problems in discrete geometry. *Combinatorica* **3**, 381–392.

TAIT P.G. (1880). Remarks on colouring of maps. *Proc. Royal Soc. Edinburgh Ser. A* **10**, 729.

TAO T. (2006). Szemerédi's regularity lemma revisited. *Contrib. Discrete Math.* **1**, 8–28 (electronic).

TAO T. and VU V. (2006). *Additive Combinatorics.* Cambridge Studies in Advanced Mathematics, Vol. 105, Cambridge University Press, Cambridge.

TARJAN R. (1972). Depth-first search and linear graph algorithms. *SIAM J. Comput.* **1**, 146–160.

TARJAN R.E. (1983). *Data Structures and Network Algorithms.* CBMS-NSF Regional Conference Series in Applied Mathematics, Vol. 44, Society for Industrial and Applied Mathematics (SIAM), Philadelphia.

TASHKINOV V.A. (1984). 3-regular subgraphs of 4-regular graphs. *Mat. Zametki* **36**, 239–259.

THOMAS R. (1998). An update on the four-color theorem. *Notices Amer. Math. Soc.* **45**, 848–859.

THOMAS R. and YU X. (1994). 4-connected projective-planar graphs are hamiltonian. *J. Combin. Theory Ser. B* **62**, 114–132.

THOMASON A.G. (1978). Hamiltonian cycles and uniquely edge colourable graphs. In *Advances in Graph Theory*, volume 3, Exp. No. 13, 3. North-Holland, Amsterdam.

THOMASSÉ S. (2001). Covering a strong digraph by $\alpha - 1$ disjoint paths: a proof of Las Vergnas' conjecture. *J. Combin. Theory Ser. B* **83**, 331–333.

THOMASSEN C. (1978). Hypohamiltonian graphs and digraphs. In *Theory and Applications of Graphs*, 557–571. Lecture Notes in Math., Vol. 642, Springer, Berlin.

THOMASSEN C. (1980). 2-linked graphs. *European J. Combin.* **1**, 371–378.

THOMASSEN C. (1981). Kuratowski's theorem. *J. Graph Theory* **5**, 225–241.

THOMASSEN C. (1983a). Infinite graphs. In *Selected Topics in Graph Theory, 2* (L.W. Beineke and R.J. Wilson, eds.), 129–160. Academic Press, London.

THOMASSEN C. (1983b). A theorem on paths in planar graphs. *J. Graph Theory* **7**, 169–176.

THOMASSEN C. (1989). Configurations in graphs of large minimum degree, connectivity, or chromatic number. In *Combinatorial Mathematics: Proceedings of the Third International Conference (New York, 1985)*, 402–412. Ann. New York Acad. Sci., Vol. 555, New York Acad. Sci., New York.

THOMASSEN C. (1990). Resistances and currents in infinite electrical networks. *J. Combin. Theory Ser. B* **49**, 87–102.

THOMASSEN C. (1994). Every planar graph is 5-choosable. *J. Combin. Theory Ser. B* **62**, 180–181.

THOMASSEN C. (1997a). Chords of longest cycles in cubic graphs. *J. Combin. Theory Ser. B* **71**, 211–214.

THOMASSEN C. (1997b). Color-critical graphs on a fixed surface. *J. Combin. Theory Ser. B* **70**, 67–100.

THOMASSEN C. (1997c). The zero-free intervals for chromatic polynomials of graphs. *Combin. Probab. Comput.* **6**, 497–506.

THOMASSEN C. (1998). Independent dominating sets and a second Hamiltonian cycle in regular graphs. *J. Combin. Theory Ser. B* **72**, 104–109.

THOMASSEN C. (2000). Chromatic roots and Hamiltonian paths. *J. Combin. Theory Ser. B* **80**, 218–224.

THOMASSEN C. (2005). Some remarks on Hajós' conjecture. *J. Combin. Theory Ser. B* **93**, 95–105.

TURÁN P. (1941). Eine Extremalaufgabe aus der Graphentheorie. *Mat. Fiz. Lapok* **48**, 436–452.

TUTTE W.T. (1946). On Hamiltonian circuits. *J. London Math. Soc.* **21**, 98–101.

TUTTE W.T. (1947a). The factorization of linear graphs. *J. London Math. Soc.* **22**, 107–111.

TUTTE W.T. (1947b). A family of cubical graphs. *Proc. Cambridge Philos. Soc.* **43**, 459–474.

TUTTE W.T. (1947c). A ring in graph theory. *Proc. Cambridge Philos. Soc.* **43**, 26–40.

TUTTE W.T. (1948a). *An Algebraic Theory of Graphs.* Ph.D. thesis, Cambridge University.

TUTTE W.T. (1948b). The dissection of equilateral triangles into equilateral triangles. *Proc. Cambridge Philos. Soc.* **44**, 463–482.

TUTTE W.T. (1954a). A contribution to the theory of chromatic polynomials. *Canadian J. Math.* **6**, 80–91.

TUTTE W.T. (1954b). A short proof of the factor theorem for finite graphs. *Canadian J. Math.* **6**, 347–352.

TUTTE W.T. (1956). A theorem on planar graphs. *Trans. Amer. Math. Soc.* **82**, 99–116.

TUTTE W.T. (1961a). On the problem of decomposing a graph into n connected factors. *J. London Math. Soc.* **36**, 221–230.

TUTTE W.T. (1961b). A theory of 3-connected graphs. *Nederl. Akad. Wetensch. Proc. Ser. A 64 = Indag. Math.* **23**, 441–455.

TUTTE W.T. (1963). How to draw a graph. *Proc. London Math. Soc. (3)* **13**, 743–767.

TUTTE W.T. (1965a). Lectures on matroids. *J. Res. Nat. Bur. Standards Sect. B* **69B**, 1–47.

TUTTE W.T. (1965b). The quest of the perfect square. *Amer. Math. Monthly* **72**, 29–35.

TUTTE W.T. (1966a). *Connectivity in Graphs.* Mathematical Expositions, Vol. 15, University of Toronto Press, Toronto.

TUTTE W.T. (1966b). On the algebraic theory of graph colorings. *J. Combin. Theory* **1**, 15–50.

TUTTE W.T. (1970). On chromatic polynomials and the golden ratio. *J. Combin. Theory* **9**, 289–296.

TUTTE W.T. (1972). Unsolved Problem 48, in Bondy and Murty (1976).

TUTTE W.T. (1976). Hamiltonian circuits. In *Colloquio Internazionale sulle Teorie Combinatorie (Rome, 1973), Tomo I*, 193–199. Atti dei Convegni Lincei, No. 17. Accad. Naz. Lincei, Rome.

TUTTE W.T. (1998). *Graph Theory as I have Known It.* Oxford Lecture Series in Mathematics and its Applications, Vol. 11, Clarendon Press, New York. With a foreword by U. S. R. Murty.

TUZA Z. (1990). A conjecture on triangles of graphs. *Graphs Combin.* **6**, 373–380.

TVERBERG H. (1982). On the decomposition of K_n into complete bipartite graphs. *J. Graph Theory* **6**, 493–494.

ULAM S.M. (1960). *A Collection of Mathematical Problems.* Interscience Tracts in Pure and Applied Mathematics, Vol. 8, Interscience, New York–London.

UNGAR P. and DESCARTES B. (1954). Advanced Problems and Solutions: Solutions: 4526. *Amer. Math. Monthly* **61**, 352–353.

VAN DER WAERDEN B. (1927). Beweis einer Baudetschen Vermutung. *Nieuw Arch. Wisk.* **15**, 212–216.

VAZIRANI V.V. (2001). *Approximation Algorithms.* Springer, Berlin.

VEBLEN O. (1912/13). An application of modular equations in analysis situs. *Ann. of Math. (2)* **14**, 86–94.

VINCE A. (1988). Star chromatic number. *J. Graph Theory* **12**, 551–559.

VIZING V.G. (1964). On an estimate of the chromatic class of a p-graph. *Diskret. Analiz No.* **3**, 25–30.

VIZING V.G. (1965). The chromatic class of a multigraph. *Kibernetika (Kiev)* **1965**, 29–39.

VOIGT M. (1993). List colourings of planar graphs. *Discrete Math.* **120**, 215–219.

VOIGT M. (1995). A not 3-choosable planar graph without 3-cycles. *Discrete Math.* **146**, 325–328.

VON NEUMANN J. (1928). Zur Theorie der Gesellschaftsspiele. *Math. Ann.* **100**, 295–320.

WAGNER K. (1936). Bemerkungen zum Vierfarbenproblem. *Jber. Deutsch. Math.- Verein.* **46**, 26–32.

WAGNER K. (1937). Über eine Eigenschaft der ebenen Komplexe. *Math. Ann.* **114**, 570–590.

WAGNER K. (1964). Beweis einer Abschwächung der Hadwiger-Vermutung. *Math. Ann.* **153**, 139–141.

WATKINS M.E. (1970). Connectivity of transitive graphs. *J. Combin. Theory* **8**, 23–29.

WELSH D.J.A. (1976). *Matroid Theory*. London Mathematical Society Monographs, Vol. 8, Academic Press, London.

WELSH D.J.A. (1993). *Complexity: Knots, Colourings and Counting*. London Mathematical Society Lecture Note Series, Vol. 186, Cambridge University Press, Cambridge.

WHITNEY H. (1932a). Congruent graphs and the connectivity of graphs. *Amer. J. Math.* **54**, 150–168.

WHITNEY H. (1932b). A logical expansion in mathematics. *Bull. Amer. Math. Soc.* **38**, 572–579.

WHITNEY H. (1932c). Non-separable and planar graphs. *Trans. Amer. Math. Soc.* **34**, 339–362.

WHITNEY H. (1933). 2-isomorphic graphs. *Amer. J. Math.* **55**, 245–254.

WHITNEY H. (1935). On the abstract properties of linear dependence. *Amer. J. Math.* **57**, 509–533.

WHITNEY H. (1992). *Collected Papers. Vol. I.* Contemporary Mathematicians, Birkhäuser, Boston. Edited and with a preface by James Eells and Domingo Toledo.

WILLIAMS J.W.J. (1964). Algorithm 232: Heapsort. *Commun. ACM* **7**, 347–348.

WILSON R. (2002). *Four Colors Suffice: How the Map Problem was Solved*. Princeton University Press, Princeton, NJ.

WINDER R.O. (1966). Partitions of N-space by hyperplanes. *SIAM J. Appl. Math.* **14**, 811–818.

WITTE D. and GALLIAN J.A. (1984). A survey: Hamiltonian cycles in Cayley graphs. *Discrete Math.* **51**, 293–304.

WONG P.K. (1982). Cages—a survey. *J. Graph Theory* **6**, 1–22.

WOODALL D.R. (1971). Thrackles and deadlock. In *Combinatorial Mathematics and its Applications*, 335–347. Academic Press, London.

WOODALL D.R. (1973). The binding number of a graph and its Anderson number. *J. Combin. Theory Ser. B* **15**, 225–255.

WOODALL D.R. (1978). Minimax theorems in graph theory. In *Selected Topics in Graph Theory* (L.W. Beineke and R.J. Wilson, eds.), 237–269. Academic Press, London.

WOODALL D.R. (2001). List colourings of graphs. In *Surveys in Combinatorics, 2001 (Sussex)*, 269–301. London Math. Soc. Lecture Note Ser., Vol. 288, Cambridge Univ. Press, Cambridge.

WOODALL D.R. and WILSON R.J. (1978). The Appel-Haken proof of the Four-Color Theorem. In *Selected Topics in Graph Theory* (L.W. Beineke and R.J. Wilson, eds.), 237–269. Academic Press, London.

WORMALD N.C. (1979). Classifying k-connected cubic graphs. In *Combinatorial Mathematics, VI*, 199–206. Lecture Notes in Math., Vol. 748, Springer, Berlin.

YOUNGER D.H. (1965). A conjectured minimax theorem for directed graphs. *Technical Report 42*, Digital Systems Laboratory, Department of Electrical Engineering, Princeton University.

YOUNGER D.H. (1973). Graphs with interlinked directed circuits. In *Proceedings of the Sixteenth Midwest Symposium on Circuit Theory*, volume II, XVI 2.1–XVI 2.7. IEEE, New York.

YOUNGER D.H. (1983). Integer flows. *J. Graph Theory* **7**, 349–357.

ZHANG C.Q. (1997). *Integer Flows and Cycle Covers of Graphs*. Monographs and Textbooks in Pure and Applied Mathematics, Vol. 205, Marcel Dekker, New York.

ZHAO Y. (2000). 3-coloring graphs embedded in surfaces. *J. Graph Theory* **33**, 140–143.

ZHU X. (2001). Circular chromatic number: a survey. *Discrete Math.* **229**, 371–410. Combinatorics, Graph Theory, Algorithms and Applications.

ZYKOV A.A. (1949). On some properties of linear complexes. *Mat. Sbornik N.S.* **24(66)**, 163–188.

General Mathematical Notation

Graph Parameters

Most notation common to graphs and digraphs is listed only for graphs

$\delta(G)$	minimum degree	7
$\delta^+(D)$	minimum outdegree	32
$\delta^-(D)$	minimum indegree	32
$\gamma(G)$	orientable genus	281
$\kappa'(G)$	edge connectivity	216
$\kappa(G)$	connectivity	207
$o(G)$	number of odd components	432
$\mu(G)$	multiplicity	463
$\nu(H)$	packing number of hypergraph	512
$\omega(G)$	clique number	188
$\pi(D)$	path partition number of digraph	298
$\pi(G)$	path partition number	481
$\overline{cr}(G)$	rectilinear crossing number	273
$\tau(H)$	transversal number of hypergraph	512
$\theta(G)$	thickness	261
$a(D)$	size of digraph	31
$c(G)$	number of components	29
$d(G)$	average degree	7
$e(G)$	size	2
$f(G)$	number of faces of embedded graph	249
m	size (of graph G or digraph D)	3
n	order (of graph G or digraph D)	3
$r(G,G)$	Ramsey number	325
$t(G)$	number of spanning trees	107
$t(G)$	number of triangles	311
$t_{xy}(G)$	number of spanning trees containing edge xy	550
$v(G)$	order	2
$w(F)$	weight of subgraph	50
$w(\mathbf{d})$	weight of outdegree sequence	387
$\binom{G}{F}$	number of copies of graph F	67
$\mathrm{aut}(G)$	number of automorphisms	16
$\mathrm{cr}(G)$	crossing number	248
$\mathrm{la}\,(G)$	linear arboricity	358
$\mathrm{pm}(G)$	number of perfect matchings	422
$|G \to H|$	number of embeddings in graph H	70

Operations and Relations

Families of Graphs

Structures

Other Notation

Index

Graduate Texts in Mathematics

(*continued from page ii*)

Printed in the United States
By Bookmasters